(*Cambridge University Aerial Photograph No. Y 84*)

SOUTHWARD AERIAL VIEW OF CHURCH STRETTON VALLEY AND NEIGHBOURING HILLS

Frontispiece PLATE 1

BRITISH GEOLOGICAL SURVEY

MEMOIRS OF THE GEOLOGICAL SURVEY OF GREAT BRITAIN

Geology of the country around Church Stretton, Craven Arms, Wenlock Edge and Brown Clee

Explanation of one-inch geological sheet 166, New Series

D C Greig, MA, BSc, J E Wright, BSc, B A Hains, BSc, PhD, and G H Mitchell, DSc, PhD, DIC, FRS, FRSE

with contributions by

V Wilson, PhD, MSc, DIC, and R H Hoare

Petrography R W Elliot, BSc, and R Dearnley, PhD

Palaeontology J D D Smith, BSc, W T Dean, BSc, PhD, M Mitchell, MA, A W A Rushton, BA, PhD, and B Owens, BSc, PhD

Geophysics M Brooks, BSc, PhD, and P J Fenning, BSc

Water supply H J Richards, BSc

LONDON HER MAJESTY'S STATIONERY OFFICE 1968

1989 reprint

HER MAJESTY'S STATIONERY OFFICE

HMSO publications are available from:

HMSO Publications Centre
(Mail and telephone orders)
PO Box 276, London SW8 5DT
Telephone orders (01) 873 9090
General enquiries (01) 873 0011
Queueing system in operation for both numbers

HMSO Bookshops
49 High Holborn, London WC1V 6HB
(01) 873 0011 (Counter service only)
258 Broad Street, Birmingham B1 2HE
(021) 643 3740
Southey House, 33 Wine Street, Bristol
BS1 2BQ (0272) 264306
9 Princess Street, Manchester M60 8AS
(061) 834 7201
80 Chichester Street, Belfast BT1 4JY
(0232) 238451
71 Lothian Road, Edinburgh
EH3 9AZ (031) 228 4181

HMSO's Accredited Agents
(see Yellow Pages)

And through good booksellers

First published 1968
Second impression 1989

Printed in the United Kingdom for Her Majesty's Stationery Office

Dd 240446 C5 3/89 16268

ISBN 0 11 884257 9

BRITISH GEOLOGICAL SURVEY

Keyworth, Nottingham NG12 5GG
Plumtree (060 77) 6111

Murchison House, West Mains Road, Edinburgh EH9 3LA (031) 667 1000

BGS London Information Office at the Geological Museum, Exhibition Road, London SW7 2DE (01) 589 4090

The full range of Survey publications is available through the Sales Desks at Keyworth and Murchison House. Selected items can be bought at the BGS London Information Office in the Geological Museum, and orders are accepted here for all publications. The adjacent Geological Museum Bookshop stocks the more popular books for sale over the counter. Most BGS books and reports are listed in HMSO's Sectional List 45 and can be bought from HMSO and through HMSO agents and retailers. Maps are listed in the BGS Map Catalogue and the Ordnance Survey's Trade Catalogue, and can be bought from Ordnance Survey Agents as well as from BGS.

The British Geological Survey carries out the geological survey of Great Britain and Northern Ireland (the latter as an agency service for the government of Northern Ireland), and of the surrounding continental shelf, as well as its basic research projects. It also undertakes programmes of British technical aid in geology in developing countries as arranged by the Overseas Development Administration.

The British Geological Survey is a component body of the Natural Environment Research Council.

Bibliographical reference

Greig, D C, Wright, J E, Hains, B A, and Mitchell, G H. 1968. Geology of the country around Church Stretton, Craven Arms, Wenlock Edge and Brown Clee. *Memoir of the Geological Survey of Great Britain*, Sheet 166 (England & Wales).

PREFACE

THE AREA of the Church Stretton Sheet (166) was included in Old Series one-inch Sheets 55 N.W., 55 N.E., 56 N.E., 60 S.E., 61 S.W., and 61 S.E. The area of these maps was originally surveyed, on the one-inch scale, by W. T. Aveline, H. W. Bristow, H. H. Howell, E. Hull, J. Phillips, A. C. Ramsay, and W. W. Smyth, and the maps were published between 1850 and 1855. A small area in the north-eastern corner and along the eastern edge was mapped on the six-inch scale between 1923 and 1932, during the survey of the adjacent districts of the Shrewsbury (152) Sheet and the Dudley (167) Sheet. In 1945 the Carboniferous rocks of the Clee Hills were surveyed on the six-inch scale by Dr. G. H. Mitchell and Dr. V. Wilson.

The systematic six-inch survey of the Church Stretton district was carried out between 1952 and 1959, under the direction first of Dr. V. A. Eyles and latterly of Dr. V. Wilson. Lists of six-inch maps and the names of the surveyors are given on p. xiii. The one-inch geological map of the area was published in 1967.

The identification of the large number of fossils collected from the area, in the past and during the recent survey, has been shared between the Palaeontological Department and outside specialists. We are grateful for the assistance of Professor O. M. B. Bulman, Dr. E. I. White, and Dr. W. T. Dean in the identification respectively of graptolites, fishes, and Ordovician fossils. Messrs. M. Mitchell and J. D. D. Smith of the Palaeontological Department have carried out most of the other identifications.

Petrographical descriptions have been prepared by Dr. R. Dearnley, Mr. R. W. Elliot, Dr. Diane C. Knill, and Mr. J. E. Wright.

On the completion of the mapping six boreholes were undertaken in the area under contract for the Geological Survey, and the Geophysics Department carried out a programme of exploration to study the sub-drift profiles of the Church Stretton valley. The results of these additional projects have been incorporated in the maps and in this Memoir.

The variety of rocks, the striking physiography, and the scenic beauty of the Church Stretton area have attracted the attention of geologists since the early years of the 19th century, and today the area is much used for student instruction and is the setting for many specialized geological studies. The stratigraphical interpretation presented in this Memoir may appear to indicate some differences of opinion between the Geological Survey and other geologists who have worked recently in the area. In fact, however, these differences are largely due to the lithological basis on which the Survey classification is built. The significance of different classifications is discussed in the appropriate parts of the Memoir. We acknowledge the benefit obtained from many discussions with other geologists who have worked in the area or nearby. In particular the late Professor W. F. Whittard kindly allowed us to peruse his large-scale unpublished field maps of the Silurian rocks between Norbury (Montgomery Sheet 165) and Little Stretton, and gave much additional information about exposures not now visible. The late Mr. T. H. Whitehead gave much assistance to the surveyors in discussion of numerous aspects of the geology of the solid rocks and the superficial deposits. He also recorded a number of trench sections through the drift in the Church Stretton valley. Valuable help has also been given by Dr. H. W. Ball and Dr. W. T. Dean, and by Dr. J. D. Lawson and his collaborators in the Ludlow Research Group. The memoir has been edited by Mr. J. E. Wright.

We thank the Committee for Aerial Photography of the University of Cambridge for permission to publish oblique aerial photographs taken by Dr. J. K. St. Joseph in Plate 1 (Frontispiece), Plate 6A and Plate 13A. We are grateful too for the friendliness and help of landowners and tenants in an area in which the geologist is by no means *rara avis*.

This Memoir received final approval from Sir James Stubblefield, F.R.S. as Director before his retirement in December 1966.

K. C. DUNHAM,
Director

Institute of Geological Sciences
Geological Survey Office
Exhibition Road
South Kensington
London, S.W.7
12th October 1967

CONTENTS

(References are listed at the end of each chapter)

ILLUSTRATIONS

TEXT FIGURES

PLATES

LIST OF SIX-INCH MAPS

Geological six-inch maps included wholly or in part in the Church Stretton (166) Sheet are listed below, together with the initials of the surveyors and the dates of survey. The surveyors were R. W. Elliot, D. C. Greig, B. A. Hains, R. H. Hoare, G. H. Mitchell, the late R. W. Pocock, T. Robertson, V. Wilson and J. E. Wright. The maps will be available in MS form for public reference at the Geological Survey Offices, London and Leeds. Those sheets marked * have been surveyed only in part.

National Grid Maps: all within the 100-kilometre square SO.

*SO 39	NE	Bridges	J.E.W.	1958
*	SE	Norbury and Wentnor	D.C.G., J.E.W.	1958–9
SO 49	NW	Ratlinghope	J.E.W.	1957–8
	NE	All Stretton and Comley	G.H.M., J.E.W.	1952, '57
	SW	Little Stretton and Minton	D.C.G., B.A.H., G.H.M., J.E.W.	1952, 1957–9
	SE	Church Stretton and Hope Bowdler	B.A.H., R.H.H., G.H.M., J.E.W.	1952–3, 1956–9
SO 59	NW	Cardington	R.H.H., G.H.M.	1952, '57
	NE	Easthope and Bourton	R.W.E., R.H.H.	1952, 1956–7
	SW	Rushbury and Longville	B.A.H., R.H.H., G.H.M.	1952, 1956–8
	SE	Shipton and Brockton	R.W.E., R.H.H.	1954–6
SO 69	NW	Much Wenlock and Acton Round	R.W.E., R.H.H., T.R.	1924, '52, 1956–7
	NE	Shirlett	R.W.E., R.H.H., R.W.P., T.R.	1924, '29, '52, '57
	SW	Monkhopton	R.W.E., R.H.H.	1952, 1954–5
	SE	Morville	R.W.E., R.H.H., R.W.P.	1929, '52, '54, '57
*SO 38	NE	Plowden	D.C.G.	1958–9
*	SE	Hopesay	D.C.G., B.A.H.	1957, '59
SO 48	NW	Horderley and Wistanstow	D.C.G., B.A.H.	1958–9
	NE	Acton Scott and Westhope	B.A.H., R.H.H.	1956, '58
	SW	Craven Arms	D.C.G., B.A.H.	1952, 1957–9
	SE	Culmington	D.C.G., B.A.H., R.H.H.	1952, 1954–9
SO 58	NW	Diddlebury and Munslow	R.H.H.	1952–6
	NE	Tugford and Abdon	R.H.H., G.H.M.	1945, 1952–3, '56
SO 58	SW	Hayton and Great Sutton	D.C.G., R.H.H.	1952–7
	SE	Stoke St. Milborough	D.C.G., R.H.H., G.H.M.	1945, '52, 1955–6
SO 68	NW	Cleobury North	R.H.H., G.H.M.	1945, 1952–3, '55
	NE	Overton	R.H.H., R.W.P.	1930, '52, '55
	SW	Wheathill and Farlow	D.C.G., R.H.H., G.H.M., V.W.	1945, 1955–7
	SE	Stottesdon and Prescott	D.C.G., R.H.H., R.W.P., V.W.	1930, '45, 1955–6
*SO 37	NE	Clungunford	B.A.H.	1957
*SO 47	NW	Aldon	D.C.G., B.A.H.	1957–8
*	NE	Onibury and Stanton Lacy	D.C.G.	1953, 1956–8
*SO 57	NW	Downton Hall	D.C.G.	1952–3
*	NE	Cleestanton	D.C.G.	1952, '54
*SO 67	NW	Silvington	D.C.G., V.W.	1945, '54, 1956–7
*	NE	Catherton	D.C.G., R.W.P., V.W.	1930, '32, '45, '54

Chapter 1

INTRODUCTION

AREA AND LOCATION

THE DISTRICT described in this memoir lies entirely in the southern part of the county of Shropshire. It is almost untouched by industrial and urban development, and agriculture remains the basic occupation of the inhabitants. In most of the district the rearing of sheep and beef-cattle is the most important type of husbandry, with some dairying and mixed farming on the lower ground. There is a low proportion of arable land except on the light dry soils of the dip slope of the Aymestry Group and Upper Ludlow Shales.

The largest town is Church Stretton, with a population of less than 3000. This market town lies in a deep valley amid some of the finest scenery in the Welsh Borderland; it attracts a large number of tourists, especially from the towns of the West Midlands, and for many years the surrounding area has been a popular field training ground for students from many university geology departments. The main agricultural centre is Craven Arms, a small settlement on the River Onny in the south-western part of the district. It is ideally situated for road and rail communication with much of Central Wales and the Welsh Borders, and the annual sheep sales are among the most important in the country.

Parts of the Clee Hills and Leebotwood coalfields fall within the district, but since active working ceased many years ago and was never on a large scale, it has made little noticeable impact on the landscape. Most of the harder rocks have been quarried locally for building or roadstone, but this development was never extensive and there now are no working quarries within the district.

B.A.H.

GEOLOGICAL SEQUENCE

The formations represented on the one-inch map and section are summarized in the following table:

SUPERFICIAL FORMATIONS (DRIFT)

RECENT AND PLEISTOCENE

Landslips	Head
Peat	Fluvio-glacial Gravel
Lacustrine Deposits	Glacial Sand and Gravel
Alluvium	Boulder Clay
Alluvial Fan	Morainic Drift
River Terraces	

Solid Formations

	Approx. thickness in ft
Carboniferous	
Coal Measures	
Upper Coal Measures	
Coed-yr-Allt or Newcastle-under-Lyme Group of the Leebotwood Coalfield: sandstones and shales with coal seams	200
Middle and Lower Coal Measures of the Clee Hills Coalfield: shales with coal seams, with a basal sandstone	350
Unconformity	
Cornbrook Sandstone	400 to 700
Unconformity	
Carboniferous Limestone Series	
Limestone with basal conglomerate (up to 20 ft thick)	0 to 150
Unconformity	
Old Red Sandstone	
Upper Old Red Sandstone	
Farlow Series: yellow and grey sandstone with bands of marl and occasional pebbles	0 to 500
Unconformity	
Lower Old Red Sandstone	
Clee Group (and Brownstones of Titterstone Clee Hill): sandstones and conglomerates with some marls ..	0 to 1200
Ditton Series: red marls with sandstones, cornstone conglomerates and cornstones. Abdon limestones at top, '*Psammosteus*' Limestones at base	2000
Downton Series:	
Ledbury Group: red marls with subsidiary cornstones and sandstones	1500
Temeside Group:	
Temeside Shales: purplish grey and green shales with thin sandstones	40 to 150
Downton Castle Sandstone: yellow fissile and thick-bedded sandstone with Ludlow Bone Bed at base	20 to 50
Undivided beds (west of Church Stretton Fault): purple silty mudstones and greyish green sandstones	600
Silurian	
WEST OF CHURCH STRETTON FAULT	
Ludlow Series (undivided)	
Grey calcareous shales and flags	at least 2000
Wenlock Series	
Edgton Limestone: nodular grey limestone with numerous bands of grey calcareous silty mudstone	125 to 200
Wenlock Shales: grey silty mudstones with thin nodular limestones	1000
Llandovery Series (Upper)	
Hughley (Purple) Shales: purple shales with subsidiary green bands	300 to 550
Pentamerus Beds: grey shales with thin limestone bands	0 to 300
Pentamerus Beds (Grits and Conglomerates): purple pebbly grits and conglomerates	0 to 180

EAST OF CHURCH STRETTON FAULT

	Approx. thickness in ft
Ludlow Series	
Upper Ludlow Shales: olive-green and buff flaggy siltstones	100 to 400
Aymestry Group: grey nodular argillaceous limestone with siltstone bands	80 to 210
Lower Ludlow Shales: olive-green and grey flaggy mudstones and siltstones	600 to 850
Wenlock Series	
Wenlock Limestone: grey nodular crystalline limestone	0 to 50
Wenlock Limestone Reef Facies	0 to 90
Tickwood Beds: alternating bands of grey argillaceous limestone and grey calcareous siltstone	50 to 180
Wenlock Shales: grey silty mudstones with occasional limestone bands	1000
Llandovery Series (Upper)	
Hughley (Purple) Shales: purple shales with subsidiary green bands	0 to 250
Pentamerus Beds: grey shales with thin limestone bands	0 to 225
Kenley Grit: sandstones, pebbly grits and conglomerates	0 to 75
Unconformity	
ORDOVICIAN	
Caradoc Series	
Onny Shales: olive-green and grey micaceous mudstones	up to 400
Acton Scott Group: grey silty mudstones with local yellow sandstones and sandy limestones	200 to 500
Cheney Longville Flags: yellowish brown flaggy siltstones with sandstone bands, *alternata* Limestone at base	350 to 850
Chatwall Sandstone: brown and purple sandstones ..	120 to 500
Chatwall Flags: yellowish brown flaggy siltstones and sandstones, with argillaceous beds towards the base	100 to 300
Harnage Shales: olive-brown silty mudstones ..	300 to 1000
Hoar Edge Grit: brown sandstones and pebbly sandstones with quartz conglomerates at the base ..	0 to 400
Unconformity	
CAMBRIAN	
Upper Cambrian	
Tremadoc Series—Shineton Shales	
Non-sequence	
Dolgelly Beds	
Black Shales (doubtfully present in the district)	
Grey '*Orusia*' Shales	?1000
?*Unconformity*	
Middle Cambrian	
Upper Comley Series: shales and sandstones ..	300 to ?over 600
Unconformity	
Lower Cambrian	
Lower Comley Series: green and greyish green sandstone with the Lower Comley Limestones at top ..	500
Wrekin Quartzite	up to 140
Unconformity	

	Approx. thickness in ft
PRE-CAMBRIAN	
Longmyndian	
Wentnor Series (Western Longmyndian)	
Bridges Group: purple laminated siltstones and sandstones	2000 to 4000
Bayston–Oakswood Group: massive purple and greenish grey sandstones with thick conglomerate bands	4000 to 8000
Unconformity	
Stretton Series (Eastern Longmyndian)	
Portway Group: green and purple siltstones and sandstones with the Huckster Conglomerate at the base	600 to 3500
Lightspout Group: greenish grey flaggy siltstone with sandstone bands, thin tuffs near top	1700 to 2700
Synalds Group: purple silty shales with sandstone bands, Batch Volcanics and thin tuffs	1600 to 2800
Burway Group: greenish grey sandstones and siltstones with the Cardingmill Grit at the top and the Buxton Rock at the base	2000
Stretton Shale Group: greenish grey shales with the Helmeth Grit at the base .. Thickness uncertain	?3000
?*Unconformity*	
(The stratigraphical relationships between the Uriconian rocks in the areas given below are not known)	
Uriconian of Caer Caradoc area	
Ragleth Tuffs	1500
Cwms Rhyolites	350
Caer Caradoc Andesites	400
Caer Caradoc Rhyolites	300
Little Caradoc Basalts	1500
Little Caradoc Tuffs	50
Comley Andesites	upwards of 100
Uriconian of Cardington area	
Hope Batch Dacites	100
Woodgate Tuffs	900
Woodgate Batch Dacites and Andesites	700
Middle Hill Andesites and Dacites	1200
North Hill Dacites	900
Stoneacton Tuffs and Andesites	uncertain
Uriconian of The Lawley, Hazler Hill, Hope Bowdler Hill and Wart Hill areas	unknown
Intrusive igneous rocks	
Dolerite in Coal Measures and Pre-Cambrian. Quartz-porphyry and quartz-microdiorite in Pre-Cambrian.	

PHYSIOGRAPHY

The district is one in which the relationship of physiography to geology is developed to a classical degree. Fig. 1 is a physiographic map of the Church Stretton District and the adjacent areas and Fig. 2 shows the main features of the geology of the district. In the north-west the broad plateau of the Long

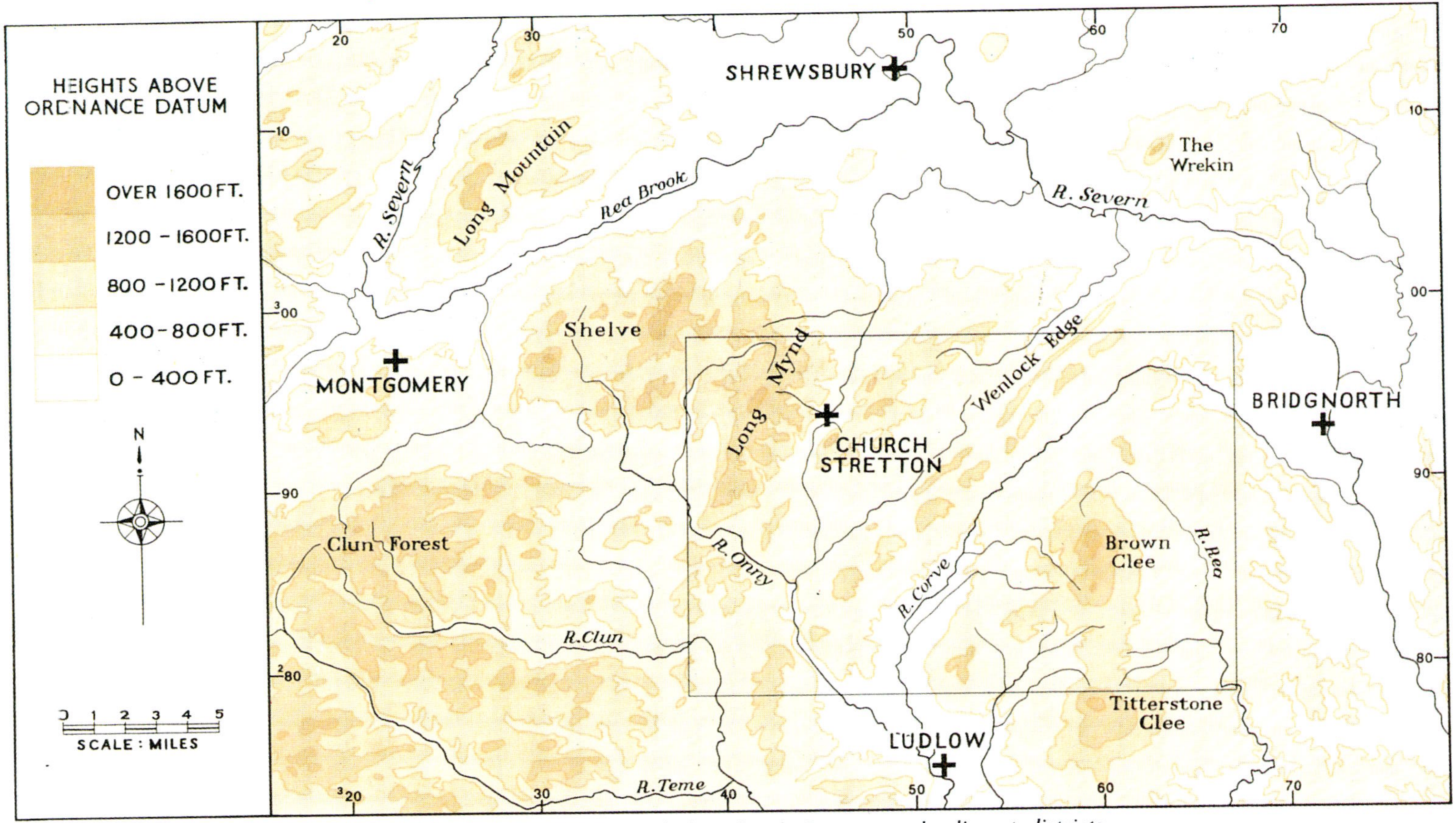

FIG. 1. *Physiographic map of the Church Stretton and adjacent districts.*

Mynd, rising to 1696 ft O.D. at Pole Bank, owes its form to the thick groups of relatively resistant sandstones and shales of which it is composed. The original peneplanation of the Long Mynd has been variously ascribed to the Silurian, Carboniferous and Triassic periods (Whittard 1952, p. 183), but sub-aerial weathering, continuous since pre-Glacial times, has produced the thick mantle of rock waste which obscures the solid rock on the high ground. The steep western escarpment of the southern part of the Long Mynd is developed along the line of a major fault but farther north, where evidence for the fault is lacking, the escarpment is much less pronounced. As far as drainage is concerned it is noteworthy that the valley of the East Onny lies within the Bridges Group, the more highly argillaceous of the Longmyndian formations in that part of the district and the relatively long immature valleys, 'hollows', and 'batches' of the eastern side of the Long Mynd originate from a watershed near the outcrop of the base of the arenaceous Bayston Group.

The Church Stretton valley, broad and relatively flat, follows the line of the most westerly component of the Church Stretton Fault Complex (F1). The rocks that crop out in the valley bottom are shales of Longmyndian, Silurian, and Carboniferous ages, much softer than the rocks on the valley sides, in the Long Mynd itself and in the Uriconian hills to the east. The line of F1 can be traced in comparatively low ground from Marshbrook to Horderley, Hopesay, and the limit of the area near Coston. West of the fault, between Horderley and Edgton, limestones in the Wenlock Series cap a steep north-facing escarpment, which reaches a height of over 950 ft O.D.

The varied topography of the hills east of the Stretton valley is in marked contrast to the smooth profile of the Long Mynd, reflecting the variety of Uriconian and Longmyndian rocks of which they are composed. The extension of Uriconian rocks to the east is manifested by the high ground of Hope Bowdler (1396 ft O.D.) and Cardington hills, and farther south the isolated knoll of Wart Hill (1065 ft O.D.) reflects the occurrence of an inlier of Uriconian igneous rocks.

East of this zone of Pre-Cambrian rocks there are several vales and 'edges' in a pattern conforming to the succession of shales and harder sandstones and limestones which make up the Ordovician, Silurian, and Old Red Sandstone of the area. The major escarpments are formed from west to east by the Hoar Edge Grit and Chatwall Sandstone of Ordovician age, the Wenlock Limestone and Aymestry Group (1112 ft O.D. at Callow Hill) of the Silurian, and the sandstones above the '*Psammosteus*' Limestones in the Old Red Sandstone. The Ludlow Shales occupy extensive and well-defined dip slopes behind the two Silurian escarpments, and the dip slopes of the Aymestry Group are characterized by the development of sharply defined narrow valleys, flat-floored and dry, or with insignificant streams. Many originate in cwm-like depressions, commonly near the crest of the escarpment and they were probably developed at a time when the water table stood higher than today. This may have happened because the escarpment was more distant and higher, because rainfall was higher, or because ground absorption was reduced by, for example, the presence of permanently frozen ground at no great depth. Similar origins have been suggested for the dry valleys on the Chalk dip slope in south-eastern England (Reid 1887, p. 364). The shales between the Cheney Longville Flags and the Wenlock Limestone form a conspicuous zone of low relief as do the red marls predominant in the Downton Series. D.C.G.

The south-eastern part of the district, and particularly the extensive ground lying south-east of the belt of Silurian escarpments, is dominated by the bold mass of Brown Clee Hill which is a high tableland, some three miles long from north to south and about one and a half miles from west to east, its general rounded form being made up of a succession of small scarps and shelves arranged one above another and culminating in two summits rather more than a mile apart. The more northerly one, Abdon Burf, 1792 ft O.D., is separated from the southern hill top, Clee Burf, 1650 ft O.D., by the col known as Burwarton Pole. Both these summits are composed of basalt, or Dhustone, probably intrusive into the Coal Measures which form two discrete outliers with their bases at about 1500 ft O.D. and which lie unconformably on Old Red Sandstone marls and sandstones. Within the Old Red Sandstone minor sandstone and cornstone bands often form well-defined features. G.H.M.

South-east of Brown Clee Hill Carboniferous sandstones and limestones form an area of high ground which is a north-eastern spur of Titterstone Clee Hill, in the Ludlow (181) Sheet.

Broad flat areas underlain by alluvial deposits are seen near Church Stretton and at places in the valleys of the rivers Clun, Onny, and Corve and the Quinny and Mor brooks, but extensive featureless slopes on the northern and eastern flanks of Brown Clee Hill owe their form to the deposition from the higher ground of a thick layer of solifluxion head. D.C.G.

The whole of the district is situated within the basin of the River Severn whose northward-flowing tributaries drain the northern parts of the Church Stretton valley and of Ape Dale; the north-eastern corner of the district is drained by the Mor Brook, which flows into the Severn south of Bridgnorth.

South of the watershed, which is at a height of about 620 ft O.D. at Church Stretton and about 590 ft O.D. in Ape Dale, the drainage is by southward flowing tributaries of the River Teme, of which one, the River Onny, flows in a wide valley to the west of the Long Mynd, turning sharply to the east at Plowden to flow through a narrow defile towards Craven Arms. Below the confluence with tributaries draining the southern part of the Church Stretton valley and the broad vale of Ape Dale, it flows southwards through a wide valley before cutting through the Aymestry Group scarp at Stokesay. The broad valley of Corve Dale, between the Aymestry Group ridge and Brown Clee Hill, is drained by the River Corve, and the eastern slopes of the Clee Hills by the River Rea. In the southern-western corner of the district the River Clun meanders in a broad vale. All these southerly flowing streams eventually join the Teme, which flows into the River Severn south of Worcester.

Geological History

The oldest rocks in the district are the Pre-Cambrian (Uriconian) volcanics, which form the range of hills along the eastern side of the Church Stretton valley. These volcanic rocks are succeeded by a great thickness of Pre-Cambrian (Longmyndian) sediments, forming the high ground of the Long Mynd and Hopesay Hill. The predominantly fine-grained rocks of the Stretton Series (Eastern Longmyndian) are overlain unconformably by coarse-grained red and purple sandstones of the Wentnor Series (Western Longmyndian), indicative of derivation from an area of high relief.

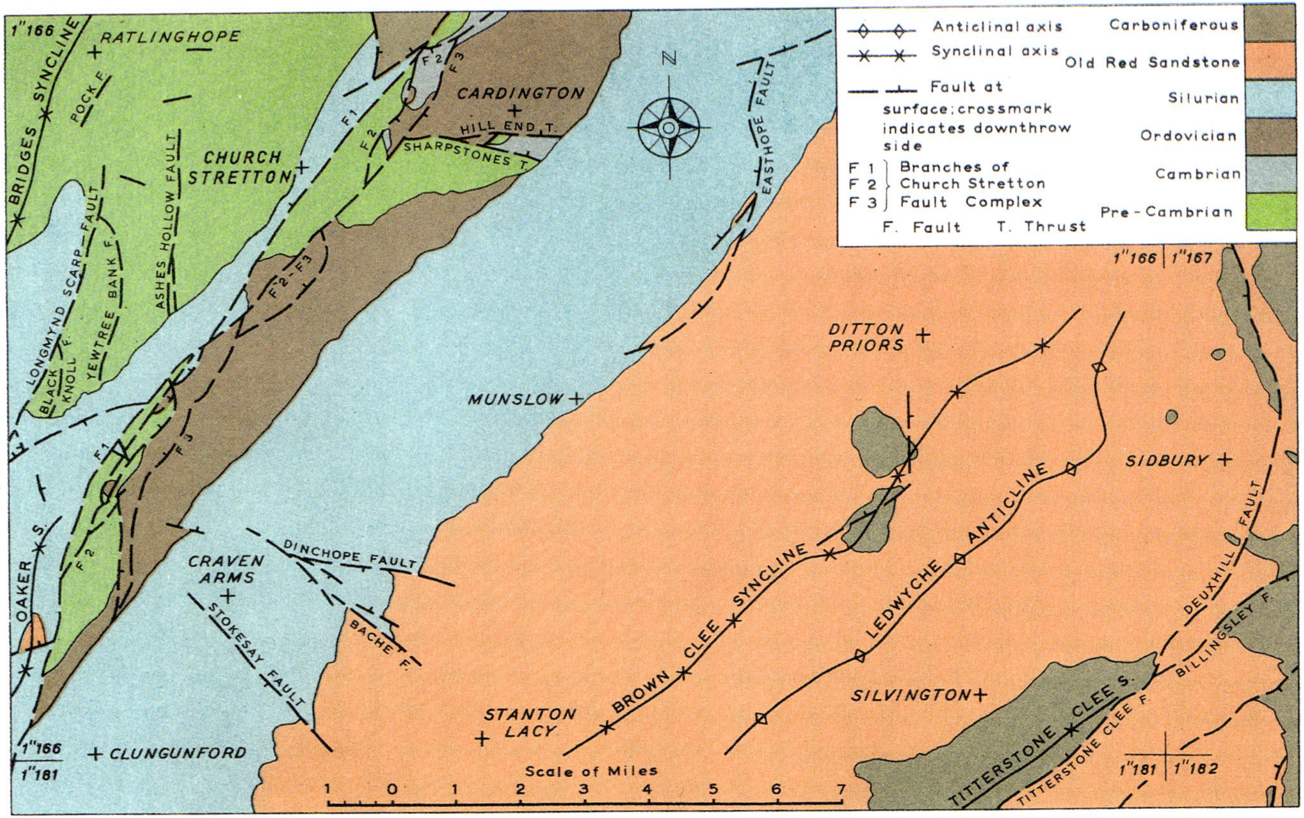

FIG. 2.–The major structures and generalized geology of the Church Stretton district

Strong folding and denudation preceded the deposition of the Cambrian sediments, which crop out in two small areas near Cardington in the north of the district. The basal beds are quartzites, laid down in a shallow current-swept sea, succeeded by shallow-water glauconitic sandstones and these Lower Cambrian beds were folded and differentially eroded before the deposition of the Middle Cambrian shales and sandstones. Steady subsidence followed in Upper Cambrian times, with the deposition of a considerable thickness of argillaceous sediments but the Cambrian sedimentation was terminated by a major uplift with some minor folding.

During most of Ordovician and Silurian times sedimentation in the subsiding Welsh Geosyncline contrasted strongly with that on the relatively stable Midland block to the east. In the Ordovician the western limit of the stable area appears to have lain to the west of the Long Mynd, and farther west, in the Shelve district (Welshpool (151) and Montgomery (165) sheets), sedimentation was continuous from Arenig to Caradoc times, probably in fairly deep water. In the Church Stretton district, however, only Caradoc strata are present, and these beds are mainly shallow-water sandstones and siltstones with an abundant shelly fauna.

Further uplift and folding affected the district at the end of the Ordovician period, and a landscape of considerable relief was developed. In Upper Llandovery times the sea invaded the district from the Welsh geosynclinal area, and shoreline deposits, such as grits and conglomerates, were laid down locally on the south-eastern margin of the Long Mynd and near Plaish. The basal grits and conglomerates were succeeded and overlapped by the shallow-water limestones and siltstones of the *Pentamerus* Beds and the mudstones of the Hughley Shales. During Wenlock and Ludlow times the line of the Church Stretton Fault Complex appears to have marked approximately the boundary between shelf sediments in the stable area to the east, and basin deposits to the west.

The 'shelf' facies is seen in the broad outcrop of Wenlock and Ludlow rocks which extends from Clungunford in the south-western corner of the district to Presthope in the north. The weakly calcareous siltstones and mudstones of the Wenlock Shales are succeeded by the shallow-water Wenlock Limestone, with reef development in the north of the district. The Lower Ludlow Shales mark a return to more argillaceous sedimentation, with at times a predominantly planktonic fauna. The higher part of these shales shows a gradual reversion to calcareous silty sedimentation with a shelly fauna, followed by a passage into the shallow-water impure limestones of the Aymestry Group. During Upper Ludlow times the sea became still more shallow, and current-bedded siltstones and fine-grained sandstones were laid down.

The sediments of the 'basin' facies to the west of the Church Stretton Fault Complex, around Edgton and Oaker are a thick monotonous series of siltstones and mudstones which include an impure limestone facies near the top of the Wenlock Series.

Sediments of Old Red Sandstone age occur extensively over the eastern part of the district, and form the slopes of the high ground of the Clee Hills. The base of the Old Red Sandstone is delimited by the Ludlow Bone Bed, which has a long sinuous outcrop over a large area of the Welsh Borders to the east of the Church Stretton Fault. This bed, which contains abundant fish remains, appears to mark a pause in sedimentation between the marine conditions of the Silurian and the oncoming of the brackish water and fluviatile deposits of the Old Red Sandstone. The change in sedimentary environment was due to the Caledonian orogeny in Wales and the consequent inception of a large delta plain over much of the Welsh Borders. The lowest beds of the Lower Old Red Sandstone, the Downton Castle Sandstone and the sandstones and grey shales of the Temeside Shales, were probably deposited in brackish water, but the oncoming of the red marls, sandstones and cornstones of the Ledbury Group marked a change to subaerial and fluviatile sedimentation, with fish remains as almost the only fossils. The red colour of the sediments may indicate derivation from areas of lateritic weathering on the rising mountain chain in Wales. In the higher beds of the Lower Old Red Sandstone, the Ditton Series and Clee Group, red sandstones become predominant, indicating a closer proximity of the Caledonian mountains, and at the end of Lower Old Red Sandstone times the earth movements extended their influence over the whole of the district, with uplift and gentle folding succeeded by extensive denudation.

No deposits of Middle Old Red Sandstone age are known in the district, and the Upper Old Red Sandstone, represented by the Farlow Series, is confined to a small area around Farlow, to the south-east of Brown Clee Hill. The Farlow Series, which consists mainly of sandstones with pebbly bands, was laid down on a fluviatile delta which was invaded by the sea at the close of Old Red Sandstone times. Marine limestones of the Lower Carboniferous with a basal conglomerate up to 20 ft thick overlie the Farlow Series, and are overlain unconformably by the Upper Carboniferous deltaic Cornbrook Sandstone. Other Upper Carboniferous sediments include the Middle and Lower Coal Measures of the Clee Hills, and the Upper Coal Measures of the southern end of the Leebotwood Coalfield. Rhythmic deltaic sedimentation took place in both areas, with the deposition of a sequence of shales, sandstones and thin coal seams.

B.A.H.

During the Pleistocene period, the area was invaded by ice from the Shropshire Plain to the north and from the Welsh hills to the west. The evidence afforded by the glacial deposits suggests only one period of glaciation which is tentatively correlated with maxima of the northern and western ice sheets but it is not known whether the maxima of the two ice sheets were coincident. It appears that the high ground of the Long Mynd, the Cardington hills, the Silurian limestone escarpments and the Clee Hills, was not covered by ice and the present distribution of boulder clay suggests that ice invaded the major valleys in the area with the exception of Corve Dale which apparently was not covered over the greater part of its length between Culmington and Stanton Long. During the retreat of the ice, melt water from the waning glaciers laid down deposits of sand and gravel, remnants of which are now found, for example, in the Church Stretton valley, near Marshbrook and near Ticklerton.

J.E.W.

References

Reid, C. 1887. On the Origin of Dry Chalk Valleys and of Coombe Rock. *Quart. J. Geol. Soc.*, **43**, 364–73.

Whittard, W. F. 1952. A Geology of South Shropshire. *Proc. Geol. Assoc.*, **63**, 143–97.

The following publications contain extensive bibliographies of works concerning the Church Stretton district and adjacent areas:

CANTRILL, T. C. 1908. Chapter I Geology, *in* A History of Shropshire. *The Victoria History of the Counties of England*, 1–46.

EDMUNDS, F. H. and OAKLEY, K. P. 1963. The Central England District. 2nd Edit. (1963 reprint). *British Regional Geology. Geol. Surv.*

LAPWORTH, C. and WATTS, W. W. 1894. The Geology of South Shropshire. *Proc. Geol. Assoc.*, **13**, 297–355.

POCOCK, R. W. and WHITEHEAD, T. H. 1961. The Welsh Borderland. 2nd Edit. (1961 reprint.) *British Regional Geology. Geol. Surv.*

WATTS, W. W. 1925. The Geology of South Shropshire. *Proc. Geol. Assoc.*, **36**, 321–63.

WHITTAKER, W. and WATTS, W. W. 1889. List of works on the Geology, Mineralogy and Palaeontology of Shropshire. *Trans. Shropshire Arch. and Nat. Hist. Soc.*, **12**, 33–62.

WHITTARD, W. F. 1952. A Geology of South Shropshire. *Proc. Geol. Assoc.*, **63**, 143–97.

Chapter 2

PRE-CAMBRIAN, URICONIAN

General Account

THE URICONIAN rocks of Shropshire are of dominantly volcanic character and form part of a discontinuous outcrop which extends from Wart Hill in the south, northwards to the neighbourhood of the Wrekin. These rocks are known as the Eastern Uriconian and are similar to the Western Uriconian which occurs in a parallel belt along the line of the Pontesford-Linley fault zone, west of the main outcrop of Pre-Cambrian sediments, the Longmyndian. Modern opinion equates the Eastern Uriconian with the Western (Whittard 1952, p. 147; James 1956, p. 323), and the Uriconian as a whole is considered to be older than the sedimentary Longmyndian rocks. J.E.W.

Uriconian rocks form the narrow, steep-sided hills known as The Lawley, Caer Caradoc, Helmeth, Hazler and Ragleth along the line of the Church Stretton Fault. To the east a short saddle links this high ground to the Hope Bowdler, Willstone and Cardington hills, likewise composed of Uriconian strata. About six miles south-west of Church Stretton is another small, faulted Uriconian area at Wart Hill. These inliers are separated either by unconformities or faulted junctions from younger strata ranging in age from Longmyndian to Upper Coal Measures.

The northern part of The Lawley is within the area shown on the Shrewsbury (152) Sheet. Little has been published concerning the Uriconian of the Church Stretton district since the papers of Allport (1877), Callaway (1879), Blake (1890) and Cobbold (1900), though Robertson (*in* Pocock and others 1938, p. 27) described the northern part of The Lawley.

It has not been found possible among the various lavas and tuffs which comprise the Uriconian to recognize a stratigraphical succession applicable to the whole area. The individual outcrops are generally bounded by faults, the throws of which cannot be precisely determined. Different portions of the Uriconian succession are found within the separate fault blocks, but cannot always be related to each other. The problem is made more acute locally by lack of exposures and by the sheared and altered state of the rocks.

It is convenient to describe the Uriconian outcrops under separate districts as follows:

1. The Lawley, 2. Caer Caradoc Hill, Helmeth Hill and Ragleth Hill, 3. Hazler Hill, 4. Hope Bowdler Hill, 5. Cardington Hill, 6. Wart Hill.

The Lawley

The Lawley inlier is bounded to the west by two major faults of the Church Stretton system, the more westerly of these bringing Carboniferous rocks against Uriconian. The inlier is limited to the east by the unconformable basal Cambrian quartzite. A number of north-east to south-west faults cross the southern part of the hill and cross-faults running eastwards from the main Church Stretton Fault (F1 of Cobbold 1927, p. 565) bring in an area of Cambrian and Ordovician rocks separating The Lawley from Little Caradoc.

Details of the structure of the Uriconian are uncertain because of a lack of exposures but it appears that the regional dip is generally steep to the east, the strike swinging from about north-north-east in the north to about north-south at the southern end of The Lawley. There is some doubt as to the correct stratigraphical succession but it appears that the Basic Tuffs (1a) of the northern part of The Lawley (Robertson *in* Pocock and others 1938, p. 28) are the highest beds exposed in the southern part of the hill where the probable downward succession is:

Basic Tuffs (probably over 500 ft)
Upper Andesites (probably over 300 ft)
Rhyolite (about 100 ft)
Lower Andesites (more than 500 ft)

Lower Andesites. The Lower Andesites are badly weathered, close-jointed, dark grey rocks, commonly amygdaloidal. The amygdaloidal infillings, mainly quartz and calcite, frequently weather out on exposed surfaces giving a scoriaceous appearance. Near the south-west corner of the hill are breccias with limonite-coated fragments less than half an inch across, the rock being similar to other Uriconian rhyolitic breccias, particularly those thought to be crush breccias (p. 14). G.H.M.

The andesite fragments (E 25142)[1] in the breccias consist of albite-oligoclase laths, about 0·1 mm long, with abundant granular leucoxene, much interstitial chlorite and occasional small hematite grains. The rock is silicified locally. R.W.E.

Rhyolite. The succeeding lava is a flow-banded, amygdaloidal rhyolite, pinkish brown to grey in colour and commonly shattered and jointed. Locally it is broken into an angular crush breccia of rhyolite fragments, up to ½-in across, in a brownish base. At 650 yd south of Penkridge Hall [490969][2] a band of this breccia several inches thick, is overlain by shear planes dipping N. 20° W. at 35° to 40°. Locally, however, the rhyolite is brecciated without any visible shear planes. G.H.M.

The rhyolite (E 25469) contains rare quartz phenocrysts and microphenocrysts of acid plagioclase. The spherulitic groundmass consists of feldspar microlites set in a mosaic of quartz, alkali-feldspar and abundant tiny flakes of chlorite. Quartz veins carry magnetite associated with limonite and sphene and there are also veinlets of limonite. A strained quartz phenocryst and brecciation of the groundmass afford evidence of the stresses to which the rock has been subjected. R.W.E.

Upper Andesites. The Upper Andesites are dark grey or purplish amygdaloidal lavas, the amygdales being filled with quartz, calcite or distinctive yellow radiating epidote. Flow breccias are recognized on weathered surfaces. Certain lavas carry noticeable feldspar phenocrysts. A few exposures of tuff suggest that the group may consist of thin andesite flows with interbedded tuffs. G.H.M.

---

[1]Such numbers refer to rock slices in the Survey collection.

[2]Such numbers are National Grid References within the 100-km square SO(32).

These andesites (E 25147, 25149) contain oligoclase laths up to 0·5 mm long, often showing flow structure. The fine-grained matrix contains chlorite, granular sphene and abundant ilmenite. Amygdales are filled with quartz, chlorite or epidote. In one specimen (E 14759) irregular amygdales, rimmed with iron ore, contain fine-grained igneous material, possibly the result of autobrecciation or derived from a succeeding flow. R.W.E.

Basic Tuffs. The Basic Tuffs are poorly bedded, greenish grey or purple in colour, fine-grained or even rather sandy in grade, and contain ragged green or pink lithic fragments. G.H.M.

These rocks are vitric or vitric-lithic tuffs (E 25145) containing abundant devitrified shards and fragments of pumice-like rock largely altered to chlorite and calcite, together with quartz and feldspar, probably albite-oligoclase. Other lithic fragments include rhyolite and andesite. The groundmass is very fine-grained with some recognizable feldspar and granular iron ore. Carbonate commonly replaces feldspar and shards. R.W.E.

Intrusions. A band of dolerite, 250 yd wide, extends across The Lawley ¼ mile north of the summit. Other smaller intrusions, all apparently transgressive, occur to the north and south. Another body, 400 yd south of the summit, is probably a sill. No dolerite intrusions are known to cut the Cambrian which suggests that they are Pre-Cambrian in age. G.H.M.

The basic intrusions include ophitic olivine-dolerite (E 14755, 14757, 14763) composed of laths and plates of extremely sericitized labradorite, ophitic plates of augite, serpentine pseudomorphs after olivine, and irregular patches of iron ore. Chlorite occurs in interstitial patches. R.W.E.

Other intrusions include a banded rhyolite dyke (E 25144) cropping out at 510 yd N. 16° W. of Shootrough [489969]. At 440 yd E. 42° S. of Penkridge Hall [493973] is a small exposure of albite-porphyrite (E 25148), probably an intrusion in the tuffs. G.H.M.

Caer Caradoc, Helmeth and Ragleth Hills

The Uriconian of Caer Caradoc Hill is limited to the north-west by the Church Stretton Fault, while on the north-eastern spur of the hill the basal Cambrian quartzite rests unconformably on, or is faulted against, the Uriconian (Fig. 3). Running south-westwards from Comley [485965], the Lawley Fault (F2 of Cobbold 1927, p. 565) separates the Uriconian from rocks of Caradoc, Cambrian and possibly Wentnor age to the south-east. To the south of Helmeth Hill, this fault joins with the Cwms–Hoar Edge Fault (F3 of Cobbold). The last named brings together two areas of Uriconian east of Hazler Hill, but farther south at Ragleth Hill it throws Harnage Shale of Caradoc age against Uriconian. Between Caradoc Coppice [468948] and Little Stretton [445916] the Uriconian of Ragleth and Helmeth hills is bounded to the north-west by the arcuate outcrop of the Helmeth Grit which rests with possible unconformity upon the Uriconian (but see Cobbold and Whittard 1935, p. 348).

On Caer Caradoc Hill the dip is steep to the south-west but to the south the regional dip is high to the west although there are some steep easterly dips. The oldest rocks are exposed on Little Caradoc and the succession in downward order is as follows:

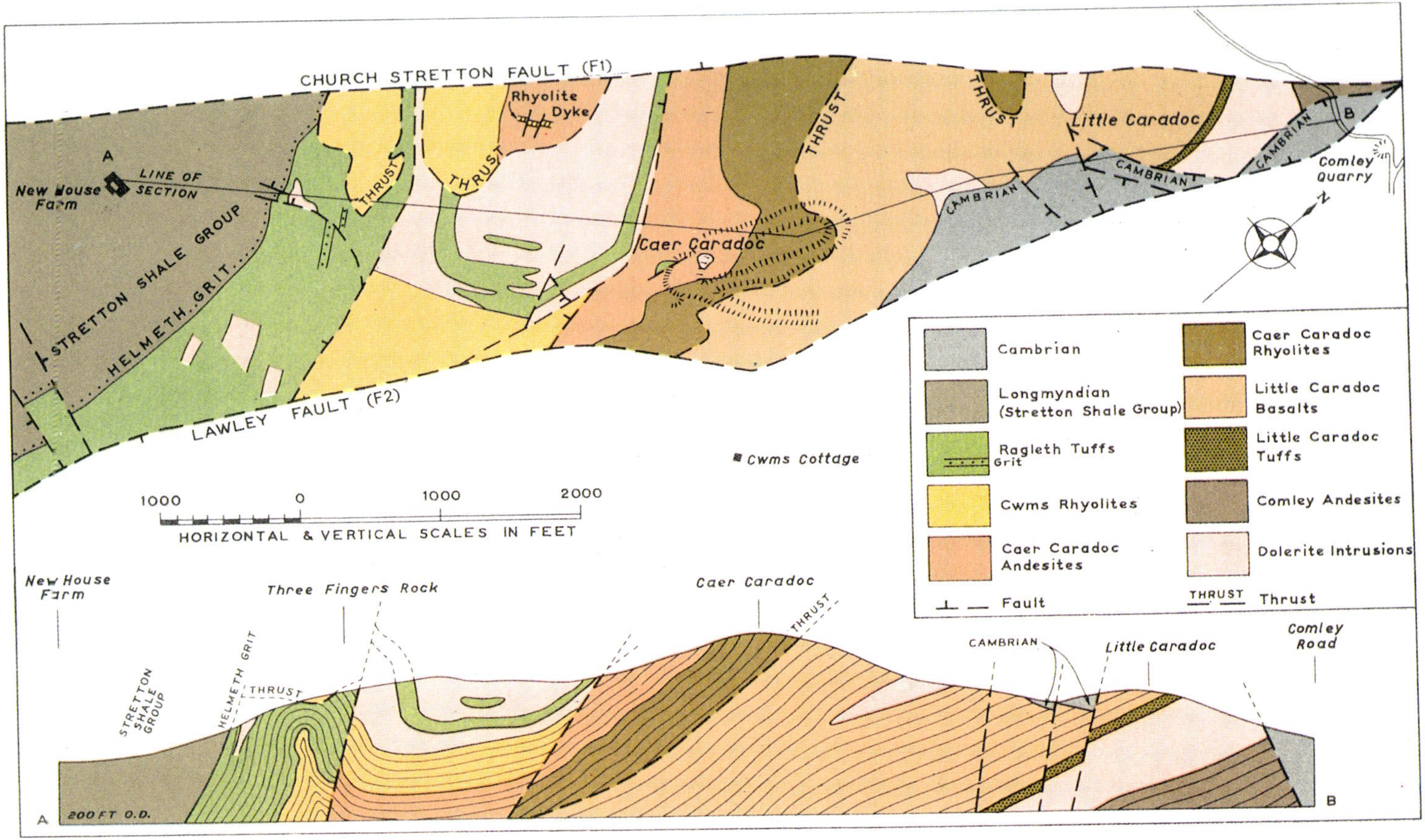

FIG. 3.–Geological plan and section of Caer Caradoc.

	Approx. thickness in ft
Longmyndian (Stretton Series)	
Helmeth Grit	100
Uriconian	
Ragleth Tuffs: fine-grained acid lithic and dust tuffs with thin rhyolite flows and grit bands	1500
Cwms Rhyolites: rhyolites, commonly amygdaloidal ..	350
Caer Caradoc Andesites: amygdaloidal andesites	400
Caer Caradoc Rhyolites: flow-banded vesicular rhyolites ..	300
Thrust junction	
Little Caradoc Basalts: albitized basalts, aphanitic, amygdaloidal and flow brecciated	1500
Little Caradoc Tuffs: well bedded, fine-grained acid tuffs ..	50
Comley Andesites: amygdaloidal andesites	100+

There are numerous intrusions of altered dolerite. Fig. 3 shows the distribution of these groups on Caer Caradoc Hill.

Comley Andesites. The Comley Andesites are exposed in the side of the lane [482964] about 750 yd north-east of Botvyle and in the fields to the north-east. They are amygdaloidal, dark blue-green, much jointed and weathered. G.H.M.

The rock (E 25426) contains scattered oligoclase microphenocrysts, up to about 0·6 mm long, in a groundmass of oligoclase microlites with abundant interstitial granular iron ore and veined by calcite and quartz with some epidote. Irregular amygdales are filled with calcite and quartz. R.W.E.

Little Caradoc Tuffs. The Comley Andesites are overlain by an apparently concordant intrusion of altered dolerite 200 yd wide. Above this there is a thin, well-bedded dust tuff which is exposed at about 500 yd S.S.W. of Comley Quarry [482961] and is faulted against Cambrian quartzite to the south. The dip is south-west at 57° and minor current bedding suggests a normal succession. At 680 yd S. 40° W. of Comley Quarry [481960] there is an outcrop of similar tuff, bounded by faults to the west and south and probably overlain by Cambrian quartzite to the north and east. A small inlier of tuff, apparently surrounded by Cambrian quartzite at 520 yd S. 22° W. of Comley Quarry [483960], may also belong to this horizon. G.H.M.

The matrix of these tuffs (E 25178, 25180, 25427) consists of a fine-grained quartz-feldspar mosaic with limonite, some hematite and abundant chlorite, and containing small fragments of quartz, rhyolite and feldspar. R.W.E.

Little Caradoc Basalts. Albitized basalt lavas overlie the Little Caradoc Tuffs and occupy the northern ridge of Caer Caradoc as far as the inner ramparts of the summit camp, also extending round the lower slopes of the hill below the camp. The lowest flows are dark grey and aphanitic, similar rocks occurring higher up the ridge, but there are also flow-brecciated and amygdaloidal basalts together with at least one band of lithic tuff. The amygdaloidal rocks are mainly greenish grey or purplish, the amygdales being filled with calcite, chlorite, quartz or hematite. There are probably several distinct flows. G.H.M.

The altered basalts (E 25179, 25439) contain microphenocrysts of albite-oligoclase up to 1·3 mm long, often partly epidotized, in a groundmass of albite-oligoclase laths, prismatic augite, interstitial chlorite and rods of magnetite. Chlorite replaces augite and may pseudomorph olivine (E 25179). Amygdales contain chlorite and epidote and, in one example (E 25439), prehnite, quartz and calcite.

The tuff band (E 25440) is a vitric-lithic tuff containing fragments of albite-oligoclase, sometimes epidotized, andesite, rhyolite, quartz, grains of magnetite and much interstitial hematite. Fragments of pumice and chloritized shards are common. R.W.E.

Caer Caradoc Rhyolites. The junction between the Little Caradoc Basalts and the overlying rhyolites is thought to be a thrust (see p. 266 for discussion) so that the precise stratigraphical relationship between the two groups is not known.

The crags forming the eastern ramparts of the Iron Age Camp on Caer Caradoc consist of pink or dull red rhyolite with marked flow banding similar to the bedding of fine-grained tuffs. In amygdaloidal varieties, the flow surfaces are marked by lines of elongated amygdales. The severe contortion of this flow banding is discussed below (p. 266). Joints and slickensided shear planes are conspicuous. Rhyolitic breccias, thought to be crush breccias, consist of fragments of reddish rhyolite, locally amygdaloidal and up to about 1 in across, set in a brownish granular matrix. Similar breccias occur on The Lawley (p. 11). The structure of these crush breccias resembles flow brecciation and an example with fragments up to 6 in across, forming the top of the crag near the camp entrance [477952], may be due to the latter cause.

The outcrop of the rhyolites extends across the hill from the Church Stretton Fault in the west to the Lawley Fault in the east, the dip being south-westerly. An isolated crag [478959] 600 yd N. of Caer Caradoc summit, is thought to be a thrust slice of these rhyolites and is much sheared and brecciated. G.H.M.

The rhyolites (E 25187, 25428) consist of microlites of albite and orthoclase in a fine-grained matrix containing chlorite, quartz and scattered granular hematite. Spherulites are common and amygdales, often elongated, are filled with quartz and chlorite. The crush breccias (E 25182) contain rhyolite fragments in a very fine-grained matrix rich in chlorite. The larger angular fragments occasionally fit into adjacent fragments but the smaller pieces of rhyolite are more rounded. Thin quartz veins cutting the fragments do not traverse the matrix of the breccia. R.W.E.

Caer Caradoc Andesites. The Caer Caradoc Andesites overlie the Caer Caradoc Rhyolites, the junction being apparently folded. The dip is south-westerly and the outcrop extends across Caer Caradoc, south of the summit. The upper limit is a strike fault, apparently with normal throw which truncates successively higher beds of the Ragleth Tuffs on the west side of the hill as it is followed westwards.

The andesites are mainly dark grey, commonly amygdaloidal and much jointed, with thin tuff bands locally.

Close to the Church Stretton Fault and north of Three Fingers Rock [471952] andesites correlated with the Caer Caradoc Andesites are, with the associated Cwms Rhyolites, thrust over the adjoining dolerite to the east.

G.H.M.

The andesites (E 25420) consist of oligoclase microlites, about 0·05 mm long, in a cryptocrystalline matrix containing granular iron ore and chlorite. Amygdales are filled with quartz, chlorite and calcite.

A specimen of a tuff band (E 25445) from 1070 yd N. 34° E. of New House Farm [467944] is a vitric tuff with shards and fragments of pumice, quartz, oligoclase and andesite in a cryptocrystalline matrix rich in chlorite. R.W.E.

Cwms Rhyolites. The junction of the Cwms Rhyolites with the underlying Caer Caradoc Andesites is exposed in a gully [475949] 550 yd S.S.W. of Caer Caradoc summit. From this section the rhyolite outcrop extends southwards on the east side of the hill as far as a fault trending about north-south near Three Fingers Rock [471947]. To the west of Three Fingers Rock there are outcrops of sheared and crushed Cwms Rhyolites, thought to be a thrust sheet overlying Ragleth Tuffs. Cobbold (1900, p. 96) equated the Cwms Rhyolites with the Caer Caradoc Rhyolites but Dr. G. Wilson appears to have been the first to suggest that they are on different horizons (personal communication). Though the resemblance is close, the Cwms Rhyolites tend to be more reddish purple and green in colour. Flow banding is less conspicuous than in the Caer Caradoc Rhyolites but amygdales and nodular structures occur, particularly near the base of the group. G.H.M.

The nodular rhyolites (E 25186) contain abundant microlites of orthoclase and albite in a very fine-grained groundmass with small irregular grains of leucoxene. The nodular areas, rimmed by thin clear zones poor in leucoxene, are partly devitrified with feldspar microlites set in irregular patches of quartz and possibly feldspar. Elongated amygdales contain chalcedony and quartz. The groundmass of the more crystalline rhyolites (E 25185, 26289) contains recognizable quartz and amygdales contain epidote. A chemical analysis of a specimen of these rhyolites is shown in Table 1, col. 8 (p. 29). R.W.E.

Ragleth Tuffs. The Ragleth Tuffs appear to follow the Cwms Rhyolites conformably and the lowest beds of the group, consisting of pink and brown, fine-grained, bedded, rhyolitic tuffs, form a narrow U-shaped outcrop (Fig. 3) lying athwart the southern ridge of Caer Caradoc Hill. To the north-east a fault separates the tuffs from the Caer Caradoc Andesites, this fault being followed by a dolerite intrusion for much of its course. Observations of dip show that this U-shaped outcrop of tuffs is the expression of a steep syncline, plunging to the north-west, the form of which is clearly seen from the west slopes of Hope Bowdler Hill. The fold core is occupied by a large dolerite sill which possesses an irregular ramifying junction with the underlying tuffs and which encloses a raft of tuff about 100 yd long. On the south-west side of Caer Caradoc Hill a thrust sheet of Cwms Rhyolites and Caer Caradoc Andesites overlies the folded Ragleth Tuffs and associated dolerite.

The main outcrop of Ragleth Tuffs is terminated to the north-east by a fault trending about north-north-west to south-south-east and running near the Three Fingers Rock [471947]. The northern end of the tuff outcrop is bounded by a thrust sheet of Cwms Rhyolites and the outcrop runs thence south-westwards to the south end of Ragleth Hill, being truncated to the south-east by The Lawley and Cwms-Hoar Edge faults of the Church Stretton Complex, and bounded to the north-west by the outcrop of the Helmeth Grit.

At the south end of Caer Caradoc Hill the regional dip is steep to the north-east, the succession in this area including grit bands with conspicuous pink grains and sill-like intrusions of dolerite. It may be that there is repetition by folding and faulting in this part. The presence of rhyolitic flows in this area and on Helmeth Hill is suggested by the unusual grey colour of some rocks together with brecciation and minor quartz veining. On the west side of Hazler Hill there are outcrops of fine-grained tuffs with intrusive dolerites associated with considerable development of epidote.

On Ragleth Hill the regional strike is about north-east to south-west but the dip is variable. A salmon-pink rhyolitic tuff is conspicuous and other types include brownish poorly bedded tuffs, locally brecciated and rough-weathered. Somewhat more basic tuffs are exposed at 450 yd E. 10° S. of Little Stretton church and grits crop out on the north-west side of Ragleth Hill. It is uncertain whether there is repetition by folding or faulting but if the succession is unbroken the rocks on Ragleth Hill must be over 1500 ft in thickness.

The group includes pink, brown and purplish, fine-grained rhyolitic tuffs, locally interbedded with darker greenish rocks of more sedimentary aspect. Bedding may be marked by a delicate colour banding and there are traces of current bedding and slump structures. Graded bedding may be seen in the coarser bands. The sequence includes variable yellowish grits with conspicuous quartz grains, pink feldspar and pink rhyolite fragments. Cobbold (1900, p. 95) has recorded quasi-fossil markings simulating '*Cruziana*' and '*Arenicolites*' from the tuffs. The rocks are much jointed and sheared, locally slickensided, and contain crush breccia similar to those in the underlying lavas (p. 14).

G.H.M.

The tuffs (E 25150, 25188–9, 25470–1) contain small scattered fragments of quartz and feldspar in a very fine-grained matrix containing granules of epidote and sphene and cut by veinlets of quartz, calcite and clear albite with epidote. Some of the tuffs (E 25453) are well banded and contain more abundant fragments of quartz and feldspar. The lithic tuffs (E 25456) contain fragments of rhyolite, andesite, quartz and plagioclase in a fine-grained quartzo-feldspathic matrix containing granular epidote. The rhyolite fragments include spherulitic varieties with perlitic fractures marked out by chlorite or with the rhyolite completely replaced by chlorite.

The coarse grits (E 25431) contain fragments, up to 9 mm across, of quartz, microcline, albite, quartzite and rhyolite in a matrix of quartz and feldspar with interstitial chlorite. These rocks resemble the Helmeth Grit.

Rocks, from near the top of the group, which may be rhyolite flows (E 25447) consist of small irregular crystals, about 0·05 mm across, of albite, orthoclase and quartz in a very fine-grained quartzo-feldspathic matrix containing flakes and small rounded amygdales of chlorite. Some specimens (E 25429) are brecciated and cut by veinlets of quartz and epidote.

A soda-trachyte lava (E 25443) consists of subhedral albite laths, occasional microphenocrysts of albite with a little interstitial quartz and accessory epidote and granular sphene. The rock is cut by veinlets of quartz and clear albite and small quartz amygdales are rimmed by a thin zone rich in hematite. R.W.E.

Intrusions. Most of the intrusions are altered dolerites which vary considerably in grain-size, the finer varieties resembling the more basic lavas. The most important is the sill-like intrusion forming the core of the syncline south-west of Caer Caradoc. Another large, coarse sill occurs at the north end of Little Caradoc [482962]. Fig. 3 shows the distribution of the intrusions in the Caer Caradoc area.

G.H.M.

The fine-grained varieties (E 25151) contain highly sericitized oligoclase, subophitic partly chloritized augite, platy and skeletal ilmenite, and intersertal patches of turbid feldspar, leucoxene and granular sphene. Chlorite occurs interstitially and with quartz in amygdales.

An altered quartz-dolerite (E 25181), from [482962] 585 yd E. 60° N. of Botvyle, contains oligoclase laths, partly chloritized subhedral augite, quartz interstitially and in myrmekitic growths around the feldspar, skeletal ilmenite associated with granular sphene, and accessory apatite. Feldspar is partly epidotized. A rock of similar type (E 25451) was mapped [463928] 1080 yd E. 42° S. of Church Stretton station. This intrusion contains large ovoid xenoliths of coarse-grained quartz-epidote-rock (E 25449, 25452), the quartz enclosing acicular fibrous chlorite.

An altered fine-grained doleritic intrusion in the Ragleth Tuffs (E 25430), from [470946] 440 yd N.E. of New House Farm contains oligoclase phenocrysts in a fine-grained matrix of oligoclase laths, interstitial chlorite with granular iron ore, and accessory anatase and epidote. Patches of chlorite and calcite may replace augite. Calcite occurs in thin veins and amygdales are filled by chlorite and carbonate.

A coarse highly altered dolerite (E 25448), from [462928] 650 yd N. 35° E. of Dryhill, contains laths of albite-oligoclase partly replaced by prehnite which is also abundant in veins. Patches of chlorite may replace original augite.

Rhyolite dykes crop out at 980 yd and 1130 yd N. 31° E. of New House Farm. They contain microlites of albite and orthoclase in a cryptocrystalline matrix (E 25446) or in a patchy matrix, locally cryptocrystalline but commonly composed of irregular micropoikilitic plates of quartz and feldspar (E 25444). The rocks contain granular hematite, occasional small spherulites and are cut by veinlets of quartz, locally epidotic. One specimen (E 25444) shows perlitic fractures and also brecciation, some quartz veinlets being brecciated while others cut the brecciated patches. G.H.M., R.W.E.

HAZLER HILL

Hazler Hill forms a triangular area of Uriconian rocks bounded on the north-west by the Cwms–Hoar Edge Fault (F3 of Cobbold) on the north-east by a fault running from Hope Bowdler to near Sandford Seat [468932], and to the south overlain unconformably by the Harnage Shales (Fig. 4).

The rocks are mainly albitized, amygdaloidal, basaltic lavas, possibly exceeding 1200 ft in thickness, locally with interbedded fine-grained tuffs. The most extensive exposures are in and around Hazler Quarry [463924], celebrated for its Neptunian dyke (Strachan, Temple and Williams 1948, p. 276), and along the lane from Sandford Seat to Hazler Barn [468926]. Sections from these two areas are compared as follows:

Hazler Quarry		Hazler Barn area	
Basalt	50 ft	Basalt	100 ft
Tuff	6 ft	Tuff	3 ft
Basalt	2 ft	Grit	unknown
Tuff	11 ft	Tuff	unknown
Basalt	18 ft	Basalt	25 ft
(Northerly dip at 50–80°)		Tuff	100 ft
		Basalt	50 ft
		(North-westerly dip at 20–35°)	

The relationship between these two sections is not known. Since the strike of both groups is similar it is suggested that a considerable dip fault may separate them.

Near the road on the west side of Hope Bowdler [473924] there are exposures of fine-grained rhyolite tuffs and more andesitic pyroclastic rocks, apparently associated with very weathered doleritic rocks. These are overlain by Harnage Shales. G.H.M.

The albitized basalts (E 25459–60, 26291) are relatively coarse, and contain albite-oligoclase laths, 0·3-0·6 mm long, subophitic augite, interstitial chlorite, granular sphene and platy iron ore. In amygdaloidal varieties (E 25460, 26291) the amygdales are filled by chlorite and prehnite, locally with quartz. The chemical analysis of a rock from this group is given in Table 1, col. 1 (p. 29). R.W.E.

Hope Bowdler Hill

The Uriconian area of Hope Bowdler Hill is completely bounded by the Sharpstones and Willstone Hill thrusts and the Cwms–Hoar Edge (F3) Fault (Fig. 4). The principal rock types are rhyolites, rhyolitic tuffs, grits and conglomerates, the latter being the most distinctive. There are intrusions of altered dolerite and locally there is much alteration to epidote, some rocks being largely composed of that mineral. The structure of the area appears to be anticlinal with a north-easterly fold axis running through the highest point of Hope Bowdler Hill. The stratigraphical sequence is uncertain but it appears to be in downward order as follows:

3. Andesites
2. Conglomerates
1. Rhyolites and andesites with tuff bands.

Rhyolites and andesites with tuff bands. Except near the Gaer Stone [472934] there are few exposures of the lowest Uriconian rocks of Hope Bowdler Hill which consist of rhyolitic and andesitic lavas and tuffs cut in intricate fashion by numerous ramifying doleritic intrusions. G.H.M.

The rhyolites (E 25201, 25205) contain spherulites, up to about 0·5 mm across, commonly more iron-stained than the groundmass, the latter comprising a fine-grained sericitic felsitic mosaic (E 25201) or a felt of orthoclase microlites with albite and interstitial quartz (E 25205). Quartz-epidote veins (E 25201) are cut by a later generation of quartz veins poor in epidote. One specimen (E 25462) is flow banded and contains albite phenocrysts up to about 2·5 mm long.

A tuff band from near the Gaer Stone (E 25204) contains fragments of quartz, albite, andesite and rhyolite in a very fine-grained matrix.

Rocks from farther north near Cwms Farm [474940] include brecciated and epidotized rhyolites [E 25474–6] with albite-oligoclase microphenocrysts, sometimes much sericitized and epidotized, in a fine-grained interlocking quartzo-feldspathic groundmass containing albite and orthoclase microlites. One specimen (E 25474) contains rare chlorite pseudomorphs, perhaps after augite. Accessory minerals include epidote, sphene and iron ore. The rocks are cut and brecciated by veins of quartz-epidote and calcite. R.W.E.

In the area of Willstone Hill, the crags of Battle Stones [486944] and those to the south-east are composed of fine-grained pink rhyolitic tuffs with some coarser beds containing grains of purple rhyolite and pink feldspar crystals. The eastward dip of 30° in the tuffs is almost exactly opposite to that of the overlying conglomerate. This anomaly may be due to local folding, or possibly

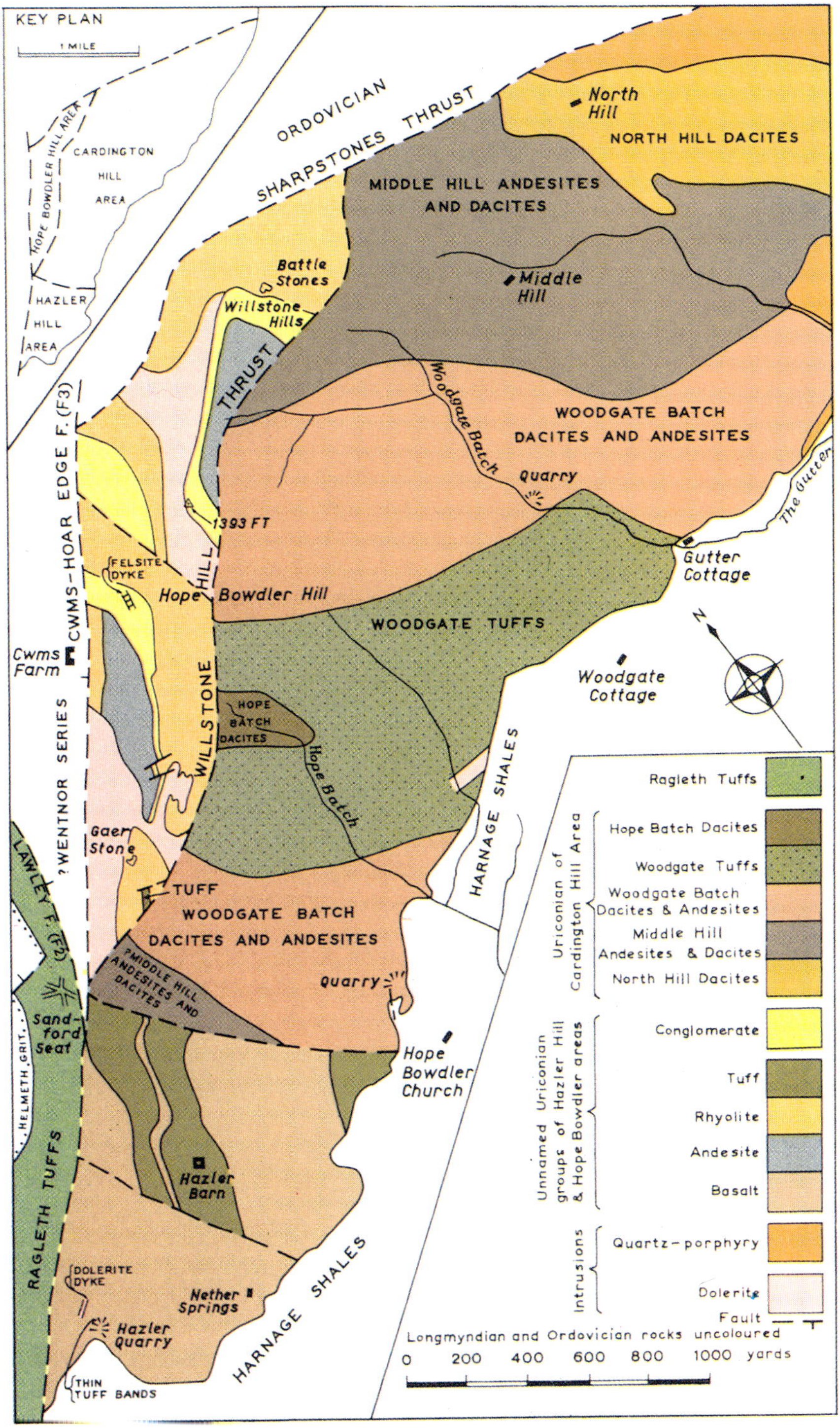

FIG. 4.- Geological map of the Uriconian rocks of the Hazler Hill, Hope Bowdler Hill and Cardington Hill areas.

a local unconformity. Elsewhere on Willstone Hill the conglomerates overlie altered dolerites. The proximity of the conglomerate at Battle Stones to acid tuffs and altered dolerites suggests correlation of these tuffs with the similar rhyolitic beds of the Gaer Stone. G.H.M.

The tuffs (E 25172, 25176) contain fragments of sericitized albite up to 3 mm across, quartz, rhyolite and andesite. The fine-grained quartzo-feldspathic groundmass is locally sericitic and chloritic. A brecciated rhyolite (E 25421) from 280 yd E. 5° S. of Battle Stones [488944] is similar to the rocks in the centre of the anticline on Hope Bowdler Hill. The rock contains microphenocrysts of altered albite-oligoclase with subordinate quartz in a fine-grained groundmass of albite, orthoclase and interstitial quartz with accessory platy and granular iron ore, granular sphene and abundant granular epidote. The rock is brecciated and cut by veins of quartz and epidote. A tuff (E 25422) from 350 yd E. 10° N. of Battle Stones [489945] is slightly brecciated and contains irregular fragments of quartz, albite-oligoclase and orthoclase up to about 0·3 mm across. The bedding is marked by chlorite wisps and the rock contains zircon and ilmenite, partly altered to leucoxene.

Fine-grained porphyritic andesites from the centre of the Hope Bowdler Hill outcrop contain partly altered oligoclase phenocrysts, up to 2·5 mm long, with hematite pseudomorphs after amphibole (E 25195), in a microlithic groundmass of albite-oligoclase with interstitial chlorite, locally rich in hematite and magnetite (E 25194) and locally highly sericitized (E 25195). Similar fine-grained amygdaloidal andesites occur close to the Gaer Stone (E 25436, 25437) and at 100 yd E. 20° N. [476940] of Cwms Farm (E 25466).

There are a few beds of tuff within this group. A vitric-lithic tuff (E 25199) from 400 yd E. 4° N. of Cwms Farm [478939] contains fragments of very fine-grained rhyolite, andesite, quartz, albite-oligoclase, accessory zircon and apatite, and flakes of altered biotite and muscovite. The fine dust matrix contains abundant large devitrified shards and the rock is cut by quartz veins. A highly epidotized tuff (E 25200) from 310 yd E. 40° S. of Cwms Farm [477938] contains fragments of quartz, albite-oligoclase and rhyolite. G.H.M., R.W.E.

Conglomerates. The conglomerates occur in two main outcrops, the more westerly running along the west side of Hope Bowdler Hill towards the Gaer Stone near which it is apparently faulted. At one place there is a westerly dip of 65°. A small faulted area of conglomerate lies north-east of the main outcrop.

The easterly occurrence of conglomerate is on Willstone Hill, where it forms an arcuate outcrop with a dip mainly to the south-east but swinging to the south-west at the eastern end of the outcrop.

The mapping suggests that these occurrences of conglomerate belong to the same bed and this provides the main evidence for an anticline running north-eastwards across the summit of Hope Bowdler Hill.

The matrix of the conglomerates is generally dark grey but in finer varieties it may be light brown. Pebbles include quartz, quartzite, pink feldspar and igneous rocks, mainly of basic composition. G.H.M.

Andesites. The conglomerate of Willstone Hill is overlain by an amygdaloidal andesite (E 25424) containing highly sericitized albite-oligoclase laths about 0·3 mm in length, chlorite pseudomorphs, probably after pyroxene and amphibole, with interstitial iron ore. The amygdales contain chlorite with some carbonate and quartz.

This andesite is succeeded by highly epidotized rocks (E 25425), possibly andesites or tuffs containing andesite lapilli.

Above these beds are epidotized andesites (E 25464) containing albite-oligoclase microphenocrysts with occasional chlorite pseudomorphs after amphibole, in a fine-grained matrix of stumpy albite-oligoclase laths with interstitial chlorite, abundant epidote, quartz and granular leucoxene. Amygdales are filled by quartz, epidote and chlorite; in some there is an outer rim of hematite. Thin quartz veins cut the rocks.

In the Willstone Hill area, higher beds are cut out by the Willstone Hill Thrust. On the western side of Hope Bowdler Hill the conglomerates are overlain by green and purple flow-brecciated andesites and altered dolerites, locally amygdaloidal and commonly epidotic. A specimen (E 25208) from 340 yd E. 40° N. of the Gaer Stone [475936] contains large fibrous epidote pseudomorphs, probably after pyroxene or amphibole and pseudomorphs of paler, more massive epidote after feldspar. The groundmass contains albite-oligoclase microphenocrysts in a fine-grained feldspathic matrix with interstitial chlorite, granular leucoxene and abundant granular epidote. Amygdales contain radiating sheaves of epidote in association with quartz and chlorite. A rock (E 25463) from 170 yd E. 36° N. of the Gaer Stone [474935] contains albite-oligoclase laths with abundant subophitic to ophitic plates of augite, partly chloritized and with interstitial chlorite and granular sphene. Amygdales are filled by chlorite, quartz and epidote and are commonly surrounded by zones rich in hematite.

An isolated exposure of rhyolite, which may possibly overlie the andesites, occurs near the Cwms–Hoar Edge Fault, 90 yd S. 12° E. of Cwms Farm [475938]. The rock (E 25193) contains a few microphenocrysts of albite-oligoclase and quartz in a fine-grained quartzo-feldspathic matrix containing disseminated sericite and chlorite and irregular crystals and aggregates of epidote. The rock is much brecciated with epidote concentrated in the very fine-grained material between the fragments.

Intrusions. A poorly exposed acid intrusion (E 25197) in the conglomerates from 250 yd E. 7° S. of Cwms Farm [477939], contains occasional albite microphenocrysts in a microlithic groundmass of albite, possible orthoclase and quartz. There are some myrmekitic intergrowths and the rock, cut by veins of quartz with epidote and chlorite, is slightly brecciated, the matrix between the fragments containing much epidote. A porphyry (E 25465), probably intrusive, from 250 yd E. 26° N. of Cwms Farm [477941], contains albite microphenocrysts in a groundmass of granular quartz, albite and orthoclase, with chlorite, epidote, and iron ore, commonly altered to sphene. G.H.M., R.W.E.

The dolerite intrusions are important, the largest outcrops occurring west and north of the Gaer Stone and on Willstone Hill. The best exposed area is north-east of the Gaer Stone where the interpenetration of dolerite and rhyolite is seen. G.H.M.

A fine-grained altered dolerite (E 25198) from 470 yd E. of Cwms Farm [479939], contains albite-oligoclase microphenocrysts in a matrix of similar feldspar with interstitial chlorite. Possible original pyroxene or amphibole is pseudomorphed by chlorite and quartz-epidote-aggregates outlined by hematite. There is abundant granular epidote and some granular sphene.

An altered dolerite (E 25206) from 200 yd N. 33° E. of the Gaer Stone [473935] contains albite-oligoclase laths, small chlorite patches, probably after a ferromagnesian mineral, interstitial quartz and chlorite and accessory sphene, apatite and ilmenite. Another part of the same intrusion (E 25207) at 215 yd E. 15° N. of the Gaer Stone [475934] is extremely altered, the feldspar being largely replaced by chlorite, epidote and prehnite. Subophitic prisms of augite occur and the rock is cut by quartz-epidote-prehnite veins.

A highly altered olivine-dolerite (E 25209) from 200 yd E. 28° N. of the Gaer Stone [474935] contains serpentine pseudomorphs after olivine, carbonate-chlorite pseudomorphs after ophitic pyroxene, and albite-oligoclase laths partly replaced by carbonate, sericite and chlorite. Accessory ilmenite shows alteration to leucoxene and sphene. R.W.E.

Near the Gaer Stone the conglomerates dip north-west at 50° while the dip of flow banding in the underlying rhyolite is in the opposite direction. The relation between conglomerate and rhyolite may be unconformable (see also Battle Stones, p. 18) or it may be a fault along which the dolerite has been intruded, in the ground about 200 yd N.E. of the Gaer Stone. In this area the dolerite locally forms distinct veins in the rhyolite while in some places xenoliths of rhyolite are all that remains of the country rock. An excellent exposure of pink and white brecciated rhyolite forms a small crag, 370 yd south-west of the summit of Hope Bowdler Hill [478938]. Farther to the north-east there are highly epidotic rocks, with an epidotized matrix and including knots and elongated vein-like masses of richly epidotic material. The latter may represent altered xenoliths of rhyolite and other country rock within the dolerite. Some of these epidotic rocks are apparently altered andesites and have been mapped as intrusive. G.H.M.

Cardington Hill

The Uriconian rocks of the Cardington Hill area extend from Hope Bowdler (Fig. 4) eastwards to Stoneacton. They are limited to the north-west by the Willstone Hill Thrust, to the north by the Sharpstones Thrust and to the south by an unconformable cover of Caradoc and Silurian sediments. The western margin of the area may be a fault but lack of exposures makes this uncertain. A small faulted inlier of Uriconian rocks occurs at Hill End.

The mapping suggests a general north-westerly trend throughout the area. A distinctive group of rocks, here called the Woodgate Batch Dacites and Andesites, occurs in two outcrops, the more westerly extending from Hope Bowdler [476926] almost to the Gaer Stone [472934], and the eastern outcrop running from near Gutter Farm [498931] to the southern flanks of Willstone Hill. The few observations of flow banding in the western outcrop suggest a north-easterly dip, whereas in the eastern outcrop, flow banding dips to the south-west. It is therefore suggested that these two outcrops lie on opposing limbs of a syncline the core of which is occupied by a belt of tuffs, the Woodgate Tuffs, separating the two outcrops of dacite (Fig. 4). These tuffs are the highest beds seen in the area apart from some thin bands of dacite poorly exposed at the head of Hope Batch [478935]. On this hypothesis, the sequence to the north-east of the synclinal axis is a descending one as follows:

	Approx. thickness in ft
Hope Batch Dacites	100
Woodgate Tuffs	900
Woodgate Batch Dacites and Andesites	700
Middle Hill Andesites and Dacites	1200
North Hill Dacites	900
Stoneacton Tuffs and Andesites	Not known

There are also intrusions of quartz-porphyry, quartz-microdiorite and altered dolerite.

Stoneacton Tuffs and Andesites. The Stoneacton Tuffs are poorly exposed at the east end of Cardington Hill [511937]. They include dark greenish grey tuffs with green lithic fragments and light brown gritty beds with pink rhyolite fragments. The outcrop is limited to the west by a considerable exposure of quartz-porphyry. In the Heath Brook [512937] there is a small development of andesite. G.H.M.

The crystal-lithic tuffs (E 25154) contain fragments of rhyolite, commonly carbonated, albite-oligoclase and quartz set in a fine-grained chloritic, quartzo-feldspathic matrix. Fragments of andesite occur in other specimens (E 25153). The porphyritic andesites of Heath Brook (E 25590) contain phenocrysts and microphenocrysts of extensively chloritized and sericitized plagioclase. The fine-grained matrix comprises altered albite-oligoclase with interstitial chlorite, granular magnetite and carbonate. Patches of turbid carbonate and granular iron-ore probably replace original pyroxene or amphibole. R.W.E.

North Hill Dacites. The outcrop of the North Hill Dacites runs south-eastwards from near North Hill [496942] towards Wall Bank [503930] and near its southern end is cut across by intrusive quartz-porphyry. The rocks are fine grained and pink or purple and grey in colour, leached to pale pink or white where weathered. G.H.M.

The dacites (E 25161) contain albite-oligoclase microphenocrysts, largely calcitized, and occasional microphenocrysts of hematite and chlorite, possibly pseudomorphing amphibole. The fine-grained quartzo-feldspathic groundmass contains irregular patches of quartz, granular magnetite and rods of hematite with chloritic alteration of the feldspar. R.W.E.

Middle Hill Andesites and Dacites. The Middle Hill Andesites and Dacites are named from their outcrop on Middle Hill [493937]. The most striking rock type is a porphyritic lava, with white feldspar phenocrysts up to 5 mm long, and occasional quartz phenocrysts, in a fine green matrix. A finer-grained variety contains pink feldspar phenocrysts and numerous chloritic amygdales. Fine-grained aphanitic rocks are also present. G.H.M.

There are two main textural types among the andesites. One (E 25157–8, 25171, 25236) contains oligoclase microphenocrysts with chlorite pseudomorphs, sometimes clearly after augite (E 25157–8), in a fine-grained matrix of oligoclase microlites with interstitial chlorite and granular iron ore. The second type (E 25159, 25235) tends to be orthophyric, with altered plagioclase microphenocrysts and pseudomorphs after hornblende, in a matrix of stumpy laths and plates of oligoclase with a dense groundmass rich in iron ore. Both types contain amygdales filled with chlorite, quartz and calcite.

The dacites (E 25156, 25164–7, 25170, 25173, 26288) contain corroded quartz phenocrysts up to about 3·5 mm long with altered albite-oligoclase phenocrysts. There are also chloritic pseudomorphs after hornblende (E 25164–5), augite, and perhaps biotite (E 25173). The equigranular quartzo-feldspathic matrix contains albite-oligoclase microlites. In some sections (E 25156) the matrix is composed largely of albite-oligoclase microlites with interstitial chlorite and quartz. A chemical analysis of a specimen of the dacites is shown in Table 1 col. 4 (p. 29). R.W.E.

To the south-east of Sandford Seat [468932] there is a small area which may be occupied by rocks of the Middle Hill group on the western limb of the syncline postulated above (p. 21). The only two exposures are of vitric tuff and andesite, 370 yd S.S.W. of the Gaer Stone [471931], occurring a short distance west of sections of Woodgate Batch rocks. G.H.M.

The vitric tuff (E 25213) is composed of pumice fragments, some apparently palagonitic, and shards. The amygdales in the pumice contain chlorite and occasionally quartz. A specimen of andesite (E 25438) collected only a few feet away from the vitric tuff is composed of oligoclase microlites in a crypto-crystalline matrix with abundant chlorite and quartz amygdales. R.W.E.

Woodgate Batch Dacites and Andesites. In their two separate outcrops the Woodgate Batch Dacites and Andesites often weather to form crags. They are well exposed in the valley, here called Woodgate Batch, to the north of Woodgate Cottage [487929] where they are brecciated in places and display conspicuous flow banding with a regional south-westerly dip of up to 60°. The upper part of a quarry section [488934] 570 yd N. 10° E. of Woodgate Cottage shows rudely spheroidal andesite, red, green and grey in colour with pink feldspars. The spheroids are about one foot across by several feet long and are surrounded by rotten amygdaloidal material, the amygdales lying parallel to the margins of the spheroids. The lower part of the quarry shows about 18 ft of massive porphyritic andesite, variably blue, greenish or pale pink in colour with red or pinkish phenocrysts. The rock is well jointed and is cut by purple and yellow veins of rotten rock up to 1½ in across. In the western outcrop of the group a section in an old quarry [476926], 250 yd N. of Hope Bowdler church shows about 25 ft of purple and pink porphyritic dacite, locally sheared. Similar rocks are exposed to the north towards the Gaer Stone. Local flow banding dips north-north-east at 30°–46°. G.H.M.

The dacites and related andesitic rocks contain albite-oligoclase phenocrysts and microphenocrysts with pseudomorphs after pyroxene, in a fine-grained microlithic albite-oligoclase groundmass, commonly with good flow structure. In the andesites (E 25210, 25290, 26287) the microlites occur in a cryptocrystalline matrix with much chlorite and granular iron ore, whereas in the dacites (E 25212, 25225, 26286) they are generally set in irregular micropoikilitic plates of quartz and feldspar. Chemical analyses of an andesite and a dacite from this group are shown in Table 1, cols. 3 and 6 (p. 29). R.W.E.

Occasional tuff bands occur within the Woodgate Batch group; a lithic-crystal tuff (E 25224) from 330 yd S. 12° W. of the Gaer Stone [472931] contains abundant dacitic fragments, much replaced by carbonate, with fragments of quartz and albite-oligoclase in a fine-grained quartzo-feldspathic matrix.
G.H.M., R.W.E.

Woodgate Tuffs. The outcrop of the Woodgate Tuffs extends north-westwards from Woodgate Cottage [487929] across the middle part of Hope Batch [477931]. The rocks are softer than the adjacent dacites and andesites, and exposures are few. The rocks comprise green and pink lithic tuffs with fragments about ½ in across and grits with grains of quartz and pink feldspar. G.H.M.

A crystal-vitric-lithic tuff (E 25215) from 670 yd N. 4° E. of Hope Bowdler church [476930] contains fragments of quartz, albite-oligoclase, some felds-pathic grit and fine-grained acid igneous rock. There are abundant recrystallized shards. There is some green biotite, partly altered to limonite, while the very fine-grained quartzo-feldspathic base is also limonitized. A lithic tuff (E 25216) from 600 yd N. 19° E. of Hope Bowdler church [478929] is composed largely of fragments of andesite with some quartz and oligoclase in a fine-grained quartzo-feldspathic matrix. A well-bedded crystal-vitric tuff (E 25227) from 1260 yd E. 36° N. of Hope Bowdler church [486931] is in part very fine-grained and chloritic, with nests of minute clear albite crystals, while the rest of the rock contains abundant devitrified shards. There are also some fragments of chloritized pumice and spherulitic rhyolite. A lithic tuff (E 25219) from 880 yd N. 40° E. of Hope Bowdler church [481929] contains fragments of andesite and dacite, similar to rocks of the Middle Hill group, together with fragments of rhyolite, quartz and acid plagioclase, in a fine-grained matrix rich in chlorite. G.H.M., R.W.E.

Hope Batch Dacites. At the head of Hope Batch [478934] there are several exposures of green dacite. Their relationships to other rocks are uncertain but it is suggested that they overlie the Woodgate Tuffs in the core of the syncline of Cardington Hill. G.H.M.

The dacites (E 25217) contain abundant albite-oligoclase microphenocrysts with chlorite-quartz pseudomorphs after augite and possibly hornblende. The fine chloritized groundmass contains albite-oligoclase, quartz and accessory leucoxene and granular sphene. R.W.E.

Intrusions. The main intrusion of this area is the quartz-porphyry which occupies the eastern end of Cardington Hill and underlies the Cambrian rocks of Hill End. Exposures are few and the mapping is dependent largely on scattered rock fragments. The intrusion appears to cut the North Hill, Middle Hill and Woodgate Batch groups but its actual shape is uncertain. The rock is pink with conspicuous phenocrysts of white feldspar, quartz and a ferromagnesian mineral. Close to the junction with the Stoneacton Tuffs there are exposures of mottled pink and green quartz-microdiorite, possibly a separate intrusion. G.H.M.

The quartz-porphyry (E 25155, 25226, 26290A and B) contains corroded phenocrysts of quartz up to 6 mm across, sericitized albite-oligoclase, ortho-clase (E 26290B) and biotite (E 25155). Chlorite may replace original amphibole. The groundmass, commonly granular, contains quartz, albite-oligoclase and orthoclase with irregular laths of albite-oligoclase and locally with myrmekite intergrowths (E 25155). In one specimen the groundmass is composed of irregular laths and plates of feldspar with interstitial quartz. Accessory minerals include ilmenite, leucoxene, sphene and rutile with occasional zircon. The quartz-microdiorite (E 25152) contains phenocrysts and microphenocrysts of sericitized albite-oligoclase with chlorite pseudomorphs after ferromagnesian minerals. The groundmass comprises albite-oligoclase laths, chlorite pseudo-morphs probably after augite, interstitial quartz and myrmekite intergrowths and accessory skeletal ilmenite. R.W.E.

A small dyke-like intrusion (E 25160) from 760 yd E. 18° N. of Middle Hill [498941] contains scattered albite-oligoclase phenocrysts, partly altered, and a few chloritic patches, possibly replacing original ferromagnesian phenocrysts, in a fine matrix of albite-oligoclase laths with patches of chlorite and calcite, perhaps after pyroxene, interstitial quartz and accessory leucoxene.

A small intrusion of altered dolerite (E 25218) in the Woodgate Tuffs, 1070 yd N. 33° E. of Hope Bowdler church [482933], shows ophitic relationships between albite-oligoclase and augite. Hematitized iron ore and granular sphene occur. Amygdales are filled by chlorite.

A small dyke from 1335 yd E. 35° N. of Hope Bowdler church [487931] is an altered fine-grained tholeiitic dolerite (E 25229) with laths and plates of albite-oligoclase, much interstitial iron ore, granular sphene, and small amygdales filled by quartz, chlorite and albite. G.H.M., R.W.E.

Wart Hill

A small inlier of Uriconian igneous rocks occurs at Wart Hill, within the zone of the Church Stretton faults. A variety of rocks is seen in a number of scattered exposures which are generally small, and the stratigraphical relationships between the different rocks are difficult to determine.

From just south of the road [400846] northwards to the summit there are several outcrops of coarse-grained tuff or tuffaceous greywacke. The more northerly beds, which appear to dip to the south-east at about 80°, are dark green and generally agglomeratic, but there are some fine-grained silicified bands which in hand specimen resemble andesite. The beds nearer the road are of a reddish colour and are evenly coarse grained. A pink andesite crops out in this area, about 170 yd S. 22° W. of the summit [400846].

Banded and brecciated rhyolites, vertical with a south-easterly strike, are exposed on the north-eastern spur of the hill [404849]. Rhyolites and tuffs crop out west and north of the summit, the banding apparently inclined at low angles towards the north-west and north. They are thought to underlie the andesites and tuffs which are exposed at many places on the steep northern slopes of the hill, apparently dipping northwards at between 45° and 80°. The rhyolites which crop out beside the footpath about 310 yd N.N.E. of the summit [402850] are thought to be the highest rocks of this group.

Altered dolerite which occurs in the wood about 400 yd E. 7° S. of Upper Carwood [406852] and in the stream 150 yd to the north [406854] is probably of Uriconian age.

The extension of the Wart Hill outcrop south-westwards is indicated by an outcrop of andesite and tuffaceous greywacke, like those seen on the southern slopes of the hill, in the stream 500 yd W. 31° S. of the summit [397845].

The junction between the Uriconian rocks and those adjacent to them is nowhere exposed and the faulted junctions are to a large extent conjectural.

D.C.G.

PETROGRAPHY OF URICONIAN OF WART HILL

Tuffs. The Uriconian tuffs include the following types: crystal tuffs, lithic tuffs with a varied assemblage of igneous and metamorphic rock fragments, and welded and non-welded tuffs (ignimbrites).

Crystal tuffs consist almost entirely of broken fragments of feldspar (mainly oligoclase) and quartz, varying in grain size from about 0·03 mm to about

2·0 mm. Coarse-grained crystal tuff or tuffaceous sediment (E 29533) occurs at 725 yd S. 14° W. of Upper Carwood [400847], and consists of fragments up to 2 mm of feldspar (mainly oligoclase and perthite, but also more rarely microcline) and some quartz. A few granophyre fragments occur consisting of micrographic intergrowths of quartz and alkali-feldspar with no coloured minerals. In addition rare fragments of crushed and sutured quartzite and quartzose schist are present. Epidote is common, and chlorite and clay-minerals together with minor amounts of finely divided hematite form the interstitial material.

Tuffs or tuffaceous sediments containing a wide variety of lithic fragments are common in the Upper Carwood area (E 29523, 29530, 29532, 29542–4, 30207). The most abundant fragments are igneous and range from porphyritic and non-porphyritic andesites (E 29543) and basalts (E 29523) through flow-banded andesitic rocks and devitrified, flow-banded (E 29523) and spherulitic rhyolites (E 29523, 29543) to granophyres (E 29523, 29532, 29543, 30207). Fragments of feldspar (mainly oligoclase with some perthite), quartz and epidote comprise a minor proportion of the tuffs. Fragments of metamorphic rocks occur including quartz-schists, sutured and foliated quartzites (E 29544), and muscovite-schists (E 29523). The groundmass of some of the specimens (e.g. E 29542, 29544) consists of sericite, chlorite, clay minerals, magnetite and microcrystalline quartz.

The tuffaceous sediment from 975 yd S. 31° W. of the west end of Upper Carwood [397845] (E 29523) is in contact with a porphyritic andesite (E 29524).

No welded tuffs are known *in situ* from the Uriconian of the Church Stretton District but both welded and non-welded tuffs occur elsewhere in the Uriconian (Dearnley 1966), on Earl's Hill, The Wrekin and The Ercall (Shrewsbury Sheet 152). Practically unwelded pumice and shard tuffs (sillars of Fenner 1948) with rare lithic fragments are found in the Eastern Uriconian of the Hope Bowdler–Willstone Hill area, two miles east of Church Stretton (E 25172, 25699, 25215, 25227–8, *see* Plate 2 A, B).

Some of these tuffs (e.g. E 25199, 25172) contain fragments of thoroughly welded tuffs, in which a (eutaxitic) flattened shard structure is clearly visible.

One specimen of crystal-vitric tuff from 440 yd S. 8° E. of the east end of Upper Carwood [402849], (E 29535) consists dominantly of a dusty quartzo-feldspathic, partly recrystallized groundmass containing relatively abundant devitrified glassy shards in which fragments of sodic plagioclase, quartz, basic to intermediate volcanic rocks and rare fragments of a devitrified welded tuff occur.

The comparatively widespread occurrence of ignimbrites in the Uriconian is indicated by the presence of fragments of welded tuff in the Longmyndian Haughmond Conglomerate (E 29939); the Oakswood Conglomerate (E 29938), the Lawn Hill Conglomerate (E 29945) and the Huckster Conglomerate (E 29955).

Basalts. The Wart Hill basalts invariably show some degree of alteration with the development of oligoclase-albite, chlorite and serpentine. The amount of alteration may vary considerably within a small area as in the scattered outcrops [406852], about 400 yd E. 8° S. of the east end of Upper Carwood (E 29519–22). The least altered basalts (E 29521–2) consist of albite-oligoclase phenocrysts with ophitic pale brown titanaugite which is locally altered to chlorite. Olivine phenocrysts (E 29521) are present and are usually pseudomorphed by serpentine and magnetite altering to hematite. Interstitial chlorite is abundant and some replacement of pyroxene by chlorite occurs. Accessory apatite and ilmenite altered to leucoxene are present. The rock is cut by veins of quartz, calcite and prehnite, and the amygdales are filled with chlorite, quartz and epidote.

The more extensively altered porphyritic olivine-basalts (E 29519–20, 29529) consist of laths of albite-oligoclase up to 1·0 mm long in a groundmass of albite-oligoclase laths with pseudomorphs of chlorite and serpentine after pyroxene and olivine (E 29529) and chlorite-filled interstitial areas. The opaque oxides are altered to hematite. The amygdales are filled by quartz or quartz, chlorite and epidote. The porphyritic olivine-basalt from 300 yd S. 11° W. of the east end of Upper Carwood [402850] (E 29529) is interbanded with rhyolites (E 29526–8).

Rhyolites. These have been examined from a number of localities in the Wart Hill area; 285 yd S. 3° W. of the east end of Upper Carwood, on the northern side of Wart Hill [402850] (E 29526–8); on the eastern slope of Wart Hill, at 500 yd S. 20° E. of the east end of Upper Carwood [404848] (E 29537–8), and 520 yd S. 17° E. of the east end of Upper Carwood [404848] (E 29539); just north of the inner camp ring, Wart Hill [401848], 490 yd S. 11° W. of the east end of Upper Carwood (E 29540–1); 285 yd S. 1° W. of the east end of Upper Carwood [402850] (E 30206) and 630 yd S. 15° W. of the east end of Upper Carwood [400847] (E 30209).

Two main varieties of rhyolite occur, a banded type (E 29527, 29537–9, 30209) and a massive homogeneous type (E 29526, 29528, 29540–1, 30206); neither type is porphyritic. The former has a pronouced banding of layers of slightly differing composition and texture or grain size, presumably due to flowage. These rhyolites are completely devitrified and the recrystallization results in a banded maculose quartzo-feldspathic mosaic showing aggregate polarization. The banding is emphasized by magnetite altering to hematite. Local bands and patches of microcrystalline quartz occur, and the rock is cut by thin irregular veins of clay minerals and microcrystalline quartz and chlorite.

Some rhyolites (E 29537–9) are pale brown in colour due to fine hematite dust. Irregular patches of opaque oxides altering to hematite are abundant. The main bulk of these rhyolites consist of flow-oriented feldspar microlites and microcrystalline quartz, but thin local bands and irregular patches show a coarser-grained (microgranitic) crystallization of feldspar laths up to 0·1 mm (with dusty sericite and clay mineral alteration) set in a patchy quartz mosaic with specks of magnetite altering to hematite. Veins of microcrystalline quartz and chlorite cut the rhyolite.

A brecciated structure is seen in one banded rhyolite (E 30209) in which there is a considerable amount of secondary sericitization.

The massive homogeneous rhyolites consist of devitrified and recrystallized quartz and feldspar showing aggregate polarization. Patches in some specimens (E 29528, 30206) show perlitic cracks. Local alteration has produced abundant sericite and (locally) felted masses of clay minerals.

Andesites. Porphyritic andesite (E 29524, 30208) interbanded with tuffs (E 29523, 30207), is exposed in the small stream south of Round Oak [397845],

975 yd S. 31° W., from the west end of Upper Carwood. The andesite consists of a groundmass of flow-textured oligoclase microlites with abundant granules of opaque ore and interstitial chlorite, with porphyritic oligoclase crystals (0·60 mm). Secondary chlorite and carbonates are abundant. Porphyritic andesite also occurs 375 yd S. 7° W. of the east end of Upper Carwood [402849] (E 29534) and 460 yd S. 24° E. of Upper Carwood [404849] (E 30204), the latter variety being coarse-grained with chlorite filled-vesicles. Intensely hematitized andesite occurs 415 yd S. 5° E. of the east end of Upper Carwood [402849] (E 30205).

Details of rocks referred to in Table 1

1. Altered dolerite, Uriconian, Hazler Quarry, 460 yd E. 21° N. of Dryhill, Shropshire: N.W. face of quarry (E 26291). Lab. No. 1700. *Sum. Prog. Geol. Surv.* for 1956 (1957) p. 57. Analysts A. D. Wilson and P. Coombs; spectrographic analysis by C. O. Harvey and K. L. H. Murray.
2. Amygdaloidal basalt, Uriconian, Hazler Quarry, 460 yd E.N.E. of Dryhill, Shropshire (E 30898). Lab. No. 1915. Analysts P. R. Kiff and G. A. Sergeant; spectrographic analysis by K. L. H. Murray.
3. Augite-andesite, Uriconian. Old Quarry, 570 yd N. 9° E. of Woodgate Cottage, Shropshire. Lower lift of quarry, 5 yd from W. corner (E 26287). Lab. No. 1697. *Sum. Prog. Geol. Surv.* for 1956 (1957), p. 57. Analysts A. D. Wilson and P. Coombs; spectrographic analysis by C. O. Harvey and K. L. H. Murray.
4. Dacite, silicified, Uriconian. Side of track immediately east of Middle Hill, Shropshire (E 26288). Lab. No. 1698. *Sum. Prog. Geol. Surv.* for 1956 (1957), p. 57. Analysts A. D. Wilson and P. Coombs; spectrographic analysis by C. O. Harvey and K. L. H. Murray.
5. Dacite, Uriconian (Woodgate Batch Dacite). Old Quarry 1700 yd N. 48° E. of Hope Bowdler church, Shropshire (E 35896). Lab. No. 1914. Analysts P. R. Kiff and G. A. Sergeant; spectrographic analysis by K. L. H. Murray.
6. Augite-dacite, Uriconian. Quarry, 250 yd N. of Hope Bowdler church, Shropshire (E 26286). Lab. No. 1696. *Sum. Prog. Geol. Surv.* for 1956 (1957), p. 57. Analysts A. D. Wilson and P. Coombs; spectrographic analysis by C. O. Harvey and K. L. H. Murray.
7. Buxton Rock. Rhyolite dust tuff at base of Burway Group, Eastern Longmyndian. Castle Hill, All Stretton. Reade and Holland (1908) Table iv, nos. 1 and 2.
8. Rhyolite, Uriconian. 780 yd E. 23° N. of New House Farm, Shropshire. (E 26289). Lab. No. 1699. *Sum. Prog. Geol. Surv.* for 1956 (1957), p. 57. Analysts A. D. Wilson and P. Coombs; spectrographic analysis by C. O. Harvey and K. L. H. Murray.
9. Devitrified pitchstone, Uriconian. Lea Rock Quarry, Wellington, Shropshire. Allport (1877, p. 457).
10. Granophyre, Uriconian, Ercall Hill. Teall (1888, p. 321).

TABLE 1

Analyses of Uriconian Rocks

	1	2	3	4	5	6	7	8	9	10
SiO_2	46·79	48·11	60·85	64·30	63·82	65·86	72·47	72·16	72·19	76·80
Al_2O_3	13·41	14·30	16·41	15·49	15·35	15·98	14·88	13·18	14·45	13·40
Fe_2O_3	7·59	7·01	3·40	2·10	2·16	3·58	0·81	2·17	1·68	0·60
FeO	6·50	5·90	3·34	3·24	3·47	0·71	2·27	0·89	0·91	0·10
MgO	6·78	6·40	3·06	2·07	2·17	0·90	0·83	0·64	trace	trace
CaO	7·11	6·27	1·45	1·79	1·44	3·37	0·62	0·68	0·93	trace
Na_2O	4·25	4·30	4·93	2·01	5·30	3·88	4·08	4·65	1·94	2·20
K_2O	0·09	0·48	2·15	2·47	1·88	2·72	1·95	2·99	6·12	6·20
H_2O+	3·31	3·53	2·72	3·51	2·24	1·33	1·79	1·37	1·47	0·50
H_2O-	0·73	0·75	0·74	1·07	0·40	0·48		0·45		
TiO_2	2·49	2·12	0·85	0·61	0·81	0·60	0·18	0·76	—	0·40
P_2O_5	0·39	0·34	0·25	0·10	0·23	0·20	—	0·18	—	—
MnO	0·18	0·16	0·16	0·11	0·16	0·05	0·09	0·05	trace	—
CO_2	0·15	trace	0·03	1·03	0·20	0·54	—	n.d.	—	—
Total S	0·01	n.d.	0·02	0·07	n.d.	0·03	—	0·01	—	—
Others	0·01	0·15	0·06	0·04	0·14	0·06	0·03	0·06	—	—
TOTAL	99·79	99·82	100·42	100·01	99·77	100·29	100·00	100·24	99·69	100·20
Trace Elements (parts per million)										
Ga	30	20	35	30	20	30	—	20	—	—
Cr	70	50	<5	<5	<5	<5	—	<5	—	—
V	400	500	70	150	50	70	—	40	—	—
Li	n.d.	30	n.d.	n.d.	60	n.d.	—	n.d.	—	—
Ni	16	40	<5	<5	<5	<5	—	<5	—	—
Co	80	35	<5	<5	<5	<5	—	<5	—	—
Zr	150	150	300	200	250	200	—	450	—	—
Sr	150	80	400	250	250	250	—	150	—	—
Ba	55	100	550	350	500	550	—	550	—	—
Norms										
Q	—	0·69	15·99	35·03	18·51	24·88	37·17	31·76	35·36	40·27
c	—	—	3·93	6·49	2·53	1·00	4·93	1·49	2·94	3·07
or	0·53	2·84	12·71	14·60	11·11	16·07	11·52	17·67	36·17	36·64
ab	35·96	36·39	41·72	17·01	44·85	32·83	34·53	39·35	16·42	18·62
an	17·25	18·30	5·56	8·23	5·64	15·41	3·08	2·20	4·61	—
di wo	6·46	4·42	—	—	—	—	—	—	—	—
di en	5·15	3·51	—	—	—	—	—	—	—	—
di fs	0·57	0·41	—	—	—	—	—	—	—	—
hy en	11·65	12·43	7·62	5·15	5·40	2·24	2·07	1·59	—	—
hy fs	1·29	1·43	2·18	3·27	3·55	—	3·37	—	0·28	—
ol fo	0·06	—	—	—	—	—	—	—	—	—
ol fa	0·01	—	—	—	—	—	—	—	—	—
mt	11·00	10·16	4·93	3·04	3·13	0·60	1·17	0·79	2·44	—
hm	—	—	—	—	—	3·16	—	1·62	—	0·60
il	4·73	4·03	1·61	1·16	1·54	1·14	0·34	1·44	—	0·21
ap	0·90	0·79	0·58	0·23	0·53	0·46	—	0·42	—	—
pr	0·02	—	0·04	0·15	—	0·06	—	0·02	—	—
ru	—	—	—	—	—	—	—	—	—	0·29

Petrological Discussion of the Uriconian

Allport (1877) was apparently the first to recognize the volcanic character of the Uriconian, and subsequently various aspects of the Uriconian rocks have been described, principally by Callaway and Bonney (1879), Blake (1890), Lapworth and Watts (1910), Watts (1925), Hirst (1925), and James (1956), and in Geological Survey Memoirs, principally by Cantrill (*in* Whitehead and others 1928) and Pocock and Whitehead (1938).

The extrusive and minor intrusive rock types which comprise the Eastern Uriconian of the Church Stretton district are petrographically similar to those of the other Eastern Uriconian areas, for example Wrockwardine and Wrekin, and are similar also to the Western Uriconian of Pontesford and Earl's Hills and the neighbouring outcrops.

The Uriconian suite of volcanic rocks and associated minor intrusions consists principally of basalt–andesite–dacite–rhyolite lavas and various types of tuff together with intrusions of granophyre, rhyolite and dolerite.

Ten chemical analyses of the Uriconian Suite (Table 1) have been plotted (Fig. 5) following the method of Larsen (1938, p. 505). One analysis (no. 10) is of the Ercall granophyre of the Wrekin and one (no. 7) is of the Buxton Rock, a rhyolitic dust tuff at the base of the Longmyndian Burway Group.

The variation curves (Figs. 5–6) indicate that the Uriconian rocks comprise a calc-alkaline volcanic province, comparable with the typical calc-alkaline province of the western U.S.A. (Larsen 1938, p. 514, figs. 6–7). These latter provinces represent the latest period of igneous activity of orogenic regions characterized by surface vulcanicity "typically separated by a lengthy period of time from the main phase of folding and plutonic activity" (Turner and Verhoogen 1951, p. 201). This phase is typically accompanied by welded rhyolitic and dacitic tuffs (Turner and Verhoogen 1951, p. 219).

The age and tectonic setting of the calc-alkaline Uriconian volcanic suite is consistent with its probable occurrence as the last phase of igneous activity of an orogenic cycle. The Uriconian tuffs contain abundant evidence, in the form of derived fragments, that they are later than a series of low grade (greenschist-facies) metamorphic rocks.

Fragments of muscovite-schist, muscovite-chlorite-schist, magnetite-muscovite-schist, quartzose schists and highly foliated (nemablastic) quartzites, muscovite-quartz-tourmaline-rocks, chlorite-schist and quartzite with vermicular chlorite inclusions occur in the Uriconian tuffs of the Church Stretton district (e.g. E 25196, 29530, 29532–3, 29543–4), in the Charlton Hill conglomerate (E 8579–81), and in the Lawrence Hill volcanic ash (E 8598). These derived fragments of metamorphic rocks are similar to those which occur in the Longmyndian sediments and are of Mona Complex and Rushton Schist type (see p. 74). Similar derived fragments of quartzose muscovite- and chlorite-schists and quartz-schists occur in some of the Pebidian tuffs (e.g. E 456, 7189, 7247–8) and in the Arvonian tuffs (Greenly 1919). It may be significant that many of the available reliable analyses of Pebidian, Arvonian and Charnian igneous rocks also fall on to the Uriconian curves of Figs. 5 and 6.

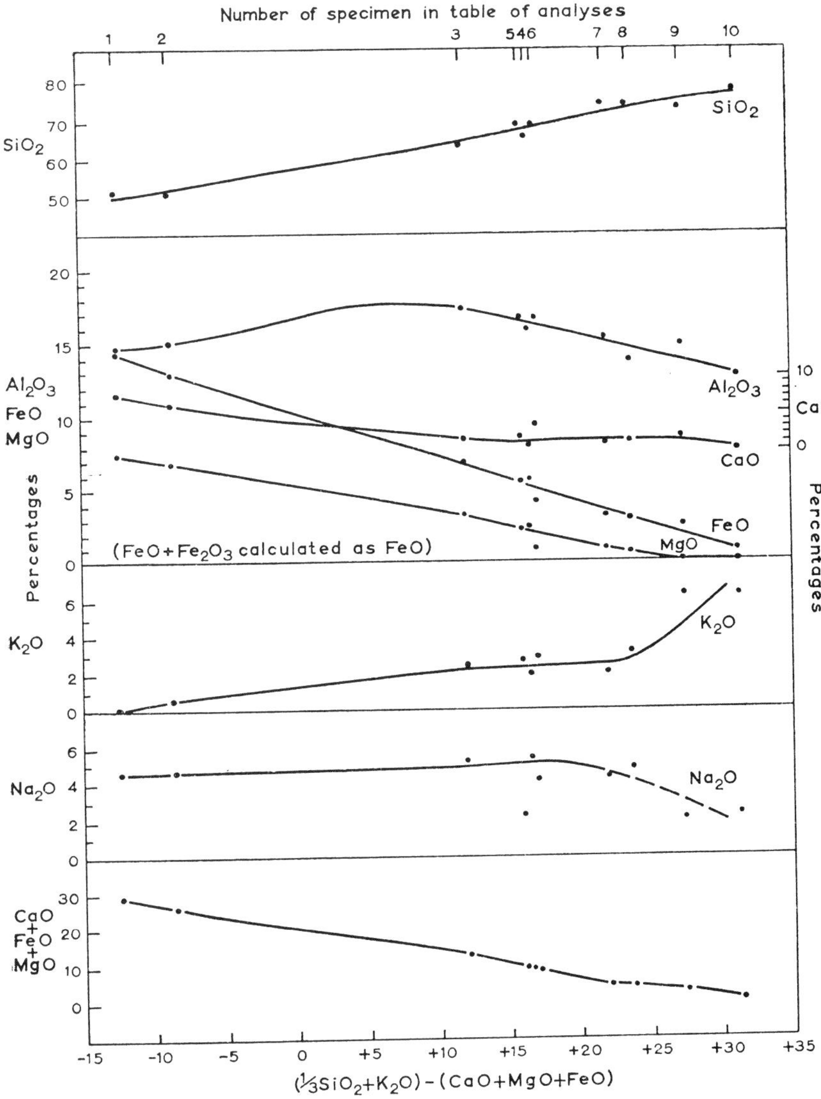

FIG. 5. Variation diagram of Uriconian igneous rocks

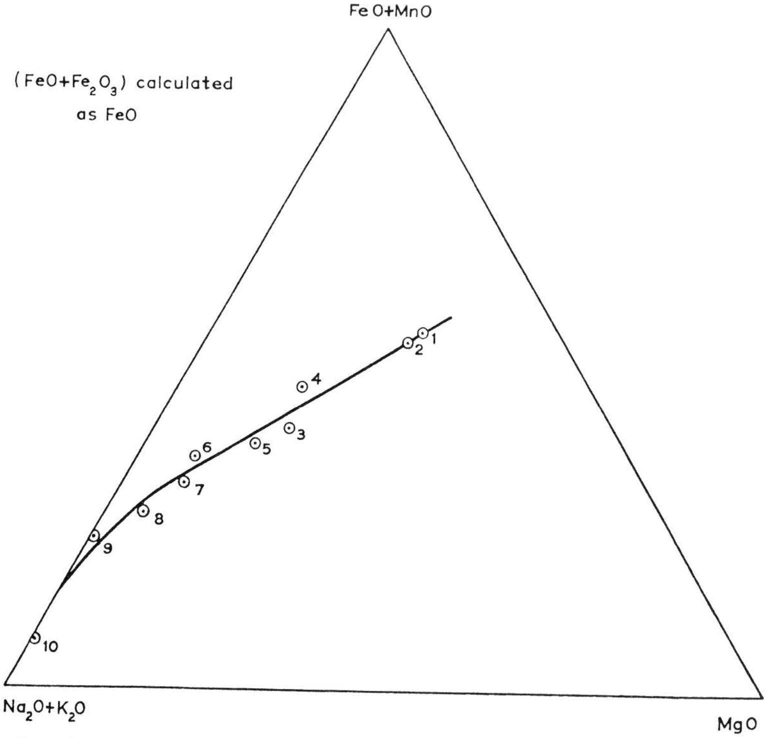

FIG. 6. (FeO+MnO)–(Na$_2$O+K$_2$O)–MgO diagram of Uriconian igneous rocks

The rock types in these scattered Pre-Cambrian volcanic groups are similar; rhyolites, andesites, basalts, granophyres and tuffs are common to the Pebidian and Arvonian; ignimbrites occur in both the Uriconian and the Arvonian; and tuffs, agglomerates, dacites and porphyries are common to the Uriconian and Charnian, since although some of the Charnian rocks have long been termed syenites the suite consists of markfieldites (quartz-diorite with micro-graphic groundmass), granodiorites, diorite-porphyries and dacites.

It seems likely therefore that these isolated volcanic areas form portions of a late Pre-Cambrian igneous episode. The suite has the characters of the supracrustal (subsequent) igneous activity of an orogenic cycle, the meta-morphic rocks of which are probably represented by the fragments, of Mona complex and Rushton Schist types, and the Molasse stage of which is repre-sented by the Longmyndian and probably parts of the Charnian, both with horizons of tuff or agglomerate.

R.D.

References

ALLPORT, S. 1877. On certain Ancient Devitrified Pitchstones and Perlites from the Lower Silurian District of Shropshire. *Quart. J. Geol. Soc.*, **33**, 449–60.

BLAKE, J. F. 1890. On the Monian and Basal Cambrian Rocks of Shropshire. *Quart. J. Geol. Soc.*, **46**, 386–420.

CALLAWAY, C. 1879. The Precambrian rocks of Shropshire. *Quart. J. Geol. Soc.*, **35**, pt. I, 643–62.

COBBOLD, E. S. 1900. in *Church Stretton*, 1. ed. C. W. Campbell—Hyslop. Shrewsbury.

—— 1927. The Stratigraphy and Geological Structure of the Cambrian area of Comley (Shropshire). *Quart. J. Geol. Soc.*, **83**, 551–73.

—— and WHITTARD, W. F. 1935. The Helmeth Grits of the Caradoc Range, Church Stretton; their bearing on part of the Pre-Cambrian succession of Shropshire. *Proc. Geol. Assoc.*, **46**, 348–59.

DEARNLEY, R. 1966. Ignimbrites from the Uriconian and Arvonian. *Bull. Geol. Surv.*, **24**, 1–6.

FENNER, C. N. 1948. Incandescent tuff flows in Southern Peru. *Bull. Geol. Soc., Amer.*, **59**, 895–917.

GREENLY, E. 1919. The Geology of Anglesey. *Mem. Geol. Surv.*

HIRST, T. 1925. Notes on the Igneous and Associated Rocks of the Wrekin Range. *MS. Thesis for D.I.C.*

JAMES, J. H. 1956. The Structure and Stratigraphy of part of the Pre-Cambrian outcrop between Church Stretton and Linley, Shropshire. *Quart. J. Geol. Soc.*, **112**, 315–35.

LAPWORTH, C. and WATTS, W. W. 1910. Geology in the Field, Shropshire. *Geol. Assoc. Jubilee Vol.*, 747–9.

LARSEN, E. S. 1938. Some New Variation Diagrams for Groups of Igneous Rocks. *J. Geol.*, **46**, 505–20.

POCOCK, R. W., WHITEHEAD, T. H., WEDD, C. B. and ROBERTSON, T. with contributions by WRAY, D. A., STUBBLEFIELD, C. J., CANTRILL, T. C. and DAVIES, W. M. 1938. Shrewsbury District. *Mem. Geol. Surv.*

READE, T. M. and HOLLAND, P. 1908. Analyses of Longmyndian Rocks. *Proc. Liverpool Geol. Soc.*, **10**, 276–87.

STRACHAN, I., TEMPLE, J. and WILLIAMS, A. 1948. The Age of the Neptunian Dyke at Hazler Hill (Shropshire). *Geol. Mag.*, **85**, 276–78.

SUMMARY OF PROGRESS. 1957 (for 1956). *Geol. Surv.*

TEALL, J. J. H. 1888. *British Petrography with special reference to the igneous rocks.* London.

TURNER, F. J. and VERHOOGEN, J. 1951. *Igneous and Metamorphic Petrology*, N. York.

WATTS, W. W. 1925. The Geology of South Shropshire. *Proc. Geol. Assoc.*, **36**, 321–63.

WHITEHEAD, T. H., ROBERTSON, T., POCOCK, R. W., and DIXON, E. E. L., with CANTRILL, T. C., DEWEY, H., SHERLOCK, R. L., PRINGLE, J. and CROOKALL, R. 1928. The Country between Wolverhampton and Oakengates. *Mem. Geol. Surv.*

Chapter 3

PRE-CAMBRIAN, LONGMYNDIAN

GENERAL ACCOUNT

Within the Church Stretton district, Longmyndian rocks form the massif of the Long Mynd. They occur also in the adjacent hills of the Wentnor and Bridges areas to the north-west, and in the Church Stretton valley. In addition to the main outcrop, faulted inliers of Longmyndian rocks are found close to the line of the Church Stretton Fault Complex, in the Cwms area south-east of Caer Caradoc, in the Brokenstones, Horderley and Wart Hill areas and on Hopesay Hill in the south-western part of the district.

Following previous practice, the Longmyndian succession has been divided into two major parts, the Stretton Series, which occupies the eastern part of the Long Mynd, and the Wentnor Series, which occupies the western part of the main outcrop. The general statement of the subdivisions adopted in this account is as follows:

	Approx. thickness in ft
WENTNOR SERIES	
BRIDGES GROUP	
Purple mudstones, siltstones and sandstones commonly laminated and flaggy, occasional massive sandstone bands	2000–4000
BAYSTON–OAKSWOOD GROUP	
Coarse purple sandstones and grits with subordinate purple mudstones and siltstones. In this district, three conglomerate bands are present, the Darnford, Stanbatch and Lawn Hill conglomerates	4000–8000
Probable unconformity	
STRETTON SERIES	
PORTWAY GROUP	
Purple and greenish grey shaly mudstones and siltstones with sandstone bands. Sandstones locally massive, particularly near base. Huckster Conglomerate at base	600–3500
LIGHTSPOUT GROUP	
Mainly flaggy, greenish grey siltstones and sandstones with massive sandstone bands, more particularly in lower part of group. Topmost 400 ft of beds mainly purple in colour and containing at least one band of tuff	1700–2700
SYNALDS GROUP	
Mainly purple shaly mudstones and siltstones with subordinate sandstone bands; locally some greenish grey mudstones and sandstones; contains several horizons of volcanic tuff, notably the Batch Volcanics	1600–2800

	Approx. thickness in ft
BURWAY GROUP	
Mainly flaggy greenish grey laminated siltstones with sandstone bands; locally some massive sandstones. Cardingmill Grit (c. 100 ft micaceous sandstone) at top. Buxton Rock (c. 15 ft siliceous dust tuff) at base	2000
STRETTON SHALE GROUP	
Mainly greenish grey shaly mudstones and siltstones; commonly well laminated; locally puckered and contorted. Helmeth Grit (includes four bands of tuffaceous grit) at base	Uncertain ?3000

This succession is essentially that recorded by J. H. James (1956, p. 316) with the modification that James's 'Mintonian' Series is, in the present account, returned to its former status as the Portway Group, a subdivision of the Stretton Series. It is not intended to give a detailed account of the history of research on the Pre-Cambrian of Shropshire as this has been done admirably by several authors e.g. Watts (1925, p. 327), Whitehead (*in* Pocock and others 1938, p. 8 and 1955, p. 465) and Whittard (1952, p. 144). However a brief resumé of the relevant points from previous literature is given below, to explain the historical development of the present classification of the Longmyndian:

(a) Callaway (1888, p. 239) gave the name Longmyndian to the sedimentary Pre-Cambrian rocks of Shropshire. He also suggested that fragments in the Longmyndian conglomerates were derived from the Uriconian.

(b) Blake (1890, p. 386) produced a map and section of the Long Mynd area. He divided the Longmyndian into an eastern part, correlated with the Monian of Anglesey, which, he concluded, was unconformably overlain by the western part, correlated with the Cambrian. He also subdivided the Eastern Longmyndian. In some localities Blake's unconformity coincided with the present base of the Wentnor Series but in other places it was extended eastwards to include the Huckster Conglomerate, which is now recognized as part of the Stretton Series.

(c) Callaway (1891, p. 109) wrote a short paper largely refuting criticisms of his work made by Blake (1890, p. 386). Callaway pointed out Blake's erroneous interpretation of evidence in favour of an unconformity within the Longmyndian.

(d) Lapworth (1894, p. 306) divided the Longmyndian into an Eastern Longmynd Series and a Western Longmynd Series corresponding with Blake's 'Monian' and 'Lower Cambrian'. At this date Lapworth adopted Blake's subdivisions of the Eastern Longmyndian.

(e) Cobbold (1900) contributed a comprehensive guide to the geology of the Church Stretton district. He adopted Blake's subdivisions of the Eastern Longmyndian but named marker horizons within this sequence as follows: Buxton Rock, Cardingmill grits and Batch volcanic rocks. In addition he attributed to Lapworth the names of Helmeth grits and Huckster Conglomerate, which names had apparently not been published previously. Cobbold (1900, p. 71) suggested that there might be "two separate unconformities in the western part of the Longmynd, each marked by a sandy and conglomeratic group of beds".

(f) Lapworth (1910, p. 748) divided the Longmyndian as follows: Wentnor Series (Western Longmyndian) subdivided into Ratlinghope Group and Bayston Group. This was considered to overlie the Stretton Series (Eastern Longmyndian), subdivided into the Stretton Shale, Burway, Synalds, Lightspout and Portway groups. The Stretton Shale Group was further subdivided into the Watling and Brockhurst shales. Within the Bayston Group Lapworth named the Haughmond, Darnford and Stanbatch conglomerates. No opinion was expressed as to a possible unconformity between the Wentnor and Stretton series.

(g) Cobbold and Whittard (1935, p. 348) demonstrated an upward sequence from the Uriconian into the Helmeth Grit, the latter being considered as the base of the Stretton Shale Group. In a review of the possible age relations between the Longmyndian and the Uriconian, these authors pointed out that (i) the Stretton Series is probably younger than the Eastern Uriconian; (ii) the Wentnor Series appears to be younger than the Western Uriconian; (iii) in view of its greater geographical extent, the Wentnor Series might be unconformable on the Stretton Series.

(h) Whitehead (*in* Pocock and others 1938, p. 42) adopted Lapworth's subdivisions of the Stretton Series but subdivided the Wentnor Series into the Bayston, Bridges and Oakswood groups. Within the Oakswood Group, Whitehead named the Radlith, Oakswood and Lawn Hill conglomerates. He also suggested the possibility of inversion in the Wentnor Series based on evidence of disharmonic folding between competent and incompetent beds (Pocock and others 1938, pp. 12 and 53).

(i) Whittard (1952, p. 149) quoted the contemporary work of J. H. James (1952, p. 198), and suggested the possibility of a deep synclinal overfold in the Wentnor Series to explain the equivalence of the Oakswood Group and the Bayston Group, the core of this fold being occupied by the Bridges Group.

(j) J. H. James (1956, p. 315) gave a detailed map and section of the Church Stretton and Linley districts. His stratigraphical succession is practically the same as that given at the beginning of this chapter (p. 34) with the exception that the Portway Group is separated as a new series, the Mintonian. James cited sedimentary structural evidence for inversion in the western (Oakswood) belt of the Bayston–Oakswood Group and demonstrated (1952, p. 198 and 1956, p. 324) that rocks of the Wentnor Series are unconformable to the Western Uriconian near Chittol, the whole sequence there being inverted. James's section (1956, fig. 4) shows the Wentnor Series as a major isoclinal syncline, overturned on the western limb with the core formed by the Bridges Group. He concluded that unconformities probably occurred at the base of the Wentnor Series and below the Mintonian (Portway Group). As indicated below (p. 43) in the present account, it is considered that James misinterpreted the evidence for a major unconformity at the base of the Portway Group.

Age of Longmyndian Sediments

Whitehead (*in* Pocock and others 1938, p. 9) noted that the Longmyndian is regarded as Pre-Cambrian, mainly because the great thickness of Longmyndian sediments cannot be fitted into the known Cambrian or Lower Ordovician sequences of Shropshire. The oldest rocks known to rest, without a tectonic break, upon Longmyndian are of Upper Ordovician age, except in the Cwms area [474944][1] east of Church Stretton, where the basal Cambrian quartzite rests on a faulted outcrop of weathered reddish grits, tentatively correlated with some part of the Wentnor Series.

Lapworth (1910, p. 747) stated that the Western Longmyndian (Wentnor Series) . . . "is probably closely related to the Torridonian rocks of Scotland". The equation of the Torridonian with the Wentnor Series has received more recent support from the work of Creer (1957, p. 214) on the natural remanent magnetization of rocks from the Wentnor Series. This author found a close comparison between the axes of magnetization of Upper Torridonian and Wentnorian rocks and suggested that the two groups are of approximately equal age.

From the suggested correlation with the Torridonian and from recent isotope geochronological work, it may be inferred that the Longmyndian sediments are older than 600 million years (lower limit of Cambrian–Holmes 1959, p. 204), definitely younger than 1600 million years (Laxfordian orogeny) and possibly younger than 1160 million years (rubidium/strontium age of biotite from Lewisian gneiss of Loch Torridon area—Giletti and others 1961, p. 249).

Stretton Series

STRETTON SHALE GROUP

The Stretton Shale Group occurs in three areas: (a) on the south-eastern flanks of the Long Mynd in the Church Stretton valley. from Minton in the south to Lower Wood in the north; (b) on the north-western flanks of Ragleth, Hazler and Helmeth Hill and separated from (a) above by an outcrop of Silurian rocks; (c) along the line of the Church Stretton Fault Complex from near Wart Hill to Cwm Head. Areas (a) and (b) above have been previously described as Brockhurst and Watling shales respectively (Lapworth 1910, p. 749) but this distinction is not made in the present account. The rocks comprise greenish grey shaly mudstones and siltstones which are commonly laminated. Several authors have reported the occurrence of calcareous nodules in these shales, some showing cone-in-cone structure (e.g. James *in* Whittard and others 1953, p. 238).

The base of the group is marked by the Helmeth Grit which overlies the Uriconian on Ragleth, Hazler and Helmeth hills. According to the present mapping the grit extends from near Little Stretton [445916] in the south to Caradoc Coppice [468948] in the north, the outcrop terminating at each end against F1 of the Church Stretton Fault Complex. The one-inch map shows the Helmeth Grit as overlying both Ragleth Tuffs and Cwms Rhyolites (both Uriconian) in the vicinity of Caradoc Coppice [468948]. This evidence and the marked divergence in strike between the Uriconian of the Caer Caradoc area and the Longmyndian, suggest that there is an unconformity at the base of the Helmeth Grit. Cobbold and Whittard, who made a detailed study of the Helmeth Grit (1935, p. 348) came to a different conclusion. They described the Helmeth Grit as four massive gritty bands within a 100-ft sequence of

---

[1]Such numbers are National Grid References within the 100-km square SO(32).

green and purple shales, and remarked (1935, p. 354) that the so-called grits are better described as lithic tuffs formed largely of fragments derived from the Uriconian by volcanic processes. Thus the Helmeth Grit may represent the dying phase of the Uriconian vulcanicity and form a passage from the Uriconian to the Longmyndian.

The typical grey shales of the group are poorly exposed in the floor of the Church Stretton valley but are better seen at the mouths of the major valleys of the Long Mynd e.g. the Cardingmill Valley. The regional dip is steep to the west-north-west but the bedding is commonly contorted by small folds, and in some localities the dip is to the east e.g. near St. Michael's Church, All Stretton, [462959] where the dip is 55° E.S.E. with a strong cleavage dipping steeply north-west. A short distance to the south near the war memorial, the true bedding is obscure and the apparent steep westerly dip here is due to closely spaced cleavage planes. J.E.W.

From the western flanks of Wart Hill [4084][1] to the vicinity of Cwm Head [4288], rocks of the Stretton Shale Group form a narrow outcrop between faults F1 and F2 of the Church Stretton disturbance. This outcrop attains a width of about ¼ mile near its southern end, where it is broken into a number of fault blocks, and narrows gradually towards the north-north-east as the two faults converge. A small outcrop east of F2 occurs in the Brokenstones area [420882]. In places a narrow zone of Wentnor Series rocks lies between the shales and F1. As might be expected the shales are commonly contorted and crushed, but in general they preserve a north-north-easterly strike with dips rarely less than 60°. D.C.G.

BURWAY GROUP

The outcrop of the Burway Group extends almost the whole length of the Long Mynd from near Hillend Farm [401881] in the south to Lower Wood [466976] in the north. Between Minton [430907] and Mount Gutter [404882] the unconformable base of the Silurian rocks trends obliquely across the strike of the Longmyndian, gradually cutting out the rocks of the Burway Group from north-east to south-west. The group consists largely of flaggy greenish grey mudstones, siltstones and sandstones, with some more massive sandstone horizons. The siltstones and mudstones are laminated locally and there are occasional bands of siltstone and sandstone which are purple or purplish grey in colour. In common with the other members of the Stretton Series, the Burway Group is best seen on the sides of the major Long Mynd valleys (e.g. Callow Hollow and Ashes Hollow) and there are only scattered exposures on the intervening spurs. Probably the most continuous section occurs in Ashes Hollow, between 400 yd and 550 yd upstream from Ashes Cottage [439926] where there are extensive exposures of uniform, vertical, flaggy siltstone and sandstone.

The regional dip is steep to the west-north-west but in several areas along the length of the outcrop the dip is south-easterly with values ranging from about 40° to about 75°. These contrary dips are found in a zone which lies between 250 yd and 450 yd north-west of the base of the group and they suggest the existence of minor anticlinal structures, possibly arranged en échelon along the outcrop of the group.

[1]Such numbers are four figure National Grid References within the 100-km square SO(32). They refer to general areas rather than specific localities.

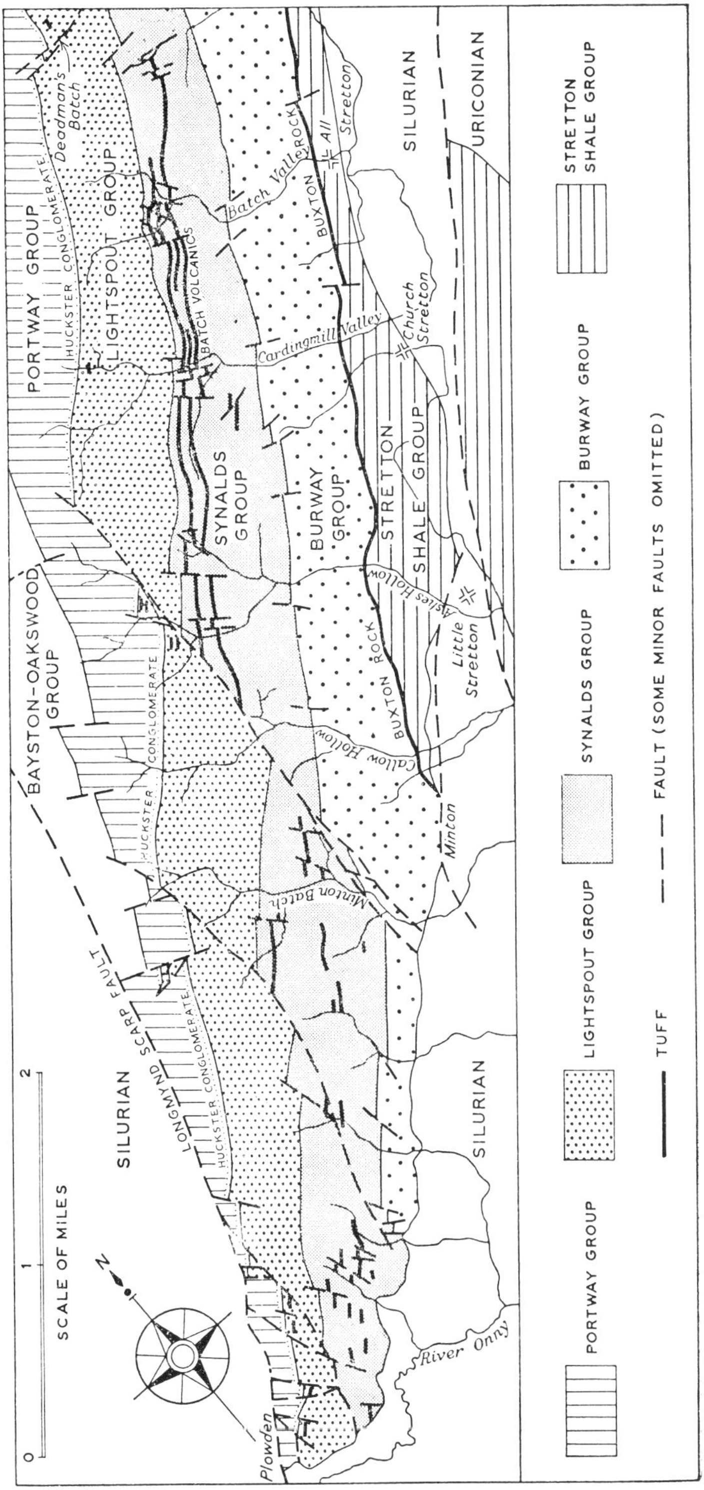

FIG. 7. Distribution of tuff bands in the Stretton Series

The lower and upper limits of the Burway Group are marked by the Buxton Rock and the Cardingmill Grit respectively. The type locality of the Buxton Rock is Buxton Quarry, All Stretton [459955], where it consists of 24 ft of massive, very fine-grained, greenish grey, cherty rock with small black spots locally and with abundant thin quartz veins on the joints. This band can be traced at intervals where it crosses the main valleys from near Minton village [430908] to near Pensylvania [468971] in the north (Fig. 7).

In thin slice the Buxton Rock is seen to consist largely of finely divided clay minerals, chlorite, quartz and feldspar and is probably best described as a silicified dust tuff. In some specimens there are scattered, rounded aggregates of albitic feldspar (E 27845)[1] and rounded aggregates of radially arranged chlorite (E 27846) which appear to be of authigenic origin. One specimen contains subhedral prisms and anhedral fragments of albitic feldspar (E 29618) and is more obviously of pyroclastic origin. This rock also contains thin strings of small sphene crystals. Most specimens are cut by thin veinlets of quartz, of intergrown quartz and albite, and of carbonate. Some specimens contain diffuse patches of carbonate, probably of secondary origin (E 27846, 29618).

The Buxton Rock divides the characteristically shaly rocks of the Stretton Shale Group from the more flaggy members of the Burway Group. This change in lithology is gradual, however, and does not always occur exactly at the level of the Buxton Rock. In Buxton Quarry, the uppermost 50 ft of the Stretton Shale Group are exposed to the east of the Buxton Rock. These beds are moderately flaggy and are similar to the characteristically flaggy basal beds of the Burway Group on the west side of the quarry. In contrast to this, the section in Ashes Hollow [439927] shows the Stretton Shale type of lithology continuing for about 50 ft into the lower part of the Burway Group.

The Cardingmill Grit is the topmost member of the Burway Group and, as described in this account, is equivalent to the Lower Cardingmill Grit of James (1956, p. 317). That author described a lower and an upper grit, the latter being purplish in colour and placed by him in the lower part of the Synalds Group. Cobbold also noted this upper grit horizon (1900, p. 79). While it was found from the present survey that purplish sandstone bands occur in the basal beds of the Synalds Group at some localities e.g. Batch valley, All Stretton [450957], it is considered that these bands are not sufficiently persistent or distinctive to be separated as an Upper Cardingmill Grit.

The Cardingmill Grit is a massive, micaceous, greenish sandstone (subgreywacke) which in the northern part of the Long Mynd averages about 80–100 ft in thickness and contains occasional very thin siltstone partings. To the south of Minton Batch the grit is less well defined and may measure up to about 190 ft in overall thickness, consisting of beds of massive sandstone up to about 30 ft in thickness with a much higher proportion of interbedded siltstone. Current bedding, younging to the west, has been observed at some localities, e.g. Batch valley, All Stretton [450957]. On the sides of the major valleys, the grit forms fairly prominent crags the shape of which is controlled by strong vertical and horizontal joints, some of which carry quartz veins. In these valley sections the line of the outcrop is commonly disrupted by minor faults giving small dextral and sinistral displacements. On the intervening ridges the grit forms a series of minor summits (e.g. Packetstone Hill, Grindle

[1]Such numbers refer to rock slices in the Survey collection.

and Yearlet hills, Devilsmouth crags and Bodbury hill) which overlook the slight hollow occupied by the Synalds Group to the west. On the higher ground, however, the Cardingmill Grit is very poorly exposed and the line of the outcrop shown on the map is generalized. In fact it is probably as much disrupted by minor faults as are the better exposed valley sections.

SYNALDS GROUP

The outcrop of the Synalds Group extends the whole length of the Long Mynd from the Onny valley in the south to near Womerton [456972] in the north. The base of the group is defined by the top of the Cardingmill Grit, and the top of the group is taken at the change in colour from the dominantly purple shales and sandstones of the Synalds Group to the greenish grey flaggy siltstones and sandstones of the Lightspout Group to the west. In well-exposed areas the change is seen to be rapid and quite well marked. The Synalds Group contains several tuffaceous horizons among which are the Batch Volcanic beds, first named by Cobbold (1900, pp. 72 and 77).

The rocks of the Synalds Group are mainly purple shaly mudstones and siltstones, locally laminated, with subordinate bands of purplish grey sandstone up to several feet in thickness. The regional dip is steep to the west-north-west but in this group the shaly members commonly exhibit a strong cross cleavage which is usually vertical or steeply dipping to the east-south-east (Plate 7A). Cleavage and bedding may have the same strike but at some localities there may be a divergence in strike of about 20° between the two. There are good exposures of this cross cleavage in the crags below the Pike [442949], 400 yd N. of the Carding Mill and on the north-eastern side of the Batch valley [449961], 300 yd S.W. of Jinlye. At some horizons there are relatively thin beds of greenish grey sediments and in the vicinity of Callow Hollow and Ashes Hollow there is a considerable thickness of green siltstone and sandstone in the lower part of the group. In Ashes Hollow there are about 400 ft of greenish grey beds separated from the Cardingmill Grit by about 200 ft of purple beds. The width of outcrop of these green sediments is increased considerably near the valley bottom by a major sigmoid flexure in strike, this fold apparently being confined to the Synalds Group. These green beds are found at other localities but cannot be mapped continuously along the Synalds outcrop.

In the Synalds Group the bedding surfaces of mudstone layers commonly carry numerous small depressions which have also been found in other groups and were first described by Salter as impressions of rain drops, gas bubbles and as marking worm burrows (1856, p. 246 and 1857, p. 199). Salter's two types of worm burrow were named *Arenicolites didymus* and *Arenicolites sparsus*, the former consisting of small paired depressions. The paired character suggests an organic origin for this variety of marking but so far no specimens have been seen which show definite burrows cutting across the laminations of the sediment. The existence of genuine worm burrows in the Longmyndian is discussed below (p. 72) under Sedimentary Structures. The supposed trilobite *Palaeopyge ramsayi*, recorded by Salter (1856) is no longer considered to be organic (Watts 1925).

Batch Volcanic beds and other tuff bands. (Fig. 7.) To the north of Ashes Hollow the two dominant tuff bands within the Synalds Group have been grouped together as the Batch Volcanic beds and have been called the Andesitic Ash (lower band) and White Ash (upper band) respectively (Cobbold 1900, p. 77). These two bands have fairly distinctive lithological characters, the lowest one-third of the Andesitic Ash being pale green with dark green patches, and the upper two thirds purple with conspicuous dark green patches. The White Ash is almost white or very pale green and carries small dark green patches. These bands can be traced at intervals from Ashes Hollow [433931] north-eastwards to the vicinity of Womerton Farm [456973] and maintain consistent positions relative to the top of the Synalds Group. In the northern part of the Long Mynd two other tuffs occur near the top of the Synalds Group, one lying between the Andesitic Ash and the White Ash and the other above the White Ash. In the type area of the Batch valley, All Stretton, all four bands are seen on the spur which rises westwards from the junction of Jonathan's Hollow and Long Batch [446961]. The section on this spur, reading from west to east, is as follows:

	Thickness ft
Base of Lightspout Group	
Purple shales with sandstone bands	70
Tuff, greenish grey, fine-grained	7
Purple shales with sandstone bands	120
White Ash: moderately coarse, pale greenish tuff	7
Purple shales	40
Tuff, pale greenish, fine-grained with some coarser bands	15
Purple shales with sandstone bands	150
Andesitic Ash: moderately coarse tuff, pale green in lower part, purple in upper part	16

The two unnamed tuffs in this section have not been traced as continuously as the White Ash or the Andesitic Ash. This is also true of a 10-ft band of tuff found only in the Devilsmouth area [439944], about 500 ft above the base of the Synalds Group.

To the south of Ashes Hollow, it is difficult to distinguish between the Andesitic Ash and the White Ash, and the tuffs in the southern part of the outcrop are rather similar in appearance, being mainly moderately coarse and varying from greenish to purplish grey in colour. It has not been possible to trace these bands so continuously nor do they appear to maintain such constant positions within the group. In this area the tuffs appear generally to occupy a lower position in the group than they do farther north, but no confident correlations can as yet be made with the Batch horizons of the type area.

LIGHTSPOUT GROUP

This group extends over the length of the Long Mynd from Plowden in the south to Deadman's Batch [448977] in the north. The base of the group is taken at the colour change from dominantly purple in the Synalds Group to mainly greenish grey in the Lightspout Group. In many sections this change is accompanied by a westward transition from shales with subordinate sandstones (Synalds Group), to more flaggy mudstones and siltstones with massive and more abundant sandstone-bands (Lightspout Group), but in the northern part of the Long Mynd the lower beds of the Lightspout Group consist mainly of flaggy siltstone with very little sandstone in the lowest 150 ft. The top of the group is taken at the base of the Huckster Conglomerate, the uppermost 400 ft (approximately) consisting of purple shaly mudstones and siltstones with bands of coarse purplish grey sandstone. These beds are very similar to those of the Synalds Group and contain one or more bands of tuff (Fig. 7) which are exposed at several localities (see discussion on James's 'Mintonian' unconformity). Between about 150 ft and 500 ft above the base of the group there are several bands of pale-weathering massive sandstone. These are well exposed in the Cardingmill Valley on the southern side of Haddon Hill [438951] and at Jonathan's Rock [446965]. One of these bands has been separated by James as the Haddon Hill Grit (1956, p. 318; see also Cobbold 1900, pp. 72 and 76). It is considered from the present survey that no single band of sandstone can be satisfactorily distinguished or traced for sufficient distance to be classed as a separate 'grit' horizon.

The regional dip is steep to the west-north-west and the cleavage in the shaly beds, where seen, is commonly vertical or very steep with a strike only slightly divergent from that of the bedding. In the type area of Lightspout Hollow [430950] the rocks exposed are characteristic of the group. The competent massive sandstone bands are cut by thin quartz veins which occupy tension cracks and joints. The veins are not seen in the siltstones and shales. There are occasional bands of purple sandstone and siltstone but these are subordinate in amount except in the topmost beds of the group.

Validity of James's 'Mintonian' unconformity. James (1956, p. 319) renamed Lapworth's Portway Group as the 'Mintonian' and raised it to a status comparable with that of the Stretton and Wentnor series. This resulted from his interpretation of a major unconformity below the Huckster Conglomerate, mainly from evidence in the Deadman's Batch section [448977]. In this area, the Huckster Conglomerate is displaced dextrally along an east–west fault, running a short distance to the north of High Park cross roads (James 1956, fig. 3). To the south of the fault the Conglomerate overlies purple and greenish siltstones and sandstones of the Lightspout Group, but to the north of the fault it rests upon purple shales and sandstones with an 8-ft band of coarse purplish tuff at 250 ft below the contact. James (1956, p. 329) classified this tuff with the Batch Volcanics and thus concluded that the purple shales and sandstones belonged to the Synalds Group. This conclusion implies that the association of tuffs with purple shales and siltstones in the Longmyndian is diagnostic of the Synalds Group. To explain the occurrence of Synalds Group rocks at Deadman's Batch James suggested that, prior to the deposition of the Huckster Conglomerate, the rocks to the north of the east-west fault were upthrown and eroded, removing the Lightspout Group. This was followed by the deposition of the Conglomerate on Lightspout Group rocks south of the fault, and on Synalds Group rocks to the north; later (post-folding) transcurrent movement on the same fault would displace the Huckster Conglomerate to its present position. This sequence indicates a major unconformity below the Huckster Conglomerate.

In the present survey bands of fairly coarse tuff were found in the purple beds near the top of the Lightspout Group at several localities south of Deadman's Batch (Fig. 7) e.g. in the Mott Road [432955] and Ashes Hollow sections [423936], and to the south of James's area near Black Knoll [392886] and Plowden [386878]. These tuffs are not lithologically distinct from those in the Synalds Group, and they occur at about the same level below the Huckster

Conglomerate as does the tuff at Deadman's Batch. Although these exposures may not represent a continuous tuff horizon, their importance lies in the demonstration that the occurrence of tuffs with purple shales and sandstones does not necessarily imply correlation with the Synalds Group.

It is considered therefore that the purple beds at Deadman's Batch belong to the upper part of the Lightspout Group and there is thus no evidence of a major unconformity at the base of the Huckster Conglomerate. The name of the Portway Group has therefore been retained and the term 'Mintonian' has not been adopted in this account.

The conglomeratic nature of the Huckster horizon suggests that it represents a depositional break but the evidence does not support the major hiatus postulated by James. Whitehead (*in* Pocock and others 1938, p. 44) expressed doubt as to whether the Lightspout Group occurred in the area of the Shrewsbury (152) Sheet to the north of the present district. This view supports James's theory of major unconformity. The ground to the north of the Long Mynd is extensively drift-covered and, with the exception of Haughmond Hill where the basal rocks of the Wentnor Series probably rest unconformably on the lowest beds of the Synalds Group, there are few exposures of rocks of the Stretton Series. It is possible that rocks of the Lightspout Group occur in the area between Deadman's Batch and Longnor Park [SJ 469005][1], but from an examination of the ground, it was concluded that exposures are not sufficiently extensive to provide a definite answer, one way or the other, to this question.

PORTWAY GROUP

The outcrop of the Portway Group extends over the length of the Long Mynd from Plowden to High Park, mainly on the poorly exposed summit plateau. The lower limit of the group is taken at the base of the Huckster Conglomerate, which is well developed and moderately well exposed south of Asterton, locally reaching a thickness of 60 ft. To the north of Asterton the conglomerate is less well developed, locally being no more than a pebbly sandstone. It is exposed only in the main valley sections and the outcrop, as shown on the one-inch map, is therefore generalized and is probably much more disrupted by minor faulting than has been indicated. The top of the group is taken at the change from purple and green shaly siltstones and sandstones (Portway Group) to the more massive, coarse, purple and grey sandstones of the Bayston–Oakswood Group. In the Church Stretton district the change is seen only at the north end of the Long Mynd in Hawkham Hollow. To the south, as far as Asterton there are no sections near the top of the Portway Group and, over this stretch, the base of the Wentnor Series has been drawn conjecturally at a constant distance below the Darnford Conglomerate. To the south of Asterton the Portway Group occurs on the western scarp slope of the Long Mynd and the outcrop is limited to the west by the Long Mynd Scarp Fault except possibly in a small area south of Myndtown [389887]. At this locality, on the east side of the Scarp Fault there is an outcrop of conglomerate and massive sandstone which may belong to the lower part of the Bayston–Oakswood Group. It is not certain whether these beds rest unconformably

[1]This National Grid Reference lies within the 100-km square SJ (33). References without initial letters lie within the 100-km square SO (32).

on the Portway rocks to the east or whether they are separated by a second fault. If in fact there be an unconformity, then the basal Wentnor rocks lie only about 700 ft stratigraphically above the Huckster Conglomerate at this locality. (See later discussion of unconformity).

The Portway Group consists largely of purple and green shaly mudstones and siltstones, with purple and greenish sandstone bands locally abundant. In the uppermost parts of Callow Hollow and Ashes Hollow there are sections in the lower part of the group showing about 100 ft of massive purplish sandstone immediately west of the Huckster Conglomerate, succeeded westwards by banded siltstones and sandstones. Siltstones and sandstones in the higher part of the group are exposed in two strike sections in Hawkham Hollow. In the southern part of the Long Mynd, purple and dark green micaceous grits are very common for some 250 ft above the Huckster Conglomerate. Dark green grits with many fragments of purple siltstone occur in places, in beds high in the group, between Plowden and Myndtown.

James (1956, p. 319) suggested that the amount of structural deformation and compaction of the Portway rocks (Mintonian) was less than that of the remainder of the Stretton Series. This was used as supporting evidence for an unconformity between the Strettonian and the Mintonian. It must be recorded that in the present survey, no tangible differences of this type were noted between the Portway Group and other members of the Stretton Series.

Evidence of unconformity between the Stretton and Wentnor series. No fresh evidence concerning the post-Strettonian unconformity was obtained from the present survey, but a brief account is given of the relevant points which have been extensively discussed in previous literature (e.g. Whittard 1952, p. 146). The first main point in favour of an unconformity is the greater geographical extent of rocks of Wentnor type in comparison with those of Stretton type. The occurrence of the latter is confined to the main outcrop stretching from Plowden to Haughmond Hill (Shrewsbury) and to a few small faulted blocks occurring close by, on the line of the Church Stretton Fault Complex. The more massive purple sandstone lithology of the Wentnor Series is found both in the type area and outside it, in the inliers of Pedwardine, Old Radnor, and Huntley in Gloucestershire. As suggested by Whittard, this implies a widespread and violent overstep of the Wentnor Series. The second point is the apparent decrease in the width of outcrop of the Portway Group when traced southwards from High Park to Asterton. In the High Park area the Portway Group measures about 3500 ft in thickness, whereas just north of Asterton there are only about 2000 ft of Portway beds present. It is admitted that the basal boundary of the Bayston–Oakswood Group is mapped conjecturally but the marker horizons of the Huckster (Portway Group) and Darnford (Bayston Group) conglomerates converge southwards in a manner which indicates a southward thinning of the Portway outcrop. Furthermore, if it be correct to interpret the section south of Myndtown [389887] as showing Wentnor Series conglomerate and sandstone resting unconformably on Portway rocks (see also p. 87), there are here only about 700 ft of Portway Group below the Bayston–Oakswood Group. Northwards from High Park, outside the Church Stretton district, there are few exposures of the base of the Wentnor Series, but at Haughmond Hill near Shrewsbury the basal Wentnor Haughmond Conglomerate appears to rest unconformably on beds which are tentatively correlated with the lower part of the Synalds

Group. Here then is the suggestion that the Wentnorian unconformity has cut downwards across the Portway, Lightspout and (upper) Synalds groups.

James (1956, p. 324) demonstrated that the basal Wentnor rocks of the Chittol area rest unconformably on Western Uriconian rocks, the whole sequence being inverted. If James's structural interpretation of the Longmyndian be correct, then on the eastern limb of his syncline the basal Wentnor rocks rest upon the Portway Group while on the inverted western limb the Wentnor Series rests on Western Uriconian. In the discussion on James's paper, (1956, p. 336), Whittard remarked that in excavations near the Brokenstones [420882] a section was found showing conglomerate and sandstone of Wentnor type apparently resting unconformably on Stretton Shales. This seems to be another indication of the unconformable overstep of the Wentnor Series upon the Stretton Series.

Wentnor Series

As noted above (p. 36) this account follows James (1956, p. 316) in a twofold division of the Wentnor Series into the Bayston–Oakswood Group and the overlying Bridges Group. This bipartite classification has developed from the earlier threefold division, used in the Shrewsbury (152) Memoir, on the hypothesis that the Wentnor Series has been folded into a large isoclinal syncline with the Bayston and Oakswood groups forming the eastern and western limbs respectively, and the Bridges Group occupying the core. James's theory of a major synclinal structure is not disputed in this account, although it is considered that the evidence so far gathered in support of the idea cannot be regarded as completely satisfactory.

Bayston–Oakswood Group

The outcrop of the Bayston–Oakswood Group is in two main parts, one on the eastern limb of the syncline (the Bayston Group of the Shrewsbury (152) Memoir) and the other on the western limb (the Oakswood Group of the Shrewsbury (152) Memoir).

Within the Church Stretton district the Bayston outcrop occupies the north-western part of the Long Mynd massif stretching from Asterton [398912] in the south to Darnford [425976] in the north. At its eastern boundary the group rests upon Portway rocks, probably with unconformity (see p. 45 above), while between Stanbatch [402937] and Ratlinghope [403970], the Bridges Group succeeds it to the west. To the south-west of Stanbatch the Bayston outcrop disappears beneath a cover of Upper Llandovery rocks in the large hollow between Wentnor [384927] and the Long Mynd.

The base of the group is seen only in Hawkham Hollow [433977] and is taken at the change from purple and green shales and sandstones of the Portway Group, to the more massive, coarse purplish sandstone of the Bayston Group. There is no obvious angular discordance at this horizon but the lithological change is well marked and the basal Bayston sandstone contains occasional small pebbles. James (1956, p. 319) suggested that this pebbly sandstone may represent the dying out of the Haughmond Conglomerate which marks the base of the Bayston–Oakswood Group in the Shrewsbury area. The top of the Bayston–Oakswood Group is taken at the change from massive sandstones to the more flaggy siltstones and laminated sandstones of the Bridges Group. The rocks are poorly exposed along this boundary and its transitional character has been noted by James (1956, p. 322), who commented that the delineation of the two groups is arbitrary.

The rocks of the eastern fold limb are mainly coarse, massive, purplish sandstones, locally gritty and with occasional siltstone bands. The eastern outcrop contains the two marker horizons of the Darnford and Stanbatch conglomerates which occur at about 800 ft and 2200 ft respectively above the base of the group. The Darnford Conglomerate takes its name from exposures in Colliersford Gutter [426973] 250 yd south of Upper Darnford [425976]. Here there are about 30 ft of coarse conglomerate with rounded pebbles, up to 2 in across, of quartz, quartzite, rhyolite and occasional pink granite. In this section the Conglomerate is succeeded westwards by flaggy green siltstones which are also found above the Conglomerate in other exposures to the south. Between Darnford and Stanbatch [402937] the Conglomerate is traced only from small scattered exposures and from debris in the soil. To the south of Stanbatch the Darnford Conglomerate is more extensively exposed on the western scarp slope of the Long Mynd and the outcrop is disrupted by a number of minor transcurrent faults.

The type locality of the Stanbatch Conglomerate is at Stanbatch [402937] on the west side of the Long Mynd, whence it can be traced northwards to Middle Darnford [421976] at the northern limit of the area under discussion. To the south of Stanbatch there are no exposures of the Conglomerate in the low ground west of the Long Mynd. It is better exposed than the Darnford Conglomerate and in most of the major valley-sections gives rise to the characteristic double-ribbed feature (Plate 7B) noted by Whitehead (*in* Pocock and others 1938, p. 49). The best section is probably that on the Ratlinghope road [416956] 700 yd W.N.W. of the Shooting Box. Here there are 200 ft of coarse conglomerate, separated from an upper 60-ft conglomerate band (occurring to the west) by 70 ft of soft, coarse purple sandstone. The conglomerate pebbles are well rounded, up to about 2 in across, and consist mainly of quartz or quartzite with subordinate amounts of rhyolite and felsite. In the eastern (lower) rib the reddish brown silty matrix is very strong, where not weathered, and pebbles can be broken across with the hammer rather than separated from the matrix. In this section there is some suggestion of a decrease in size from east to west in accordance with the recognized view of an east-west upward stratigraphical sequence. This gradation however is not sufficiently well marked to be used alone as evidence of age sequence. As the map shows, the outcrops of the Stanbatch and Darnford conglomerates are disrupted by a number of transcurrent faults trending approximately east-west or east-south-east to west-north-west. These have been mapped mainly in the exposed valley sections and the generalized outcrop on the unexposed intervening spurs may in fact be more disrupted than has been shown.

The rocks of the Bayston outcrop west of the Stanbatch Conglomerate are mainly massive purple sandstones, locally with thin quartz veins. Concerning the section on the Ratlinghope road between Belmore Ring [408960] and Ratlinghope [402967] James (1956, p. 320) suggested that there is a gradual westward decrease in grain size with an incoming of shaly bands, marking a passage to the Bridges Group to the west. This idea was not supported by the results of the present survey, no marked progressive change being observed in this section.

In the district under discussion the western (Oakswood) part of the Bayston–Oakswood Group forms only a small area in the vicinity of Stedment Farm [388969], north-west of Bridges. Exposures are poor and scattered and consist mainly of massive purple sandstone, locally gritty. The outcrop of the Lawn Hill Conglomerate just enters the area but is nowhere exposed within it, being represented only by two small sections a short distance west of the limit of the district. These sections are in a conglomerate composed mainly of well-rounded pebbles of quartz and quartzite up to about 1 in across. According to the positions of these two outcrops the Lawn Hill Conglomerate is here about 600 ft in thickness. In better exposed ground to the north, the Lawn Hill Conglomerate is described as being up to 500 ft thick, consisting of grits with coarser pebbly or conglomeratic bands (Pocock and others 1938, p. 55). This description may also apply in the Stedment area although exposures are insufficient to confirm it.

The Oakswood outcrop is considered to belong to the western, inverted, limb of the major Longmyndian syncline which implies that there is an upward stratigraphical sequence into the Bridges Group to the east. As with the eastern limit of the Bridges Group, its western boundary with the Oakswood outcrop is poorly exposed and in most places must be regarded as conjectural.

BRIDGES GROUP

The Bridges Group is considered to overlie the Bayston–Oakswood Group and to occupy the core of the Longmyndian syncline. Accordingly, the eastern and western limits of the Bridges Group outcrop mark the base of the Group, but it is unfortunate that both boundaries are poorly exposed. Within the Church Stretton district the Bridges Group outcrop extends from near Criftin [380915] in the south to Ratlinghope Hill [405977] in the north. To the east and south of Wentnor [384927] the Bridges Group rocks are unconformably overlain by sediments of Upper Llandovery age. There is a marked decrease in the width of the outcrop from south to north and this decrease continues northwards in the area of the Shrewsbury (152) sheet. This was noted by Whitehead (*in* Pocock and others 1938, p. 52) and James (1956, p. 333) who suggested that the decrease was due to a southerly plunge of the longitudinal axis of the syncline.

The rocks of the Bridges Group consist mainly of purple siltstones and sandstones, commonly laminated and moderately flaggy. The well-marked bedding together with a higher proportion of silty sediment serves to distinguish these rocks from the coarser, more massive beds of the Bayston–Oakswood Group. At a number of localities specimens have been obtained showing minor sedimentary features, such as current and graded bedding and load cast structures, which indicate the age sequence in these beds (Plate 3). These specimens are more fully discussed below (p. 50) but it may be noted here that a great deal of the evidence for stratigraphical inversion has been obtained from the laminated beds of the Bridges Group. Such evidence is not readily obtained from the more massive sandstones of the Bayston–Oakswood Group. The regional dip in the Bridges Group is steep to the west-north-west. At some localities the amount of dip may vary to 50°–60° and some sections may show a steep dip to the east, but in general it may be said that the attitude of the Bridges rocks conforms with the steep isoclinal west-north-west dip of the rest of the Longmyndian.

FAULTED INLIERS OF WENTNOR SERIES

In addition to the outcrops described above there are a number of small fault-inliers of presumed Wentnor Series rocks, situated along the line of the Church Stretton Fault zone. In all instances the rocks have been assigned to the Wentnor Series solely on the grounds of general lithological similarity and it has not been possible to make correlations with either the Bayston–Oakswood or Bridges groups. J.E.W.

Cwms area. In the Cwms area [474940], where the Lawley and Cwms–Hoar Edge faults converge, there is a triangular outcrop assigned to the Wentnor Series which is bounded by these faults and by an unconformable cover of Cambrian rocks to the north. The rocks comprise grits and conglomerates which Cobbold (1927, p. 551) suggested might belong to the Wentnor Series. They are poorly exposed and are locally much weathered.

Little Stretton area. At the south end of Ragleth Hill [446915] near Little Stretton, there is an outcrop, measuring about 300 yd by 50 yd, of red and pink conglomerate and grit, assigned by Cobbold and Whittard (1935, p. 349) to the Wentnor Series. There is some doubt as to the nature of the junction of these beds with the Uriconian to the north. The southern limit of the outcrop is formed by an unconformable cover of Harnage Shale. G.H.M.

Aston-on-Clun to Cwm Head. Between Aston-on-Clun [394818] and Wart Hill [400847] there is a large outcrop composed mainly of purple micaceous sandstone and siltstone which is tentatively correlated with the Wentnor Series. Exposures are poor and scattered but this outcrop appears to be fault-bounded in most places except in the vicinity of Aston-on-Clun where there is an unconformable cover of Hoar Edge Grit. B.A.H.

Purple grits referred to the Wentnor Series crop out all round Wart Hill and north-north-eastwards to Cwm Head [423886]. They lie mainly to the east of fault F2 and are succeeded eastwards by Ordovician sediments, which locally rest unconformably on them and elsewhere are faulted against them. Narrow outcrops lie in three places between fault F1 and the Stretton Shale Group. Sandy shale and blocky mudstone are locally developed, possibly belonging to the Bridges Group, but no general stratigraphical pattern has been observed within the outcrop as a whole. Near Brokenstones [4288] the Wentnor and Stretton rocks are seen in contact in several outcrops, at one of which (p. 77) the former seem to be resting unconformably on the latter. Whether this relationship obtains elsewhere in this area is not clear, all the exposures being complicated by faulting. D.C.G.

Evidence of Stratigraphical Sequence in the Longmyndian

Within the Stretton Series, minor sedimentary features such as graded and current bedding, load casts and flame structures indicate a general upward succession from east to west. At several localities in the Long Mynd there are exceptions to the general statement made above, e.g. (1) In exposures north and south of St. Michael's Church, All Stretton [462959] the Stretton Shales dip to the east-south-east at 55° and also young in this direction. (2) In the Burway Group, in a zone lying between 250 yd and 450 yd north-west of the Buxton Rock, easterly dips have been recorded along the strike (see p. 79). (3) In the discussion on James's paper (1956, p. 337) Whitten described features from the Synalds Group of the Cardingmill Valley which he considered as younging to the east. These examples indicate local reversal of dip and in some instances, inversion of the beds, but this small number of exceptions can be accounted for by the existence of minor folds on the steeply dipping eastern limb of the major syncline. The evidence so far obtained favours the idea of a general east-west stratigraphical sequence in the Stretton Series.

According to Whittard's and James's theory of a syncline in the Longmyndian, the eastern (Bayston) outcrop of the Bayston–Oakswood Group and the eastern part of the Bridges Group exhibit a normal westward succession. In the western part of the Bridges Group and the western (Oakswood) outcrop of the Bayston–Oakswood Group it is considered that the beds are inverted and young to the east. The rocks of the Bayston–Oakswood Group are mainly coarse massive sandstones with conglomerates, which do not in general provide good evidence of age sequence. However, Whitehead (1955, p. 466) described a section in the quarry east of Haughmond Abbey, near Shrewsbury, where graded and cross bedding in grits of the Bayston outcrop young to the west. From the Oakswood outcrop, James (1952, p. 198 and 1956, p. 324) described localities in the Chittol–Coldyeld area, west of the Church Stretton district, where tuffaceous beds near the western edge of the Longmyndian outcrop show inverted sedimentary structures younging towards the east. Apart from these observations, most of the evidence of age sequence in the Wentnor Series comes from the Bridges Group.

Fig. 8 shows localities within the Bridges Group outcrop from which definite evidence of age sequence has been obtained. The dip of the bedding is shown at each locality together with the direction in which the beds young. The sedimentary features which have been used as evidence include current and graded bedding, erosion, load cast and pull apart structures. These commonly occur on a small scale and are best seen on a smoothed surface with a hand lens or binocular microscope. Plate 3 illustrates various specimens which have been examined in this way.

James (1956, p. 322) noted only two localities within the Bridges Group, one being at Drive Coppice [397964] (Fig. 8 locality 2) where he observed current bedding younging to the west. The second locality is in the stream section 150 yd north of the Bridges road junction [393967] (Fig. 8 locality 4) where he observed graded bedding younging to the east. From this evidence alone, it might be inferred that the axial plane of the Longmyndian syncline lay between these two localities. It appears more probable that the Bridges Group contains an axial zone (Fig. 8) in which there are reversals of sequence due possibly to small, tightly packed isoclines. Localities 2–8 lie within this axial zone, while of the remaining three, locality 1 lies on the normal eastern limb of the syncline and localities 9 and 10 lie on the inverted western limb. It is admitted that the numerous small isoclines postulated for the axial zone were not in fact seen in the limited exposures available. However, the possible existence of such folds affords a reasonable explanation of the observed reversals of age sequence.

It may be concluded from the above statements that the evidence obtained to date cannot be regarded as definite confirmation of the theory of a major synclinal structure in the Longmyndian. On the other hand, nothing has yet been found in strong opposition to the idea and for this reason, Whittard's and James's proposal is, in this account, regarded as tenable.

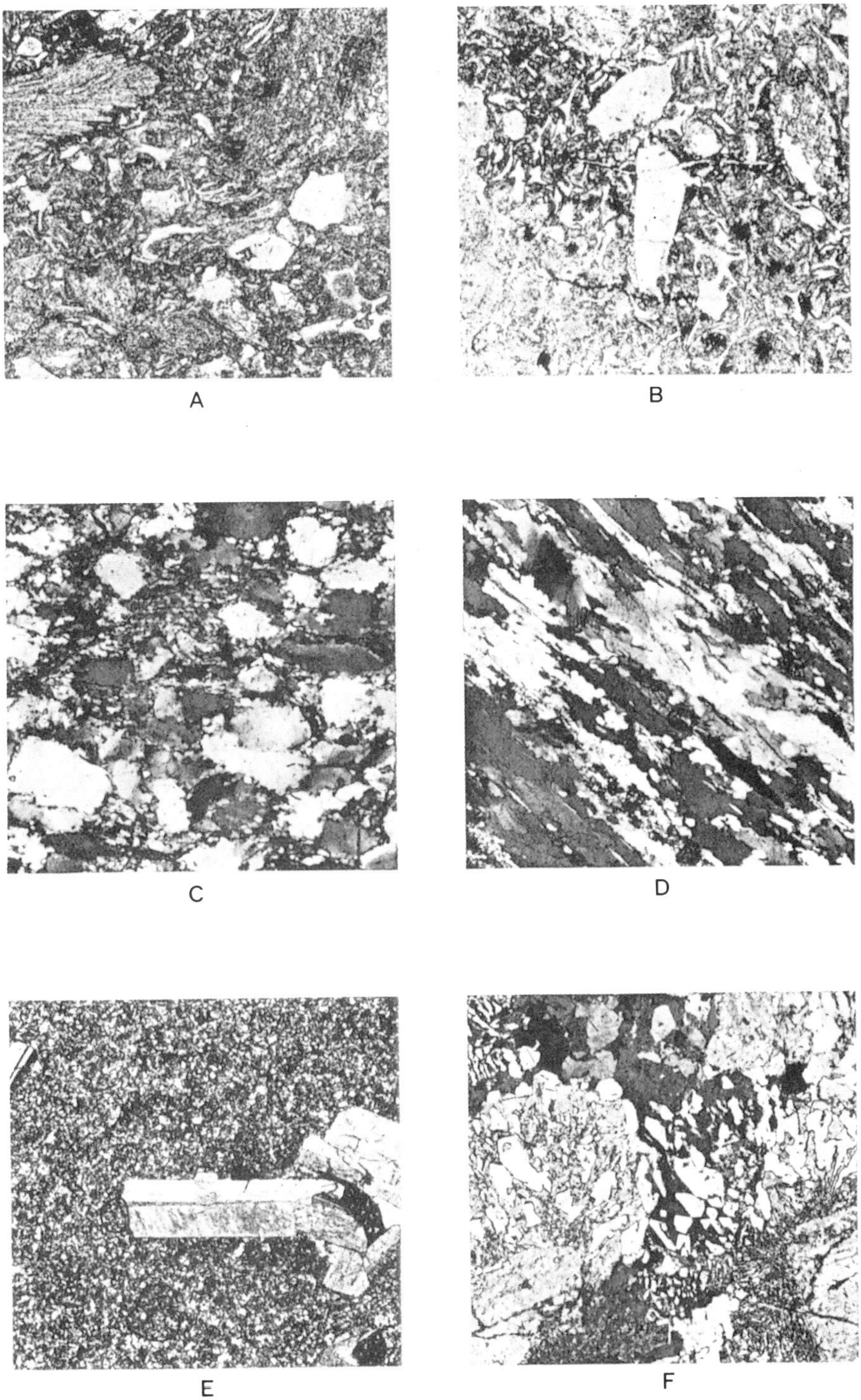

PHOTOMICROGRAPHS OF URICONIAN TUFFS AND PEBBLES FROM LONGMYNDIAN CONGLOMERATES

(For detailed explanation see p. ix)

A

B

C

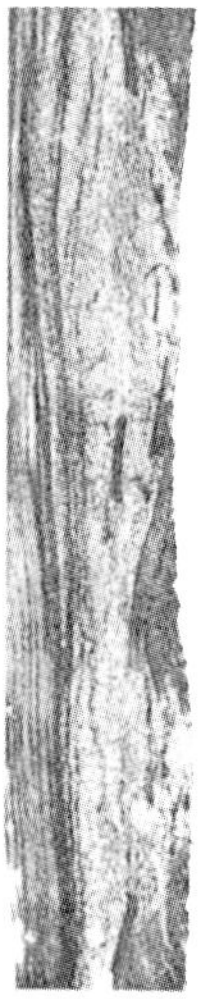

D

E

F

G

POLISHED SURFACES OF LONGMYNDIAN SEDIMENTS
SHOWING SEDIMENTARY STRUCTURES
(For detailed explanation see pp. ix-x)

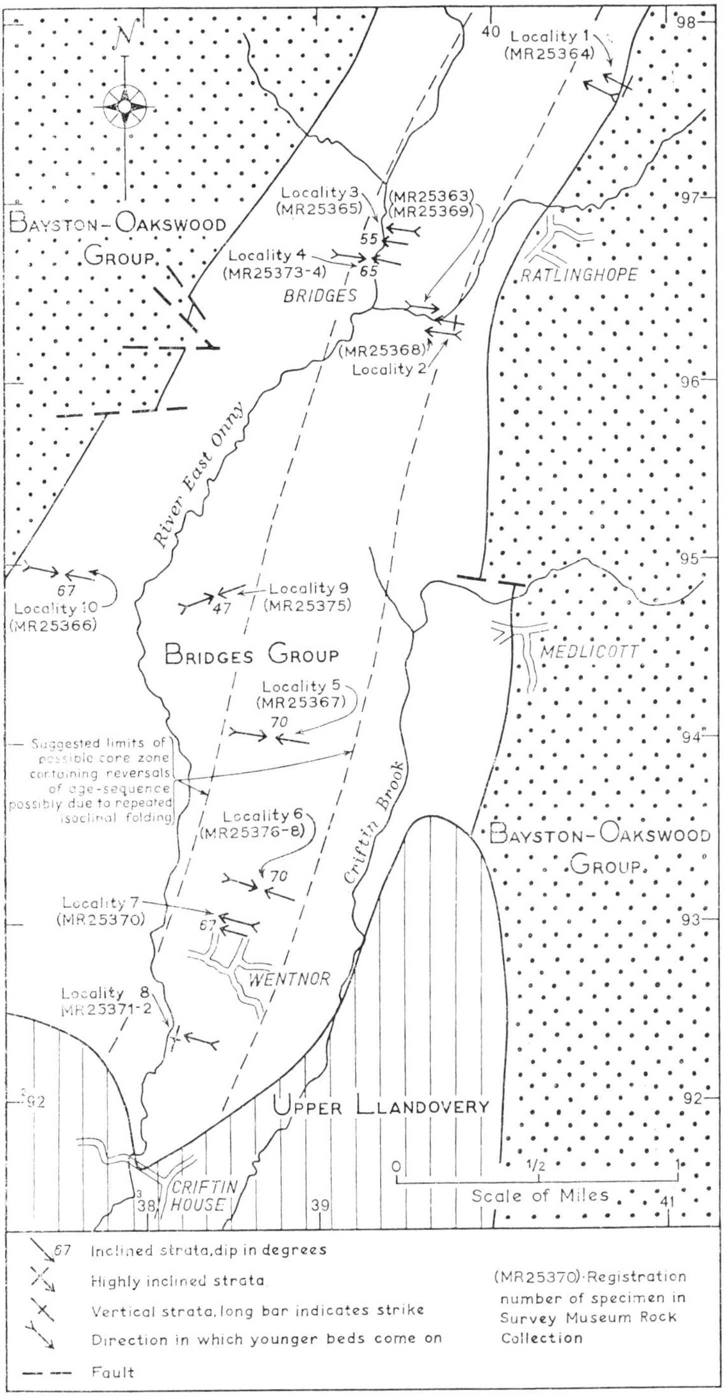

FIG. 8. Geological map of part of the outcrop of the Bridges Group, showing localities from which evidence of age sequence has been obtained

Petrography and Petrology

TUFFS IN SYNALDS GROUP

The tuffs in the Synalds Group are mostly moderately coarse, some with a rough foliation, and varying in colour from purple to pale green or greenish grey. Dark green epidotic aggregates up to about 5 mm across are seen in the hand specimen in addition to white or flesh-coloured fragments, probably feldspar, of similar size.

No satisfactory petrological distinction can be made between one band of tuff and another. They are all epidotic crystal-lithic tuffs, generally very similar in appearance. On average the phenoclasts make up about one quarter to one third of the rock but one or two specimens contain a much higher proportion of large fragments.

The most abundant crystal phenoclasts are of feldspar, mainly albite or sodic oligoclase, occurring as stout subhedral prisms and as more irregular anhedral fragments, about 0·5 mm to 1 mm across. In some specimens there is partial alteration of the feldspar to sericite, chlorite and epidote, the degree of alteration varying considerably between specimens. Some rocks contain albite phenoclasts with slightly distorted twin lamellae (E 28753, 28998, 29015), presumably a result of their pyroclastic origin. Subangular quartz phenoclasts occur in most specimens but quartz is less abundant than feldspar and the fragments are usually smaller, up to about 0·2 mm across. In some specimens there are anhedral flakes of altered greenish brown biotite up to about 1 mm across (E 27859, 29016, 29642). The biotite commonly shows considerable alteration to chlorite and white mica and there is some development of very small crystals of turbid sphene, presumably an alteration product. The cleavage of these altered biotite flakes is commonly slightly contorted. They are thought to be of pyroclastic origin. Small euhedral prisms of apatite occur as phenoclasts in some specimens (E 28994, 29039, 29612), but apatite is not a prominent constituent of these rocks.

Lithic phenoclasts are of three main types; siltstone or mudstone, crypto-crystalline igneous rocks and irregular aggregates of chlorite, micaceous aggregate, and epidote. The last type of phenoclast may represent earlier tuff deposits which have been redeposited in the bands of the Synalds Group. Siltstone and mudstone fragments are roughly tabular in shape and average about 1 mm to 2 mm across, although in some specimens they occur up to about 10 mm in length. In some of the tuffs the foliation or lamination of the siltstone fragments is contorted or even disrupted (E 27853). Igneous rock fragments are probably of rhyolitic or andesitic composition and consist of a finely crystalline or cryptocrystalline aggregate of chlorite, quartz and feldspar, which in some instances encloses small prisms of plagioclase. These fragments are usually subrounded, polygonal in outline, and average about 0·2 mm across although they may reach 2 or 3 mm. In a few specimens these igneous fragments show poor trachytic texture (E 27859) or sub-spherulitic structure (E 29065). As noted above, the third type of lithic phenoclast comprises aggregates of chlorite, micaceous aggregate and epidote, in some instances with feldspar and carbonate. These aggregates are commonly elongate and rather irregular in outline. They may possess a rough foliation and usually have sharp junctions with the matrix of the tuff. There is some doubt as to their origin; if they are pyroclastic they may represent reworked earlier tuff deposits.

On the other hand they could be of authigenic origin. Other lithic types found as phenoclasts include greywacke (E 27843) and quartzite (E 28737, 29624, 29628). Some of the quartzite fragments show strained extinction and sutured crystal boundaries; these may be of metamorphic origin.

The matrix of the tuff bands is finely divided and is composed mainly of clay minerals, micaceous aggregate and subordinate chlorite, with abundant epidote, usually rather turbid and occurring in irregular aggregates throughout the rock. In some instances a rough foliation may be marked by the sub-parallel arrangement of wisps of micaceous aggregate (E 27851).

DOLERITE INTRUSIONS IN THE LONGMYNDIAN

The dolerite intrusions of the Long Mynd occur mainly as dykes which cut the steeply folded Longmyndian sediments. In addition to the obviously transgressive dykes there are a number of small intrusive bosses of dolerite and also some steeply inclined concordant sheets, which could be regarded as sills folded with the sediments, but which more probably belong to the post-folding phase of dyke intrusion. The dykes vary in thickness from a few feet to 200 ft but are commonly up to about 10 ft in thickness. Trends are difficult to establish because of the generally restricted nature of the exposures but in some instances dykes have been traced at intervals for distances of about ½ mile. The direction of these is rather variable but is dominantly easterly. From a number of other less well-defined dykes there appears to be another major trend running approximately northwesterly. Fig. 9 shows the distribution and directions of dykes and other intrusions in the Longmyndian.

In hand specimen the fresher dolerites are massive, eucrystalline, bluish grey rocks with clear feldspar and with small amounts of visible pyrite. In the more altered varieties, or on weathered surfaces, the ferromagnesian minerals are dark green while the feldspar is opaque and white or pinkish in colour.

The specimens examined all appear to belong to the quartz-dolerite suite as opposed to the olivine-bearing variety. Typically they show good ophitic texture and are composed of interlocking subhedral prisms of plagioclase-feldspar which may be wholly or partly enclosed by anhedral plates of augite. This latter mineral is commonly very pale brown or almost colourless. In some specimens quartz occurs as a relatively abundant accessory (E 27830) in anhedral crystals interstitial to the feldspar, while in other rocks the original magmatic acid residuum may be represented solely by chloritized patches of mesostasis. The dolerites invariably contain chlorite, commonly penninitic and weakly pleochroic, which may partly or wholly replace augite, and which also occurs in polygonal areas replacing original fine-grained acid mesostasis. Accessory minerals include ilmenite or titanomagnetite, apatite in small acicular prisms, and occasional small anhedral aggregates of pyrite.

In the specimens examined the amount of deuteric or subsequent alteration varies considerably. The least altered rocks contain clear labradorite with only a very little sericitization (E 29616, 29635) and in some instances the feldspar shows some zoning. In addition to pale augite these rocks may contain small amounts of pleochroic pale brown hornblende, closely associated with the pyroxene (E 29616, 29617), or subordinate greenish brown biotite, again closely associated with pyroxene (E 29635). Two dykes in this group contain abundant accessory quartz (E 29616, 29617).

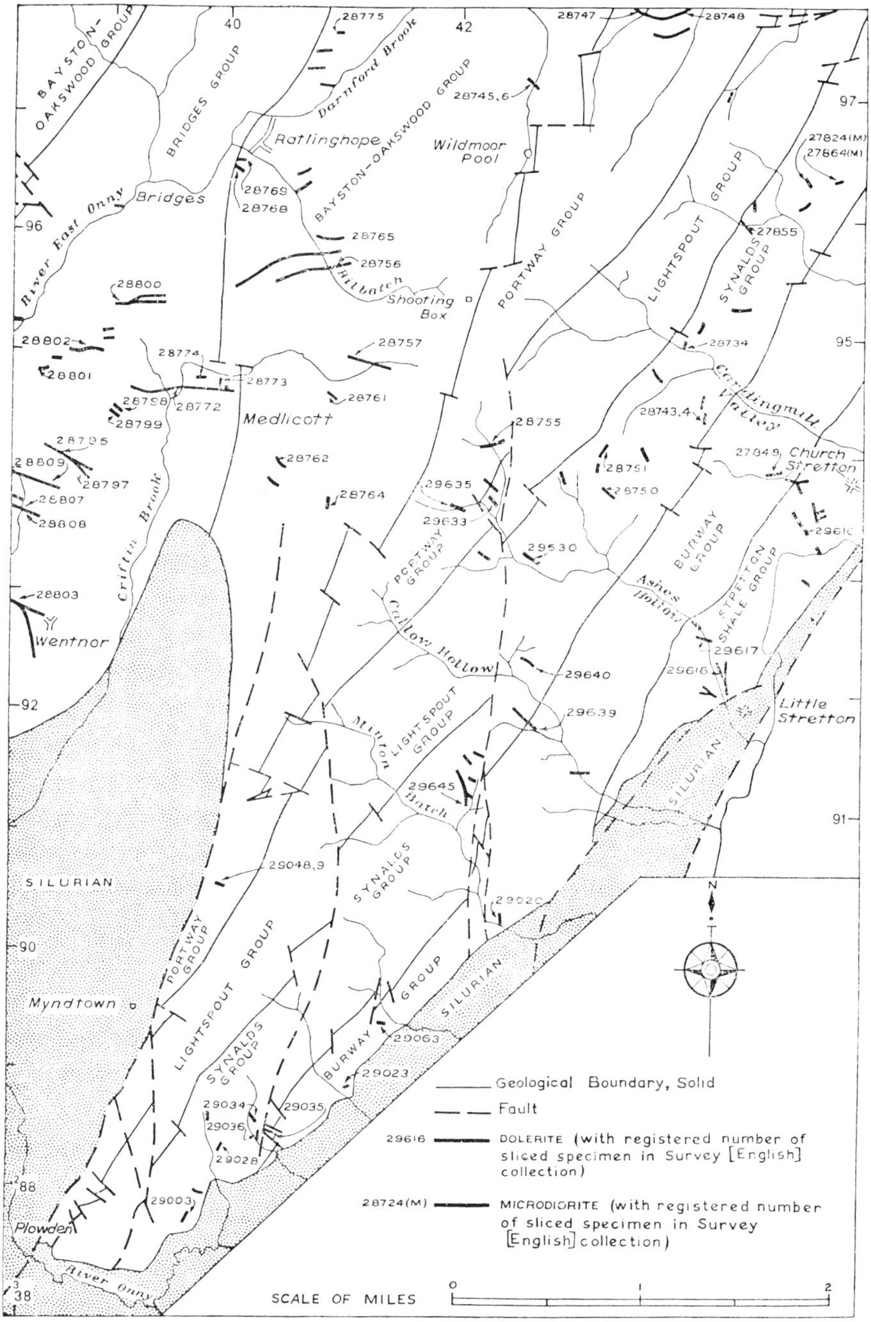

FIG. 9. Distribution of minor intrusions (mainly dolerites) in the Longmyndian

In the more noticeably altered dolerites the feldspar varies in composition from labradorite to albite and shows more extensive development of small flakes of sericite (E 28768), which in some instances is accompanied by the growth of small flakes and veinlets of pale green chlorite (E 28803). In some rocks thin films of pale green chlorite are developed on cleavage planes within the feldspar (E 28806). Several specimens in this group contain plagioclase with turbid sericitized centres and rather poorly defined clear margins, probably more albitic in composition than the centres (E 28803). There is considerable growth of epidote within the feldspar in a few specimens (E 27830) but epidotization is not conspicuous in the partly altered dolerites. Some rocks of this group contain accessory quartz (E 28730, 28751, 28764). Augite is partly altered to chlorite, commonly penninitic, and in some instances the original ophitic texture between augite and plagioclase may be preserved between chlorite and plagioclase (E 28748, 28751). Chlorite commonly contains small turbid anhedral crystals of sphene, especially where the chlorite is in contact with ilmenite or titanomagnetite (E 28751, 28807). This suggests that the sphene is a product of the chloritization process, particularly where the ore minerals have provided titanium. Some specimens contain a small amount of pleochroic greenish brown hornblende (E 29630, 29633) which is closely associated with augite and chlorite, and appears to have developed from augite. In other rocks there is accessory greenish brown biotite as fairly large anhedral flakes and as fine micaceous aggregate intergrown with chlorite aggregate (E 28765, 28794).

In the more extensively altered dolerites the original ophitic texture may still be preserved but augite may be almost completely replaced by chlorite and, in some instances, partly replaced also by calcite (E 28717). In some members of this group, however, there is still a considerable amount of augite remaining (E 28757). Feldspar is oligoclase or albite and shows some variation in the amount of sericitic alteration. Epidote is more common in this group than in the less altered rocks, and occurs as rather turbid aggregates and as smaller clear crystals within plagioclase and chlorite (E 29020, 29639). Ilmenite or titanomagnetite shows extensive alteration to leucoxene (E 27826) and may also carry narrow rims of rather turbid sphene (E 28757). One altered dyke with poor doleritic texture contains abundant accessory quartz with which is associated small amounts of micropegmatite, some of which appears to replace feldspar (E 28743, 28744).

In addition to the dolerite intrusions described above, there are two small bosses in the Synalds Group, close to Jinlye, which are better described as microdiorites (E 27824, 27864). These rocks consist largely of stout, interlocking albite prisms with subordinate chloritic aggregate interstitial to the feldspar. The albite contains some sericitic mica and there is a little accessory quartz. In one specimen (E 27864) the chlorite partly replaces pale greenish brown biotite. These are the only minor intrusions so far found in the Longmyndian, which have not been referred to the dolerite suite.

Age of dolerite intrusions. In the Long Mynd and Shelve areas of south Shropshire dolerite intrusions cut Uriconian, Longmyndian, Cambrian and Ordovician rocks. No intrusions have been found in the Silurian rocks of this area and it should be noted that intrusions in the Ordovician occur only to the west of the Church Stretton Fault Complex. The only igneous rock so far found in the Ordovician to the east of the fault (p. 121), is a chloritized and carbonated basalt (E 29498–29500), probably a lava, in the Harnage Shales near Sibdon Carwood [407834] (see also Whittard 1952, p. 162).

The basaltic lavas and doleritic intrusions in the Uriconian of the Church Stretton district are extensively albitized and chloritized and contain considerable amounts of epidote. In their degree of alteration they are not unlike some of the more altered intrusions of the Long Mynd although many of the Uriconian dolerites contain conspicuous pseudomorphs after olivine. The basaltic lavas are clearly of Uriconian age while the altered doleritic intrusions are probably pre-Cambrian and possibly pre-Longmyndian in age, as is suggested by the following evidence. According to the Shrewsbury (152) Memoir (p. 32) "The main intrusion of the Lawley terminates abruptly on the eastern side of the hill, under the Cambrian quartzite, which does not seem to be affected by it". This suggests a probable pre-Cambrian age for the intrusion. On Caer Caradoc a large doleritic intrusion, enclosing a large raft of Ragleth Tuffs, apparently forms the core of a major synclinal fold in the Uriconian and may be contemporaneous with, or earlier than the folding. Also, doleritic intrusions in the Caer Caradoc area are affected by thrusts within the Uriconian, which may be contemporaneous with the folding. In the discussion on the possible unconformity at the base of the Longmyndian it is suggested that the folding of the Uriconian may be of pre-Longmyndian age (p. 277). This points to the conclusion, admittedly with some uncertainty, that the doleritic intrusions in the Uriconian may be of pre-Longmyndian age.

From a study of the basic intrusions of the Shelve area, Blyth (1944, p. 169) concluded that these rocks were emplaced during or after the folding of the Ordovician (Arenig to Caradoc) sediments but prior to the deposition of the basal Silurian (Upper Llandovery) rocks. Blyth described the Shelve intrusives as having affinities with the olivine-basalt magma type although the Corndon intrusion is a quartz-dolerite. In their general appearance in thin slice, many of the dolerites of the Longmynd are similar to rocks from the Shelve, described by Blyth.

The dolerite intrusions of the Long Mynd are clearly younger than the intense folding (of probable pre-Cambrian age, see p. 277) which affected the Longmyndian sediments. They are therefore certainly younger than the Uriconian basaltic lavas and are possibly also younger than the doleritic intrusions in the Uriconian (of possible pre-Longmyndian age). So far it has not proved possible to distinguish more than one suite of dolerites within the Longmyndian intrusions and accordingly they are regarded as belonging to one phase of intrusion. In the light of the above arguments and in the absence of more conclusive evidence it is suggested that the dolerites of the Longmynd are approximately contemporaneous with those of the Shelve area and are of post-Caradocian and pre-Upper Llandoverian age. The apparent absence of dykes in the Caradoc Series east of the Church Stretton Fault Complex suggests that this structure may have acted as a barrier to the parent magma.

J.E.W.

LONGMYNDIAN SEDIMENTS

General description of the rocks. The Longmyndian sediments include all types of clastic sediment from banded argillaceous rocks and siltstones to relatively coarse conglomerates. Subgreywacke-sandstones are common in both series and are especially abundant in the Wentnor Series. Calcareous rocks are represented only by a few thin nodular limestones and slightly calcareous grits in the Stretton Shales, and by rare slightly calcareous grits in the Bayston Group.

The colour variation is from dark greenish to dark purple although the markedly conglomeratic horizons tend to be reddish.

Interbedded with these clastic sediments are tuffs of various types and tuffaceous sandstones or subgreywackes. There is apparently a complete lithological gradation from normal subgreywackes to epidotic tuffs with increasing amounts of interstitial fine-grained sericitic and chloritic cement and epidote.

A general petrographical similarity is evident between the different lithological types. Throughout the Longmyndian sediments the detrital grains are predominantly subangular or angular, and the same suite of derived lithic fragments is found in all grades of sediments from siltstones to conglomerates.

Conglomerates

The main conglomeratic horizons in the Longmyndian, which occur in the Portway and Bayston–Oakswood groups, are interbedded with conglomeratic and coarse-grained (subgreywacke) sandstones and siltstones.

In general the conglomerates are poorly sorted and contain a variable proportion of sand-grade fragments in which the pebbles are embedded. The pebbles are commonly 2–3 cm and in places up to 12 cm or more in size. The largest pebbles are predominantly metamorphic quartzites or vein-quartz and generally these are subrounded, although the smaller fragments tend to be subangular or angular. The relative proportions of the sandy and pebbly constituents varies within wide limits; one extreme is represented by pebbly sandstones such as the Huckster Conglomerate from the crags on the north side of Deadman's Batch [448977], 300 yd E. 30° N. of High Park crossroads (MR 22231)[1] and the crags on the south-west of Calf Ridge [428953], about 1100 yd N. 44° E. of the Boiling Well (MR 22232). In these specimens pebbles up to 2·5 cm across occur in a reddish medium to coarse-grained sandstone either as distinct thin beds (2–3 cm) or randomly scattered throughout the rock. The medium to coarse-grained sandstone is a massive rock lacking any stratification and may comprise up to 90 per cent of the rock as a whole. At the other extreme, conglomerates occur which are almost entirely composed or subrounded pebbles (0·5 cm–3·0 cm), for example the Huckster Conglomerate from the crags above the road [393894], about 330 yd E. 35° S. of Myndtown church (MR 22234) and the Oakswood Conglomerate from the hilltop [372973], about 600 yd W. 5° S. of Squilver Farm (MR 22170), (Montgomery Sheet 165).

The derived lithic fragments and pebbles in the conglomerates consist mainly of igneous and metamorphic rocks, although rare sedimentary fragments also occur.

Of the igneous rock fragments, acid to intermediate volcanic types predominate. Rhyolite-fragments are found in almost every specimen of conglomerate, occasionally as pebbles up to 10 cm across. In many instances these

[1]Such numbers prefixed by MR refer to registered specimens in the Museum Rock Collection of the Geological Survey.

rhyolites are cut by thin quartz veins. Flow-banded rhyolites are abundant in the Darnford Conglomerate east of the road [426974], 280 yd south of Darnford (E 29958). An 8 cm × 5 cm pebble of flow-banded rhyolite in the Stanbatch Conglomerate from a roadside exposure at Bayston Hill (E 29943), (Shrewsbury Sheet 152) is a particularly good example. It consists of bands of recrystallized devitrified rhyolite alternating with bands and patches of quartz and oligoclase-feldspar crystals. In places it appears that the partly solidified bands have been subsequently broken and bent, probably due to flow brecciation. Accessory magnetite occurs in small grains and as dusty particles which are frequently altered to hematite, giving a pink colour to the rock.

Fragments of rhyolite in which the groundmass is more or less devitrified are common and occur for example in the Oakswood Conglomerate at Habberley Brook, 2200 yd S. 35° E. of Pontesbury church (E 29938) (Shrewsbury Sheet 152). Typical pebbles of similarly devitrified rhyolite occur in the Stanbatch Conglomerate at a roadside exposure [416956], 1900 yd E. 36° S. of Ratlinghope Manor House. In one such (4 cm × 2·5 cm) pebble (E 29973), in which a few oligoclase phenocrysts occur, the groundmass has recrystallized to form a mosaic showing patchy and streaky aggregate polarization, although showing clear evidence of the original fluidal texture. Alteration of magnetite to hematite imparts a pink colour to the rock.

Spherulitic texture is developed in some rhyolite fragments for example in the (5cm × 3 cm) pebble from the Stanbatch Conglomerate at the previous locality (E 29974) in which a fine-grained quartz-feldspar groundmass is crowded with well developed spherulites associated with oligoclase phenocrysts. In some instances the spherulites have grown around previously crystallized oligoclase phenocrysts or groups of small phenocrysts; elsewhere the feldspar laths have a stellate arrangement and are intimately intergrown with the spherulites. At the margins of the phenocrysts a micrographic texture is locally developed. A patchy polygonal pattern is developed in places by the mutual interference during growth of adjacent spherulites. The groundmass in which the spherulites are embedded consists of a microcrystalline aggregate of quartz and feldspar, and small scattered epidote crystals occur. Finely divided hematite gives the rock a pronounced red coloration.

Less well-developed spherulites occur within a devitrified recrystallized groundmass in a rhyolite pebble from the Darnford Conglomerate east of the road [426974], 280 yd south of Darnford (E 29964–5, 29967). One (5 cm) pebble (E 29967) consists of a devitrified recrystallized groundmass showing aggregate polarization in which flow banding can be seen around the phenocrysts of quartz and oligoclase. Some of the phenocrysts show euhedral outlines, but others are highly corroded and have deeply embayed margins. A similar fragment but with a finer-grained groundmass occurs in the Oakswood Conglomerate (E 29940) from the east bank of Habberley Brook, 2200 yd S. 35° E. of Pontesbury church (Shrewsbury Sheet 152). Another (10 cm × 5 cm) pebble of porphyritic rhyolite or dacite from the same location (E 29965) shows almost complete replacement of the original phenocrysts and groundmass by microcrystalline silica and some feldspar, although the original amygdaloidal, porphyritic texture is still perfectly preserved. Other porphyritic rhyolite pebbles from the same locality show similar alteration with the presence also of clay minerals in the groundmass and partly replacing the original oligoclase phenocrysts (E 29971).

A porphyritic rhyolite or dacite (E 29947) occurs as pebbles in the Lawn Hill Conglomerate from an outcrop in Squilver Plantation [379970] (Montgomery Sheet 165). This contains partly sericitized phenocrysts of oligoclase in a felsitic groundmass of microcrystalline quartz and feldspar (Plate 2E).

Rhyolites such as those described above constitute a relatively large proportion of the igneous rock fragments in the conglomerate; a much smaller proportion of andesitic fragments occurs and basaltic types are practically absent.

The only conglomerate sliced in which andesite fragments are relatively abundant and form a significant portion of the rock is the Haughmond Conglomerate (E 29939) from 100 yd east of the Castle, Haughmond Hill (Shrewsbury Sheet 152). The andesite fragments are generally subrounded and up to 8 mm in size. They are mainly composed of laths of oligoclase-feldspar associated with a minor proportion of albite and interstitial magnetite altering to hematite. In some fragments irregular interstitial patches of glass occur. Similar fragments occur in the Huckster Conglomerate (E 29955) from High Park cross roads [445976] and the Lawn Hill Conglomerate (E 29945) from an outcrop in Squilver Plantation [379970], 490 yd S. 32° E. of Squilver Farm (Montgomery Sheet 165). Basalt fragments in the conglomerates are rare and have only been found in the Oakswood Conglomerate (E 29938) from Habberley Brook, 2200 yd S. 35° E. of Pontesbury church (Shrewsbury Sheet 152).

Fragments of volcanic tuff are a relatively minor but widespread constituent of the Longmyndian conglomerates. A number of different varieties of tuff are found, namely devitrified dust tuffs, tuffs composed of partially devitrified glassy shards, crystal tuffs (E 29955) and crystal-lithic tuffs together with fragments of microcrystalline quartz with recrystallized clay minerals or fragments composed almost entirely of clay minerals which may be of a tuffaceous nature. In the Oakswood Conglomerate at Habberley Brook 2200 yd S. 35° E. of Pontesbury church (E 29938) and in the Haughmond Conglomerate, 100 yd east of the castle, Haughmond Hill (E 29939) (Shrewsbury Sheet 152) small fragments of a glassy shard tuff occur. These fragments are devitrified but are clearly composed of shards which exhibit either a flattened (welded) texture (E 29938), or a random orientation of broken shards with occasional unbroken and nearly circular devitrified glass bubble walls (E 29939). Similar fragments of welded tuff occur in the Huckster Conglomerate (E 29955) from High Park cross roads [445976], in the Lawn Hill Conglomerate (E 29945), from Squilver plantation [379970], which also contains fragments of crystal tuff, and in the Stanbatch Conglomerate 260 yd N. 14° W. of Darnford [425978] (E 29966) (Shrewsbury Sheet 152).

A 4-cm banded (?welded) tuff pebble from the Stanbatch Conglomerate at a roadside exposure [416956] 1900 yd E. 36° S. of Ratlinghope Manor House (E 29972) consists of corroded fragments of quartz and kaolinized feldspar in a banded fine-grained groundmass. Discernible in the groundmass are outlines of what appear to be completely devitrified flattened glassy shards, closely moulded around the feldspar and quartz xenocrysts. In places the groundmass material has corroded the quartz fragments producing deep lobate embayments which may extend irregularly into the centre of the grain.

Fragments of crystal-lithic tuff in the Huckster Conglomerate (E 29955) from High Park cross roads [445976] contain smaller lithic fragments of quartz-muscovite-schists.

The most distinctive and relatively abundant plutonic igneous rock occurring as fragments in the conglomerates is a pink granophyre. Particularly large rounded pebbles of this type, up to 10 cm by 6 cm, occur in the Stanbatch Conglomerate (E 29970), 260 yd N. 14° W. of Darnford [425978] (Shrewsbury Sheet 152), and up to 7 cm in length in the Darnford Conglomerate (E 29963) east of the road [426974] 280 yd south of Darnford. These pebbles consist almost entirely of a fine (E 29970) or relatively coarse (E 29963) granophyric intergrowth of quartz and albitic feldspar, together with scattered phenocrysts of albite and quartz. The pink coloration is due to oxidation of accessory opaque ore to hematite. Smaller pebbles of granophyre are present in almost every conglomerate specimen (e.g. E 29938, 29945, 29958, 29964, 29966), (*see* Plate 2F).

A 4-cm rounded fragment of granophyric quartz-porphyry occurs in the Darnford Conglomerate (E 29960) east of the road [426974] 280 yd south of Darnford; the groundmass consists of quartz and alkali-feldspar, much of which shows granophyric intergrowth. Larger phenocrysts of albitic feldspar and quartz and a few xenoliths of a muscovite-bearing quartzite occur. The rock is slightly altered (sericitized) and the red colour is due to finely disseminated hematite.

Subangular to rounded fragments of granitic rock are relatively rare constituents of the conglomerates. A 4-cm leucogranite pebble from the Stanbatch Conglomerate 600 yd east-south-east of Wildreck (E 14723)[1] (Shrewsbury Sheet 152) consists of quartz, albite-oligoclase and perthitic feldspar with muscovite and accessory apatite and opaque ores altering to hematite. The rock has a crushed granular texture and the quartz shows suturing and strain shadows. A similar sodic leucogranite pebble (6 cm × 5 cm) has been sliced from the Oakswood Conglomerate (E 15922) from the east bank of Habberley Brook, 2080 yd E. 29° S. of Pontesbury church (Shrewsbury Sheet 152).

Smaller fragments of similar, slightly crushed quartz-albite-oligoclase-muscovite-leucogranites occur in the Darnford Conglomerate (E 29958), Stanbatch Conglomerate (E 29964, E 29966) and the Huckster Conglomerate (E 29955). In these conglomerates feldspar fragments are rather common and consist dominantly of albite-oligoclase, perthite, and less commonly of microcline and microcline-perthite (E 29966).

Low-grade metamorphic rock fragments are widespread and relatively common in the Longmyndian conglomerates. These fragments comprise: (i) quartzites and quartzose schists which exhibit suturing of the grain boundaries, strain shadows and not infrequently a foliated (elongated or 'nemablastic') texture produced by the extreme crushing and elongation of a dominantly quartzose mosaic, (ii) quartz-muscovite-schists, with or without chlorite and epidote, (iii) schistose grits and (iv) a peculiar type of fragment comprising small green chlorite vermicules within quartz grains.

The most commonly occurring metamorphic fragments are quartzites and quartz-schists. One such (6 mm) pebble in the Huckster Conglomerate from High Park cross roads [445976] (E 29955) consists of a crushed and partly or wholly recrystallized quartz mosaic crossed by parallel streams of minute fluid pores. Similar fragments are common in the Lawn Hill (E 29945), the Stanbatch (E 29966), Haughmond (E 29939) and Darnford conglomerates

[1](E 14723). This specimen, incorrectly quoted as E 14273, was noted by Whitehead in the Shrewsbury (152) Memoir, (1938, p. 48).

(E 29959). Some quartz-schist pebbles also contain relatively minor amounts of albite-oligoclase (E 29945).

Small fragments and pebbles of extremely crushed, strained and recrystallized quartzose schist, in which the texture is characterized by long thread-like deformed quartz rods (nemablastic texture) are relatively common in the conglomerates (E 29059, 29955).

Pebbles and small fragments of two types of quartz-mica-schist are common constituents of the conglomerates. Schists of the first type are distinctly foliated and consist dominantly of crushed and elongated quartz and relatively minor albite-oligoclase grains together with abundant oriented blades of muscovite. Such fragments are common for instance in the Darnford (E 29958–9), Oakswood (E 29938), Haughmond (E 29939) and Stanbatch conglomerates (E 29964) and in an isolated large exposure [421883], 460 yd S. 33° W. of Cwm Head church (E 29059). Fragments of quartz-muscovite-schist occur in crystal-lithic tuffs which are found as small pebbles in the Huckster Conglomerate (E 29955). Pebbles of this type in the Oakswood Conglomerate (E 29938A) up to 2 cm across contain scattered augen of quartz and oligoclase with some chlorite in a schistose, highly foliated rock consisting dominantly of elongated quartz grains (Plate 2D). Similar fragments in the Oakswood Conglomerate (E 29938) also contain oriented albite flakes associated with the muscovite-mica. Rarer fragments of quartz-muscovite-magnetite-schist are found, for instance in the Huckster Conglomerate (E 29955), in which magnetite constitutes up to about a fifth of the rock and occurs in irregular long and narrow lenses parallel to the foliation.

A second variety of quartz-muscovite-schist, in which parallel oriented muscovite plates occur in an essentially granoblastic quartz rock with minor oligoclase, is less common than the foliated variety but is found for instance in the Stanbatch Conglomerate (E 29964) and the Oakswood Conglomerate (E 29938). A 3-cm pebble from the Darnford Conglomerate (E 29959) consists of a foliated, crushed and sutured granular aggregate of quartz with minor oligoclase, abundant scattered muscovite plates and accessory opaque ores partly altered to hematite.

Schistose grit pebbles are not common, but particularly good examples have been sliced from the Oakswood (E 29938) and Huckster conglomerates (E 29955) (Plate 2C). The former is a 1-cm rounded pebble consisting of 0·1mm strained quartz grains with partly granulated margins. The interstitial sericite and chlorite have recrystallized giving the rock a slightly schistose texture. One of the clastic grains in this schistose grit is a devitrified rhyolite showing a patchy aggregate polarization. A 1-cm schistose grit pebble from the Huckster Conglomerate (E 29955) contains much less recrystallized matrix, which in this rock consists almost entirely of oriented chlorite flakes winding between the elongated and granulated original clastic quartz grains. The fragment is cut by a thin chlorite-magnetite vein.

Fragments of quartz containing vermicular rouleaux of chlorite occur in the Lawn Hill Conglomerate (E 29945). The green chlorite vermicules may be partly replaced by secondary iron oxides altering to hematite.

Pebbles and small fragments of sedimentary rocks are relatively rare in the Longmyndian conglomerates. Where they occur, as for instance in the Stanbatch (E 29966, 29968), Darnford (E 29961) and Lawn Hill (E 29945) conglomerates they consist of tuffaceous quartzitic sandstones or subgreywackes

(E 29966) similar to the Longmyndian sediments (see Table 2, analysis 1). Similar fragments occur in the Oakswood Conglomerate (E 29938) associated with crushed silicified rocks resembling the Buxton Rock of the Burway Group.

Subgreywackes

Sandstones of subgreywacke type make up one of the largest lithological groups of the Longmyndian. The term subgreywacke is used in accordance with the classification of Pettijohn (1957, pp. 316–9) to define a sandstone which is characterized by an abundance of lithic fragments and feldspar (with a considerable excess of lithic fragments over feldspar) with a sparse matrix which generally constitutes less than 10 per cent of the rock. Subgreywacke differs from greywacke in the grain-matrix ratio, greywacke having more than 15 per cent matrix.

The Longmyndian subgreywackes, which range from dark greenish grey to dark purple in colour, consist essentially of abundant subangular to subrounded lithic fragments and grains of quartz, with subordinate feldspar fragments in a sparse matrix composed essentially of clay minerals, small amounts of chlorite and some secondary silica. Approximate modal analyses (Table 2) show that lithic fragments constitute about half the rocks and that the matrix is limited to about 10 per cent or less. The feldspars are chiefly sodic plagioclase (albite to sodic oligoclase). Microcline is rare but perthitic feldspars are relatively common. Detrital flakes of muscovite are present in most specimens, but biotite is rare and is almost exclusively pale green. Tourmaline, garnet, zircon, sphene, opaque iron ores and apatite are widespread accessory minerals.

The lithic fragments comprise acid to intermediate and rare basic volcanic rocks, acid plutonic rocks, low-grade metamorphic types and sediments, in the relative proportions shown in Table 3. The abundance of volcanic rocks, especially rhyolite and acid tuffs is notable, as also is the relatively high percentage of metamorphic fragments. Granitic and sedimentary rock fragments are rare.

Thin section grain-size analyses of subgreywackes (Fig. 10, and Table 4) indicate that the median grain size ranges from 0·21 mm to 0·48 mm (average 0·31 mm). The degree of sorting of the subgreywackes is good (coefficient of sorting averages 1·24) and there is only a slight variation throughout the Eastern and Western Longmyndian (Fig. 10).

The grains are generally subangular to subrounded, and the coarser-grained types contain well rounded fragments up to pebble size. Graded bedding occurs relatively rarely, and current bedding has been noted in the Cardingmill Grit by James (1956, p. 318).

The sandstones throughout the Longmyndian are almost exclusively of subgreywacke type and are relatively common in the Burway (E 27827, 27829, 27862), Synalds (E 27860–2, 28989, 29024), Lightspout (E 29001, 29029), Portway (E 29002, 30224) and Bayston–Oakswood groups (E 28758, 28760, 28763, 28776–7, 29059–60, 29069, 29944).

The lithic fragments contained in the subgreywackes are dominantly of igneous origin. Fragments of rhyolite, usually recrystallized into a patchy mosaic or with spherulitic or banded structure, and fragments of acid tuffs are almost always present (e.g. E 27861, 28760, 28763, 28776–7, 29001, 29059), but other volcanic rock fragments of intermediate (andesitic) to basaltic composition are less common.

TABLE 2

Approximate Modal Analyses of Longmyndian Subgreywackes Expressed as Percentages

	(1)	(2)	(3)	(4)	(5)	(6)
Lithic Fragments	47	49	55	52	60	48
Quartz	31	39	28	30	26	35
Feldspar	12	5	8	6	3	7
Matrix	10	7[1]	9[2]	12	11	10
	100	100	100	100	100	100

1. Subgreywacke. 260 yd N. 14° W. of Darnford (1″ 152, 6″ 48 S.E.). MR 22457, E 29966. Pebble in Stanbatch Conglomerate.
2. Subgreywacke. 1200 yd E. 37° S. of Myndmill Farm (1″ 166, 6″ 63 N.W.). E 29019. 25-ft bed in Lightspout Group.
3. Subgreywacke. 1040 yd S. 21° E. of Myndmill Farm (1″ 166, 6″ 63 N.W.). E 29001. Bed below Huckster Conglomerate, Lightspout Group.
4. Subgreywacke. Hilltop, 680 yd W. 7° S. of Stedment (1″ 166, 6″ 55 N.W.). E 29944. Bayston–Oakswood Group.
5. Subgreywacke. Lightspout Hollow, crags 1000 yd N. 42° E. of Boiling Well. (1″ 166, 6″ 55 N.E.). E 30224. Portway Group.
6. Subgreywacke. Crags on north side of All Stretton Batch, c. 100 yd north-east of Batch cottage. (1″ 166, 6″ 56 N.W.). E 30223. Cardingmill Grit.

[1] Includes a small amount of calcite.

[2] Includes epidote.

Secondary alteration is widespread in the intermediate to basic volcanics; pyroxenes are invariably replaced by chlorite and feldspars are sericitized or albitized, although traces of original ophitic and flow-banded textures still remain.

Metamorphic rocks, of similar types to these described from the conglomerates, make up the next largest group of detrital lithic fragments. These may be found in practically every specimen of subgreywacke (e.g. E 27834, 27861–2, 28024, 28740, 28760, 28767, 28776–7, 29619).

Small subangular fragments of granophyre and granophyric felsite occur in many subgreywackes (e.g. E 28758–9, 28777, 29001, 29019, 29047, 29068, 29516). No ferromagnesian minerals have been noted from these granophyres, which consist almost entirely of quartz and alkali-feldspar in micrographic intergrowth.

Fragments of sedimentary rocks, usually siltstones and mudstones, are relatively rare. A minor amount of calcite cement, perhaps amounting to about one per cent, occurs in a few subgreywackes (e.g. E 30210–11). Heavy mineral layering is occasionally developed for example in the Synalds Group (E 28989, 29064) and consists mainly of concentrations of magnetite, garnet and tourmaline.

TABLE 3

Approximate Proportions of Lithic Fragments in Longmyndian Subgreywackes Expressed as Percentages

	(1)	(2)	(3)	(4)	(5)	(6)
Rhyolite, Tuff	42	60	55	53	50	60
Intermediate to Basic Volcanic	25	20	15	26	18	16
Metamorphic	25	20	20	15	24	17
Granite and Granophyre	4	<1	5	4	5	3
Sedimentary Rocks ..	4	<1	5	2	3	4
	100	100	100	100	100	100

1. Subgreywacke. Crags on north side of All Stretton Batch, c. 100 yd north-east of Batch Cottage (1″ 166, 6″ 56 N.W.). E 30223. Cardingmill Grit.
2. Subgreywacke. Lightspout Hollow, crags 1000 yd N. 42° E. of Boiling Well (1″ 166, 6″ 55 N.E.). E 30224. Portway Group.
3. Subgreywacke. Hilltop, 680 yd W. 7° S. of Stedment (1″ 166, 6″ 55 N.W.). E 29944. Bayston–Oakswood Group.
4. Subgreywacke. 1040 yd S. 21° E. of Myndmill Farm (1″ 166, 6″ 63 N.W.). E 29001. Bed below Huckster Conglomerate, Lightspout Group.
5. Subgreywacke. 640 yd S. 31° W. of Jinlye (1″ 166, 6″ 56 N.W.). E 27860. Synalds Group.
6. Subgreywacke. c. 1130 yd S. 12° W. of west end of Myndtown church (1″ 166, 6″ 63 N.W.). E 29069. Bayston–Oakswood Group.

Passega (1957, p. 1952) has shown by an extensive analysis of grain size distribution in sediments of varying depositional agents that the "sample point patterns representing the varieties in the deposit of two parameters (C, an approximation of the maximum grain size, and M the median) are characteristic of the depositional agent . . . These CM patterns are generally sharply defined and vary considerably with the type of depositional agent". The one per cent value, C, of the cumulative grain size curve is used as an approximation to the maximum size and is plotted against the median, M, for each sample. The CM curve, based on a study of Mississippi sediments (Passega 1957, fig. 2) was found to be typical of river-transported sediments in general. Shallow marine, tidal-flat sediments also formed a similar pattern.

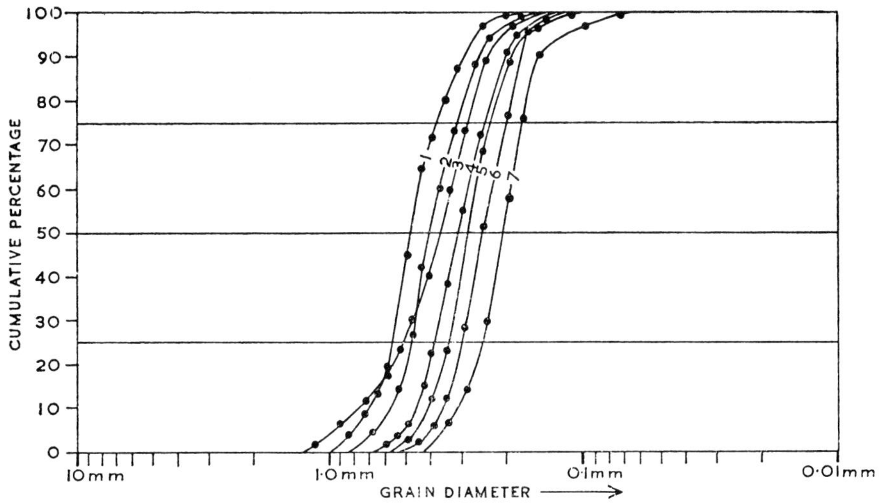

FIG. 10. Grain-size distribution of typical Longmyndian subgreywackes
1. Oakswood Group (E 29944)
2. Bayston Group (E 29069)
3. Lightspout Group (E 29001)
4. Cardingmill Grit (E 27862)
5. Cardingmill Grit (E 30223)
6. Lightspout Group (E 29019)
7. Synalds Group (E 27860)

TABLE 4

Average Grain Size and Coefficient of Sorting of Longmyndian Subgreywackes

	(1)	(2)	(3)	(4)	(5)	(6)	(7)	(Average)
Median (mm)	0·48	0·37	0·31	0·29	0·28	0·25	0·21	0·31
One per cent value (mm) ..	0·98	1·10	0·64	0·56	0·52	0·47	0·42	0·67
Coefficient of sorting Q3/Q1	1·27	1·33	1·25	1·17	1·19	1·25	1·20	1·24

1. Subgreywacke. Hilltop, 680 yd W. 7° S. of Stedment (1″ 166, 6″ 55 N.W.). E 29944. Bayston–Oakswood Group.
2. Subgreywacke. c. 1130 yd S. 12° W. of west end of Myndtown church (1″ 166, 6″ 63 N.W.). E 29069. Bayston–Oakswood Group.
3. Subgreywacke. 1040 yd S. 21° E. of Myndmill Farm. (1″ 166, 6″ 63 N.W.). E 29001. Bed below Huckster Conglomerate, Lightspout Group.
4. Subgreywacke. 650 yd S. 22° W. of Jinlye (1″ 166, 6″ 56 N.W.). E 27862. Cardingmill Grit.
5. Subgreywacke. Crags on north side of All Stretton Batch, c. 100 yd north-east of Batch cottage. (1″ 166, 6″ 56 N.W.). E 30223. Cardingmill Grit.
6. Subgreywacke. 1200 yd E. 37° S. of Myndmill Farm (1″ 166, 6″ 63 N.W.). E 29019. 25-ft bed in Lightspout Group.
7. Subgreywacke. 640 yd S. 31° W. of Jinlye. (1″ 166, 6″ 56 N.W.). E 27860. Synalds Group.

The CM pattern of the Longmyndian subgreywackes is closely similar to the basic CM river pattern of Passega. The inference therefore, is that the subgreywackes are a result of river or near-bottom shallow water currents. The depositional environment of typical subgreywackes (Pettijohn 1957, p. 320) is generally considered to be of the floodplain deltas of the coastal plains, and the closely associated marine environment. Consequently the combined available evidence of the lithology of the subgreywackes, the occurrence in them of current bedding together with the associated fine-grained sediments showing shallow water (tidal flat) sedimentary features (e.g. pit-and-mound structures, rill marks, wave-foam bubble impressions, rain prints etc., see below) strongly suggest a shallow water environment.

Chemical Composition of the Longmyndian Sediments

A notable feature of the chemical composition of the subgreywackes (Table 5, analyses 1–3) grits (Table 5, analyses 4–5) and conglomerates (Table 5, analysis 6) is, with one exception, the marked dominance of soda over potash. This is a reflection of the predominantly sodic detrital feldspar fragments which in the subgreywackes comprise 5–10 per cent of the rock, together with the predominantly sodic feldspars present in the lithic fragments. Potash-feldspar is practically confined to the granophyre and rhyolite fragments.

The relatively high alumina content (10–16 per cent) of the subgreywackes varies between the average alumina content of arkose (Pettijohn 1957, p. 324 gives 10 per cent), and the average alumina content of greywacke of about 15 per cent, (Pettijohn 1957, p. 307). The alumina content of arkose is a reflection of the presence of relatively abundant detrital feldspar and in the greywackes the alumina is largely present in the interstitial clay mineral 'paste'. The relatively high alumina content of the subgreywackes may be accounted for by a combination of that present in the sparse clay mineral and chloritic matrix, together with that of the detrital fragments of sodic feldspars and also the alumina of the feldspars within the abundant igneous (mainly volcanic) rock fragments.

The subgreywackes, grits and conglomerates have a comparatively high Fe_2O_3: FeO ratio; which contrasts with the low Fe_2O_3 : FeO ratio of average greywacke (Pettijohn 1957, p. 307). The relatively high degree of oxidation of the iron in the (generally purple) Synalds Group shales (Table 5, analysis 8) contrasts with the low Fe_2O_3: FeO ratio in the (predominantly grey-green) Stretton Shales (Table 5, analysis 7).

An approximate estimate of the relative proportions of the arenaceous deposits (subgreywackes, grits, conglomerates) to the shaly horizons in the Longmyndian as a whole suggest that the ratio is about 1 : 1. The average bulk composition of the sediments based on this estimate (Table 5, col. III) indicates a composition close to that of an intermediate igneous rock. This is probably a reflection of the average composition of the source area, which on the evidence of the lithic fragments, was largely composed of Uriconian volcanics.

Sedimentary Structures

Several varieties of sedimentary structures are found in the Longmyndian and the presence of some of these together with some supposed organic impressions (e.g. *Arenicolites*) was first recorded by Salter (1856, 1857).

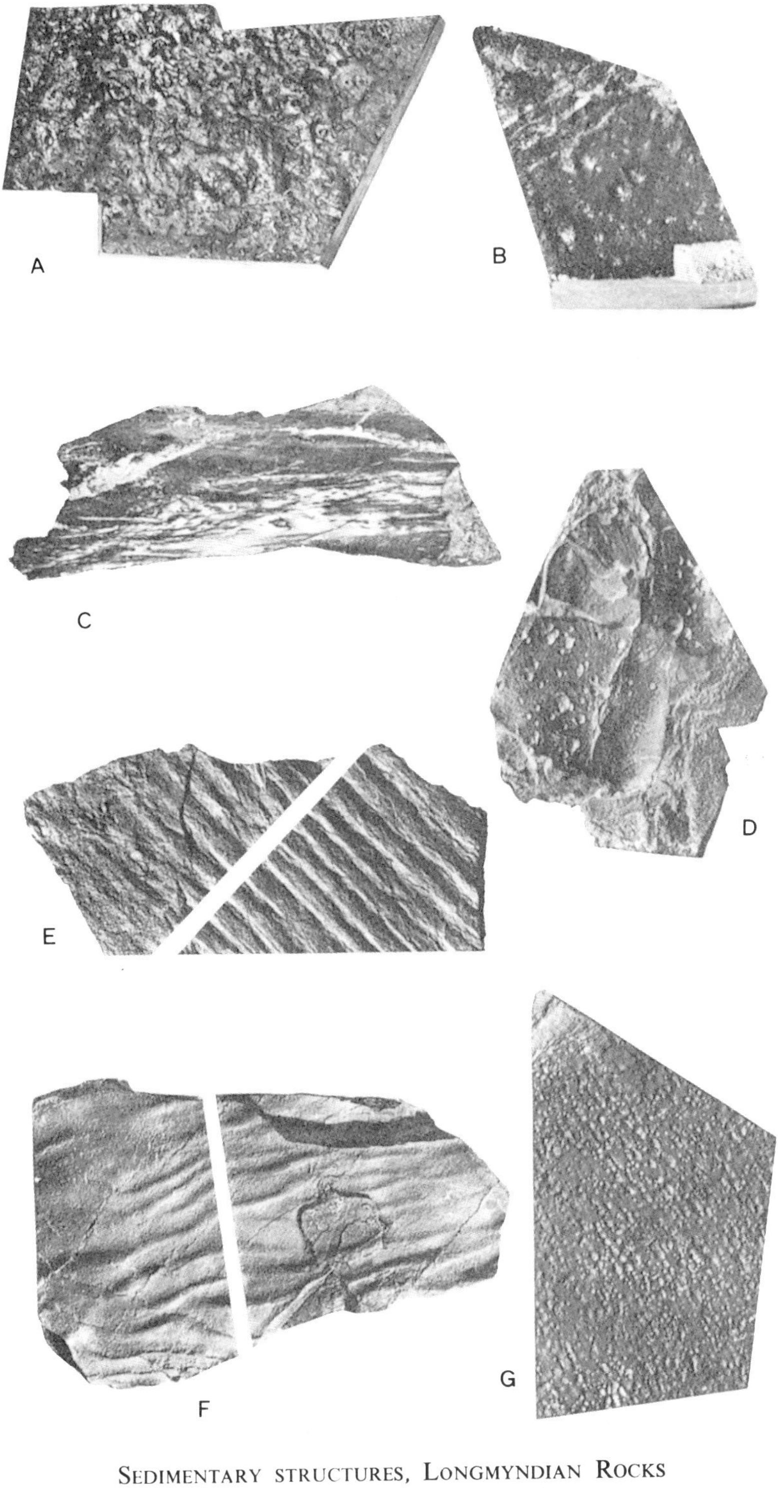

SEDIMENTARY STRUCTURES, LONGMYNDIAN ROCKS

(For detailed explanation see p. x)

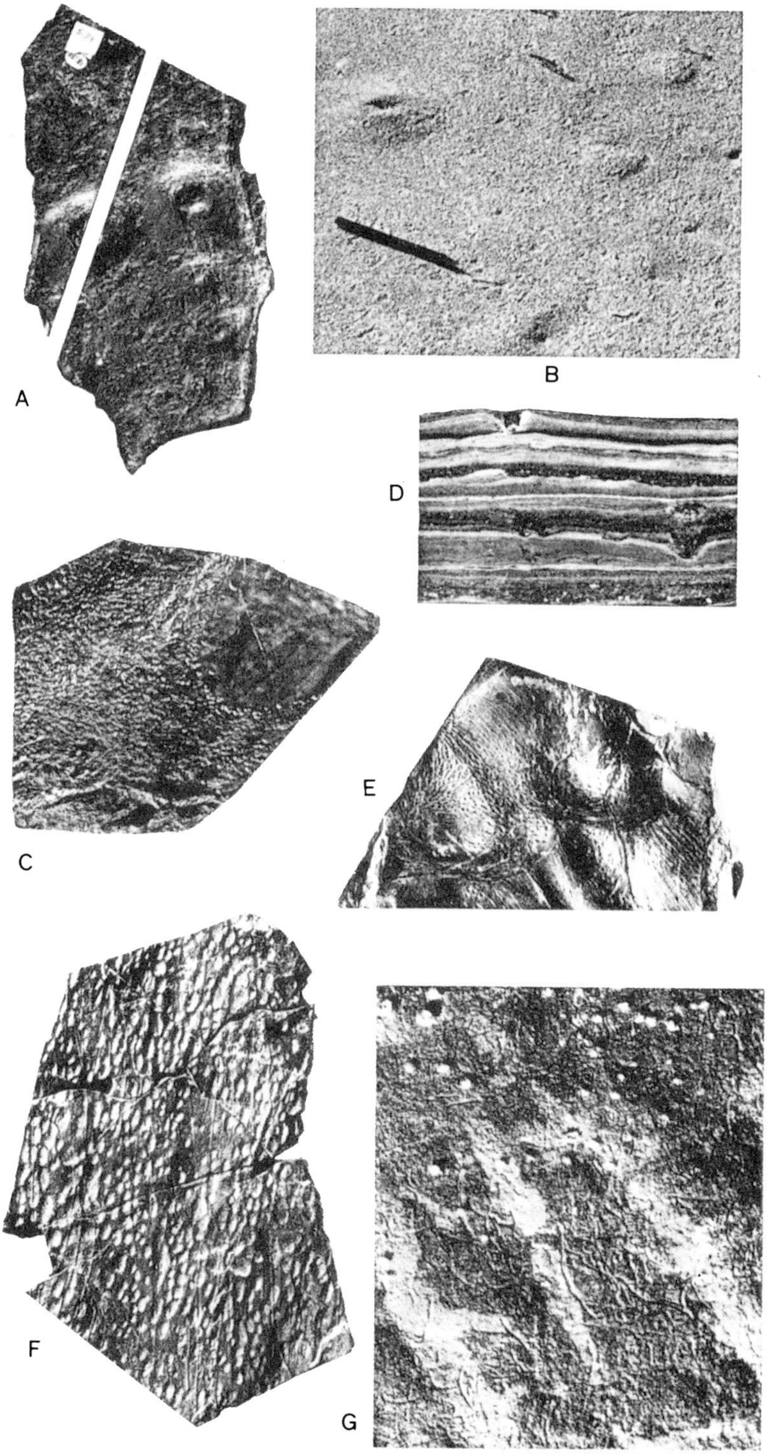

SEDIMENTARY STRUCTURES, MAINLY LONGMYNDIAN ROCKS

(For detailed explanation see p. xi)

TABLE 5

Analysis of Longmyndian Sediments
(Percentages)

	Subgreywackes			Grits		Conglo-merate	Shales: Stretton	Shales: Synalds			
	1	2	3	4	5	6	7	8	I	II	III
SiO_2	76·27	67·26	62·61	62·05	61·56	68·90	60·98	62·93	66·44	61·95	64·20
Al_2O_3	10·39	14·15	15·98	18·66	18·74	14·32	19·27	17·88	15·32	18·57	16·95
Fe_2O_3	3·61	2·82	5·37	6·09	4·62	5·45	2·58	6·24	3·76	4·41	4·09
FeO	0·56	3·09	1·42	1·02	1·44	1·08	4·69	1·23	1·43	2·96	2·20
MgO	1·56	1·67	2·49	2·32	2·03	2·02	2·15	2·10	2·02	2·13	2·08
CaO	0·63	2·55	3·75	1·06	3·32	0·33	0·22	1·02	1·95	0·62	1·29
Na_2O	2·37	3·07	2·91	2·12	2·15	2·39	2·03	1·77	2·50	1·90	2·20
K_2O	1·38	1·24	1·01	2·80	1·95	1·46	3·00	3·38	1·47	3·19	2·33
H_2O^+	1·82	2·38	2·71	} 3·27	3·45	3·24	4·25	2·69	3·05	3·47	3·26
H_2O^-	0·53	0·43	0·45								
TiO_2	0·46	0·83	0·75	0·49	0·44	0·72	0·71	0·62	0·62	0·66	0·64
P_2O_5	0·05	0·09	0·11	none	0·17	trace	none	trace	0·11*	trace	0·05
MnO	0·06	0·15	0·17	0·12	0·13	0·08	0·12	0·14	0·12	0·08	0·10
CO_2	0·06	nd.	nd.	none	none	none	none	none	0·06†	—	0·03
S	0·04	0·02	nd.	—	—	—	—	—	0·04‡	—	0·02
Others	0·05	0·09	0·07								
TOTAL	99·80	99·82	99·80	100·00	100·00	99·99	100·00	100·00	99·89	99·94	99·44

N.B. *Average of 4 analyses
†One analysis only
‡Average of 2 analyses

I Average of analyses 1–6
II Average of analyses 7–8
III Average of 50 : 50 of I and II

1. Subgreywacke. Hilltop, 680 yd W. 7° S. of Stedment, (1″ 166, 6″ 55 N.W.). Bayston–Oakswood Group. E 29944. Lab. No. 1893. Analysts: P. R. Kiff and W. H. Evans.
2. Subgreywacke. Crags on north of All Stretton Batch, c. 100 yd north-east of Batch cottage, (1″ 166, 6″ 56 N.W.). Cardingmill Grit. E 30223. Lab. No. 1894. Analysts: P. R. Kiff and G. A. Sergeant.
3. Subgreywacke. Lightspout Hollow, crags 1000 yd N. 42° E. of Boiling Well, (1″ 166, 6″ 55 N.E.). Portway Group. E 30224. Lab. No. 1895. Analysts: P. R. Kiff and G. A. Sergeant.
4. Gritty Slate. Beds in dingle below High Park. Portway Group. Reade and Holland 1907, *Proc. Liverpool Geol. Soc.*, Table iv, anal. 4, p. 14.
5. Grit. Beds in dingle below High Park. Portway Group. Reade and Holland 1907, *Proc. Liverpool Geol. Soc.*, Table iv, anal. 5, p. 14.
6. Huckster Conglomerate. Corner of roads below High Park. Portway Group. Reade and Holland 1907, *Proc. Liverpool Geol. Soc.*, Table iv, average of analyses of two specimens (analysis 6).
7. Shale. House excavation above Halescroft, Madeira Walk, Church Stretton. Stretton Shale Group. Reade and Holland 1907, *Proc. Liverpool Geol. Soc.*, Table iv, anal. 8.
8. Purple Shale or ?Slate. Above Cardingmill. Synalds Group. Reade and Holland 1907, *Proc. Liverpool Geol. Soc.*, Table iv, anal. 11, p. 14.

The following account is based partly on a re-examination of Salter's original specimens[1] and partly on new material.

Current bedding and graded bedding. Current bedding has previously been recorded from the Lightspout Group by Westoll (in discussion of Cobbold and Whittard 1935, p. 358); from the Stretton Series by Whittard (1952, p. 146); from the Cardingmill Grit by James (1956, p. 318) and Taylor (1958, p. 1080) and from the Bayston–Oakswood and Bridges groups by Taylor (1958, p. 1082). It occurs on a small scale in some of Salter's specimens, for example in the Lightspout Group (GSM 49141) associated with ripple marks, and from 'Yearling Hill', ?Yearlet hill, (Burway–Synalds Group, GSM 49151, 49156–7) associated with pit-and-mound structures and rain prints. A few examples also occur in the Bridges Group (e.g. MR 25366, 25376).

According to Shrock (1948, p. 1083) "graded bedding and current bedding do not ordinarily occur together, one may give way to the other or the two may alternate in successive layers". The occurrence of current bedding certainly indicates a relatively shallow water environment (Bailey 1930, p. 88) although the presence of associated graded bedding does not necessarily indicate a change to deep water, but only to relatively still water such as may be found in lagoonal areas. This is well illustrated by the relatively close association in the Burway–Synalds groups of Yearling Hill of current bedding (e.g. GSM 49151, 49157) and specimens showing graded bedding together with rain pits and mud cracks.

Ripple marks. Salter (1857) was apparently the first to note and figure ripple marks from the Longmyndian (Lightspout Group). They have since been recorded by James (1956) and Taylor (1958) from the Synalds Group and the Wentnor Series respectively.

Although Whitehead (1938, p. 43) has shown that structures resembling ripple marks are locally produced by puckering between small shear planes the ripple marks from the Lightspout Group (Salter 1857) are clearly defined and include both current-ripple marks (GSM 49142) associated with wispy current bedding, and symmetrical (wave) ripple marks (GSM 49141, 82971) in current-bedded undeformed subgreywackes (Plate 4, E, F). The association of symmetrical ripple marks and current bedding strongly indicates formation in relatively shallow water.

Mud cracks. These were first figured by Salter (1857, pl. v, figs. 9–10) from Yearling Hill (Burway–Synalds Group) and described (1857, p. 202) as dividing the surface "into areolae of various sizes and shapes . . .; and when, as is most common, the superficial layer of mud is darker than the stone, show themselves well in relief by exposing the lower stratum. The edges of the areolae are most generally curved upwards, the heat of the sun having caused shrinkage".

A specimen from Yearling Hill (GSM 82964) shows cracked and upturned edges in a greenish grey mudstone layer which on its undersurface has clear casts of rain drop impressions. The overlying subgreywacke-siltstone grades from coarse- to fine-grained through a vertical distance of 1 cm.

Other good examples occur of mud-cracked surfaces associated with undoubted rain prints in the Burway–Synalds Group of Yearling Hill (e.g. GSM 49155, Plate 4 D (cf. Salter, 1857, pl. v., fig. 10), and GSM 49145). Vertical

[1]Nos. GSM 10656–8, 49141–64, 49884–5 in the palaeontological collections of the Geological Survey.

polished faces clearly show the fine- to medium-grained sand infilling the cracks and penetrating underneath the upturned edges of the cracks in the grey-green mudstone.

Rain imprints. Salter (1857, pp. 202–3 and pl. v., fig. 10) first described and figured rain imprints in the Burway–Synalds Group of Yearling Hill. James (1956, p. 318) mentioned similar features from the Synalds Group and stated that "the more circular depressions are surrounded by a raised rim and these, without doubt represent rain pits". Specimens showing undoubted rain imprints (e.g. GSM 49145, 49149, 49155–6, and 82964) and rather shallower impressions which are possibly slightly eroded rain imprints (e.g. GSM 49148) are relatively common in the Synalds and Burway Groups. The rain imprints occur in thin mudstone layers (1 to 3 mm) overlying fine-grained sand and siltstone in which current bedding, graded bedding and small scale wash outs occur. Clearly defined ragged raised margins, in some instances higher on one side than the other, surround the 1 to 7 mm diameter depressions and in places later depressions are superimposed on the earlier ones. These characteristics are similar in all respects to the features described by Twenhofel (1921, pp. 360–3) and Shrock (1948, pp. 139–42) which are considered to be diagnostic of falling drops of water (compare pl. v., fig. 1 of Salter 1857, and Plate 4 D, of this memoir with Shrock 1948, p. 142, fig. 101).

Air-heave structures. The sedimentary structures grouped under this heading, are apparently all formed in one way or another by subsequent movements of air trapped in the pore spaces of the sediment. Criteria for the recognition of air-heave structures have been given by Stewart (1956, p. 159) who concluded that "the occurrence of air-heave structures in the geologic column may be interpreted as indicative of a tidal area exposed at low tide with a continuous source of sand . . . the necessary conditions develop most commonly near the entrances to tidal estuaries and lagoons". This type of environment is supported by the sand-hole type of air-heave structure described from beaches by Palmer (1928, pp. 176–80) and Emery (1945, p. 39) and the association of desiccation features with air-heave structures described by Paborski (1954, p. 995).

Pit-and-mound structures. Features which resemble pit-and-mound structures occur in some of the specimens collected by Salter from the Burway–Synalds Group of Yearling Hill (GSM 49152–4, 82961). The structures occur in finely bedded green mudstones with thin (1 mm) silt partings. Graded bedding is locally present in small units of 1 to 2 mm. On the upper surfaces of the specimens small (1 to 2 mm diameter) craters occur (GSM 49153–4, 82961, see Plate 4 A) or in some specimens extremely shallow mounds with rather larger (2 to 3 mm diameter) central depressions (GSM 49152; figured by Salter 1857, pl. v., fig. 2, syntype of *Arenicolites sparsus* Salter).

On the under surfaces of some specimens (e.g. GSM 49153, Plate 4 B) these features are represented by casts consisting of a central small mound set in a surrounding moat-like depression.

In vertical polished faces (Plate 5 D) these structures are seen in cross section; one specimen from Yearling Hill (GSM 49152) consists of graded bedding units, less than 1 mm in thickness, of grey siltstone (average grain size 0·05 mm) grading upwards to green silty mudstone and mudstone. Individual layers of siltstone and mudstone average 0·2 mm in thickness. Erosional features at the upper surfaces of the green mudstone layers are common and the siltstone forms small washouts and occasional 'sharpstone' conglomerates of disoriented fragments of finely banded consolidated mudstone in a siltstone matrix. A common feature is the downward protruberances of the siltstone layers into the green mudstone, which are about 1 to 2 mm in diameter and about 2 mm in depth. The bedding-surface pit-and-mound features, which are directly related to these structures are abundant in some specimens (GSM 49153–4, 82961) and may reach a density of 25 to 30 per square inch. Sections through pits show a vertical core of silt penetrating the banding of the green mudstone, which is slightly contorted towards the base of the infilling. Similar downward protruberances of coarser sand cores into finely banded mudstone occur in the shales of the Burway Group (e.g. MR 8673). The bedding surface is studded with 1 to 3 mm diameter pits and mounds corresponding to the sandstone core infillings.

These features are presumed to have been formed while the siltstones and mudstones were in an unconsolidated plastic state, either owing to the unequal loading of the sand or silt on the underlying mud, in a similar manner to the formation of flow casts, where water-filled sand sinks into a viscous hydroplastic mud, or by movement of entrapped air bubbles followed by slumping in the wake of the rising bubbles. Similar features have been produced experimentally by Kindle (1916) during the settling of clay in water by small upward-moving water currents.

Pit-and-mound surface features of a similar nature are found in the Burway–Synalds Group of Ashes Hollow (e.g. GSM 85185; E 30234) in finely laminated mudstone and silty mudstone (Plate 5 A). The underside of the specimen shows numerous circular or slightly elliptical moat-like depressions with a central mound about 2 mm in diameter. These features are the counterparts of small slightly elevated mounds or domes with small central craters, similar to those which are seen on the top side of the specimen. They occur singly, in pairs, or in groups of three or more. The two largest and best developed domes have a diameter of 3 cm, and a maximum height of about 3 mm; the central core being about 1 cm across. These larger structures clearly pass through the 7-mm thick specimen, the axis of the central core being inclined at about 50° from the bedding. In thin section (E 30234) the disturbance of the otherwise regular bedding planes corresponds to the peripheral surface updoming; the central surface depression is underlain by broken and deformed bedding planes suggesting a collapsed structure. The crater infilling of apparently down-slumped sand partly truncates the underlying bedding planes. These structures are similar to experimental air-heave structures caused by rising air bubbles (Stewart 1956, p. 157, fig. 4 etc.) a notable similarity being the central slump in the wake of the rising air pocket. These structures (Plate 5 A) are similar to the sand domes found on present day beaches (Plate 5 B). This plate also shows numerous small holes (see Palmer 1928; Emery 1945) with random arrangement, although occasionally in pairs or groups of three or more, which may be similar to the '*Arenicolites*' holes of the Longmyndian.

Groove casts and rill-like markings. Groove markings have previously been noted from the Stretton Shale Group by Taylor (1958, p. 1079).

Well developed groove casts occur in the red siltstone and mudstone of the Synalds–Lightspout Group of the 'Rabbit Warren, Choulton Lodge', ?near Plowden, (e.g. GSM 82970, see Plate 4 C) and also in the Burway–Synalds

Group of the Cardingmill valley (e.g. GSM 49163, see Plate 5 F). Salter (1856, p. 250) attributed the raised lines (on specimen GSM 49163) to "mineral structure rather than anything organic" and suggested also that the hollows on the specimen which were cut across by these ridges may have been due to gaseous bubbles or decomposition of small concretions rather than rain drops. The way-up evidence of the small-scale graded bedding, however, indicates that the structures occur on the underside of the slab and the raised lines are consequently groove casts which cut across the small mounds (up to 3 to 4 mm in diameter) some of which are elongated parallel to the grooves. The casts are preserved in a red silty mudstone overlain by a grey-green fine-grained subgreywacke-sandstone containing fragments of the red mudstone. Irregular contorted bedding suggests turbulent movements of trapped air within the sediment. The grooves are up to 5 cm in length and are approximately 0·3 mm in depth and width, but where they commence or die out the ends are gradually sloping, suggesting that they were caused by current-drifted material which occasionally touched bottom. The grooves are commonly branched and the consistent direction of branching suggests movement in one direction only. Specimens from the Lightspout Group figured by Salter (1856, pl. iv., figs. 5–6) show probable rill markings (see also Salter 1856, p. 251, fig. 2).

Irregularly banded green siltstones and mudstones from Yearling Hill show a number of features including rill-like markings. Small ridges occur (GSM 49143, 49147) down the sides of which are groups of subparallel curving grooves (see Plate 5, E). The combination of ridges, irregular mounds and depressions with rill-like grooves suggests intertidal current action.

One specimen (GSM 49157) from Yearling Hill has a large number of 1 to 5 mm long curving irregular grooves, approximately 0·1 mm wide and in depth (Plate 5 G). Casts of these impressions are preserved on the counterpart (GSM 49151). These have a random orientation and shape and may be only slightly curved, or U or S-shaped, but others are branching and Y-shaped. The ends of the grooves may be either sharp or gradually shelving and the grooves may possibly represent the impressions made by drifting objects, floating bubbles or seaweed-like material (cf. Twenhofel 1921, p. 367).

Swash-like markings. Small curvilinear grooves and blister-like protuberances occur on the underside of a specimen of current-bedded siltstone and mudstone (GSM 49155) from Yearling Hill (Salter 1857, pl. v., fig. 10). Undoubted rain prints occur on the upper surface. The undersurface features are casts of original small bubble-like impressions and tiny linear ridges about 0·1 mm in height and width, which occur on more than one bedding-plane parting. On the surface one large undulating ridge occurs 0·5 to 3 mm in height, separating concave-upward segments which may be due to drying out and cracking of the mud surface. Other fine parallel ridges occur of about the same cross section as those which are found as casts on the underside of the specimen. The fine ridges and grooves are approximately parallel and concave in the same direction. In one instance a short groove lies at right angles between, and terminates sharply against, two long subparallel grooves. These curvilinear ridge features may possibly represent overlapping and truncating swash marks, that is the "lobate fronts of waning waves, advancing over sandy beaches . . . [forming] . . . a system of tiny imbricating sand ridges that are convex landward and mark the line of furthest advance" (Shrock 1948,

p. 128). The observed patterns could well be produced in such a way and it is difficult to imagine any other mode of formation.

Bubble impressions. These may be produced in modern deposits "by bubbles floating and becoming anchored to the bottom through sediment becoming attached to the films or stranding in shallow water" (Twenhofel 1921, p. 365). The impressions of these bubbles are usually less than a hemisphere; there are no raised rims and "the edges of the depressions are sharply defined and regular" (Twenhofel 1921, p. 366).

Examples of impressions probably made by floating wave-foam bubbles occur in the Portway and Bridges groups. In the East Onny River 500 yd W. 20° S. of Wentnor church specimens of red siltstone and mudstone (MR 25371–2) showing small scale current bedding, washout structures and intraformational erosional features have oriented parallel lines of small shallow depressions on the upper surfaces. These depressions up to 2 mm × 4 mm do not have raised rims and are not associated with any structures in depth in contrast to pit-and-mound surface features. The impressions are similar to those which may be made by "bubbles floating in very shallow waters, which are gently rising and falling a few millimetres because of wave activity" (Twenhofel 1921, p. 367). The arrangement of the impressions in oriented lines or streams (Plate 4 G) may be due to current action.

The lower surface of a specimen of mudstone from the 'upper fork of the stream Portway' (GSM 49164) is the counterpart of a surface (Plate 5 C) covered with small circular (1 to 2 mm) depressions having a density of up to 100 per square inch, which do not occur above the level of a curved 'contour' running round the base of a small mound. It is suggested that these are bubble impressions due to the stranding of wave foam around the small mound.

Other features, which are somewhat similar to the casts of the presumed bubble impressions described above, but which occur on the upper surfaces as small blister-like mounds, are found in the Burway–Synalds Group of Yearling Hill and at the Carding Mill (GSM 49143, 49147, 49157, 49163) and in the Bridges Group (e.g. MR 9999, 25369).

Two of these specimens (GSM 49147 and the counterpart of GSM 49143) show irregular low (ripple) ridges with rill-like markings. The low ridges are studded with small blister-like mounds (about 1 to 2 mm across) commonly slightly elongated parallel to the rill-like markings (Plate 5 E). Some appear to be slightly eroded. Exceptionally clear clusters of similar 1-mm blisters occur along the slight ridges on specimen GSM 49157 (seen in counterpart on GSM 49151) and as impressions on the lower surface of a banded red mudstone-siltstone with sandstone veining from the Bridges Group (MR 9999). Similar elongated mounds occur together with groove marks on a specimen from the Carding Mill (Plate 5 F), and slightly smaller elongated mounds on banded red siltstone and mudstones, with current bedding and erosional features in the Bridges Group (MR 25369). The origin of these features is obscure, but may be either mud bubbles or blisters, perhaps akin to pit-and-mound structures.

Arenicolites. It seems that many of the identifications of *Arenicolites* originally made by Salter (1856, 1857) refer to inorganic bubble holes and pit-and-mound structures, and in some instances to the counterparts of original mounds or small blister-like features. As early as 1884 L. F. James pointed out the inorganic origin of similar 'mud bubbles' which had previously been considered organic and had been given names.

Specimens described and figured by Salter (e.g. GSM 49151, 49157; Salter 1857, pl. v., fig. 1, pp. 202–4) as showing annelid holes (syntype of *Arenicolites sparsus* Salter) have been re-examined, and these depressions (GSM 49151) are seen to lie on the underside of the specimen and are consequently moulds of small blister-like mounds on the top surface of the counterpart (GSM 49157 see Plate 5 G). The 'way up' evidence is provided by the small-scale current bedding, graded bedding and occasional small wash-out features seen in vertical polished sections. Thus the 'annelid holes' on this particular specimen do not occur 'sheltered in the hollows' (Salter 1857, pp. 203, 204) but the features are moulds of small mounds concentrated on the low ridges.

A characteristic feature of the 'annelid holes' was stated to be their paired nature (Salter 1857, p. 203) but on close examination except on portions of one specimen (see Salter 1856, pl. iv., fig. 1a, b; GSM 49160, *Arenicolites didymus* Salter), the arrangement of moulds appears to be quite random; they occur singly, in pairs or in groups of three or more (cf. Salter 1857, pl. v, figs. 1 to 4; and compare with Plate 5 B of this memoir).

Another (syntype) specimen of '*Arenicolites sparsus* (adult)' (Salter 1857, pl. v, fig. 2) with upper-surface depressions shows features which suggest pit-and-mound structures (GSM 49152, cf. p. 69).

Salter used the term '*Arenicolites*' for 'all worm burrows with double openings' (1857, p. 204), but since on the basis of the above evidence some of the features described as '*Arenicolites*' are not paired, some are small depressions, others are small mounds, and as no burrows have yet been found in any of the specimens examined, including Salter's specimens, it is suggested that the term '*Arenicolites*' should not be applied to any of the structures so far found in the Longmyndian.

Source of the Longmyndian sediments

Detrital pebbles and fragments in the Longmyndian conglomerates and sandstones fall into two characteristic suites. The first suite which also occurs in the Uriconian tuffs comprises fragments of metamorphic rocks such as foliated quartz-schist and muscovite-chlorite-schists, which fall within the muscovite-chlorite and biotite-chlorite subfacies of the greenschist facies (Fyfe, Turner and Verhoogen 1958), corresponding to regional metamorphic rocks of the chlorite and biotite grades. Garnet occurs locally but is not a common constituent of the lithic fragments.

Metamorphic rocks of these types are exposed in three areas not far removed from the Long Mynd; (i) Rushton Schists and Primrose Hill Gneisses; (ii) Malvernian gneisses, and (iii) the Mona Complex of Anglesey.

The Rushton Schists (E 8572, 8574, 8594, 8935–6, 12956, 12985, 14249–50, 14252) include quartz-muscovite-chlorite-(epidote)-albite-oligoclase-schists and quartz-biotite-chlorite-(garnet)-albite-oligoclase-schists of greenschist metamorphic facies. It was considered possible by Pocock that "the Rushton Schists represent Eastern Longmyndian rocks in a sheared and hornfelsed condition", and also that the later granophyres "might be responsible for the thermal element in the metamorphism of the Rushton Group and also of the rocks of Primrose Hill" (1938, p. 31).

It is clear, however, from the mineral assemblages that the Rushton Schists are low-grade regional metamorphic types, not contact-metamorphosed varieties. Their most obvious similarities lie with the metamorphic rocks of the Mona Complex and with the Malvernian gneisses.

One of the characteristic textures in the metamorphic fragments is an elongated highly sheared (or 'nemablastic') texture produced by extreme crushing and elongation of quartz. This texture is common in the Mona Complex, especially in the Gwna greenschists of the Penmynydd zone (Greenly 1919). The Malvernian is in general not so finely foliated as the Mona schists and nemablastic texture is rare.

Another feature of the lithic fragments which can be matched with the Mona Complex is the occurrence of quartz grains with inclusions of pale green vermicular chlorite. Vermicular chlorite occurs in the Penmynydd zone (Greenly 1919, p. 372; e.g. E 10745).

Schistose grits, which have not been recorded from either the Rushton Schists or the Malvern Gneisses, but which are common in parts of the Mona Complex occur as lithic fragments or pebbles in the Longmyndian.

The predominant detrital feldspars in the Longmyndian sediments are perthite and sodic plagioclase. Microcline is rare. Greenly (1919, p. 414) has emphasized that "the most salient fact in the composition of the Mona Complex is the extraordinary predominance of albite-feldspar". Microcline occurs only in some of the Penmynydd mica-schist and then only rarely.

It appears then that this metamorphic suite of fragments in the Longmyndian was derived from a metamorphic complex similar to that now exposed in the Rushton Schists and the Mona Complex.

The second suite of lithic fragments comprising rhyolite, andesites, basalts, tuffs and granophyres is clearly of Uriconian aspect and derivation.

Depositional Environment of the Longmyndian

The conditions of formation and the tectonic setting of the Longmyndian may be deduced from the lithological types present, the sedimentary structures. the degree of sorting of the sediments and the relationships of the Longmyndian sediments to the underlying Uriconian.

The sandstones throughout both the Eastern and Western Longmyndian are almost exclusively of subgreywacke type. Although the precise conditions of formation of subgreywacke are not thoroughly understood, "those better known from close study appear to be a product of 'paralic sedimentation', i.e. they accumulated on floodplains or deltas in the coastal plains or in clearly associated marine environments", (Pettijohn 1957, p. 320).

The mode of deposition of the Longmyndian subgreywackes is also suggested by comparison with the lithologically very similar subgreywackes of the Downton Series (e.g. Downton Castle Sandstone, see p. 216) and also with the only slightly different characters of the Pennant Measures subgreywackes of the South Wales coalfield. All three groups of subgreywackes contain the same suite of detrital lithic fragments. The depositional environment of both the Downton Series and the Pennant Measures is usually considered to be essentially deltaic and hence it seems that the Longmyndian subgreywackes, common to both the Stretton and the Wentnor Series, may also be products of essentially deltaic sedimentation.

The sedimentary structures which are found in the Stretton Series in particular, such as current bedding, ripple marking, some graded bedding, rain pitting, desiccation cracks, probable wave-foam bubble impressions and pit-and-mound structures indicate a generally shallow-water environment with intermittent exposure, such as would occur in a tidal estuary or delta.

The grain-size distribution of the subgreywackes is consistent with their formation as river-transported sediments or as shallow marine, tidal-flat deposits.

Since these features are characteristic of the Longmyndian subgreywackes as a whole, it seems that they reflect an essentially similar sedimentary environment throughout the Longmyndian. Heavy mineral layering associated with current bedding which is occasionally present in the subgreywackes (e.g. Cardingmill Grit) implies fairly extensive reworking to produce the 5-mm bands of garnet-magnetite-tourmaline concentrates. It has been suggested that the transition from the Eastern Uriconian lithic tuffs through the Helmeth Grits to the Stretton Series (Stretton Shales) is essentially conformable (Cobbold and Whittard 1935, p. 351), and hence, since the Uriconian is an essentially terrestrial formation the lower portion at least of the Stretton Shales is probably a relatively shallow-water series.

The continuation of volcanic activity as represented by the tuffs of the Buxton Rock and the Batch Volcanics from the Uriconian through into the Longmyndian suggests that the latter is a post-tectonic Molasse-type deposit following on the Uriconian (subsequent type) volcanicity.

Taylor (1958) discussed the tectonic environment of the Longmyndian sedimentation and concluded that two distinct phases may be recognized. The Eastern Longmyndian or Strettonian was regarded as typically geosynclinal or Flysch-like sedimentation (1958, pp. 1090–1), and the Western Longmyndian or Wentnorian as Molasse, the Portway Group constituting a transitional stage.

The Burway and Synalds Groups in particular were regarded by Taylor (1958, p. 1090) as synorogenic (Flysch) sediments, but it is within these groups that the evidence of the sedimentary structures is strongest in suggesting a shallow water or intermittently emerging environment rather than geosynclinal conditions.

Both the Stretton and Wentnor series have a common feature in that the sandstones (subgreywackes) of each series are similar. These subgreywackes are not geosynclinal sediments and consequently a division of the Longmyndian into a 'geosynclinal' Eastern Series and a 'Molasse' Western Series is questionable. The whole sequence appears to be a product of Molasse-type (paralic) sedimentation, showing similarities of tectonic setting, lithology and thickness with the Old Red Sandstone of Scotland. R.D.

It is considered that the Longmyndian was deposited in a narrow crustal depression which originated by subsidence of the underlying basement between major fault lines. It is possible that the Pontesford–Linley and Church Stretton lines of disturbance were the two controlling fault systems and the apparent absence of the Stretton Series to the south-east of the Church Stretton line may be due to original non-deposition. This picture invites comparison with the taphrogeosyncline of Kay (1951, p. 61) which arises from block faulting in the basement rather than from broad crustal downwarp. The form of the

Longmyndian trough is not identical with that illustrated by Kay, the latter being wedge-shaped in cross section and fault-bounded on one side only. The shallow-water character of the Longmyndian sediments may be compared with the non-marine, fluvatile and estuarine deposits of the Triassic taphrogeosyncline of North America (Kay 1951, p. 60). The interpretation of an unconformity below the Wentnor Series requires slight, pre-Wentnorian folding of the Stretton Series, followed by limited erosion and subsequent deposition of the Wentnor Series over an area more extensive than that of the original trough. It is further suggested that the deep overfolding of the whole Longmyndian sequence was restricted approximately to the limits of the original depositional trough and was due largely to renewed compressive block movements in the underlying basement. Similar intense folds, apparently resulting from movements of crustal blocks are known from other areas, e.g. the overfolding and thrusting of the Palaeozoic rocks to the west of the Malvern line of disturbance. J.E.W.

DETAILS

STRETTON SERIES

Stretton Shale Group

EAST OF CHURCH STRETTON FAULT

Helmeth Grit

The outcrop of the Helmeth Grit extends along the western flanks of Ragleth, Hazler and Helmeth hills from near Little Stretton [445916] in the south to Caradoc Coppice [468948] in the north, and is cut off at each end by the fault F1 of the Church Stretton Complex. There are some exposures near the south end of Ragleth Wood [448920], and at about 700 yd north-east of Little Stretton [449922] the grit dips westward at angles of from 40° to 65°. To the north-east the outcrop is traced mainly by feature and debris, and is disrupted by a series of small faults. There is a section in the Helmeth Grit in the Hazler Old Road [465933], 1000 yd E. 16° S. of Church Stretton station. This and a trench section dug near the south end of Ragleth Wood, have been measured in detail by Cobbold and Whittard (1935, p. 348) who give the thickness of the grit group as about 100 ft, comprising four tuffaceous grit bands, varying from about 2 ft to 13 ft thick, with interbedded shales. North of Hazler the grit is traced across Helmeth Hill, being displaced westwards between two east-west faults near the summit [469939], and is then followed to Caradoc Coppice [468948] where it is cut off by F1. In this area the grit is mapped as overlying both the Ragleth Tuffs and the Cwms Rhyolite and is thus considered to be unconformable on the Eastern Uriconian (see also p. 37).

Stretton Shales

The area enclosed by the outcrop of the Helmeth Grit and F1 of the Church Stretton Fault Complex is underlain by Stretton Shales (≡ Lapworth's Watling Shales). Along the north-western flanks of Ragleth Hill there are scattered exposures of well laminated greenish grey shales with steep westerly or north-westerly dips. In a sunken lane [464940] 430 yd S. 23° W. of New House Farm there is a section in weathered greenish grey shales with vertical bedding. To the north-east of New House Farm [467943] there are scattered outcrops of greenish shales on the south-western flanks of Caer Caradoc Hill.

WEST OF CHURCH STRETTON FAULT

To the west of F1 the Stretton Shale Group occupies part of the floor of the Church Stretton valley and the adjacent lower slopes of the Long Mynd, the Helmeth Grit being absent in this area. Lapworth called these rocks the Brockhurst Shales but, as noted above (p. 37), the Brockhurst and Watling Shales are now combined as the Stretton Shale Group. It is not known whether part of the Watling outcrop may be equivalent to part of the Brockhurst outcrop due to fault repetition.

At Minton [430908] the Stretton Shale Group is cut out by overstep of the Upper Llandovery rocks but the outcrop widens to the north. There are exposures of typical grey shales, commonly weathered, in the lower parts of Callow Hollow, Ashes Hollow, Cardingmill Valley, and along the line of the road from All Stretton to Inwood. In the south the shales contain occasional thin bands of sandstone or siltstone, but these were not found north of All Stretton where the shales are well laminated. Calcareous nodules have been reported from some localities (James *in* Whittard and others 1953, p. 238). The regional dip is steep to the west-north-west and the cleavage, where distinct, may dip steeply to the west-north-west or east-south-east. Minor variations of dip are seen at some localities, e.g. in the Burway Road section [451940] there is a small anticline with a horizontal longitudinal axis striking N. 15° E. At an exposure [462959] 50 yd N. of St. Michael's Church, All Stretton, the shales dip east-south-east at 55° with a steep cleavage dip to the north-west. The intersection of cleavage and bedding gives rise to a lineation which plunges E. 38° N. at 20°. Lineations of similar character and attitude are seen about 300 yd to the north-east on the east side of Mousehill Wood [464960]. J.E.W.

Wart Hill to Cwm Head. Contorted olive-green shale, dipping steeply to W. 25° S., is well exposed in the yard at Upper Carwood [402853], the attitude of the beds being probably influenced by a nearby fault. Another fault, between Stretton and Wentnor Series rocks, is seen in the stream 215 yd south-west of the farm [401851]. Green shales, often finely laminated, are exposed in another stream below the bend north-west of Upper Carwood [399856]. They dip at varying high angles on a generally north-north-easterly strike. At the River Onny the outcrop is interrupted by a faulted wedge of Silurian rocks (p. 171), but green and purple Stretton Shales are seen immediately to the east in the cart track [408867] 250 yd W. 24° S. of Horderley bridge, and elsewhere nearby. Green shales are well exposed in the Onny for some 300 yd above the bridge.

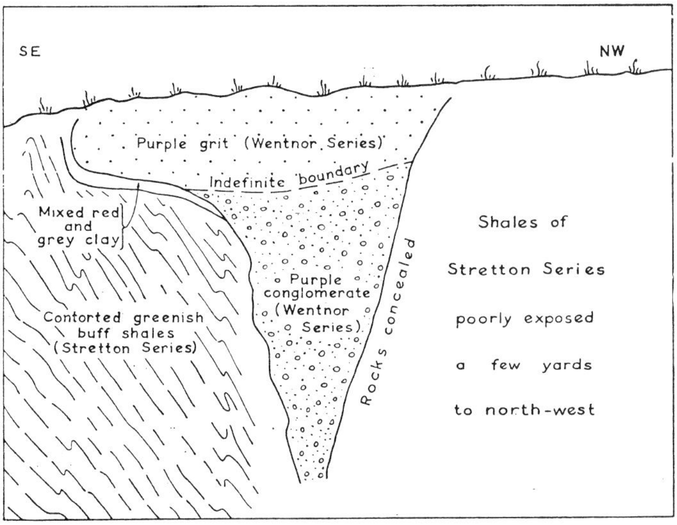

FIG. 11. Sketch of outcrop about 8 ft high, in Stretton Series and Wentnor Series rocks, beside a small stream [420882] 680 yd S. 30° W. of Cwm Head Church

Shales, generally green and contorted, are widely exposed beside the road leading from Horderley to Marshbrook and in places in the stream just west of it. Calcite veins and small faults occur locally. In general the dip is steep, on a north-north-easterly strike. Between 900 and 950 yd south-west of Cwm Head church crushed green shale appears to be faulted against Wentnor Series grits on the right bank of the stream. Whittard (1953, p. 247) and Dean (1964, p. 270) consider that the boundary here and in the plantation about 300 yd to the east-north-east [420882] is essentially an unconformity. In the plantation Wentnor Series grits are seen in contact with Stretton Shales along an undulating boundary which dips steeply to the north-west (Fig. 11). A thin band of red and grey clay at one place appears to have originated by faulting and a considerable element of doubt remains. On the steep right bank of the valley 250 yd south-west of Cwm Head church [421886] the evidence for faulting is more convincing, the shales lying at a higher level than the adjacent Wentnor Series grits. D.C.G.

Burway Group

Buxton Rock

The base of the Burway Group is marked by the Buxton Rock which is a silicified dust tuff, varying in thickness from about 8 ft to about 24 ft. Near Minton [430908] the Buxton Rock is overstepped by Upper Llandovery rocks and the most southerly exposure is at 330 yd N. 17° W. of the Manor House, Minton [430910], where there is 10 ft of very fine-grained, greenish grey cherty rock. On the southern and northern flanks of Wern Hill [431911], near the mouth of Callow Hollow, there are sections showing about 9 ft of greenish grey cherty rock with thin quartz veins on joints. There are no further exposures of the Buxton Rock south of Ashes Hollow and over this stretch the mapped outcrop is conjectural. In a small quarry [439927] 70 yd north-north-west of Ashes cottage there is 14 ft of massive cherty rock which appears to thin slightly when traced up the steep valley side. In the Cardingmill Valley the Buxton Rock is again exposed in a small quarry [451943] 900 yd E. 22° S. of the Carding Mill. Here there is 18 ft of massive greenish grey cherty rock containing some thin shaly bands. About 200 yd to the north-north-east a section in the road bend [452944] near the Church Stretton Golf Club House shows 13 ft of cherty rock, rather more flaggy than usual and with thin quartz veins. A farther 200 yd to the north-north-east the Buxton Rock is only 8 ft in thickness and is here probably displaced dextrally for a short distance along a minor fault. At the type locality, Buxton Quarry, All Stretton [459955], the Buxton Rock is 24 ft in thickness and is a massive greenish grey cherty rock with small dark spots locally and with thin quartz veins on the joints. In this section the Buxton Rock stands out from the adjacent shales and siltstones by reason of its massive and quartz-veined appearance. Northwards from All Stretton the Buxton Rock is again seen in the south side of Castle Hill [461959], where it is about 22 ft thick. The most northerly exposure in the district occurs in a field [466967] 1000 yd N. 27° E. from St. Michael's Church, All Stretton.

Main Part of Burway Group

The Burway Group consists largely of greenish grey sandstones, siltstones, and mudstones, commonly flaggy but with some more massive sandstone bands which carry thin quartz veins locally. At some horizons there are bands of purplish sediment but the greenish grey material is far more abundant. J.E.W.

Hillend to Minton Batch. Greenish grey shale, siltstone, and massive coarse sandstone, approximately vertical on a strike of E. 40° N., crop out on the slopes 1050 to 1100 yd north-east of Hillend Farm [403885], below the buttress in the Cardingmill Grit (p. 79). South-westwards these rocks are covered by solifluxion deposits until they disappear beneath Silurian rocks. They form the steep south-eastern slope of the Long Mynd from here to Callow Hollow but are best exposed on the sides of the deeply cut transverse valleys, for example in Woolers Batch, the valley about 1 mile north-east of Hillend Farm [407889]. Some of the finer beds here are bluish grey or, in the highest beds, faintly purple. They dip mainly south-eastwards at from 60 to to 80 degrees, with a general steepening in the upstream outcrops. Current bedding developed locally suggests that these more north-westerly beds are the younger. An outlying spur of Burway rocks, projecting east-south-eastward from the hillside some 1600 yd north-east of Hillend Farm [409887], is interpreted as an isolated stack which was surrounded by the Llandoverian Sea (Whittard 1932, p. 893; pl. lxii, (ii)).

Massive and shaly green siltstones are exposed in the old quarry 800 yd N. 13° E. of Churchmoor Hall [413893] and at the roadside to the east. The rocks of the group are also occasionally exposed in the valley north-west of Priors Holt Farm, and on the slopes south-east of Priors Holt Hill. About 960 yd N. 7° E. of Hamperley cross roads [421901] cleaved, brecciated and veined, green and purple siltstones mark the position of the Ashes Hollow Fault. In the lower part of Minton Batch the rocks of the Burway Group often exhibit anomalous dips. Massive quartz-veined sandstones and shales are well exposed on the summit and south-western slopes of Shooter's Knoll [4290], generally nearly vertical on a north-easterly strike. D.C.G.

Minton Batch to Lower Wood. To the north of Minton Batch the best exposures occur in the steep valley sides and of these the most extensive is probably in Ashes Hollow [436929] about 450 yd upstream from Ashes cottage, where there is a good section in vertical, flaggy siltstones and sandstones. Other good sections occur in Callow Hollow [429915] and on the north-eastern rim of Minton Batch [423906] near Shooter's Knoll.

In several areas along the length of the outcrop the dip is south-easterly with values varying from about 40° to about 75°, this being contrary to the regional steep dip to the west-north-west. These dips are found along a zone which lies between 250 yd and 450 yd north-west of the outcrop of the Buxton Rock, e.g. on the north side of Ashes Hollow [440930] 400 yd north of Ashes cottage, on the south-eastern flanks of Burway Hill [446940], in Gogbatch [463968] 1050 yd north of St. Michael's Church, All Stretton, and near Lower Wood [466973] 300 yd N. 20° W. of Pensylvania. These easterly dips suggest the presence of minor flexures or possibly anticlinal folds arranged en échelon along the outcrop of the group.

Cardingmill Grit

The upper limit of the Burway Group is taken at the top of the Cardingmill Grit (≡ Lower Cardingmill Grit of James). This is a massive, micaceous, greenish grey sandstone of moderately coarse texture. J.E.W.

Hillend to Minton Batch. The outcrop of the Cardingmill Grit can be followed north-eastwards from about 450 yd north-north-east of Hillend Farm [400880]. The rock is well exposed, and much faulted, on the banks of Mount Gutter between 800 and 1100 yd north-north-east of the farm. It forms a prominent buttress 1100 yd N. 30° E. of the farm [403885], where it comprises about 190 ft of massive, gritty, green sandstones, with green and purplish shale and siltstone interbedded. The sandstone, locally current-bedded, constitutes at least 75 per cent of the total thickness, in beds up to 36 ft thick. Quartz veins up to 4 inches thick cut the sandstones, commonly along joints dipping at 17° to N. 20° W. A good section through the Grit is seen in a series of outcrops on the left bank of Woolers Batch, some 1750 yd N. 33° E. of Hillend Farm [407890]. Here it appears to consist of three bands of massive, quartz-veined, greenish grey sandstone, 20, 30, and 20 ft thick, with a total of about 125 ft of interbedded purple shales, all dipping steeply to S. 40° E.

Similar massive sandstones exposed on the top of Old Churchmoor Hill and in the valley to the north-east, where they are much faulted, have been assigned to the Cardingmill Grit. Between 1150 and 1240 yd N. 10° E. of Churchmoor Hall [414897] greenish grey massive sandstone and purple shale and siltstone are exposed to a thickness of 79 ft, of which sandstone, in three bands, makes up 50 ft. The Grit is not exposed again west of the Ashes Hollow Fault. D.C.G.

Minton Batch to Lower Wood. In Minton Batch, between Ram's Batch and Windy Batch [421908], there are exposures of about 60 ft of massive quartz-veined sandstone, mapped as the Cardingmill Grit. The beds are vertical and strike about north-south, the outcrop here being displaced sinistrally for 350 yd along the Ashes Hollow Fault. The outcrop can be traced northwards from near the Packet Stone [422913] across Sleekstonebank Hollow to Callow Hollow [426920] on the northern slope of which 110 ft of Grit are exposed. There are few sections on Grindle hill to the north but on the northern slope of Ashes Hollow [434931] there are good exposures of about 120 ft of massive, micaceous, quartz-veined sandstone with occasional thin silty partings. This rock can be followed north-eastwards towards the summit of Yearlet Hill, the outcrop being disrupted by a series of minor faults. The prominent crags near the Devil's Mouth [441942] on the Bur Way are formed by the Cardingmill Grit, which is here about 100 ft thick and is cut by prominent vertical and inclined joints. The rock is a massive, coarse micaceous sandstone, with subordinate flaggy bands up to about 2 ft in thickness. The type locality is close to the Carding Mill [443946] where the Grit is about 120 ft thick and carries thin quartz veins. It can be traced from the small quarry at the valley bottom up the hill towards Bodbury Ring [445948], the outcrop being dextrally faulted for a short distance at about halfway up the slope. In the Batch valley, All Stretton, the Grit forms a prominent crag feature running up the northern valley side from near Batch cottage [450957]. The massive quartz-veined sandstone is about 90 ft thick and is dextrally faulted on the upper part of the slope. Current bedding, younging to the west, can be seen in some exposures. About 80 ft of the Grit crop out on the road between All Stretton and Jinlye [453961], and it can be traced thence along the north-western side of a prominent glacial drainage channel running northwards from Worsley [455962]. Over this stretch only about 20 ft of the Grit are exposed and it is affected by a number of minor dextral and sinistral fault displacements. On the northern slope of Gogbatch [458971], 250 yd E. 25° S. of Womerton, the Cardingmill Grit is about 80 ft thick. This is the most northerly exposure of the Grit in the present district except for two small excavations in a field [461976] 600 yd E. 38° N. of Womerton.

SYNALDS GROUP

The Synalds Group consists largely of purple shaly mudstones and siltstones with subordinate bands of purplish grey sandstone. In the lower part of the group there are greenish grey sediments in some localities but these are generally subordinate in amount to the purple rocks. J.E.W.

Onny Valley to Minton Batch. In the southern part of the Long Mynd the lowest beds of the Synalds Group are mainly of a pale greenish grey colour. More uniformly fine-grained than the higher beds and including very few sandstones, they weather to a very pale colour and are usually distinguishable from the rocks of the Burway Group. Estimates of their thickness vary from about 350 to about 700 ft but the figure may be invalidated in places by undetected faults. Particularly good exposures occur on the left bank of the valley 900–1000 yd north-north-east of Hillend Farm [402884] and at the top of the valley-side some 850 yd north-north-west of Churchmoor Hall [407892].

The purple shales and thin massive sandstones with quite common greenish bands, which constitute the bulk of the Synalds Group, are well exposed over the whole of their outcrop south-west of Mount Gutter, the valley about ½ mile north-east of Hillend Farm. They form many crags at the southern end of the Long Mynd between the valleys north of Hillend Farm and the escarpment about 1000 yd west-south-west of it [389874]. The top of the group follows the foot of the scarp and, from considerations of stratigraphy, structure, and topography, the boundary here is thought

A. Aerial view along western scarp slope of the Long Mynd (*Cambridge University Aerial Photograph No. Z3*)

B. The Long Mynd and the Church Stretton valley from Wart Hill (A9463)

Geology of Church Stretton
(*Mem. Geol. Surv.*)

PLATE 7

A. Bedding and cleavage in the Synalds Group (A9564)

B. The double topographic feature formed by the Stanbatch Conglomerate (A9560)

C. Unconformity between base of Harnage Shales and Uriconian, Hope Bowdler (A8908)

to be faulted. Towards the north-east there is a gradual clockwise rotation of strike and increase of dip, the dip being generally about 70° to west-north-west.

The numerous and often extensive outcrops and the frequently alternating lithologies afford many examples of minor structures, both sedimentary and tectonic. Thus an 8-ft band of massive sandstone is dextrally displaced by some 15 ft in a buttress 525 yd W. 13° S. of Hillend Farm [393876]. A sinistral dislocation of about 30 ft and several smaller faults, on lines just east of south, are seen in well exposed ground about 265 yd N. of the farm [398882]. About 55 yd farther south elongate rounded ribs, about 1½ inches wide and 1 inch deep, are developed on the upper surfaces of some bedding planes in purple shale. They are disposed en échelon and are sub-parallel. Their origin is uncertain but they may be a form of load cast or rill mark (Shrock 1948, fig. 92). The Synalds rocks are very well exposed along the valleys north of Hillend Farm and along Mount Gutter. Barytes veins up to 4 inches wide cut a dolerite intrusion in Mount Gutter, 920 yd N. 25° E. of Hillend Farm [402884]. Barytes veins up to 2 inches wide are exposed, associated with brecciated siltstone, in an old quarry 455 yd N. 25° E. of the farm [400880]. Very fine linear markings on some bedding planes in the quarry may be interpreted alternatively as drainage lines (Salter 1856, pp. 250–1, pl. iv and 1857, p. 202) or groove casts (Shrock 1948, pp. 162–6).

An unusually coarse dark green grit forms a lens up to 4 inches thick in purple shale 860 yd N. 36° E. of Hillend Farm [402883]. It contains many pebbles of Longmyndian rock and grades upwards into finer rock. Small monoclines, with southerly-trending axial planes are visible 885 yd N. 20° E. of the farm [400885] and elsewhere nearby. Cleavage is well developed in the shales here, and fluting and pitting, resembling Salter's *Arenicolites spp.* (1856, pl. iv and 1857, pl. v), are locally seen.

The Synalds rocks are less well exposed in Woolers Batch, and Nut Batch, respectively north-west, and about ¾ mile north, of Churchmoor Hall, but there are nevertheless many extensive outcrops. There are large buttresses of purple shale and sandstone at the head of Hens Batch, about 2000 yd north by east of the Hall.

D.C.G.

Minton Batch to Womerton. Greenish grey shales and sandstones occur near the base of the Synalds Group in Sleekstonebank Hollow [423918]. Callow Hollow [425922], and Ashes Hollow [430931], where they reach a thickness of about 400 ft and are separated from the Cardingmill Grit by about 200 ft of purple beds. Between Ashes Hollow and the All Stretton–Jinlye road there are practically no green sediments near the base of the group but at about 500 yd N. of Worsley [455966] there are about 250 ft of greenish grey siltstones and sandstones separated from the Cardingmill Grit by about 20 ft of purple beds.

In the dominantly purple beds to the west of the greenish grey rocks the banded arrangement of shale and sandstone gives rise to extensive craggy exposures in tributary hollows of the main Long Mynd valleys e.g. in Callow Hollow, Ashes Hollow, at the Pike [442949] in the Cardingmill Valley, and in the Batch valley [449961] south-west of Jinlye. In many sections the difference between cleavage and bedding is well seen, for example near the Pike [442949] the cleavage in the shaly bands is vertical, striking north-north-east, whereas the interbedded sandstones dip steeply to the west-north-west. '*Arenicolites*' hollows (see p. 72) are common at many localities e.g. Cardingmill Valley [439950], 900 yd E. 4° S. of Lightspout waterfall, and on the northern side of the 'swimming pool' [441949], 1150 yd E. 9° S. of Lightspout waterfall. They are also well exposed in New Pool Hollow on the south-eastern side of the reservoir [436945]. In every example found so far these hollows face to the west.

J.E.W.

Batch Volcanic Beds and Other Tuff Bands

Onny valley to Minton Batch. Tuff apparently occupies two principal horizons within the Synalds Group south-west of Mount Gutter, about 1000 yd north-north-east of Hillend Farm (Fig. 7). The bands, which are about 700 and 1150 ft below the top of the Group, cannot be distinguished by any obvious lithological criterion, both being composed of pale greenish grey and pale and bright purple tuff with many dark green aggregates of epidote. Creamy white feldspar fragments are common in the greenish grey bands. The recognizable tuff is generally overlain by pale greenish siltstone which may also be a tuffaceous deposit.

The lower band is well exposed at three points along its outcrop. It is 10 ft thick at a locality [394878] 460 yd W. 20° N. of Hillend Farm, and is exposed to 8 ft some 920 yd to the north-east [399885], 880 yd N. 10° E. of the farm. On the crags on the right bank of Mount Gutter, at the sharp bend in the stream [400885], it is about 6 ft 6 in thick. Some 500 ft lower in the succession a band of green tuff, coarse at the base and at least 2 ft 9 in thick, is exposed 625 yd N. 10° E. of Hillend Farm [399882]. It may be tentatively correlated with a band of fine-grained green tuff which occurs some 600 ft above the Cardingmill Grit on the southern side of Hens Batch, 1320 yd N. 14° W. of the Hamperley cross roads [417904].

The upper band crops out at the southern extremity of the Long Mynd, 815 yd W. 22° S. of Hillend Farm [391873]. About 80 yd to the north-north-east it is exposed to 6 ft 6 in. About 8 ft of tuff are exposed at a point 610 yd N. 25° W. of the farm [395882]. A band of tuff some 400 ft higher in the succession, crops out at about 150 yd to the north-west [395883]. It is of the same lithology as the others and is exposed to a thickness of 4 ft in a second outcrop about 100 yd to north-north-east. The band beneath can be followed, across several tear faults, to a point 1000 yd N. 1° E. of Hillend Farm [398886].

Two bands of tuff, about 4 ft and 7 ft 6 in thick are exposed on the left bank of Woolers Batch, 1120 yd N. 30° W. [406895] and 1200 yd N. 34° W. of Churchmoor Hall. They may be the two main bands mapped farther south. Two bands, only 130 ft apart, are exposed on the Forestry Commission road 1670 and 1685 yd north of Churchmoor Hall [412902]. The lower, 9 ft of pale greenish tuff, is not exposed elsewhere. The upper occurs again about 50 yd south-west of the road, where it is at least 6 ft thick, and in the crags 1550 yd N. 28° W. of Hamperley cross roads [414905], where it is about 13 ft 6 in thick. It is also exposed at the roadside nearby and can be followed north-eastwards towards Minton Batch. D.C.G.

Minton Batch to Womerton. From Stony Batch [411908] an 8-ft band of purple and greenish tuff can be traced north-eastwards almost to the foot of Minton Batch, but is not found on the northern side of this valley. This band lies at about 400 ft below the top of the Group. Farther east in Minton Batch [417911] two bands of tuff are exposed near the stream, the lower band being purple and the upper mainly green. They appear to lie at about 1600 ft and 1200 ft respectively below the top of the Synalds Group. The upper, green, band can be traced north-eastwards to a locality [420915] close to the path leading from Minton to the Portway, the outcrop being broken by small transcurrent faults. Across Sleekstonebank and Round Hill, the outcrop of the Synalds Group is displaced sinistrally for about 1400 yd along the Ashes Hollow Fault, to the east of which the most southerly exposures of tuff occur on Barrister's Plain [424927]. The section here is of a coarse purplish grey tuff between 5 ft and 10 ft thick. It lies at about 700 ft below the top of the group, is correlated with the Andesitic Ash of the Batch succession and can be traced north-eastwards across Ashes Hollow, the outcrop showing several small transcurrent fault displacements. A similarly faulted tuff band occurs on the north side of Ashes Hollow [425934], lying about 400 ft below the top of the Synalds Group. This is pale green in colour and is correlated with the White Ash.

In a tributary valley north of Ashes Hollow [430938], 950 yd E. 22° N. of Narnell's Rock, the coarse, purple Andesitic Ash is 15 ft thick and forms a good crag feature trending north-eastwards up the hillside. Near the top of the slope [429939] at about 300 ft north-west of the Andesitic Ash, there is an outcrop of 9 ft of pale green coarse tuff correlated with the White Ash. This band is cut off to the south-west by a north-south fault, on the west side of which there are 12 ft of fine-grained green tuff on the line of strike of the White Ash. This fine-grained band is tentatively correlated with a band in the Batch succession lying between the Andesitic Ash and the White Ash. The Andesitic Ash crops out on the Burway road [433941], 1440 yd E. 20° S. from the Boiling Well, and can be traced northwards to the upper slopes of New Pool Hollow [434942]. Here there is about 20 ft of coarse tuff, purple above and green near the base. The White Ash, 5 ft thick, crops out at about 350 ft to the north-west. At about 600 yd to the east there is a section on the Burway road, near the Devil's Mouth [439943], showing 11 ft of coarse green and purple tuff, which can be traced north-eastwards for about 200 yd into Devilsmouth Hollow. This band is about 600 ft above the base of the Synalds Group and is much lower than the typical Batch horizons.

There are some exposures of the Andesitic Ash and White Ash on the rocky spur north of New Pool Hollow [438949] and in the Cardingmill Valley [439949], but here the outcrops are considerably disrupted by minor transcurrent fault displacements. On the north-east side of Haddon Hill [443956], at the head of a tributary of the Batch valley, there is an outcrop of 8 ft of green tuff, the lower 5 ft being coarse and the upper part fine. This is correlated with the White Ash. At about 70 yd east of this section there is an exposure of 6 ft of green tuff, considered to be the equivalent of the band lying between the Andesitic Ash and the White Ash in the Batch section.

The type area of the Batch Volcanics lies at the junction of Long Batch and Jonathan's Hollow [446961] in the Batch valley. The full sequence of four tuff bands is found on the rocky spur between Long Batch and Jonathan's Hollow and has been described above (p. 42). The tuffs are well exposed on the rocky valley sides in the vicinity and the outcrops are much disrupted by minor fault movements. The Andesitic Ash, about 16 ft thick, forms a good crag feature extending north-eastwards from the valley bottom up the spur towards Plush Hill. The Andesitic Ash and White Ash are both exposed on the flat ground about 300 yd west of Plush Hill [448963], but there are no further exposures between this point and the head of Gogbatch 150 yd E. of Bullocks Moor Cottage [453969]. Here there is a good section showing 13 ft of Andesitic Ash with a characteristic purple upper part and pale green lower part. The outcrop is broken by minor faulting and can be traced as far as the road leading westwards from Womerton [453971], where it is again displaced on an east-west fault. At 420 yd W. of Womerton [453972] there is an exposure of 15 ft of coarse pale green tuff correlated with the White Ash, and at about 200 yd north and east of Womerton there are two small exposures of tuff which have not been correlated with the Batch succession. These are the two most northerly exposures of tuff in the Synalds Group.

Lightspout Group

The Lightspout Group consists mainly of greenish grey shales, flaggy siltstones, and sandstones, interbedded with occasional massive sandstone bands. The uppermost part of the group (approximately 400 ft) is mainly purple in colour and similar in appearance to the rocks of the Synalds Group. J.E.W.

Plowden to Asterton. On the long southern slope of the Long Mynd about ½ mile east of Plowden, are many relatively small, low outcrops of the green shales and sandstones of the Lightspout Group. They are well exposed in an old quarry just north of the road, 350 yd N. 29° E. of Plowden Mill [387875], where they are cut by barytes veins and carry some low-angle slickensiding. Purple and green shales and green massive sandstones are commonly seen along the fault-scarp about 850 yd west of Hillend Farm. Minor faults and irregularities of strike are common here. Beyond the Rabbit Warren to the west, where green shales and sandstones are widely exposed, purple shales probably indicate proximity to the top of the Group, as, for example, in two old quarries about 1300 yd S. 8° E. of Myndmill Farm [384875].

About 150 ft of purple shales and grits are exposed below the Huckster Conglomerate about 1070 yd S.S.E. of Myndmill Farm [386878]. They include, 62 ft below the Conglomerate, a 3-ft band of pale purple tuff which occurs again about 350 yd to the north-east [388879]. Purple shales about 270 ft below the Conglomerate are exposed 1080 yd S. 34° E. of the farm [388878], beside a tarn.

On the valley side 1180 yd E. 38° S. of Myndmill Farm [391880] is a small disused quarry in green massive sandstone in which a barytes vein 3 to 9 inches thick fills a joint dipping 70° to S. 10° W. In this area the dip is commonly about 70° to S. 20° E. the beds being probably inverted. Outcrops of green shale and sandstone just south-west of the summit of Black Knoll are clearly faulted against Portway rocks to the south-west.

Within purple shale and grit south-east of a crag 950 yd E. 6° S. of Myndmill Farm [391886] pale purple tuff, 2 ft 6 in thick, occurs about 140 ft below the Huckster Conglomerate. At 940 yd E. 14° S. of the farm [391885], 5 ft 6 in of similar rock are exposed apparently about 100 ft below the Conglomerate. About 6 ft of pale purple and green tuff crop out 570 yd S. 12° E. of Myndtown church [392890], with much purple grit and shale nearby. No tuff is seen in the topmost 100 ft of the Group which are well exposed on the northern side of a gully 1130 yd S. 16° W. of Handless Farm [394893]. Here only the top 70 ft are purple. Green shales and massive, quartz-veined sandstones are commonly exposed on the steep slopes above.

North-east of Black Knoll the Group crops out on the relatively flat summit plateau where there are very few exposures. Green shales and sandstones, apparently near the base of the Group, are exposed in and near the stream about 1250 yd W. 12° N. of Priors Holt [402896]. D.C.G.

Asterton to High Park. There are good exposures of typical Lightspout rocks in Minton Batch, Callow Hollow, Lightspout Hollow and Long Batch. The westward change from the purple beds of the Synalds Group into the greenish Lightspout rocks is well exposed on the north side of Ashes Hollow [425935], on the south side of the Cardingmill Valley [436949] and in Long Batch [444961]. Massive sandstone bands in the lower part of the Group occur on the south side of Haddon Hill [438951], in Long Batch [442961], and at Jonathan's Rock [446965]. As noted above (p. 43) James distinguished one of these bands as the Haddon Hill Grit.

The type area is near the Lightspout waterfall [431951], where there are good sections in massive greenish grey sandstones up to about 20 ft thick, interbedded with greenish grey shales and siltstones. There are some beds of purple sediment but these are subordinate in amount. The beds dip to the west-north-west at about 70° and the cleavage in the shales is mainly vertical, or nearly so, with a strike coincident with that of the bedding. The massive sandstones carry conspicuous thin quartz veins, probably filling tension-gashes in these competent beds. Probable ripple marks occur in a bed of purple sandstone on the steep rock face near the top of the Lightspout waterfall.

The topmost purple beds of the Lightspout Group are important in the consideration of James's 'Mintonian' unconformity at the base of the Huckster Conglomerate (see p. 43 for discussion). These beds are well exposed in the upper parts of Minton Batch [409919], Callow Hollow [415926], Ashes Hollow near Narnell's Rock [422934], Lightspout Hollow [428952], Mott Road [432935] and Deadman's Batch [448977]. At Narnell's Rock the Huckster Conglomerate rests immediately upon a small thickness of greenish sandstones and siltstones which are underlain to the south-east by 400 ft of purple siltstones and sandstones containing two bands of coarse, epidotic, crystal-lithic tuff, each about 3 ft in thickness and lying about 300 ft below the Conglomerate (Fig. 7). The same conformable succession occurs on the rocky spur [422937] 400 yd north of Narnell's Rock as a result of lateral movements along the Ashes Hollow Fault and a north-westerly branch of this fault. The occurrence of the same horizons at these two localities disproves Blake's supposition (1890, p. 394) that here the Huckster Conglomerate (base of Blake's 'Cambrian') is violently unconformable across different horizons in the Stretton Series (Blake's 'Monian').

In the upper part of the Lightspout Hollow [428952] the section for about 200 yd south-east of the Huckster Conglomerate is in banded purple sandstone and shaly siltstones. The bands of tuff seen at Narnell's Rock are not exposed in this section, but in the valley to the north [432955], which is followed by Mott Road, there is debris of coarse purple tuff associated with purple sandstones and shaly siltstones at the top of the Lightspout Group (Fig. 7). The purple Lightspout beds are seen again in the upper part of Long Batch [436961], but here there are no exposures of tuff nor is the Huckster Conglomerate exposed. A short distance north of the Church Stretton district in the lower part of Deadman's Batch [448977], 1040 yd W. 40° N. of Womerton, there is a section extending eastwards from the Huckster Conglomerate across 250 ft of purple sandstones and shales to an 8-ft band of coarse purplish tuff (Fig. 7). These beds were assigned by James (1956, p. 329) to the Synalds Group but, as noted above (p. 43), it is considered that this section is more correctly compared with other similar successions at the top of the Lightspout Group. The Deadman's Batch section is the most northerly exposure of known Lightspout rocks.

J.E.W.

Portway Group

Huckster Conglomerate

Plowden to Asterton. Numerous isolated outcrops of the Huckster Conglomerate occur north-east of a point 1050 yd S. 17° E. of Myndmill Farm, [385878], where 45 ft of quartz-conglomerate, with rounded pebbles up to about 1 inch in diameter, crop out on either side of the Port Way. To the south-west the conglomerate is cut off by a fault, and north-eastwards too it is cut by several tear faults, which might, in the absence of such a good marker, have been undetected. Large outcrops occur 880 yd E. 15° S. of Myndmill Farm [390885] where 35 ft of conglomerate, with two 30-inch grit bands, are exposed, and 950 yd E. 6° S. of the farm [391886] where 60 ft are visible. The conglomerate extends a farther 125 yd to the north-east, as indicated by debris, but beyond it is faulted against Lightspout rocks.

The Conglomerate crops out, but is not well exposed, low on the Long Mynd escarpment, about 600 yd south by east of Myndtown church [391890]. It is exposed to about 40 ft on the sides of a gully 1130 yd S. 16° W. of Handless Farm [394893]. Small rounded pebbles of quartz predominate but there are also some of purple grit. In a prominent crag 740 yd S. 11° W. of Handless [395897] about 27 ft of conglomerate are exposed, the pebbles mainly of white vein quartz with patchy red staining. About 30 ft of conglomerate and pebbly grit are exposed at the Huxter (Huckster) Stone, 440 yd S. 20° E. of Handless [397900], the type locality of the formation. D.C.G.

Asterton to High Park. Close to the Port Way [402907], at 740 yd S. 26° E. of the chapel, Asterton, there are several large blocks of conglomerate, probably very little displaced from the outcrop. The conglomerate is composed of rounded pebbles of quartz and quartzite averaging ½ in. across in a reddish matrix, and is mapped as the Huckster Conglomerate. About 650 yd to the north [402913] on the upper slopes above Asterton, there are exposures of about 10 ft of similar conglomerate probably displaced westwards for about 200 yd by lateral fault movement from the original line of strike. There is no section of the Huckster Conglomerate in Minton Batch [407919], its absence here being probably due to sinistral movement along the north-south Yewtree Bank Fault. In the upper part of Callow Hollow [414926] there is an outcrop of 15 ft of massive conglomerate, composed of rounded pebbles of quartz and siltstone averaging about ½ in. across in a purple sandy matrix. This is succeeded by a massive purplish grey coarse sandstone with some quartz veining.

There are good exposures of the Huckster Conglomerate in the upper part of Ashes Hollow at Narnell's Rock [422934] and on the spur 400 yd to the north [422938].

The basal conglomerate in these sections is about 10 ft thick and is composed of rounded quartz and purple siltstone pebbles up to about 1 in across. This is overlain by a massive purplish sandstone with occasional pebbly layers up to about 1 ft thick. These two localities are separated by a north-westerly branch of the Ashes Hollow Fault, with a transcurrent sinistral movement of about 250 yd. From the more northerly outcrop [422938] the Conglomerate is traced about halfway down the west side of Ashes Hollow, but is not exposed on the east side of the valley where its absence is attributed to sinistral transcurrent movement on the north-south Ashes Hollow Fault.

In the upper part of Lightspout Hollow [427953] and in the Mott Road valley to the north [431956] the horizon of the Huckster Conglomerate is represented only by a coarse massive sandstone with occasional small pebbles of quartz and purple siltstone. At High Park crossroads [445976] there are two small exposures of conglomerate in the grass verge at each side of the road. The Huckster horizon is next exposed 400 yd to the north-east in Deadman's Batch [448977], a short distance north of the limits of the Church Stretton district. Here the Conglomerate is poorly developed and is represented by a massive purplish gritty sandstone with occasional thin pebbly lenses in the basal part. J.E.W.

Main Part of Portway Group

Plowden to Asterton. Greenish grey shales, with purple shales locally predominant, crop out widely at the south-western limit of the Long Mynd, some 1100 yd S. 10° E. of Myndmill Farm. They are referred to the Portway Group, but their distinction from the Lightspout rocks to the east is uncertain. Behind the cottages [384879], about 900 yd S. 10° E. of Myndmill Farm, they dip at about 80° to W. 15° N. in several large buttresses. Similar rock, contorted into minor folds of every degree of intensity, forms several outcrops from 850 yd [385879] to 760 yd south-south-east of the farm.

The basal purplish grits and shales of the Group are well exposed for about 180 ft some 980 yd S. 18° E. of Myndmill Farm [385878]. Purple grits, apparently about 250 ft above the base, are seen 855 yd S. 40° E. of the farm [387881]. They are exposed in several outcrops some 550 yd S. 10° E. of Myndtown church [391890] and 530 yd S. 7° W. of Handless [396899], about 220 ft above the Huckster Conglomerate at the latter locality.

Green and purple shales with some bands of dark green grit are well exposed just east of the road 1200 to 1600 yd south-south-west of Myndtown church. From south to north the dip direction gradually swings from north-eastwards to east-south-east. These beds are probably all inverted but no conclusive evidence was observed. Green and purple shale with malachite in some joint and bedding planes is exposed 790 yd S. 6° W. of Myndtown church [390888]. Green and purple shales, with many thin bands of green sandstone, form many outcrops on the slopes north-east of Handless [398904]. They dip steeply to N. 40° W. and exhibit well developed cleavage, inclined at 75° to E. 30° S. D.C.G.

Asterton to High Park. The outcrop of the Portway Group occurs mainly on the summit plateau of the Long Mynd and is, to the north of Asterton, poorly exposed. On the steep slopes to the east and south-east of Asterton there are scattered exposures of purple and greenish grey shales, siltstones, and sandstones, wtih some more massive sandstone bands. In this area the dip is mainly steep to the east-south-east. There are sections in similar rocks from the lower part of the Group in the upper parts of Callow Hollow [412928] and Ashes Hollow [417937]. Here the bedding dips steeply to the west-north-west while the cleavage in the shaly beds is inclined steeply in the opposite direction. In High Park Hollow [438972] there are strike sections in purple and greenish siltstones and sandstones from the middle part of the Group, dipping to the west-north-west at angles of 50° to 75°. These beds are underlain by massive purple and greenish sandstones on the eastern side of the hollow. In Hawkham Hollow [434976] there are strike sections in the highest part of the Portway Group. These are again purple shales, siltstones, and sandstones dipping westwards at angles of 45° to 65°, the cleavage in the shaly beds dipping steeply in the same direction.

J.E.W.

WENTNOR SERIES

Bayston–Oakswood Group

bayston (eastern) outcrop

Basal Beds

Myndtown. Massive, coarse-grained, purple and green grits form a line of crags which extends northwards from an old quarry 1120 yd S. 12° W. of Myndtown church [388885]. At the northern end of the crags, 880 yd S. 11° W. of the church [389887], conglomerate about 50 ft thick, dipping steeply to the south-east, forms a narrow fault-bounded, and fault-broken rib. It resembles the Huckster Conglomerate but for its relatively high content of igneous pebbles. These rocks are referred to the base of the Bayston–Oakswood Group but their horizon is uncertain. They may be faulted against the Portway rocks to the east or rest unconformably on them and they are faulted against Portway shales about 100 yd north-east of the conglomerate rib.

D.C.G.

Asterton to Darnford. A few scattered sections in the lower part of the Group occur on the steep hillside 800 yd north-north-east of Asterton [401918]. The rocks exposed are massive grey sandstone with occasional gritty bands. These lower beds are scarcely exposed between this point and the western slopes of Hawkham Hollow [433977], where Portway siltstones and sandstones are overlain with apparent angular conformity by a massive, coarse, purple sandstone carrying occasional pebbles. In this area the beds dip westwards at about 50°-60° and the junction of the two groups is displaced slightly by two minor faults trending east-south-east. This is the only exposure of the base of the Group within the Church Stretton district.

Darnford Conglomerate

The most southerly exposures of the Darnford Conglomerate occur on the lower part of the western slope of the Long Mynd [401919], 630 yd N. 19° E. of the chapel, Asterton. The Conglomerate is 60 ft thick and is composed of rounded pebbles of quartz and quartzite, with some rhyolite, up to about 2 in across, in a greenish gritty matrix. This band can be traced north-eastwards, crossing obliquely up the hill slope with minor dextral fault displacements to a point [405924] 1300 yd N. 31° E. of the chapel, Asterton, where there is an apparent sinistral displacement of about 200 yd, possibly on the line of the Yewtree Bank Fault. The Conglomerate is displaced almost to the bottom of the hill again whence it is traced north-north-eastwards to Stanbatch. Over this stretch there are a number of exposures of coarse, well rounded conglomerate, with a greenish gritty matrix and with pebbles occasionally up to 6 in across. The outcrop suffers a number of minor dextral and sinistral displacements in this section. At Stanbatch [407935] the Darnford Conglomerate crops out close to the Stanbatch Conglomerate as a result of lateral fault movements, but the two bands are distinguished by the greenish gritty matrix of the former and the characteristic double-ribbed character of the latter.

The Darnford Conglomerate appears to be dextrally faulted at Stanbatch for about 350 yd, and is traced thence to the north mainly by conglomerate debris on the upper slopes of the Long Mynd. There is one small exposure of conglomerate in a gully [413947], 300 yd N. 28° W. of the summit (1696 ft) of Pole Bank, and two more outcrops at the head of Bilbatch [418955], 340 yd W. 23° N. of the Shooting Box. These are identified with the Darnford Conglomerate by their association with overlying greenish grey shaly siltstones, which are also found in the type area to the north. From Bilbatch to Colliersford Gutter [426973] there are no exposures of the Darnford Conglomerate, and between Pole Cottage [413938] and Colliersford Gutter the mapped fault shifts in the outcrop are inferred largely from similar displacements in the better exposed Stanbatch Conglomerate to the west. The type section on the west side of Colliersford Gutter [426973], 250 yd south of Upper Darnford, is described in the general section above (p. 47). This is the most northerly exposure in the Church Stretton district, and to the north of this section the Conglomerate is thought to be dextrally faulted to the east side of the valley.

Stanbatch Conglomerate

The most southerly exposure of the Stanbatch Conglomerate occurs on the south side of Stanbatch [406936], on the loop in the hill road leading from Pole Cottage to Wentnor. Here the Conglomerate is not wholly exposed but probably measures about 300 ft in thickness and is composed of two conglomeratic bands separated by a central band of soft-weathering purple grit. To the south of this locality the Conglomerate is probably faulted out against the Long Mynd Scarp Fault and does not reappear in the lower ground of Wentnor Prolley Moor. The Stanbatch Conglomerate is exposed on the north side of Stanbatch and again to the north in the valley leading from Pole Bank to Medlicott [410948]. Here there is the typical double feature and in the valley bottom there is a dextral displacement of about 100 yd along a west-north-west to east-south-east transcurrent fault. There are good exposures of the Conglomerate on both sides of Bilbatch [414955], separated by a dextral displacement of 150 yd on an east-west fault in the valley bottom. The double crag feature is seen on both sides of the valley and there is an almost continuous section across the Conglomerate on the road [416956] leading from the Shooting Box to Ratlinghope. This section is described above in the general account (p. 47). There are no exposures of the Conglomerate across Wild Moor but in the upper part of Catbatch Brook [418967], there is a strike section in the lower rib showing about 120 ft of coarse purplish conglomerate. This is underlain by a hard, dark grey, siliceous breccia containing small angular fragments of feldspar, quartz and grit. On the north side of this valley the Conglomerate is displaced dextrally for about 400 yd on an east-west fault, and is thence traced only by debris and feature to the vicinity of Middle Darnford [423976]. At 200 yd east of this farm there are exposures in the lower part of the Conglomerate, separated by dextral fault displacements from the more extensive exposures on the northern side of the valley where, as noted by Whitehead (*in* Pocock and others, 1938, p. 51), the double ribbed feature is well displayed (Plate 7B).

Upper Part of Bayston Outcrop

There are no very extensive sections in the upper part of the group but scattered outcrops occur on the north-western slopes of the Long Mynd from Stanbatch [403939] to Ratlinghope Hill [411976]. The rocks are mainly massive, coarse, purple sandstones, locally with quartz veins and with occasional bands of greenish grey sandstone. In the vicinity of the Pock [400952], Rock Cottage [403955] and Shaw Wood [406962] near Belmore Farm [410962], there are exposures of hard grey brecciated and silicified sandstone with abundant ramifying quartz veins. These veins show slickensides which plunge in various directions. These outcrops are considered to lie along the line of the Pock Fault (see p. 263).

OAKSWOOD (WESTERN) OUTCROP

Within the Church Stretton district the area occupied by the outcrop is small and so poorly exposed that the general description given above (p. 48) cannot usefully be further amplified in detail.

Bridges Group

The most southerly exposures of the Bridges Group occur just west of the Church Stretton district in the River East Onny, 250 yd W. 40° N. of Criftin House [378917]. Here there are flaggy purple siltstones and sandstones dipping steeply to the west-north-west. In sections along the roads radiating from Wentnor village [384928] there are exposures of purple sandstone and siltstone, commonly laminated and moderately flaggy. The dip is mainly steep to the west-north-west and at several localities in this area specimens showing evidence of age sequence have been collected (see above p. 50 and Fig. 8). Between Wentnor and Bridges there are scattered outcrops along the course of the River East Onny and along the flanks and summit of Adstone Hill [389948]. These sections are mainly in flaggy purple siltstones and sandstones, locally well laminated, but at some localities, notably along the top of Adstone Hill, there are more massive purple and grey sandstones.

Near the Criftin Brook [394937] 900 yd west of Stanbatch there is an outcrop of purplish grey silty Longmyndian sandstone. This indicates a northern limit to the tongue of Upper Llandovery rocks which occupies the hollow of Wentnor and Asterton Prolley Moors to the south. At Medlicott [400946] there are roadside exposures of laminated purple siltstone and sandstone dipping steeply to the west. These rocks occur near the base of the group and may be contrasted with the massive gritty sandstones of the Bayston outcrop occurring about 300 yd to the east on the hill road rising to Pole Bank.

At 100 yd north of the inn at Bridges [393966] there are extensive roadside exposures of well jointed, flaggy, purple siltstone and sandstone dipping westwards at about 50°. Similar rocks are found in the stream section [393969] about 500 yd north of Bridges and at the south end of Drive Coppice [397964], 350 yd east of Bridges. These last two localities have yielded specimens showing evidence of age sequence (see p. 50). In Rookery Wood [402970] west of the mill at Ratlinghope, there are exposures of vertical or steeply dipping laminated purple siltstones and sandstones, and to the north there are scattered outcrops of similar rocks with some more massive sandstone bands, on the north-western part of Ratlinghope Hill.

The western limit of the Bridges Group is very poorly exposed, but outcrops of banded purple siltstone and sandstone, considered to lie near the base of the Group (on the inverted western synclinal limb) occur at Stedment farm [388969] and in a small quarry [392978] 1000 yd N. 25° E. of Stedment. J.E.W.

Wentnor Series–Undifferentiated

Coston to Hopesay Hill. The most southerly exposure of the Wentnor Series lies 430 yd E. 21° S. of Beambridge [392811], near Aston-on-Clun, where purple micaceous sandstones crop out in the River Clun. The basal beds of the Hoar Edge Grit (basal Caradoc) are exposed a few yards farther downstream. Purple sandstone with bands of rusty brown siltstone is seen in a track section [396818], 350 yd N. 30° E. of the Kangaroo Inn, Aston-on-Clun and, a little farther north, 8 ft of slickensided and disturbed purple micaceous sandstone dipping at 30° W. were seen in a temporary section 425 yd N.E. of Aston Hall. A water borehole 60 yd N.W. of Aston Hall proved 123 ft of sandstone.

There are no exposures on the high ground of Aston and Hopesay Hills, but there is much debris of purple micaceous sandstone in the soil. Purple sandstone is exposed in a stream section [405831] 180 yd N.W. of Oldfield, with the Hoar Edge Grit cropping out about 30 yd downstream.

An obscure and intermittent track section extends for 600 yd northwards from a locality 475 yd N. 33° E. from Hopesay church [392836]. This shows coarse purple sandstones, purple siltstones and occasional beds of hard grey siltstone overlain by boulder clay and head up to 5 ft in thickness. The beds are strongly sheared and slickensided due to the proximity of the Church Stretton Fault to the west. B.A.H.

Wart Hill to Cwm Head. Purple grits, with subsidiary shales and conglomerates, crop out in places in the streams on the flanks of Wart Hill [4084], and debris of these rocks is abundant over the northern end of Hopesay Hill [403843]. A conglomerate with well-rounded pebbles up to 2 inch in diameter is exposed in the stream [403847] 285 yd E. 7° S. of the top of Wart Hill. The pebbles are generally of grit or igneous rock of Uriconian aspect. At 410 yd N. 11° E. of Wart Hill [402851], a band of dark green tuff (E 29525) lies within a group of purple and dark green grits and shales.

Purple and green grits are locally exposed between Upper Carwood and the River Onny. On the steep southern bank of the river [408864], some 815 yd N. 39° E. of Lower Carwood, grits, siltstones, and shales, heavily veined and locally crushed, are faulted against Silurian rocks, exposed immediately to the west and south (p. 171). In the old quarry [412861], 150 yd west-south-west of Glenburrell, Ordovician rocks rest on soft weathered purple grit, all dipping steeply to the east. Buff-coloured leached zones can be followed from the hard purple grit through the 4 ft of soft rock at the top, and show none of the disturbance which might be expected were the junction between the two formations faulted. About ¼ mile north-east and south-east of the old railway station at Horderley [382875], purple grits with interbedded siltstones and blocky mudstones are exposed in old quarries and a stream. Purple grit is exposed on the hill top at Brokenstones about 1000 yd S. 21° W. of Cwm Head church, separated from the adjacent Ordovician outcrops by a north-easterly fault.

Purple pebbly grit and shale in the stream [421881] 710 yd S. 23° W. of Cwm Head church, indicates the presence of a fault-bounded outcrop of Wentnor Series rocks. The north-eastern limit of this outcrop is unknown.

An outcrop of purple grit in the stream [404864] 575 yd N. 6° E. of Lower Carwood is interpreted as part of a narrow strip of Wentnor Series rocks faulted between the Church Stretton Fault (F1) and the Stretton Shale Group. Two similar strips occur along F1 between Horderley and Cwm Head. In a tributary valley [414877] 1450yd S. 44° W. of Cwm Head church, purple and green crushed micaceous rock and slickensided clay are exposed immediately south-east of F1 and north-west of the Stretton Shale Group. The Wentnor Series grits are better exposed on the western bank of the main valley from 480 to 550 yd to the north-east where they are cut by narrow andesitic intrusions. Here and in a third narrow strip, south of Pillocksgreen [422887], and also in the wood [420882] 680 yd S. 30° W. of the church, rocks of the Wentnor and Stretton series are seen in close contact (Fig. 11). In every case the junction is complicated by faulting but Whittard (1953, p. 247) believed that, in the first and third localities at least, it is essentially an unconformity. The field evidence is not considered to be at all conclusive.

Wentnor Series grits crop out in places at the roadside from 200 to 850 yd south-west of Cwm Head church, in the lane 740 yd S. 30° W. of it [420881], and in the stream from 230 to 480 yd south-west of it. These outcrops form two fault-bounded strips trending north-north-east. D.C.G.

References

Bailey, E. B. 1930. New light on Sedimentation and tectonics. *Geol. Mag.*, **67**, 77–92.

Blake, J. F. 1890. On the Monian and Basal Cambrian Rocks of Shropshire. *Quart. J. Geol. Soc.*, **46**, 386–420.

Blyth, F. G. H. 1944. Intrusive Rocks of the Shelve Area, South Shropshire, *Quart. J. Geol. Soc.*, **94**, 169–204.

Callaway, C. 1888. On some Ancient Salopian Conglomerates. *Trans. Shropshire Archaeol. Nat. Hist. Soc.*, **9**, 239.

—— 1891. On the Unconformities between the Rock-systems underlying the Cambrian Quartzite in Shropshire. *Quart. J. Geol. Soc.*, **47**, 109–25.

COBBOLD, E. S. 1900. in *Church Stretton 1*. ed. C. W. Campbell-Hyslop. Shrewsbury.

—— 1927. The Stratigraphy and Geological Structure of the Cambrian Area of Comley (Shropshire). *Quart. J. Geol. Soc.*, **83**, 551–73.

—— and WHITTARD, W. F. 1935. The Helmeth Grits of the Caradoc Range, Church Stretton; their Bearing on Part of the Pre-Cambrian Succession of Shropshire. *Proc. Geol. Assoc.*, **46**, 348–59.

CREER, K. M. 1957. The Natural Remanent Magnetization of Certain Stable Rocks from Great Britain. *Phil. Trans. Roy. Soc.* (A), **250**, 111–29.

DEAN, W. T. 1964. The Geology of the Ordovician and adjacent strata in the Southern Caradoc District of Shropshire. *Bull. Brit. Mus. (Nat. Hist.) Geol.*, **9**, 257–96.

EMERY, K. O. 1945. Entrapment of Air in Beach Sand. *J. Sed. Pet.*, **15**, 39–49.

FYFE, W. S., TURNER, F. J. and VERHOOGEN, J. 1958. Metamorphic reactions and metamorphic facies. *Mem. Geol. Soc. Amer.*, **73**.

GILETTI, B. J., MOORBATH, S. and LAMBERT, R. St. J. 1961. A Geochronological Study of the Metamorphic Complexes of the Scottish Highlands. *Quart. J. Geol. Soc.*, **117**, 233–72.

GREENLY, E. 1919. The Geology of Anglesey. *Mem. Geol. Surv.*

HOLMES, A. 1959. A Revised Geological Time-Scale. *Trans. Edin. Geol. Soc.*, **17**, 183–216.

JAMES, J. H. 1952. Notes on the Relationship of the Uriconian and Longmyndian Rocks near Linley, Shropshire. *Proc. Geol. Assoc.*, **63**, 198–200.

—— 1956. The Structure and Stratigraphy of part of the Pre-Cambrian Outcrop between Church Stretton and Linley, Shropshire. *Quart. J. Geol. Soc.*, **112**, 315–37.

JAMES, L. F. 1884. The fucoids of the Cincinnati group. *Cincinnati Soc. Nat. Hist. J.*, **7**, 124–32, 151–66.

KAY, M. 1951. North American Geosynclines. *Mem. Geol. Soc. Amer.*, **48**.

KINDLE, E. M. 1916. Small pit and mound structures developed during sedimentation. *Geol. Mag.*, **3**, 542–47.

LAPWORTH, C. and WATTS, W. W. 1894. The Geology of South Shropshire. *Proc. Geol. Assoc.*, **13**, 297–355.

—— 1910. Geology in the Field, Shropshire. *Geol. Assoc. Jubilee Vol.*, 739–69.

PABORSKI, S. J. 1954. Virgin Formation (Triassic) of the St. George, Utah, Area. *Bull. Geol. Soc. Amer.*, **65**, 971–1006.

PALMER, R. H. 1928. Sand holes in the Strand. *J. Geol.*, **36**, 176–80.

PASSEGA, R. 1957. Texture as characteristic of clastic deposition. *Bull. Amer. Assoc. Pet. Geol.*, **41**, 1952–84.

PETTIJOHN, F. J. 1957. *Sedimentary Rocks*. 2nd Edit. New York.

POCOCK, R. W., WHITEHEAD, T. H., WEDD, C. B. and ROBERTSON, T. with contributions by WRAY, D. A., STUBBLEFIELD, C. J., CANTRILL, T. C. and DAVIES, W. M. 1938. Shrewsbury District. *Mem. Geol. Surv.*

READE, T. M. and HOLLAND, P. 1908. Analyses of Longmyndian Rocks. *Proc. Liverpool Geol. Soc.*, **10**, 276–87.

SALTER, J. W. 1856. On Fossil Remains in the Cambrian Rocks of the Longmynd and North Wales. *Quart. J. Geol. Soc.*, **12**, 246–51.

—— 1857. On Annelide-Burrows and Surface-Markings from the Cambrian Rocks of the Longmynd. *Quart. J. Geol. Soc.*, **13**, 199–206.

SHROCK, R. R. 1948. *Sequence in Layered Rocks*. New York.

STEWART, H. B. 1956. Contorted sediments in Modern Coastal horizons Explained by Laboratory Experiments. *Bull. Amer. Assoc. Pet. Geol.*, **11,** 153–61.

TAYLOR, J. H. 1958. Pre-Cambrian Sedimentation in England and Wales. *Eclogae Geol. Helv.*, **51,** 1078–92.

TWENHOFEL, W. H. 1921. Impressions made by bubbles, rain drops, and other agencies. *Bull. Geol. Soc. Amer.*, **32,** 359–72.

WATTS, W. W. 1925. The Geology of South Shropshire. *Proc. Geol. Assoc.*, **36,** 322–93.

WHITEHEAD, T. H. 1955. The Western Longmyndian Rocks of the Shrewsbury District. *Geol. Mag.* (6), **92,** 465–70.

WHITTARD, W. F. 1952. A Geology of South Shropshire. *Proc. Geol. Assoc.*, **63,** 143–97.

—— with BALL, H. W., BLYTH, F. G. H., DINELEY, D. L., MITCHELL, G. H., POCOCK, R. W., and STUBBLEFIELD, C. J. 1953. Report of Summer Field Meeting in South Shropshire, 1952. *Proc. Geol. Assoc.*, **64,** 232–50.

Chapter 4

CAMBRIAN

GENERAL ACCOUNT

CAMBRIAN ROCKS crop out in two areas in the Church Stretton (166) Sheet. The classic area of Comley and Cwms was made famous first by Lapworth (1888), whose discoveries provided the clue to the presence of Lower and Middle Cambrian rocks in Britain, and later by the detailed palaeontological and stratigraphical work of Cobbold. A second Cambrian outcrop is situated at Hill End, about half a mile south of Cardington. The gradual recognition of the presence and extent of the Cambrian System in Shropshire has been summarized in the Shrewsbury Memoir (Pocock and others 1938), and need not be repeated here.

The general sequence in the Cambrian is as follows (Cobbold's notation is given in square brackets):

UPPER CAMBRIAN

		Approx. thickness in ft.
Tremadoc Series		
Shineton Shales [Cx] (Shoot Rough Wood Shales)	Undivided on map	? 1000
Non-sequence		
Dolgelly Beds		
Black Shales (doubtfully present in this district)		
Grey (*Orusia*) Shales [Ca] (Shoot Rough Road Shales)		

Non-sequence

MIDDLE CAMBRIAN

		Approx. thickness in ft.
Upper Comley Series		
Billingsella Beds [Bc] } (Shoot Rough Road Flags)	Undivided on map	300 to ?over 600
Paradoxides davidis Flags [Bb$_5$] } (Shoot Rough Road Flags)		
'*Paradoxides rugulosus*' Sandstone [Bb$_4$] (Shoot Rough Road Sandstone)		
Hill House Shales [Bb$_2$]		
Paradoxides intermedius Grit [Bb$_1$] (Hill House Grits, Comley Breccia Bed)		
'*Dorypyge*' *lakei* Flags [Ba$_3$] (Hill House Flags)		
Quarry Ridge Shales [Ba$_2$]		
Paradoxides groomi Grits [Ba$_1$] (Quarry Ridge Grits)		

Unconformity

Lower Cambrian

		Approx. thickness in ft
Lower Comley Series		
Lower Comley Limestones [Ac_1-Ac_5, Ad]	Undivided on map	up to 6
Lower Comley Sandstone [Ab_1-Ab_4]	Undivided on map	500
Wrekin Quartzite [Aa]		up to 140

Comley area.

The Cambrian rocks of the Comley area extend along the eastern side of the Uriconian mass of Caer Caradoc Hill from Cwms to Comley. The outcrop is divided into two areas by the overstep of Ordovician rocks on the south-east side of the hill. The basal Wrekin Quartzite rests unconformably on the Uriconian in places near Comley, but for much of the way the junction is faulted, the most important fault being the Lawley Fault (F2 of Cobbold 1927a, p. 565). To the east the Cambrian is overlain unconformably by the Hoar Edge Grit (local basal Ordovician).

North of Comley the Cambrian rocks extend into the area covered by the Shrewsbury (152) Sheet (Pocock and others 1938). There the Wrekin Quartzite rests unconformably on the Uriconian of the Lawley.

The detailed stratigraphy and structure of the Cambrian rocks of the Comley area is complex, and cannot be determined from natural exposures. Our knowledge of it is due to the painstaking work of Cobbold who, between 1909 and 1933, presented a series of reports to the British Association for the Advancement of Science describing 63 excavations made at critical places on the Cambrian outcrop. His paper of 1927a summarizes most of this work. With the exception of a few which are badly degraded, the excavations are now completely overgrown or filled in.

Hill End area.

Cambrian rocks occur over an area of about a quarter of a square mile to the east of Hill End, half a mile south of Cardington. They are bounded to the north and south by the Hill End and Sharpstones thrusts, to the east by the unconformable Silurian, and to the west by their unconformable junction with Uriconian rocks.

B.A.H.

DETAILS

Comley Area

Wrekin Quartzite

The Wrekin Quartzite is exposed [491969][1] on the south-east side of The Lawley, 800 yd south of Penkridge Hall. It dips at about 38° to the south-south-east, and closely adjoins outcrops of Uriconian rocks. At the north end of Caer Caradoc Hill its junction with the Uriconian appears to be faulted for some distance. In several exposures [481960] 770 yd N. 28° E. of the summit of Caer Caradoc Hill, the Quartzite dips westward into the fault. 500 yd N. 28° E. of the summit of Caer Caradoc Hill, the Quartzite, which dips eastwards at about 45°, appears to rest unconformably

[1]Such numbers are National Grid References within the 100-km. square SO (32).

on the Uriconian. The rock is a white or blue sugary quartzite, pebbly near the base, with quartz and pink rhyolitic fragments.

To the east of the Lawley Fault, on the eastern slopes [483960] of the northern end of Caer Caradoc Hill, there is an obscure area in which quartzite blocks on the ground suggest the presence of an outcrop of the Quartzite. This is interpreted by Cobbold (1927a) as an anticline with a northerly trend bringing up the Quartzite between outcrops of Lower Comley Sandstone. In Cwms [475947–476942] the Quartzite appears to rest on red grits, probably of the Wentnor Series, but exposures are very poor. An excavation [475943] (No. 56 of Cobbold 1927b), 310 yd N. 8° E. of Cwms Farm, showed the basal part of the Lower Comley Sandstone resting on 25 ft of Wrekin Quartzite, which was apparently faulted against red sandstone of the Wentnor Series ('Torridonian'). The best section is in an old quarry [475947] 720 yd S. 17° W. of the summit of Caer Caradoc Hill in which 12 ft of quartzite with black and white patches pass up into 14 ft of fine-grained glauconitic sandstone forming the bottom of the Lower Comley Sandstone.

LOWER COMLEY SANDSTONE

The outcrop of the Lower Comley Sandstone closely follows that of the Wrekin Quartzite. No exposures were found in the ground south-east of The Lawley. The best known section is at Comley Quarry [48459647], 1400 yd N. 33° E. of the summit of Caer Caradoc Hill, where green glauconitic sandstone with an easterly dip of 73° forms the western face. About 14 ft of the Sandstone were exposed in 1953. Other exposures occur on the slope immediately to the south. Elsewhere the Sandstone is badly exposed, although a few scattered outcrops are found in and near the stream about 450 yd south-south-east of Comley Quarry. In Cwms, the Lower Comley Sandstone in the old quarry [475947] 720 yd S. 17° W. of the summit of Caer Caradoc Hill has yielded *Micromitra* (*Paterina*) *phillipsi* (Holl), *M.* (*P.*) *rhodesi* Cobbold, *Obolella? groomi* Matley, *Walcottina lapworthi* Cobbold, *Hyolithellus sp.* and *H.? sinuosus* Cobbold (Cobbold 1921). Clearing and deepening of this quarry, as described by Cobbold (1916, p. 118. Excavation No. 53) formerly allowed a more detailed description of the section than is now possible. G.H.M.

A Geological Survey inclined borehole (Church Stretton No. 7) [48899645] at Shootrough proved 565 ft 8 in of Lower Comley Sandstone (depth 239 ft 6 in to 805 ft 2 in, true thickness about 420 ft) under an unconformable cover of Middle and Upper Cambrian strata. The Sandstone was greenish grey with local purplish coloration, medium-grained and glauconitic in the upper part, becoming fine-grained and silty towards the bottom of the borehole. B.A.H.

The Lower Comley Sandstone of the Church Stretton No. 7 Borehole has yielded the following fossils:

Depth	
276 ft 10 in	*Hyolithellus micans* Billings, cf. *Indiana minima* Wiman
375 ft 2 in to 383 ft 4 in	*Hyolithellus micans*, *Torellella laevigata* (Linnarsson)
454 ft 10 in	*Indiana exigua* (Cobbold)
656 ft 8 in	Bradoriid (compare '*Bradoria* cf. *nitida* (Wiman)' of Cobbold 1936)
722 ft 1 in and 727 ft 11 in	*Micromitra?* fragment, *Helcionella sp.*

Poorly preserved ostracods have been assigned to species described by Cobbold from his divisions Ab_2 and Ab_4. It would be hazardous to correlate levels in the borehole with Cobbold's divisions on this basis alone as the ranges of these Comley Bradoriida are unknown.

The range of *Hyolithellus micans* is here extended down into Cobbold's division Ab and the species of *Helcionella* which occurs near the bottom of the borehole is probably from a horizon much lower than those of the other species of *Helcionella* recorded in the Comley area.

At several horizons between 298 ft 2 in and 505 ft 3 in there were collected casts of worm tracks of a type not previously recorded from the Lower Comley Sandstone but which resemble specimens from the glauconitic Hollybush Sandstone of the Malvern area. A.W.A.R.

LOWER COMLEY LIMESTONES

The type section of these beds is at Comley Quarry [48459647], where the following divisions of the Limestones are seen:

	ft	in
Lapworthella Limestone ('Ad' of Cobbold's notation)		
Very dark grey, composed of phosphatic material and occasional quartz pebbles. Locally thinning out .. up to	–	6
Protolenus Limestone (Ac_5)		
Pale grey fossiliferous limestone, dark and phosphatic where fossils are rarer .. .. about	–	6
Strenuella Limestone (Ac_4)		
Red to purple sandy limestone with well-rounded grains of quartz and a phosphatic matrix .. .. .. about	–	9
'*Eodiscus*' *bellimarginatus* Limestone (Ac_3)		
Phosphatic limestone .. .. about	1	9
Red *Callavia* Sandstone (Raw 1936, p. 238) (Ac_2)		
Nodules of red or purplish, micaceous and glauconitic calcareous sandstone .. .. about	2	6
Green *Callavia* Sandstone (Raw 1936, p. 238) (Ac_1)		
Bright green glauconitic sandstone		

The Lower Comley Limestones were still poorly seen by re-excavating the old trench (No. 2 in Cobbold 1927a, p. 553) 200 yd south of Comley Quarry. Full details of this section are given by Cobbold (1933, p. 475). The faunas of the Lower Comley Limestones have been described by Cobbold (1910, 1921, 1931, 1936) and Raw (1936).

UPPER COMLEY SERIES

The Upper Comley Series is not seen on the south-eastern end of The Lawley, as it is faulted out. The lower part of the Series is seen in Comley Quarry [48459647] where about 20 ft of coarse pebbly grits form the eastern face. These rocks, the *Paradoxides groomi* Grits, are the 'Quarry Ridge Grits' of Cobbold's earlier papers. The lowest bed of the Series in Comley Quarry is a dark phosphatic layer with calcite, which contains Middle Cambrian trilobites including *Paradoxides spp.* and *Kootenia* [*Dorypyge*] *lakei* (Cobbold). These grits are well exposed on the ridge south of the quarry, where they dip eastwards at 70° to 85°.

At the southern end of a second ridge [485959], 660 yd S. 4° E. of Comley Quarry, an exposure for roadstone was described by Stubblefield (*in* Whittard and others 1953, pp. 236–7). This shows the Comley Breccia Bed ('Bb_1' of Cobbold's notation), consisting of blocks of '*Eodiscus*' *bellimarginatus* Limestone (Ac_3) and fragments of soft green glauconitic sandstone (Ac_1), embedded in a glauconitic gritty matrix with moulds of Middle Cambrian fossils and occasional derived Lower Cambrian fossils. The fauna comprises the Middle Cambrian fossils *Bailiella cobboldi* Resser and *Paradoxides sp.*, and the following derived from the Lower Cambrian: *Obolella atlantica transversa* Cobbold, *Obolus parvulus* Cobbold, *Callavia callavei* (Lapworth),

Hebediscus attleborensis (Shaler and Foerste), *Serrodiscus* [*Eodiscus*] *bellimarginatus* (Shaler and Foerste) and *Strenuella pustulata* (Cobbold).

Other exposures of the Series occur in the lane [486963] 230 yd E. 35° S. of Comley Quarry, on the hill to the south, and also in the stream 400 yd east of the quarry, where a pebble bed is seen. Beds in the upper part of the Series are seen in the road 400 yd E. 5° N. of Comley Quarry (Cobbold 1927a, p. 553. Excavations 20 and 21).

The Quarry Ridge Grits are seen, though indifferently, on Robin's Tump, [483955] 650 yd E. 11° N. of the summit of Caer Caradoc Hill, where they dip eastwards at 40° and rest on the Lower Comley Sandstone. Excavations (Cobbold 1911, pp. 117–9; 1912, pp. 112–4. Excavation 29) at the summit of this hill showed that the Upper Comley Series rests unconformably on the Lower Comley Sandstone and that the Lower Comley Limestones are absent.

The southernmost outcrop of the Series, in Cwms, though established without doubt by Cobbold (1927a, p. 554. Excavation 54), can now only be followed by the feature it makes. G.H.M.

The Geological Survey inclined borehole [48899645] at Shootrough proved 118 ft 1 in (depth 121 ft 5 in to 239 ft 6 in, estimated true thickness 104 ft 6 in) of Middle Cambrian beds below the *Orusia* Shales, the sequence being as follows (fossils detailed below):

	Thickness in borehole ft in	Depth ft in
Billingsella Beds (Bc)		
Strongly sheared very calcareous grey sandstones with scattered glauconite grains	4 3	125 8
Faulted ground (sheared sandstone)	4 11	130 7
? Hill House Shales (Bb, probably Bb_2)		
Sandstones and shales	96 7½	227 2½
? Comley Breccia Bed (Bb_1)		
Very coarse-grained pale grey sandstone with angular fragments of dark greenish grey sandstone and light grey limestone	12 3½	239 6

The Breccia Bed rested unconformably on the Lower Comley Sandstone, this unconformity cutting out the lower part of the Middle Cambrian sequence and the Lower Comley Limestones. B.A.H.

The *Billingsella* Beds contain many specimens of *Acrotreta* (s.l.), including *A. schmalenseei* Walcott, *A. schmalenseei matleyi* Cobbold and *A. sagittalis* (Salter), together with *Acrothele coriacea* Linnarsson and *Obolus linnarssoni* Cobbold, the last being sufficiently well preserved to show that Cobbold was correct in assigning the species to *Obolus*.

The sandstones and shales (? Hill House Shales) below the *Billingsella* Beds have yielded *Hyolithellus micans* Billings, *Acrotreta socialis* von Seebach, *Lingulella ferruginea* Salter, *Hyolithes socialis* Linnarsson, *Paradoxides intermedius* Cobbold, *Parasolenopleura dubia* (Cobbold), *Beyrichona sp.*, *Hipponicharion* aff. *confluens* Ulrich and Bassler, *Indiana sp.* and *Hymenocaris? sp.*

These fossils indicate that the beds are middle Middle Cambrian but the precise age is uncertain; Cobbold (1921, table facing p. 326) suggested that horizons Bb_1 and Bb_2 correspond to the Zone of *Paradoxides hicksii*. *Paradoxides intermedius*, which recalls the early Middle Cambrian species *P. eteminicus* Matthew and *P. insularis* Westergård, was also recorded from the base of the Menevian Series (= base of the Zone of *P. hicksii*) in the St. David's district, Pembrokeshire, by Lake (1935, p. 202); although the evidence is far from conclusive it suggests that Cobbold's correlation is approximately correct.

Derived fragments found within the shales contain fossils indicative of the horizons Ba, Ad? and Ac, as follows:

Depth			Supposed horizon
171 ft	*Acrothele* (*Redlichella*) *granulata* Linnarsson		Ba
218 ft 8 in	*Protospongia?* *Hyolithellus micans* Billings *Coleoloides?* *Acrothyra* or *Acrotreta* sp. Tribolite fragments	in very dark brown glauconitic and phosphatic limestone	Ad?
226 ft 2 in-226 ft 4 in	*Rhombocorniculum* [*Helenia*] *cancellatum* (Cobbold). (Apparently identical with *Rhombocorniculum comleyense* Walliser, 1958). *Callavia sp.*	in greyish green calcareous sandstone	Ac_2-Ac_4

The breccia from the borehole yielded the following derived fossils:

c. 228 ft	Trilobite and brachiopod fragments in a pale limestone fragment		Ac_3-Ac_5
c. 237 ft	*Micromitra?* (s.l.) *sp.* Trilobite fragment	in soft greyish green sandstone	Ac_1?
238 ft 3 in	*Hyolithellus micans* in fine-grained green sandstone		Ab_4?
239 ft 3 in-239 ft 5 in	*Rhombocorniculum cancellatum* cf. *Lingulella viridis* Cobbold cf. *Micromitra* (*Paterina*) *labradorica* (Billings) *Callavia sp.* (fragment)	green very calcareous sandstone	Ac_1-Ac_2

Besides these, the following fossils occur in the breccia but are not obviously in derived rock fragments: *Rhombocorniculum* [*Helenia*] *cancellatum*, *Hyolithellus micans*, *Micromitra?* and *Callavia sp.* (fragments). Both the matrix and the phenoclasts vary in composition, but the bulk of the breccia consists of coarse glauconitic sandstone with many derived fragments of fine-grained green sandstone. In places the matrix is of finer composition, in which cases it compares well with the green sandstone of the phenoclasts. Some phenoclasts consist of limestone or calcareous sandstone and their lithology and fossils suggest derivation from horizons Ab to Ac_3 or higher, indicating that, at the earliest, the erosion which formed the breccia was later than the period of formation of the '*Eodiscus*' *bellimarginatus* Limestone (Ac_3). From its position in the sequence, the breccia should be assigned to the *Paradoxides intermedius* Grit (Bb_1), but at the type locality that formation contains a Middle Cambrian fauna in the matrix. In the borehole, the matrix of the breccia yielded no exclusively Middle Cambrian fossils and it appears to contain Lower Cambrian forms, particularly *Rhombocorniculum cancellatum*. However, as Cobbold (1913, p. 41) has pointed out, it may be very difficult to distinguish between matrix and phenoclasts in the *P. intermedius* Grit, and accordingly these fossils may be taken as derived. (Compare also the occurrence described by Stubblefield (*in* Whittard and others 1953, p. 237) of the Comley Breccia Bed in which Lower Cambrian fossils appear to be isolated in a Middle Cambrian matrix.) A.W.A.R.

UPPER CAMBRIAN

The soft, easily weathered shales of the Upper Cambrian form lower ground, in which they are seldom exposed. Consequently the Shineton Shales (Tremadoc Series) and Grey (*Orusia*) Shales (Dolgelly Beds) have been mapped as one unit. Although the Black Shales of the Dolgelly Beds have been recorded (Stubblefield 1930) from Bentleyford Brook, 2½ miles north-east of Comley, in the Shrewsbury (152) Sheet, they have not been seen in the Comley area. Evidence from an excavation [489967] 380 yd N. 31° W. of Shootrough (Cobbold 1927a, p. 555. Excavation 62) where the Shineton Shales succeeded the *Orusia* Shales, suggests that the absence of the Black Shales there may be due to faulting.

The outcrop of the Upper Cambrian shales extends through the gap between The Lawley and Hoar Edge, and southwards along the valley to the west of The Wilderness ridge. Isolated exposures of striped grey silty shale are seen in the wood north of Shootrough. A section [489967] in the stream in this wood, 370 yd N. 24° W. of Shootrough, has yielded *Lingulella nicholsoni* Callaway and *Dictyonema flabelliforme* (Eichwald); *Acrotreta sp.* and *Adelograptus* cf. *hunnebergensis* (Moberg) were obtained from another section 80 yd upstream, both faunas being indicative of the Shineton Shales.

Other exposures of micaceous grey shale with striped siltstone bands, folded into zig-zags, are seen in the stream 250 to 350 yd north-north-east of Comley Quarry. Calcareous concretions in these beds have yielded a fauna including *Orusia lenticularis* (Wahlenberg), *Acrothele* cf. *coriacea*, *Parabolina spinulosa* (Wahlenberg), cf. *Parabolinoides bucephalus* (Belt) and *Lingulella* (Cobbold 1927a, pp. 556–7; Stubblefield *in* Whittard and others 1953, p. 236), indicative of the *Orusia* Shales. Contorted *Orusia* Shales are seen in the stream 300 yd E. 15° N. of Comley Quarry, and there are other poor exposures in the stream [486958] 1050 yd E. 19° S. of Botvyle. G.H.M.

The Geological Survey borehole [48899645] at Shootrough proved 94 ft 11 in (true thickness not determinable) of Grey (*Orusia*) Shales resting with non-sequence on the Middle Cambrian. The beds were dark grey micaceous silty shales with many thin calcareous layers, and were strongly sheared and slickensided. The Shales contained *Orusia lenticularis*, cf. *Lingulella lepis* Salter and trilobite (olenid?) fragments.
B.A.H., A.W.A.R.

Hill End Area

The Wrekin Quartzite rests on Uriconian quartz-porphyry, and forms a prominent strike ridge extending for a quarter of a mile to the south-west from an old quarry [507945] 540 yd S. 42° W. of Mill Farm, Cardington. Near the Stoneacton–Cardington road the strike swings to east-west, almost parallel with the Sharpstones Thrust which terminates the outcrop to the south. Dips range from 40° to 50°, to the south-east on Hill End and a little west of south on the Sharpstones [502942]. The Quartzite forms crags of white rock, in places sugary and commonly stained yellow or brown. The beds are slightly coarser grained towards the base and some less well-cemented bands are sandy in appearance. The rock often appears brecciated when weathered owing to the irregular jointing. In the old quarry at the north-eastern end of the outcrop, where about 60 ft of beds are exposed, there are horizontal slickensides striking S. 20° W. on a surface inclined eastwards at about 50°. Along the ridge the principal joints dip 25° to N. 25° E. In an outcrop [505942] 950 yd S. 42° W. of Mill Farm there is a layer of brecciated quartzite 4 in thick on a vertical joint striking W. 20° N. G.H.M.

Between the outcrops of the Wrekin Quartzite on Hill End and the Shineton Shales at the eastern end of the area there are no natural exposures. Fragments of Lower Comley Sandstone are seen in several places immediately to the east of the Quartzite outcrop, but most of the intervening ground is obscured by drift. Blocks of quartzite are abundant, probably derived by solifluxion from the outcrop on Hill End.

A number of excavations were made in this unexposed ground by Dr. J. W. Cowie of Bristol University. These have shown that the Lower Comley Limestones are almost certainly absent. One of his excavations [508942], 740 yd E. 39° S. of The Villa, Cardington, showed dark grey and black shales, and weathered and disturbed black shales were revealed in an excavation made by the Geological Survey 860 yd E. 18° S.

of The Villa. These results may indicate the presence of the Black Shales of the Dolgelly Beds, although neither excavation yielded any definite palaeontological evidence. A programme of augering, together with the results from Dr. Cowie's excavations, has resulted in the more accurate delimitation of the outcrops of the Lower Comley Series, Upper Comley Series and Upper Cambrian shales. B.A.H.

The Shineton Shales are exposed at intervals in the brook for about half a mile south of the Mill, Cardington [512945]. They are mainly green or greyish green shales dipping to the east-south-east at 20° to 45°. Stubblefield and Bulman (1927, p. 116) stated that these sections have yielded *Acrotreta sabrinae* (Callaway), *Lingulella nicholsoni* Callaway, *Lingulella sp.*, *Shumardia curta* Stubblefield, *Clonograptus tenellus* (Linnarsson) and *Dictyonema flabelliforme*. In addition to these, three new trilobite species, *Asaphoon pithogastron*, *Dichelepyge phylax* and *Myindella crux*, from one of these sections, have been described by Hutchison and Ingham (1967). An abortive attempt to sink a shaft, presumably for coal, was made in 1911 on the west bank of the brook [512944], 470 yd S.S.E. of Mill Farm, Cardington, but the tipped material from the sinking is now badly weathered and overgrown. Traces of shale with *Dictyonema sp.* were seen [509941] 850 yd E. 37° S. of The Villa, and further poor exposures were seen 250–300 yd to the south-east. G.H.M., B.A.H.

References

COBBOLD, E. S. 1910. On some small Trilobites from the Cambrian Rocks of Comley (Shropshire). *Quart. J. Geol. Soc.*, **66**, 19–50.

—— 1911. Third Report on Excavations among the Cambrian Rocks of Comley, Shropshire, 1909. *Rep. Brit. Assoc.* for 1910, 113–22.

—— 1912. Fourth Report on Excavations among the Cambrian Rocks of Comley, Shropshire, 1910. *Rep. Brit. Assoc.* for 1911, 111–5.

—— 1913. The Trilobite Fauna of the Comley Breccia-Bed (Shropshire). *Quart. J. Geol. Soc.*, **69**, 27–44.

—— 1916. Sixth Report on Excavations among the Cambrian Rocks of Comley, Shropshire (1912, 1913, 1914). *Rep. Brit. Assoc.* for 1915, 117–23.

—— 1921. The Cambrian Horizons of Comley (Shropshire) and their Brachiopoda, Pteropoda, Gasteropoda etc. *Quart. J. Geol. Soc.*, **76**, for 1920, 325–86.

—— 1927a. The Stratigraphy and Geological Structure of the Cambrian Area of Comley (Shropshire). *Quart. J. Geol. Soc.*, **83**, 551–73.

—— 1927b. Seventh Report on Excavations among the Cambrian Rocks of Comley, Shropshire. *Rep. Brit. Assoc.* for 1927, 275–6.

—— 1931. Additional Fossils from the Cambrian Rocks of Comley, Shropshire. *Quart. J. Geol. Soc.*, **87**, 459–512.

—— 1933. Notes on Comley Quarry, near Church Stretton, Shropshire. *Rep. Brit. Assoc.* for 1933, 473–6.

—— 1936. The Conchostraca of the Cambrian Area of Comley, Shropshire, with a note on a new Variety of *Atops reticulatus* Walcott. *Quart. J. Geol. Soc.*, **92**, 221–35.

HUTCHISON, R. and INGHAM, J. K. 1967. New trilobites from the Tremadoc Series of Shropshire. *Palaeontology*, **10**, 47–59, pl. 8.

LAKE, P. 1935. A Monograph of the British Cambrian Trilobites, pt. 9. *Palaeont. Soc.*, 197–224.

LAPWORTH, C. 1888. On the discovery of the *Olenellus* Fauna in the Lower Cambrian Rccks of Britain. *Geol. Mag.* (3), **5**, 485.

POCOCK, R. W., WHITEHEAD, T. H., WEDD, C. B. and ROBERTSON, T. with contributions by WRAY, D. A., STUBBLEFIELD, C. J., CANTRILL, T. C. and DAVIES, W. M. 1938. Shrewsbury District. *Mem. Geol. Surv.*

RAW, F. 1936. Mesonacidae of Comley in Shropshire, with a discussion of classification within the family. *Quart. J. Geol. Soc.*, **92**, 236–93.

STUBBLEFIELD, C. J. 1930. A new Upper Cambrian Section in South Shropshire. *Summ. Prog. Geol. Surv.* for 1929, pt. 2, 54–62.

—— 1958. *Lexique Stratigraphique International.* **1**, fasc. 3a III, Cambrien, Paris.

—— and BULMAN, O. M. B. 1927. The Shineton Shales of the Wrekin District; with notes on their development in other parts of Shropshire and Herefordshire. *Quart. J. Geol. Soc.*, **83**, 96–146.

WALLISER, O. H. 1958. *Rhombocorniculum comleyense* n. gen., n. sp. (Incertae sedis, Unterkambrium, Shropshire). *Paläont. Z.*, **32**, 176–180.

WHITTARD, W. F. with contributions by BALL, H. W., BLYTH, F. G. H., DINELEY, D. L., JAMES, J. H., MITCHELL, G. H., POCOCK, R. W. and STUBBLEFIELD, C. J. 1953. Report of Summer Field Meeting in South Shropshire. 1952. *Proc. Geol. Assoc.*, **64**, 232–50.

Chapter 5

ORDOVICIAN

GENERAL ACCOUNT

THE ORDOVICIAN ROCKS of the Church Stretton district form a narrow outcrop within and east of the Eastern Uriconian axis, extending north-north-eastwards from the neighbourhood of Coston [3880][1], where they are overstepped by Silurian rocks, to the northern limit of the area north of Cardington. The outcrop is broken by the Uriconian and Cambrian rocks of Hope Bowdler and Cardington hills. The rocks belong entirely to the Caradoc Series, a formation which takes its name from the outcrops which lie east of Caer Caradoc Hill, and it is in this area, centred on the village of Cardington, that they reach their maximum thickness, some 2700 ft. In addition the Onny Shales, absent at Cardington, are present to a thickness of about 400 ft near Henley. In their most westerly outcrops the rocks generally dip steeply towards the east or south-east and are in places overturned towards the west. Eastwards the dips decrease, usually abruptly, to angles of 10 to 20 degrees.

The classification which has been adopted in the preparation of the map is as follows:

CARADOC SERIES

		Approx. thickness in ft
Onny Shales: pale olive-green and grey micaceous mudstone		up to 400
Acton Scott Group: green and grey micaceous mudstone, often calcareous; impure limestones and sandstones locally important		200 to 500
Cheney Longville Flags: green and yellowish, flaggy, fine-grained sandstones and shales; some thin shelly limestones, forming the *alternata* Limestone, in basal 100 ft or so		350 to 850
Chatwall Sandstone: greenish brown, purple-banded, fine-grained sandstone, mainly massive; thin shelly limestones, especially common near top	Mapped together on one-inch sheet	120 to 500
Chatwall Flags: greenish brown and buff, micaceous flaggy, fine-grained sandstone; some shale bands; crinoid ossicles often abundant		100 to 300
Harnage Shales: grey and green micaceous mudstone, often shaly; thin sandstones; pebbly sandstone locally at base; a thin lava flow locally		300 to 1000
Hoar Edge Grit: buff coarse-grained sandstone and conglomerate, some shell bands; thin limestones locally abundant		0 to 400

Lateral variations in thickness are illustrated in Fig. 12.

[1]Such numbers are four figure National Grid References within the 100-km square SO (32). They refer to general areas rather than to specific localities.

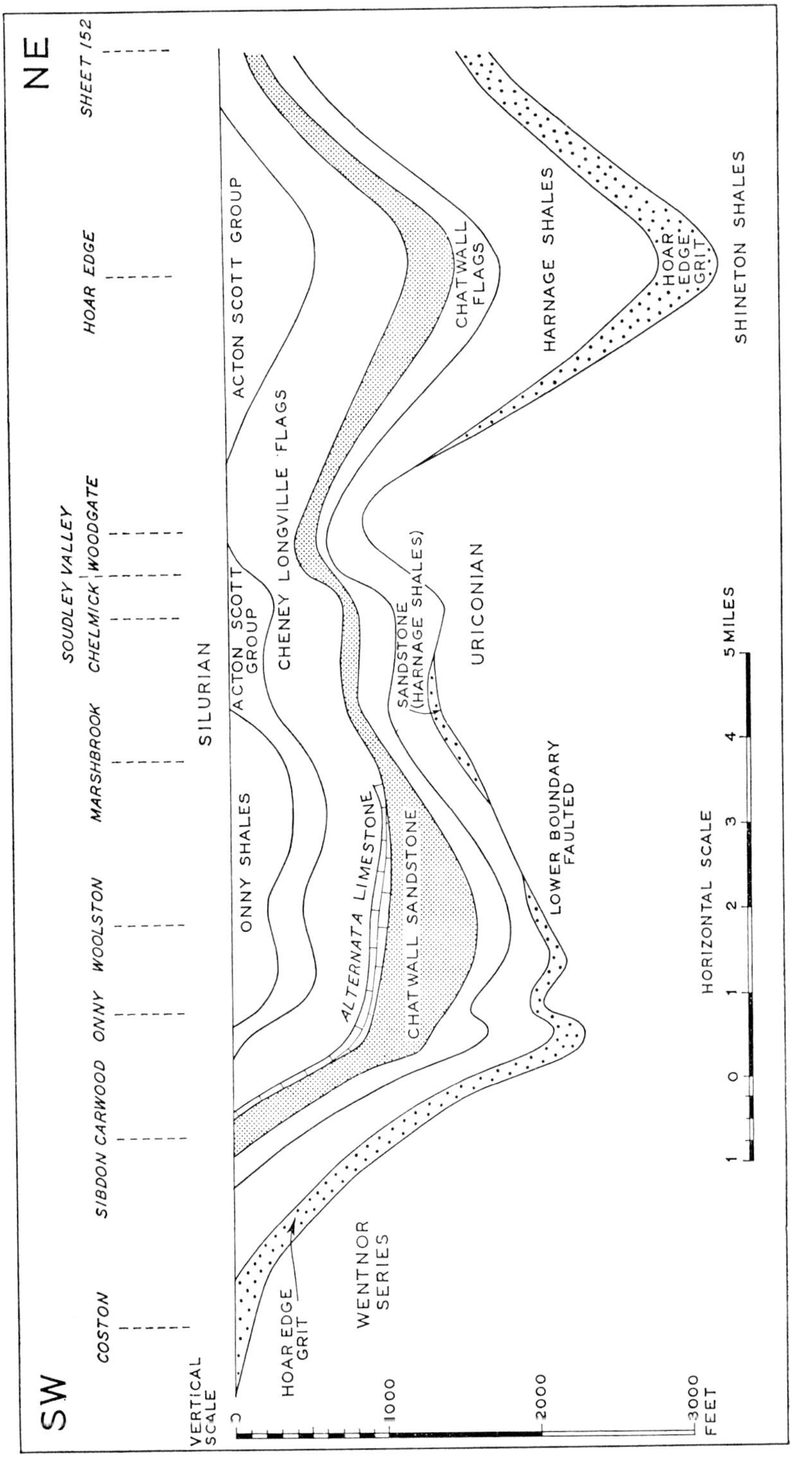

FIG. 12. *Variations in thickness of the formations of the Caradoc Series*

The Series constitutes a succession of generally calcareous, shallow-water deposits, with an abundant fauna consisting predominantly of brachiopods and trilobites. In these respects it contrasts markedly with the thicker, deeper-water deposits of the Shelve area, some 10 miles to the west, where a mixed fauna of graptolites and shelly fossils is prevalent. An additional difference is that in the Shelve area a complete succession of pre-Caradoc Ordovician rocks is also developed. The difference between the Ordovician histories of the two areas is attributed to the influence of the Church Stretton Fault Complex, which formed the hinge-like boundary between the stable shelf area to the east and the more continuously subsiding basin to the west. No indication has been observed of an approach towards the eastern shore line.

The historical development of the classification of the Caradoc Series has been described in detail by Dean (1958, pp. 193–4; 1964, pp. 260–2) and Whittard (1960, pp. 22–4). The recognition by Sedgwick (1853) of a faunal distinction between the upper and lower beds of Murchison's Caradoc Sandstone (1839, pp. 216–22), and the realization by Ramsay (1853) and Salter and Aveline (1854) that the two divisions were separated by an unconformity, were important steps in the controversy which was developing between Sedgwick and Murchison as to the stratigraphical extent of the Cambrian and Silurian systems. To Sedgwick and his supporters the break within the Caradoc Sandstone was the most significant in the whole Palaeozoic succession. As Sedgwick (1853, p. 229) pointed out, only the upper division is 'true Silurian' and is now classified as Upper Llandovery. The lower, now the Caradoc Series, is locally the uppermost formation of Lapworth's (1879) Ordovician System.

The lithological classification adopted during the recent survey follows in outline that used in the Shrewsbury district (Pocock and others 1938) and introduced for the Caradoc area by Lapworth (1916, p. 38, fig. 3). Many of the lithological units are diachronous, but in so far as they are units they are considered to constitute a classification which is useful to the layman and to the geologist in the field. The relationship between this and the palaeontological classification is considered below (p. 105).

The graptolite zones recognized in the Caradoc Series of the Anglo-Welsh area are in ascending order, *Nemagraptus gracilis*, *Diplograptus multidens*, *Dicranograptus clingani* and *Pleurograptus linearis*. The lowest of these was shown to be present in Shropshire by the identification by Stubblefield (1930, p. 87) of the eponymous fossil in collections from the Hoar Edge Grit. According to Dean (1958, pp. 229–30) the shelly faunas and the few graptolites of the highest Caradoc formations in Shropshire indicate that they belong to the *D. clingani* Zone, the overlying zone being absent at the surface but possibly occurring beneath the Silurian rocks to the east.

The palaeontological classification of the Caradoc Series of the Caradoc area is based on the work of B. B. Bancroft. In a series of papers published between 1928 and 1949 he developed a system dependent on his recognition of a succession of distinctive brachiopod-trilobite faunas. His final classification divided the Series into seven stages, to which geographical names were applied and which were related to lithological divisions. Bancroft's work has been amplified by Dean (1958) and no important change in classification has been found to be necessary, but a new correlation between the shelly and graptolitic faunas has been worked out. In particular Dean (1958 and

TABLE 6

Graptolite zone	Stage	Lithological Succession: Geological Survey	Lapworth (1916)	Lapworth (1916)	Dean (1958): Coston–Horderley	Dean (1958): Marshbrook–Henley	Dean (1958): Soudley–Wallsbank	Dean (1958): Cwms–Gretton
Dicranograptus clingani	Onnian	Onny Shales	Upper *Trinucleus* Beds	Acton Group	Upper Llandovery ~~~ *Onnia* Beds	Upper Llandovery ~~~ ? *Onnia* Beds	Upper Llandovery ~~~	Upper Llandovery ~~~ ? *Onnia* Beds
	Actonian	Acton Scott Group	Acton Calcareous Beds		Grey mudstones and limestones	Acton Scott Beds *sensu stricto*	Grey mudstones of Ticklerton	Yellow sandstones = Acton Scott Beds
	Marshbrookian	Cheney Longville Flags	Lower *Trinucleus* Shales - - - Birrels Wood Flags	Longville Group	Upper Cheney Longville Flags	Upper Cheney Longville Flags	Upper Cheney Longville Flags	Cheney Longville Flags
	Upper Longvillian		Chelmick Flags and Shales		Lower Cheney Longville Flags	Lower Cheney Longville Flags	Lower Cheney Longville Flags	
		alternata Limestone	*alternata* Limestone	Chatwall Group	*alternata* Limestone	*alternata* Limestone ~~~	*alternata* Limestone ~~~	*alternata* Limestone —?—
	Lower Longvillian	Chatwall Sandstone	Upper Chatwall Sandstone		Upper and Middle Horderley Sandstone	Probable break ~~~ Horderley Sandstone (part)	Faunal break ~~~	Upper Chatwall Sandstone
Diplograptus multidens	Soudleyan	Chatwall Flags	Lower Chatwall Sandstone or *Glyptocrinus* Flags		Lower Horderley Sandstone	Not seen owing to faulting	Soudley Sandstone —?—	Lower Chatwall Sandstone —?—
	Harnagian	Harnage Shales	Harnage Shales	Harnage Group	Glenburrell Beds		Harnage Shales and Neptunian dykes ~~~	Shales
			Transition Bed	Hoar Edge Group	Smeathen Wood Beds			'Hoar Edge Grits' of the Cwms ~~~
Nemagraptus gracilis	Costonian	Hoar Edge Grit ~~~	Upper Hoar Edge Sandstone - - - Hoar Edge Limestone - - - Lower Sandstone and Conglomerate ~~~	~~~	Coston beds including basal conglomerates ~~~			

1960a) has shown that north of the Onny valley there is a non-sequence between the *alternata* Limestone and the underlying Chatwall Sandstone, by whatever local name it is known, and, at Chatwall, between the Sandstone and the Chatwall Flags. No major diachronism is observed in the beds above the Chatwall Sandstone, but this formation occupies a lower horizon at Soudley than elsewhere, and the base of the underlying Chatwall Flags is at a lower horizon in the north-east than in the south-west. The basal Ordovician rocks in the Hope Bowdler area are younger than elsewhere and are strongly diachronous over an irregular floor of Pre-Cambrian and Cambrian rocks. With the provisos given above Table 6 relates the graptolite and brachiopod-trilobite classification to Dean's lithological units, to those employed in this Memoir and to Lapworth's (1916) classification.

Hoar Edge Grit

The outcrop of the Hoar Edge Grit is in two main parts. The more southerly extends from New House [383796][1], near Coston, to west of Woolston [420873], the more northerly from Cwms Farm [475939] to the northern limit of the district on Hoar Edge [507977]. At Hoar Edge it attains its maximum thickness of 400 ft. In the intervening ground the oldest Ordovician rocks exposed belong to the Harnage Shales. In the south the Grit rests on, or is faulted against, Pre-Cambrian rocks lying to the west, except that in the extreme south it transgresses them to lie adjacent to Ludlow rocks, from which it is separated by the Church Stretton Fault. To the east it is succeeded by the Harnage Shales, except south of Aston on Clun where the overstepping Wenlock Shales rest directly on it. The Grit in its turn is completely overstepped by the Wenlock to the south of New House. This southern outcrop is itself broken by a gap of about 1½ miles between Oldfield [406831] and the wood east of Upper Carwood [406854], within which the Grit is cut out by a fault. At the northern end of the gap a small down-faulted outlier lies within the Pre-Cambrian about 250 yd west of the main Grit outcrop. Another outlier, apparently completely fault-bounded, forms high ground at Brokenstones [4288].

The failure of the Grit outcrop north-west of Woolston is shown on the map as being due to faulting but the evidence is inconclusive. North of Brokenstones much of the lower Caradoc succession is cut out by faulting, but on Hazler Hill [4692] and nearby it has been shown (Strachan and others 1948) that rocks of Harnagian age rest unconformably on the Pre-Cambrian. Thus the outcrop failure at Woolston may be due to Harnagian overlap.

A much-faulted outcrop of Hoar Edge Grit lies on the eastern slopes of Caer Caradoc Hill, extending for about 1¼ miles to the north-north-east from Cwms Farm. The Grit rests on Cambrian rocks or is bounded to the west by the Lawley Fault. It continues on the eastern limb of a minor syncline to form a long narrow outcrop at The Wilderness [4895], immediately west of the Cwms–Hoar Edge Fault. The outcrop continues again, on the eastern side of this fault, from Shootrough [490964] to the northern limit of the district, the Grit resting unconformably on Tremadoc shales. D.C.G.

[1]Such figures are National Grid References within the 100-km square SO (32).

In the Coston area the Grit has been divided by Dean (1958, p. 197) as follows:

Costonia ultima Beds	over 35 ft
Harknessella Beds	over 25 ft
Basal Conglomerates	over 45 ft

The total thickness of the Grit may reach 200 ft but structural complications and intermittence of exposures do not permit an accurate measurement.

The Basal Conglomerates consist of white and buff quartz-conglomerates with occasional pebbles of pinkish fine-grained (Uriconian) volcanic rocks. The pebbles are rounded and generally less than ½ inch in diameter; the matrix is a coarse quartz-grit. This division is not recognized north of the Coston area. There is a passage up into the *Harknessella* Beds which are brown coarse-grained sandstones with rare quartz pebbles and local thin bands of quartz-conglomerate. Lenticular fossil bands contain abundant brachiopods, particularly *Harknessella vespertilio* (J. de C. Sowerby) and *Heterorthis patera* (Davidson). In the Onny valley beds which appear to belong to this division are highly calcareous and include many thin beds of shelly limestone. The *Costonia ultima* Beds are rather more evenly bedded medium- and coarse-grained sandstones, with scattered quartz pebbles and an abundant fauna of brachiopods and trilobites. They are doubtfully recognized in the Onny valley quarries and are also distinguishable at Brokenstones (Dean 1958, p. 198).

D.C.G., B.A.H.

In the northern outcrop the Hoar Edge Grit is a coarse sandstone with well-rounded sand grains, pebbly at the base, and usually pale grey or even white. Some beds are feldspathic with either white or pink grains of feldspar. Most of the pebbles, as in the southern outcrop, are composed of quartz, but igneous fragments, probably of Uriconian derivation, have been observed. G.H.M.

In the Harnage–Cressage area (Shrewsbury Sheet 152) Dean (1958, pp. 198–9) recognized a three-fold subdivision of the Grit:

Rhynchonellid Grits
Sandy limestones with *Harknessella subquadrata*
Sandy shales and limestones with *Harknessella subplicata*

He pointed out that the trilobite faunas indicate a correlation of the two lower divisions here with the two upper divisions at Coston, that the Rhynchonellid Grits are Harnagian, and that the Hoar Edge Grit in the Cwms area includes these beds and the underlying beds with *H. subquadrata*. The fauna of the latter beds shows certain Harnagian elements, which led Dean to the conclusion that no important break occurs at about this level in the Caradoc Series, as postulated by Robertson (*in* Pocock and others 1938, p. 86). D.C.G.

HOAR EDGE GRIT: PALAEONTOLOGY

Although the lowest strata of this subdivision in the Horderley–Coston area comprise unfossiliferous conglomerates, those of the middle portion contain numerous lenticular, originally calcareous, bands made up essentially of large brachiopod shells generally found weathered and in the form of internal and external moulds. One of the most common species is *Harknessella vespertilio* (J. de C. Sowerby), but *Dinorthis flabellulum* (J. de C. Sowerby) and *Heterorthis patera* (Davidson) are also locally abundant. Trilobite remains are rare in these beds but in the overlying subdivision known sometimes as the *Costonia ultima* Beds (Dean 1958, p. 197), they are more common, the eponymous trilobite being of some stratigraphical value. *Costonia*, represented by a different species, is known also from the corresponding horizon farther north-east, in the neighbourhood of Evenwood (Shrewsbury Sheet 152), and *C. ultima* has been found in the Spy Wood Grit of the Shelve Inlier (Whittard 1966, p. 281). The remainder of the trilobite fauna shows little variety but includes *Flexicalymene* cf. *acantha* Bancroft, a species more abundant in the basal portion of the succeeding Harnage Shales. The brachiopods are more varied and include, among others, *Horderleyella plicata* Bancroft and *Smeathenella strophomenoides* Bancroft with, less commonly, *Salopia salteri* (Davidson) and the characteristically ribbed plectorthid genus *Macewanella*. W.T.D.

HARNAGE SHALES

The Harnage Shales form an outcrop which extends from Aston on Clun [3981] to east of Hoar Edge [5097], interrupted by faulting between Brokenstones [4288] and Ragdon [4591] and again between Wall Bank [5093] and Willstone Hill [4894]. Small outcrops, much disturbed by faults, occur in the ground south-west of Ragleth Hill [4592]. Normally they lie to the east of the Hoar Edge Grit, but between Oldfield and Upper Carwood, and again just south of Brokenstones, they are faulted against Pre-Cambrian rocks (p. 90). From Ragleth Hill to Wall Bank, Harnage Shales, commonly arenaceous or rudaceous, succeed the Pre-Cambrian unconformably. The Shales are in general poorly exposed and commonly form a vale between the Hoar Edge Grit and the Chatwall Flags. This is well seen in the north of the area where they crop out between the scarps of Hoar Edge and Yell Bank. D.C.G.

In the Sibdon Carwood area the Harnage Shales appear to be about 500 ft thick, but some of the beds may be repeated by faulting. They are flaggy, micaceous, silty mudstones, locally shaly. Their colour is generally buff but orange staining is common along joint and cleavage surfaces and bedding planes. The junction with the Hoar Edge Grit is not exposed. Beds of siltstone and fine-grained sandstone occur towards the top of the Shales, and there is a gradual upward passage into the Chatwall Flags. A thin lava flow in this area is the only definite volcanic rock in the Caradoc Series of the shelf facies in South Shropshire (p. 121). B.A.H.

Between Long Lane and Brokenstones the lower beds are soft, yellowish and green, micaceous shales, often blocky, and the upper mainly blocky with interbedded flags and thin sandstones. Orange staining is again quite common. Two lithological divisions have been recognized by Dean (1958, pp. 201–2) in the Onny valley, the Smeathen Wood Beds below and, following Bancroft (1933), the Glenburrell Beds above. The lower beds of the latter are mudstones and shales like the beds below them, and it is only at their highest levels, of Soudleyan age, that beds of sandstone occur, leading by degrees into the Chatwall Flags. The thickness of the Harnage Shales in the Onny valley is between 400 and 450 ft. West of Woolston it appears to be only about 150 ft, possibly due to facies change in the lower and/or upper beds of the Onny succession. D.C.G.

In the Ragleth Hill–Wall Bank area the Harnage Shales consist largely of grey and brown shales and mudstones. Calcareous pebbly grits and sandstones commonly occur at the base and occasional thin bands of grit or flags occur at higher levels. Basal deposits include neptunian dykes in the Uriconian on Hazler Hill (p. 17). The thickness of the Shales varies between about 200 and 400 ft. North of Willstone Hill the Shales comprise grey or greenish mudstones, commonly shaly, with bands of sandstone and more calcareous bands. Between Hoar Edge and Yell Bank they appear to be about 1000 ft thick. G.H.M.

HARNAGE SHALES: PALAEONTOLOGY

Although the succession in the type area (in the vicinity of Evenwood, on the adjacent Shrewsbury (152) Sheet) consists mainly of grey shales, the succession farther south is more variable and includes also conglomerates and mudstones as well as the well-known neptunian dykes of the Church Stretton hills. The faunas are correspondingly variable but those of the strata immediately overlying the Hoar Edge Grit in the Horderley district include the diagnostic Trinucleid *Reuscholithus reuschi* Bancroft, uncommon here but fairly abundant at the same horizon farther north, for example at Coundmoor Brook (Shrewsbury Sheet 152). The other trilobites at this level show a greater variety than those of the Hoar Edge Grit and include a genus described from Bohemia, *Kloucekia*, together with *Flexicalymene acantha* Bancroft, *Proetidella fearnsidesi* Bancroft and *Acanthoparypha stubblefieldi* (Bancroft), the last-named genus occurring also in certain North American Middle Ordovician faunas.

The brachiopod faunas of the Harnage Shales exhibit, on the one hand, certain resemblances to those of the topmost Hoar Edge Grit, and on the other, affinities with certain extra-British faunas, particularly those of Bohemia and the Baltic region. *Heterorthis*, *Smeathenella*, *Sowerbyella* and *Salopia* represent genera already occurring in the Hoar Edge Grit, but it is in the lowest Harnage Shales that the last-named genus attains its greatest abundance. Plates of cystids and machaeridians (*Lepidocoleus*) are not uncommon in the Harnage Shales, particularly the lowest beds, whilst the ostracods *Primitia*, *Tallinnella* and *Ulrichia* are often extremely abundant. In this formation the genus *Tallinnella* includes species which were formerly referred to *Tetradella*. The members of these groups underwent little modification throughout much of Ordovician time and have not yet proved to be of zonal value.

Slightly higher in the succession the strata comprise mostly shales and mudstones in which Trinucleid trilobites are the only common fossils. *Salterolithus caractaci* (Murchison) appears, locally in sufficient profusion to have been used as a zonal index, but it does range upwards into what are here interpreted as the topmost Harnage Shales. The name Glenburrell Beds has been applied to these strata and they are well defined faunally by the Trinucleids *Broeggerolithus broeggeri* (Bancroft) and *S. caractaci* together with the Homalonotid *Brongniartella* and the Asaphid *Parabasilicus*. This fauna is also represented in neighbouring parts of Wales and the Welsh Borders, a particularly interesting example being that of the Pontesford district, north-northwest of the Long Mynd (Shrewsbury Sheet 152; see Dean and Dineley 1961). There, graptolitic shales closely resembling the Harnage Shales of the Harnage district are succeeded by Glenburrell Beds exhibiting the typical lithology and fauna. The occurrence is significant in that the Pontesford outcrop is not far from the lithologically distinct strata of the Shelve Inlier, from which it is separated by the Pontesford–Linley Fault, and the beds transgress the Pre-Cambrian rocks for some distance to the west of the Church Stretton Fault Complex.

Graptolites are less common in the Harnage Shales of the southern outcrops than they are in the north, around Evenwood (Shrewsbury Sheet 152). At one time the fauna was held to indicate the zone of *Dicranograptus clingani*, a supposition which demanded the presence of an appreciable disconformity between the Hoar Edge Grit and the overlying Harnage Shales (Pocock and others 1938, pp. 81, 86, 87; fig. 13). Recent revisions of the stratigraphy indicate an unbroken succession in which the intervening zone of *Diplograptus multidens* is probably fully represented (Bulman 1948; Dean 1958 pp. 227–8). W.T.D.

CHATWALL FLAGS

Because their outcrop is locally very narrow, due to its position on the face of the Chatwall Sandstone scarp, the Chatwall Flags are not separated from the Sandstone on the one-inch map. The two formations have, however, been separately mapped in the field, although the boundary between them is often difficult to determine. The outcrop follows that of the Harnage Shales, lying immediately east or south-east of the older formation, and in places it forms an escarpment separate from that of the overlying Sandstone.

In the area south of Whittlingslow [4389] the Flags consist largely of green and buff micaceous flags with some bands of shale and of more massive sandstone. The higher beds are locally dark reddish brown, many of the bedding planes carrying abundant impressions of columnals of the crinoid *Balacrinus* [*Glyptocrinus*] *basalis* (McCoy). Some bands containing many Sowerbyellids also occur in the topmost beds. The thickness varies between 120 ft and about 250 ft, being least in the valley of the Onny. D.C.G., B.A.H.

In the vicinity of Hope Bowdler the rocks are fine-grained flags, brown or green, and commonly crinoidal, varying in thickness between 75 and 225 ft. North of the Cardington hills they are yellow flaggy sandstones and sandy mudstones, with lenses of impure limestone and bands of crinoid columnals. The thickness here is about 300 ft. G.H.M.

According to Dean (1958, fig. 4) the Chatwall Flags ('Lower Chatwall Sandstone') in the northern part of the district are co-extensive with the Soudleyan Stage, including beds of the same age as the upper part of the Glenburrell Beds of the Onny valley (pp. 105 and 106). Their upper part in the north is of the same age as the Chatwall Sandstone ('Soudley Sandstone') in the Hope Bowdler area. In general the Chatwall Flags appear to begin earlier in the north-east than in the south-west. D.C.G.

CHATWALL FLAGS: PALAEONTOLOGY

The shelly fauna of the Chatwall Flags, though it may be locally abundant, is not very varied in its composition. Brachiopods are the commonest fossils, *Sowerbyella soudleyensis* Jones and *Macrocoelia* [*Rafinesquina*] *expansa* (J. de C. Sowerby) being particularly abundant whilst *Reuschella horderleyensis* Bancroft is a useful zonal index. Trilobites are not well represented but include *Parabasilicus* together with the Trinucleid *Broeggerolithus soudleyensis* (Bancroft), in which some fringe-pits of the outermost concentric row (E_2) are absent in front of the glabella. *B. soudleyensis* occurs less commonly in higher strata but has been recorded as high as the *alternata* Limestone at the base of the Lower Cheney Longville Flags. The term *Glyptocrinus* Flags was once applied to the Chatwall Flags because of the large number of remains of the crinoid *Balacrinus* [previously *Rhaphanocrinus* or *Glyptocrinus*] *basalis* (McCoy). These are found mainly as disarticulated columnals but entire calices occur in places. W.T.D.

Chatwall Sandstone

The Chatwall Sandstone appears at Sibdon Carwood [4183] from beneath the unconformable Silurian cover and continues to crop out northwards almost to Whittingslow [4389]. It is known in this area as the Horderley Sandstone. It is faulted out from there to Marshbrook [4489], between which place and Ragleth Hill [4591] it occurs in several small faulted outcrops. Another major outcrop, to which the local name Soudley Sandstone is applied, extends east-north-eastwards for about 3 miles from Ragdon [4591] towards Wall under Heywood [5092], and the most northerly crop in the Church Stretton district extends from the northern side of Willstone Hill [4894] to Chatwall Hall [5197] and into the area of the Shrewsbury (152) Sheet. Along most of its outcrop the Sandstone gives rise to a distinct topographic feature, and it commonly forms the capping to a prominent escarpment, notably at Briar Edge [4286], east of Hope Bowdler, and at Yell Bank [5097].

In the southern outcrop the formation consists predominantly of massive dark green and brown sandstone, commonly current-bedded, but occasionally flaggy. The middle beds show a characteristic purple banding, and a uniform purple colour is locally developed. Sowerbyellid brachiopods are scattered through the Sandstone and in the uppermost beds they form a number of shelly limestones up to about 3 ft thick. Other brachiopods and trilobites are also common in these limestones. Specimens from Woolston (E 28990–1)[1] consist of fine subangular grains of quartz and feldspar. Glauconite and phosphatic grains are common, particularly in the higher green sandstone at that locality. The thickness of the Sandstone varies between about 300 and 500 ft, being at its maximum in or near the Onny valley. D.C.G.

The Sandstone is seldom well exposed between Marshbrook and Ragleth Hill, but it is lithologically similar to that exposed farther south, mainly massive, but with thin bands of shelly limestone near the top. Farther east, at Soudley [477918], it is a well banded, fine-grained, massive sandstone with some soft bands of shelly, crinoidal debris. Its thickness in this area is only about 150 ft. B.A.H., G.H.M.

In the north of the district the Chatwall Sandstone consists of greenish grey and purple argillaceous sandstone, with bands containing small pebbles of quartz, quartzite, and grit, as well as green shale. It attains a thickness of about 300 ft on Yell Bank, east of Hoar Edge. G.H.M.

Bancroft (1929, p. 40) showed that at Soudley the Chatwall (Soudley) Sandstone belonged to his Soudleyan Stage and the *alternata* Limestone to the Upper Longvillian. Thus there are no sediments at that locality corresponding to the Lower Longvillian Stage. The Chatwall Sandstone at Chatwall

[1]Such numbers in brackets refer to rock slices in the Survey collection.

has been shown by Dean (1960a, p. 169) to include beds of the Upper and Lower Longvillian and Soudleyan stages. The absence of the faunas of the zones of *Bancroftina typa* and *Dalmanella horderleyensis*, the highest and lowest zones of the Lower Longvillian, led him to conclude that non-sequences occur both above and below the rocks of that stage. Lithological changes at the appropriate horizons add weight to this view. Dean suggested finally that the name Chatwall Sandstone should be applied only to the 26 ft of Lower Longvillian beds, those above and below being assigned to the *alternata* Limestone and Chatwall Flags respectively. The important section just south-west of Chatwall Hall is referred to in the Shrewsbury Memoir (Pocock and others 1938, pp. 87–8). In that publication the *alternata* Limestone is included within the Chatwall Group ('Chatwall Beds' on the one-inch map), but it is not clear at what precise horizon within the section the base of the Limestone was taken. The base of the Chatwall Sandstone was taken, on lithological criteria, at a horizon apparently some 80 or 90 ft below that suggested by Dean.

In the present survey the *alternata* Limestone has been grouped with the Cheney Longville Flags, and not, as in the Shrewsbury Memoir (Pocock and others 1938), with the Chatwall Group. South of Marshbrook it has been mapped locally as a separate unit. D.C.G.

CHATWALL SANDSTONE: PALAEONTOLOGY

The most abundant fossils in the Chatwall Sandstone are brachiopods, the shells of which generally occur in enormous numbers, forming lenticular bands of sandy limestone. Perhaps the most prolific of them is *Sowerbyella soudleyensis* Jones, described originally as a variety of the slightly later *Sowerbyella sericea* (J. de C. Sowerby) but now regarded as a distinct species. Members of the family Dalmanellidae are also common, most of them being at one time assigned to the genus *Wattsella* Bancroft which is now held to be a synonym of *Dalmanella* Hall and Clarke. Among the most important forms are *Dalmanella horderleyensis* (Whittington), *D. indica* Whittington, *D. lepta* (Bancroft) and *Howellites antiquior* (McCoy), all of which have proved useful in correlations with successions in parts of North Wales. Less common though fairly characteristic are the Triplesiid brachiopod once known as *Cliftonia spiriferoides* (McCoy), but now referred to the Bohemian genus *Bicuspina* Havlíček, and the small Strophomenid *Kjaerina jonesi* Bancroft.

Trilobites include the stratigraphically useful Trinucleid *Broeggerolithus nicholsoni* (Reed), a species widespread at this level through much of North Wales and parts of Northern England, together with *Brongniartella*, *Flexicalymene*, *Parabasilicus* and *Kloucekia*, the last-named making its return after being unrecorded from the district since the deposition of the lowest Harnage Shales. W.T.D.

ALTERNATA LIMESTONE

The group of beds known as the *alternata* Limestone forms a narrow outcrop immediately east or south-east of the Chatwall Sandstone from Sibdon Carwood to Whittingslow, near Hope Bowdler, and from Willstone Hill to Chatwall and beyond. Because of its comparative thinness, it is often difficult to trace with any accuracy in the field and it has been separately mapped only between

Cheney Longville and Whittingslow. The group has elsewhere been included with the Cheney Longville Flags.

The *alternata* Limestone consists of a succession of lenticular shelly limestones, characterized by the abundance of the brachiopod *Heterorthis alternata* (J. de C. Sowerby) accompanied by various trilobites and other brachiopods, interbedded with green micaceous flags and shales which are also highly fossiliferous. The limestone lenticles are up to 2 ft thick but are generally less than 1 ft. Near Woolston the formation attains a thickness of about 100 ft. Elsewhere, where it can be measured, it averages about 70 ft.

Cheney Longville Flags

Except near Wall under Heywood [5092], where it is cut out by Silurian overstep, the outcrop of the Cheney Longville Flags is continuous from Long Lane [416839] to the northern limit of the district near Chatwall. In the Marshbrook area it is much broken by faults but otherwise it is relatively uncomplicated.

The formation consists predominantly of greenish grey flags with interbedded shales and rubbly siltstones. Thin beds of fine-grained massive sandstone occur, and thin shelly limestones, usually decalcified, are common. The flags are generally finely bedded, the laminae being often current-bedded, slumped or otherwise irregular. A form of decalcified shelly grit, bearing some resemblance to gingerbread, occurs in places in the higher beds. In addition to the fossils in this grit and in the thin limestones, brachiopods and trilobites occur commonly throughout the formation. Amongst the other fossils which are present the most conspicuous and characteristic is *Tentaculites*.

Over most of the outcrop the Flags, excluding the *alternata* Limestone, are between 400 and 500 ft thick, but in the Cardington area about 600 ft is reached, and near Cheney Longville, at the southern end of the fully-developed outcrop, the thickness appears to be about 800 ft.

The relation of the *alternata* Limestone to the base of the Cheney Longville Flags has been discussed above (p. 111). Although Bancroft's (1933) choice of the *alternata* Limestone as the lowest member of the Upper Longvillian Stage may have been founded on his recognition in places of a non-sequence at its base, the lists published by Dean (1958, pp. 220–2) show that, in addition, the fauna of the Limestone has more in common with that of the beds above than with that of the Lower Longvillian. Dean (1958, pp. 206–7) pointed out that in the Onny valley there is a lithological gradation from the flaggy beds at the top of the Lower Longvillian Chatwall Sandstone, through the flags and limestones of the Upper Longvillian *alternata* Limestone, to the flags in the lower part of the Upper Longvillian Cheney Longville Flags.

The upper part of the Flags constitutes Bancroft's Marshbrookian Stage. Although there is no marked lithological difference between these beds and those below, there is in them a general increase in the proportion of blocky and rubbly mudstones and siltstones and, as in the beds just above the *alternata* Limestone, thin shelly limestones are more common than in the higher beds of the Upper Longvillian. The very highest beds of the Cheney Longville Flags show a transition to the mudstone facies of the Acton Scott Group, and Dean (1958, p. 219) pointed out that the faunal change from the Marshbrookian to the Actonian stages is partly controlled by the change in lithology.

However the persistence of Cryptolithids, which do not occur in the basal Actonian, suggests that these transitional beds should be included in the Marshbrookian. D.C.G.

LOWER CHENEY LONGVILLE FLAGS AND ALTERNATA LIMESTONE: PALAEONTOLOGY

The basal subdivision of the Cheney Longville Flags, the so-called *alternata* Limestone, is aptly though loosely named on account of its containing immense numbers of valves of *Heterorthis alternata* (J. de C. Sowerby). Generally these are found in lenticular limestone bands separated by greenish siltstones and mudstones, but locally they occur in sandy limestones which weather to a yellow-brown colour and exhibit both internal and external moulds of the shell. The limestone bands contain also large numbers of the Plectambonitid *Sowerbyella sericea* (J. de C. Sowerby), for which this is the type horizon, together with less common specimens of *Trematis punctata* (J. de C. Sowerby), an inarticulate brachiopod with distinctive reticulate ornamentation. The Strophomenids *Kjaerina bipartita* (Salter) and *Strophomena grandis* (J. de C. Sowerby) are abundant, particularly in the less calcareous parts of the *alternata* Limestone. The former species is characteristic of this part of the Ordovician succession, but *S. grandis* ranges upwards into the Acton Scott Group.

Above the *alternata* Limestone the succession is characterized particularly by numerous large specimens of *Kjaerina*, particularly *K. typa* Bancroft and *K. geniculata* Bancroft, in places accompanied by *Dolerorthis duftonensis* (Reed), a species first described from the Cross Fell Inlier and found also in North Wales. It is in these strata that the trilobite *Chasmops extensa* (Boeck), first described from Norway, makes its appearance, thereafter ranging upwards into the Onny Shales. Trinucleid trilobites are uncommon in the Lower Cheney Longville Flags but *Broeggerolithus longiceps* (Bancroft) is characteristic. Bedding planes throughout much of the succession are commonly found covered with the remains of *Tentaculites* (tentatively referred by some authors to the Scaphopoda), whilst large lamellibranchs, for example *Modiolopsis obliqua* (J. de C. Sowerby), are fairly common.

UPPER CHENEY LONGVILLE FLAGS: PALAEONTOLOGY

Although these strata follow conformably upon the Lower Cheney Longville Flags, they contain a distinctive and sometimes abundant shelly fauna. Brachiopods are particularly common, especially the Dalmanellids. *Dalmanella wattsi* (Bancroft) and *D. unguis* (J. de C. Sowerby) characterize successively the lower two-thirds of the succession whereas the topmost beds contain *Onniella reuschi* Bancroft. Strophomenids are generally less common than in the Lower Flags but the genera *Kjerulfina* Bancroft and *Hedstroemina* Bancroft appear for the first time in Shropshire, particularly in the highest Cheney Longville Flags, where *K. polycyma* Bancroft and *H. fragilis* Bancroft are of zonal value.

Of the trilobites, *Brongniartella bisulcata* (McCoy) and *Chasmops extensa* (Boeck) are common but long-ranging, whilst the Trinucleid *Broeggerolithus transiens* Bancroft is most abundant at this level, though it is known very rarely from the Acton Scott Group. *Flexicalymene caractaci* (Salter) is to be found throughout most of the subdivision, and the type material came from this horizon. As in the Lower Cheney Longville Flags *Tentaculites* occurs in vast numbers and lamellibranchs are not uncommon. W.T.D.

Acton Scott Group

The most southerly outcrops of the Acton Scott Group are in the banks of the River Onny [423854], and the overstep of the Group by Wenlock Shales is thought to be completed just south-east of Cheney Longville [4285]. The outcrop continues north-north-eastwards to Acton Scott [4589] and Ticklerton [4890], where it is again cut out by Silurian overstep. Near Marshbrook and Acton Scott it is much faulted, as a result of which an outlier is developed on a line between Marshbrook and Ragdon. North of Cardington Hill [5094] the Group forms a very broad outcrop between the Cheney Longville Flags and the Silurian rocks. It extends into the neighbouring district (Shrewsbury Sheet 152) as far as Church Preen [5498], but is insufficiently exposed there to have been separated from the Flags below by the surveyors of that sheet.

In the Onny valley the Group, between 200 and 300 ft thick, consists largely of grey, rubbly, micaceous mudstones with, characteristically, an abundance of calcareous shells of the brachiopod *Onniella*. Thin bands of shelly and crystalline limestone also occur. D.C.G.

Near Acton Scott the lower beds are olive-green and buff, micaceous, silty mudstones with siltstone bands, in places orange-stained and weathering shaly. Locally the beds are weakly calcareous. There is a passage up into the Acton Scott Limestone, a hard, splintery, sandy limestone or calcareous sandstone which caps the high ground around Acton Scott village, and has been quarried for local building stone. The Limestone, which appears to be absent to the west of the Marshbrook valley, reaches a maximum thickness of about 80 ft at Acton Scott, and dies out to the east between there and Hatton. The beds above the Limestone include olive-green micaceous siltstones, but are very poorly exposed. At Hatton the basal beds of the Group are steel-grey silty mudstones with occasional bands of grey argillaceous limestone up to 6 inches thick. These beds pass up into pale yellow flaggy siltstones and sandstones. In this area the greatest thickness, 200 ft, is attained in the vicinity of Acton Scott. B.A.H.

A succession like that at Hatton, with some shell bands, persists north-eastwards towards Ticklerton.

North of Cardington Hill the Group includes two thick beds of flaggy shelly sandstone, the lower about 50 and the upper about 75 ft thick. These sandstones, like the associated shales and mudstones, are of a yellow or fawn colour. The thickness of the Group is estimated to be in excess of 500 ft.

G.H.M.

The name 'Acton Scott group' was first used in print by Cobbold (1900, p. 47), who defined it in terms of the rocks exposed in the River Onny. From Cobbold's description it appears that both limits of the Group, as mapped in the recent survey, are about 75 ft higher in the succession than those he had chosen. Lapworth (1916, p. 38) applied the name Acton Group to the highest beds of the Ordovician in the Caradoc area, dividing them into the Acton Calcareous Group below and the Upper *Trinucleus* Beds above. The same beds were called the Acton Scott Group by Pocock and Whitehead

(1948, p. 51). The Group as now defined appears to be co-extensive with Bancroft's Actonian Stage (1945, pp. 183, 186), at least in the Onny valley. Dean (1958, pp. 211–2) stated that the Acton Scott Beds (Lapworth and Watts 1894, p. 320) at Acton Scott form only part of the middle Actonian, but their limits have never been defined and in any case Dean pointed out that neither the lowest nor the highest beds had been fully examined. D.C.G.

ACTON SCOTT GROUP: PALAEONTOLOGY

The strata of this subdivision consist in the main of grey calcareous mudstones and shales, and the shelly faunas exhibit marked differences from those of the preceding beds. The trilobites are more varied and genera appearing for the first time in South Shropshire include *Calyptaulax*, *Ampyxella*, *Illaenus*, *Lonchodomas*, *Onnicalymene*, *Platylichas*, and *Tretaspis* whilst *Chasmops extensa* (Boeck) ranges through the Acton Scott Group and persists into the lowest Onny Shales.

Brachiopods are abundant and often well preserved, the genus *Onniella* being particularly common. Numerous species of *Onniella* were erected by Bancroft, in both this Group and the Onny Shales, but are in need of detailed revision. *Cryptothyris paracyclica* Bancroft occurs plentifully and is restricted to the middle portion of the Group. The genus *Reuschella* has its maximum development in Shropshire at this horizon and *Strophomena grandis* (J. de C. Sowerby) persists from earlier strata, as does *Hedstroemina*, though represented by a different species, *H. robusta* Bancroft. Part of the Acton Scott Group occurs in an arenaceous facies near Acton Scott, Plaish and Cardington, and in these beds *Tentaculites* commonly occurs in abundance with brachiopods, particularly Dalmanellids. W.T.D.

ONNY SHALES

The Onny Shales form a restricted lenticular outcrop between Cheney Longville [4285] and Hatton [4690]. They are extensively covered by drift and there are but few good exposures. Lithologically they are not readily distinguishable from the underlying mudstones of the Acton Scott Group and their differentiation is to a large extent palaeontological, being based on the abundance in them of Trinucleid trilobites, a group almost unknown in the older formation. The rocks are mainly grey and buff micaceous siltstones and mudstones but in the Onny exposures they appear to be somewhat more greenish and better stratified than the underlying beds. They reach a thickness of about 400 ft near Henley but it is not known whether any horizons present there are higher than those in the River Onny, where only about 150 ft of beds crop out to the west of the Silurian.

The Onny Shales as here described and mapped are believed to be equivalent to the Onnian Stage of Bancroft (1933) and the *Onnia* Beds of Dean (1958, p. 213). The name as first used by La Touche (1884, p. 12) was applied to all the Ordovician beds above the Cheney Longville Flags. In Bancroft's nomenclature (1933) they comprised only the topmost zone of his Onnian Stage, the two lower zones being included in his Acton Scott Beds. As Dean (1958, p. 213) pointed out, there is no justification for this upward extension of the Acton Scott Beds nor for separating lithologically the highest Onnian beds from those below. The Onny Shales appear to be approximately equivalent to the *Trinucleus* or Onny Shales of Lapworth (1894, p. 320) or his Upper *Trinucleus* Beds (1916, p. 38), although from the data provided by Lapworth the base of his formations would appear to be at a horizon within our Acton Scott Group. Dean, however, (*in* Whittard 1960, p. 269) deduced, from his own investigations (1958), that the Upper *Trinucleus* Beds are equivalent to only the upper part of the Onnian. D.C.G.

ONNY SHALES: PALAEONTOLOGY

These strata, to which the name *Onnia* Beds has also been applied (Dean 1958, p. 213), mark the appearance in large numbers of the Trinucleid trilobite genus *Onnia* Bancroft, a widely-distributed form known also from Bohemia and North Africa. The beds were divided by Bancroft into three zones on the basis of successive species: *Onnia? cobboldi* (Bancroft), *O. gracilis* (Bancroft), and *O. superba* (Bancroft). Of these *O. gracilis* is the most widespread, occurring at the same level at Welshpool and in the Cross Fell Inlier of Northern England. Trilobite genera persisting from the Acton Scott group include *Ampyxella*, *Lonchodomas*, *Calyptaulax*, *Chasmops*, *Illaenus*, *Onnicalymene* and *Platylichas*, whilst newcomers include *Remopleurella* and *Triarthrus*. The combined trilobite faunas of the Acton Scott Group and the Onny Shales show marked Scandinavian affinities and enable the beds to be correlated with the Upper *Chasmops* Limestone of southern Norway and equivalent strata in parts of Sweden.

Brachiopods are represented mostly by *Onniella*, a genus which becomes less common in the topmost beds where it is accompanied by large numbers of a small inarticulate brachiopod which is usually, though inaccurately, termed '*Strophomena*' *holli* Davidson, a species in need of revision.

Graptolite remains occur rarely in both the Acton Scott Group and Onny Shales. The evidence suggests that both subdivisions belong to the zone of *Dicranograptus clingani*. W.T.D.

DETAILS

HOAR EDGE GRIT

Coston to Sibdon Carwood. Augering has revealed an outcrop of Hoar Edge Grit up to 60 yd in width along the east side of the Church Stretton Fault, extending for about 600 yd from the southern edge of the one-inch sheet. It is probably faulted against the Wenlock Shales to the east. The only exposure [379787] is 820 yd S. of Hope, near Coston, where buff, coarse-grained, slickensided sandstone is seen.

To the north the Grit is exposed in an old quarry [385798] 570 yd S. 13° W. of Coston where about 38 ft of the Basal Conglomerates are intermittently exposed.

The structure of the Coston area appears basically to be anticlinal, the beds adjacent to the Church Stretton Fault dipping to the west, and those on the eastern side of Coston Hill dipping to the south or south-east. This anticlinal structure may be due to drag by the Church Stretton Fault. Most of the exposures are on the western side of the outcrop, and are in the higher part of the Hoar Edge Grit sequence.

Immediately south of Coston farmhouse [386803] a north-south track section, 150 yd long, shows coarse-grained sandstones with occasional thin bands of quartz pebbles. The beds dip at about 60° to the west and are strongly slickensided and shattered.

The old quarry [38608008][1] (Locality 1)[2] 200 yd S. of Coston farmhouse gives the only good section in the *Costonia ultima* Beds. About 22 ft of medium-grained brown sandstone are visible in the quarry, dipping 30° W. Bancroft (1945, p. 236) named this quarry as one of the localities for *Horderleyella plicata*, and it is also the type locality for *Costonia ultima* Bancroft sp. (Dean 1960b, p. 88). At the entrance (Locality 2) to the quarry there is a small exposure of brown medium-grained sandstone with scattered quartz pebbles, used by Bancroft (1945, p. 235) as the type locality for *Smeathenella strophomenoides*. Both exposures have yielded an abundant fauna of trilobites and brachiopods.

About 18 ft of coarse sandstone and conglomerate of the *Harknessella* Beds, dipping 32° W., are exposed in the track (Locality 3), 50 yd E. of Coston farmhouse. The beds are affected by minor faulting. This exposure was assumed by Bancroft (1945, p. 231) to be the type locality of *Harknessella vespertilio*.

Similar beds, rather disturbed, crop out in an old quarry immediately south of the trackway, and also in an exposure 80 yd to the north. Two old quarries on the western edge of the wood to the east of the Hollies [387806] show beds that are probably at about the same horizon. The northerly quarry shows 18 ft of coarse sandstone with conglomerate bands with a variable westerly dip, while the southern quarry shows about 8 ft of similar beds, strongly sheared and slickensided. No fossils were found during the resurvey although Bancroft (1928, p. 184) designated one of the quarries as the type-locality for *Harknessella jonesi* Bancroft.

The Basal Conglomerates are exposed in an old quarry [390804] 400 yd N. of Coston Hall, and 48 ft of similar beds with a south-easterly dip varying from 20° to 45° are exposed at the roadside 150 yd farther north. 10 ft of the Conglomerates, with an easterly dip, crop out behind Coston Cottage [392811], 750 yd S. 5° E. of The Hall, Aston on Clun. Similar beds crop out in the south bank of the River Clun, 50 yd to the north-west. A few yards farther west purple grits of the Pre-Cambrian are visible, but the contact is not exposed.

North of Coston Cottage the outcrop is largely obscured by drift. A small exposure of buff, coarse-grained, pebbly grit is present in an old working 390 yd E. 27° S. of The Hall, Aston on Clun. Eight feet of 'unconsolidated rock' (drift) overlying 12 ft of coarse brown grit (E 21275) were recorded from a well [395816] in the angle between the two roads, 390 yd E. 35° S. of The Hall, Aston on Clun.

In an old quarry [404828] 1020 yd W. 22° S. of Sibdon Carwood church, 5 ft of the Basal Conglomerates are exposed. The beds are purple-stained and weathered, and appear to dip at about 80° to the west. Similar beds occur in an old gravel pit [405830] at Oldfield, 250 yd to the north-north-east. These beds also have a steep westerly dip, which may be due to the proximity of a north-south fault through Oldfield. The most northerly exposure of the Hoar Edge Grit in this area is in the stream 900 yd W. 2° S. of Sibdon Carwood church, where the Basal Conglomerates are again exposed. Purple sandstones of the Pre-Cambrian crop out a few yards farther upstream, but the nature of the contact cannot be determined. North of Oldfield the Hoar Edge Grit is cut out by a north-south fault. B.A.H.

Wart Hill to Brokenstones. A down-faulted outlier of Hoar Edge Grit lies between 150 and 600 yd north-north-east of Upper Carwood. The best exposure is of green and buff, coarse-grained, pebbly grit with scattered brachiopod moulds in an 'old gravel pit' [403854] 170 yd N. 43° E. of the farm. The dip cannot be determined in this outlier but it appears to be at a low angle towards the east-south-east.

The main outcrop begins [406854] about 450 yd E. 12° N. of Upper Carwood against a fault which throws the Grit against the Uriconian rocks of Wart Hill. Buff grit, dipping steeply to E. 10° S., is exposed in the roadside 510 yd E. 25° N. of Upper

[1]Such numbers are eight figure National Grid References within the 100-km square SO (32).

[2]Such numbers refer to Ordovician fossil localities listed in Appendix 1.

Carwood, and there are small exposures of buff quartz-conglomerate along the steep wooded slope which extends from there to the north-east. Buff grit is exposed in the lane [409857] 940 yd E. 30° N. of Upper Carwood and, more extensively, in the wood about 130 yd farther to the north-north-east. At the latter locality the rock dips steeply to the S.E. In outcrops between this point and the northern bank of the River Onny, bands of limestone and pebbly bands are common. The beds dip at between 60 and 70 degrees towards about E. 35° S., the dip becoming more easterly near the river. In all the ground south of the Onny the Grit forms a ridge, of which the south-eastern limit is particularly clearly defined. At some places its base is indicated by the presence in the soil of debris of the underlying purplish grits of the Wentnor Series.

The Hoar Edge Grit is fully exposed in an old quarry [41188614] (Locality 4) on the southern bank of the Onny, 180 yd W. 22° S. of Glenburrell. According to local information the rock was burnt here for lime, an undertaking to which Murchison also referred (1839, p. 219). At the western end of the quarry the Ordovician rocks rest unconformably on weathered grits of the Wentnor Series (p. 90). The basal Hoar Edge Grit dips at about 70° towards a point somewhat north of east whereas the underlying rocks are nearly vertical on a north-north-easterly strike. It is to be expected that some movement would be induced at the surface of unconformity by the forces which up-ended the Ordovician rocks. This may explain the broken buff grit, up to 1 ft 4 in thick, and the contained 2-in band of pale buff clay, which, in the quarry, underlie the lowest hard grit of the Ordovician. The broken grit was thought by Whittard (1953, pp. 243, 245) to be Pre-Cambrian, but it more closely resembles a crushed form of the Hoar Edge Grit. As Whittard pointed out, the Hoar Edge Grit is only about half as thick here as it is a short distance to the south, but it seems as likely that this may be due to a reduction in deposition as to a fault. The absence of disturbance in the weathered Pre-Cambrian grits, and the course of the Hoar Edge Grit northwards across the Onny, both point to the improbability of such a fault. On the other hand the minor overthrusting of the Hoar Edge Grit towards the south-east indicates that some movement has taken place at this horizon.

The quarry is divided into two parts, the rocks of the eastern part appearing to succeed those in the western. Some 30 ft of beds are exposed in the western quarry, consisting chiefly of brown calcareous grits, with rounded quartz pebbles at one horizon near the base. The topmost 18 ft are rather flaggy but the lower beds, up to 4 ft thick, are massive. Lenses and bands of grey crystalline limestone, commonly weathered back, and in places current-bedded, are interbedded with the grits. The thickest band, near the base of the section, is 1 ft 6 in thick. In the eastern quarry there are about 57 ft of beds, of which the topmost 4 ft are thought to belong to the succeeding Harnage Shales. Flaggy and massive calcareous grits are predominant. Thin limestones are developed as below, though less frequently, but shelly and pebbly bands are more common. Two thin bands of pale grey and brownish grey clay occur about 2 ft below the top of the Grit.

Another quarry [412862] in the calcareous Hoar Edge Grit is situated on the opposite bank of the Onny about 100 yd to the north. About 50 ft of buff, flaggy grit are exposed here with some massive bands of bluish grey limestone near the middle. A 2-ft band of calcareous quartz-conglomerate, the pebbles of which are rather angular, occurs near the top of the section. The bedding is somewhat undulatory, but the general dip of 83° towards the east-south-east is maintained throughout the exposure. Buff grit with some shell bands is exposed in an old quarry [41368658] (Locality 5) 400 yd N. of Glenburrell. The rock here dips at a low angle in a rather variable easterly direction and the extent of the quarry indicates that the outcrop of the Grit is here much wider than in the Onny valley.

North-eastwards for about half a mile there are no outcrops and little debris of the Hoar Edge Grit, but it is exposed at the north-eastern end of a line of old narrow qarries [419872], 1270 yd N. 26° E. of Glenburrell and about 600 yd west of the road junction at Woolston. The rock here is a buff shelly grit, dipping vertically on a strike of N. 35° E., and cannot be more than about 70 ft thick. Within about 250 yd to the north-east it has disappeared and it is thought to be faulted out. The fracture extends in a south-south-westerly direction to a place immediately west of the southern end of the outcrop of the Hoar Edge Grit north of Wart Hill.

A small area of high ground at Brokenstones is occupied by a fault-bounded outcrop of Hoar Edge Grit. The rock is well exposed [41938785] (Locality 7), 1020 yd S. 24° W. of Cwm Head church, where it forms a distinct knoll on the hill top. It is a buff quartz-grit, conglomeratic in part, and with scattered moulds of brachiopods in certain bands. The dip is difficult to determine but it seems to be 40° to E. 30° S. About 80 yd farther south, in a small old quarry, brown gritty sandstone 10 ft thick is overlain by 4 ft of brown quartz-conglomerate. These beds dip at 42° to W. 40° N., an attitude possibly induced by the proximity of a fault to the south-east.

The Hoar Edge Grit is again well exposed in an old quarry [41888797] (Locality 6) in the plantation about 130 yd north-north-west of the knoll just described. Brown gritty sandstone with some brachiopods forms a cliff some 20 ft high. The exposure is difficult to examine and the dip was not determined with any certainty. From this point the Grit, separated by the fault F2 (p. 267) from Longmyndian rocks to the west, forms a well-defined escarpment, covered by vegetation, which extends for some 600 yd to the south-west. At a place on the escarpment [415875] 1510 yd S. 35° W. of Cwm Head church, buff pebbly grit is exposed, dipping at about 50° to the east. The break in the escarpment south of this place, and the distribution of Longmyndian debris to the south and east, suggest that the Grit is pinched out by the convergence with F2 of a fault from the north-east. Some 800 yd south-south-west of the church the outcrop of the Grit ends against a north-westerly fault following approximately the line of a valley which cuts across the general grain of the country. It is next exposed near Cwms Farm [4793], some 5 miles to the north-east. D.C.G.

Cwms to Hoar Edge. From a place [476941] on the outcrop of the Cwms–Hoar Edge Fault, about 200 yd E. 35° N. of Cwms Farm, the Hoar Edge Grit forms a conspicuous feature which extends northwards for about 800 yd. Exposures are confined to the sides of the stream [477944] which breaches the escarpment 530 yd N. 28° E. of the farm, where coarse-grained, yellow, flaggy sandstone dips to the east at 56 degrees. These outcrops lie on the western limb of a southward-plunging syncline. On the eastern limb an outcrop of coarse-grained calcareous sandstone, with well-preserved shells occurs in a small old quarry [483951] on the northern side of the bridle road, 1520 yd N. 35° E. of Cwms Farm. This rock dips at 30° to the south-west. The outcrop of Hoar Edge Grit which closes the syncline is limited to the north-west by the Lawley Fault. To the north-east it rests unconformably on the Cambrian, but small faulted outcrops lie just east of the Lawley Fault as far as 1¼ miles north-north-east of Cwms Farm.

An exposure [48439520] (Locality 8) 1710 yd N. 38° E. of Cwms Farm shows coarse-grained and finer-grained sandstone bands, dipping at 55° to the south-east. From here a narrow outcrop of the Grit extends for some 600 yd both to south-south-west and to north-north-east. In the former direction the outcrop ultimately turns north-westward under the influence of the syncline mentioned above. In the opposite direction it is pinched out between the underlying Cambrian and the Cwms–Hoar Edge Fault which forms its south-eastern limit throughout its length.

North-eastwards from Shootrough [491964], where it re-appears on the south-eastern side of the Cwms–Hoar Edge Fault, the Grit forms the fine escarpment of Hoar Edge. Thick-bedded, coarse-grained sandstone is exposed locally, dipping at between 60 and 70 degrees to the south-east. Prominent joints dip at 60° towards S. 50° W. G.H.M.

Harnage Shales

Coston to Sibdon Carwood. The southern limit of the Harnage Shales outcrop is uncertain. The beds are obscured by drift and presumably are overstepped by the Wenlock Shales between Long Meadowend and Coston. The most southerly exposure of the Shales is in an intermittent stream section extending for 150 yd to the north-west from a place [40698284] (Locality 10) 740 yd W. 27° S. of Sibdon Carwood church. At this place 14 ft of buff, flaggy, silty, micaceous mudstones are exposed, dipping 50° to E. 5° S.; 40 yd upstream 24 ft of similar beds crop out, with a further 7 ft 20 yd to the north-west. These beds are orange-stained and dip at 60° to E. 40° S. There is a further unexposed gap of 60 yd, and then orange-stained, micaceous, silty mudstone is seen [40648295] (Locality 9) with an abundant fauna. The dip is 60° to W. 5° S., and the beds may be overturned. Similar beds continue to the north-western end of the section, with siltstone and sandstone bands locally.

Northwards from about 200 yd north of Oldfield, the Harnage Shales appear to be faulted against the Pre-Cambrian of Hopesay Hill. The line of the fault is marked by a strong break in slope and seepages along the east margin of Oldfield Wood [406832] and northwards towards the south-east corner of Sibdon Wood [407839].

Olive-brown, micaceous mudstones are exposed in the entrance to the old gravel pit [407832] 700 yd W. 8° N. of Sibdon Carwood church, and also in the north-west face of the pit, apparently with a steep dip to the north-west. Dark grey basalt is seen in the central part of the pit. Another pit 300 yd to the north shows separate obscure exposures of vertical shaly mudstones and of basalt. In neither pit can the field relations of the basalt now be made out. However, Whittard appears to have had a clearer view of the section in the northern pit, for he stated (1952, p. 162) that "A thin volcanic agglomerate and a lava flow of weathered trachybasalt, a few feet in thickness, are the only direct evidence known to me of vulcanicity in the Caradoc area. The rocks, which are associated with graptolitic Harnage Shales, are to be seen in an old 'gravel pit' . . . ". Murchison (1839, p. 229) described three dykes intruded into vertical and contorted shale "on the south-eastern face of Sibdon Hill and a little west of Sibdon House", but did not specify the exact locality. Thin-section examination of the basalt by Dr. Diane C. Knill (E 29498 and E 29499 from the north pit and E 29500 from the south pit) showed it to be a carbonated and chloritized porphyritic basalt, in which carbonated phenocrysts of olivine and pyroxene lie in a ground mass of feldspar laths (albite), interstitial chloritized augite and granular iron oxides. A few rounded amygdales and irregular cavities are filled with calcite.

Buff, micaceous, silty mudstones with occasional bands of fine-grained sandstone, probably near the top of the Harnage Shales, are exposed in a track section [408836] 700 yd W. 42° N. of Sibdon Carwood church. B.A.H.

Long Lane to Brokenstones. Green shales with flaggy bands crop out in the stream [40738424] (Locality 11) 935 yd E. 35° S. of the summit of Wart Hill, dipping at 85° to E. 15° S. They are overlain to the east by Chatwall Flags and are thought to be the highest beds of the Harnage Shales. Green blocky shales dipping at about 35° to the east crop out downstream [40958418] (Locality 12), 1150 yd E. 31° S. of Wart Hill. They are separated from the Chatwall Flags to the west by the fault F3, which has a downthrow to the west. Shales with subsidiary flags, evidently from a much lower horizon, are exposed in the stream [406849] 675 yd E. 18° N. of Wart Hill, immediately east of the fault which separates the Pre-Cambrian outcrop from the Ordovician. At this locality the shales have a blocky structure and the beds dip very steeply towards E. 15° N. Green and brown blocky shales overlain by flags, cropping out near the right bank of this stream [408850] 840 yd E.N.E. of Wart Hill, and in a ditch [409853] 1085 yd E. 34° N. of Wart Hill, are considered to be the topmost Harnage Shales. Small outcrops of weathered shale from near the base of the group occur in the track [40678538] (Locality 13) 950 yd N.E. of Wart Hill, and in a ditch about 35 yd to the south-east.

The shales are not exposed between this locality and the River Onny but their outcrop forms a depression about 150 yd wide between the Hoar Edge Grit and the Chatwall Flags. Just within the wood [411858] 1560 yd N.E. of Wart Hill is an apparently artificial mound of debris of dark green vesicular igneous rock, very like that exposed some 800 yd to the south-west (p. 25). The origin of this mound is unknown but it is not thought to indicate the occurrence of an outcrop of the, presumably Uriconian, igneous rock at this locality.

On the right bank of the Onny the basal 4 ft of the Harnage Shales are exposed at the eastern end of the old quarry [41188614] in the Hoar Edge Grit (p. 119). The rock is a yellowish buff, micaceous shale, soft at the outcrop, dipping eastwards at 65° and it is exposed again in the steep bank about 30 yd to the south-east. Some low outcrops occur in the river, mainly of green shale, with bands of green flaggy sandstone up to 4 in thick, dipping at about 75° to the east-south-east. Green shale and flags, dipping steeply to the south-east, form two small outcrops [413862] at Glenburrell, 35 and 70 yd west-north-west of the farmhouse. Higher beds are well exposed in a rather crumbling cliff immediately behind the farmhouse [414862], taken by Dean (1958, p. 203) to be the type locality of Bancroft's 'Glenburrell Beds' (1933). The rock consists of khaki-weathered knobbly shales with occasional beds of massive sandstone of an average thickness of some 2 in, but exceptionally reaching 7 in, and dips at about 80° to the east-south-east.

Interbedded green micaceous flags and shales are exposed in two small outcrops [418869] on the steep slope 1090 yd N. 35° W. and 1120 yd N. 34° W. of Upper Barn. They are considered to belong to a horizon near the boundary between the Harnage Shales and the Chatwall Flags. The shale crops out beside a spring [421874] 1380 yd S. 10° W. of Cwm Head church, some 450 yd W. 32° N. of the Woolston road junction, but it is better exposed on the northern end of the grit ridge [42088753] (Locality 14) about 100 yd farther north. The shale, which dips at about 60° to S. 30° E., is green in colour, flaggy in places, and has occasional thin bands of shelly grit.

The triangular fault-bounded area [422879] about half a mile south of Cwm Head church is devoid of outcrops and it is on this negative basis that it is shown on the map as Harnage Shales. Shale debris which is probably of this group occurs at a well [421879] 890 yd S. 17° W. of the church, and this may be from Harnage Shales, not elsewhere exposed, which are thought to overlie the Hoar Edge Grit of Brokenstones. In the track-side bank [422880] 770 yd S. 12° W. of the church green shale of Harnage type forms a small outcrop. Slickensided grey shales in the stream [421881] 550 yd W. 6° S. of Blakemoor are tentatively assigned to the Harnage Shales on the basis of lithological similarity. No rocks of the group are seen between this locality and the eastern side of the valley north of Marshbrook. D.C.G.

Little Stretton. Shaly mudstones with thin flaggy bands, dipping at 38° to E. 30° S., occur in the stream [443909] 910 yd W. 20° S. of The Hough. The small outcrop of probable Harnage Shales which lies in this valley is fault-bounded to the west and north. B.A.H.

A faulted area of Harnage Shales has been mapped south-west of The Hough, Within it, at a place 450 yd W. 17° S. of The Hough, the rocks dip to the south-west at 20°.

An outcrop of the Shales extends along the southern end of Ragleth Hill. An excellent section is visible in a small old quarry [44549136] (Localities 15 and 16), 620 yd W. 20° N. of The Hough. Steeply dipping grits, probably belonging to the Wentnor Series, rise up in the centre of the floor of the eastern part of the quarry. Their western boundary is obscured, but banked against this eastern side is a hard, gritty, pebbly sandstone, bluish grey, fine-grained, and calcareous, with calcite-veins. This rock, which dips about 22° to S. 10° W., is the basal part of the Harnage Shales. The pebbles include quartz, pink Uriconian rocks, and green fragments, probably volcanic.

The gritty calcareous sandstone seems to be continuous with, and passes upwards into, the grey to brown, calcareous, shaly mudstones which form the bulk of the Harnage sediments. The latter include certain more gritty bands as well as calcareous fossiliferous beds, and the full section is about 35 ft thick. The shales are buckled and sheared, and it seems certain that one at least of the anticlines in the lower part of the group is due to the squeezing of the beds over the mass of supposed Wentnor grit underlying them. This mass probably stood up as a small crag during the deposition of the lowest Harnage Shales.

The Shales can readily be traced eastwards from the quarry, where a band of coarse-grained, gritty, fossiliferous sandstone at or near the base makes a strong feature. The sandstone is exposed [45059148] (Locality 17) 340 yd N. 10° W. of The Hough, and again about 75 yd to the west-south-west. The Shales are also exposed in the stream to the south of this ridge. About 450 yd N.N.E. of The Hough their outcrop is terminated by faults.

Hope Bowdler. Neptunian dykes of Harnage Shales age occur at the quarry [463925] in Uriconian rocks 475 yd E. 20° N. of Dryhill. They are up to several inches thick and consist of sandy mudstone with a shelly fauna which was described in detail by Strachan and others (1948). Harnage Shales are exposed in places on the slopes south and south-west of Hope Bowdler, dipping generally south-eastwards at 12 to 15 degrees, and they were turned out of a well [471924] 500 yd W. 5° S. of Hope Bowdler church. In the banks of the road [47399244] (Locality 18) to Church Stretton, 200 yd W. 10° N. of the church, a foot of conglomerate at the base of the Shales rests on Uriconian tuff. The fossiliferous shale above the conglomerate contains an extensive fauna. A comparable section at Upper House (Plate 7c) [475926], 200 yd N. 15° W. of the church, has been described by Cobbold (1900, pp. 52, 103). Here about 2 ft 3 in of conglomerate rest on the Uriconian, and a fault with a small north-westerly downthrow cuts the section. The fragments of Uriconian rock forming the conglomerates reach 1½ inches in diameter at Upper House. An outlier of conglomeratic Harnage Shales with ostracods occurs on the north side of the road [473925] 290 yd W. 15° N. of the church, evidently an isolated pocket in the Uriconian. Poor exposures of Harnage Shales occur in the bed of the stream [479925] about 350 yd E.N.E. of the church and in an old lane [484930] 1000–1100 yd E. 40° N. of the church.

Harnage Shales are exposed in the stream flowing past Gutter Farm [498931]. The soft, weathered shales are commonly fossiliferous. The actual contact with the Uriconian occurs at the cottage [49019304] (Locality 19) 870 yd W. 4° S. of the farm, and the Shales are again exposed in the stream [49679301] (Locality 20) 140 yd above the farm. Steel-grey mudstones from a slightly higher horizon are exposed in the stream [502931] 420 yd east of the farm, and greenish grey striped silty shales crop out 100 yd farther downstream [50299309] (Locality 21). The dip hereabouts is 20° to the south-east.

Cwms to Hoar Edge. Harnage Shales form a horseshoe-shaped outcrop in the syncline north-east of Cwms Farm (p. 265). The northern closure of this outcrop [481952] lies about 1500 yd N. 30° E. of the farm but an isolated outcrop [48049550] (Locality 22) of the Shales occurs immediately east of the Lawley Fault 1800 yd N. 20° E. of the farm. In the stream [484948] about 1300 yd E. 43° N. of the farm, yellow micaceous shales are exposed. The dips recorded in this stream section vary from 45° to 90° to the north-west, the beds being presumably under the influence of the Cwms–Hoar Edge Fault which truncates them to the west.

Being relatively soft and readily eroded the Harnage Shales are not well exposed except in stream sections. There are no exposures of note along The Wilderness nor in the northern part of the outcrop, between Hoar Edge and Yell Bank. Blocky-jointed, olive-green, micaceous mudstones, near the top of the group, are exposed in the cutting of the old bridle-road [490960] 400 yd S. of Shootrough. Between 250 and 300 yd east-north-east of Shootrough [493965] an impersistent bed of flaggy, coarse-grained, feldspathic sandstone within the group makes a minor feature, with exposures showing dips of upwards of 55° to the south-east.

An inlier of shales, possibly of Harnage age, occurs within a triangle of faults some three-quarters of a mile south-west of Cardington. Behind a small garage [49949431] (Locality 31) 1160 yd S. 37° W. of the church there is an exposure of green shelly sandstone, with a westerly dip of 20°. The fauna from this sandstone, although poorly preserved, suggests that it may be part of the Chatwall Flags. G.H.M.

Chatwall Flags

Sibdon Carwood. South of Sibdon Carwood the Chatwall Flags are concealed by drift. Buff, fine-grained sandstone with flaggy siltstone bands containing *Balacrinus* [*Glyptocrinus*] *sp.* is exposed in an old quarry [409832] 450 yd W. of Sibdon Carwood church. The beds dip 53° to W. 15° N. and may be inverted. 200 yd to the north there is a track section [40868335] (Locality 23) in shaly-weathering, buff, micaceous siltstone, locally orange-stained, with fine sandstone bands towards the top. The beds dip steeply to the east. Both of these exposures are affected by the proximity of the fault F3, which runs approximately north-south along the outcrop of the Chatwall Flags, causing partial repetition of the beds. In an old quarry [410835] 500 yd W. 40° N. of Sibdon Carwood church 21 ft of fine-grained purple sandstone, dipping 34° to E. 10° S. are exposed, these beds being rather higher in the sequence than those just described. North of this quarry the Flags are unexposed for nearly half a mile, but give rise to a small feature and dip slope. B.A.H.

Long Lane to Cwm Head. Green micaceous flags, dipping to E. 10° N. at 60° to 68°, are continuously exposed in the stream for about 90 yd to the east of a place [408842] 940 yd E. 34° S. of the summit of Wart Hill. Green, micaceous, flaggy sandstone crops out on the southern side of the coppice [409844], 980 yd E. 23° S. of Wart Hill. The dip here is easterly, steepening eastwards from 65° to about 85° presumably due to the proximity in that direction of the fault F3. For 600 yd to the north this fault has the effect of widening the outcrop of the Flags, but they are not exposed. Beyond this area the outcrop is displaced to the west by an east-south-easterly fault and is no longer cut by F3.

Green micaceous flags are exposed in the stream bank [409852] 1010 yd E. 26° N. of Wart Hill, dipping at 70° to E. 35° S., and again in the ditch a farther 140 yd to the north, where the underlying shales are also exposed (p. 121). In the large old sandstone quarry [410852] 1150 yd E. 27° N. of Wart Hill the lowest beds are distinctly flaggy, and in the ground immediately to the west there is much debris of crinoidal flags. The beds in the quarry dip at 68° to E. 35° S. (p. 126). The Flags form a distinct feature over the ridge [409855] some 1250 yd north-east of Wart Hill, and in an old quarry [411856] just south of the road, 1420 yd N. 40° E. of Wart Hill, there are greenish brown, flaggy sandstones with brachiopods and crinoids, dipping at 42° to E. 5° S. Steeply dipping brown flaggy sandstone is exposed in the road [412857] between 500 and 600 yd S. 20° W. of Glenburrell and in an old quarry [413859] 320 yd S. 11° W. of that farm.

The Flags are poorly exposed in the River Onny [41328610] (Locality 24) and near Glenburrell, where they appear to be attenuated at the expense of the overlying sandstone. They occupy the steep western slope of Briar Edge and are visible in small outcrops 970 yd W. 40° N. [416866], and 1080 yd N. 35° W. [418868], of Upper Barn. Debris of green flags is common on the lower part of a steep wooded slope [423878] about 1000 yd south of Cwm Head church, and dark green sandstone and flags crop out near the top of the escarpment [424878] 920 yd S. 5° E. of the church. Bands containing an abundance of *Sowerbyella sp.* occur at the latter locality and the beds exposed appear to be at a horizon near the junction between the Flags and the Chatwall Sandstone.

The outcrop of the Flags is very broad north of the fault complex at Brokenstones, the rock being apparently repeated by a strike fault. At the side of a track [42158805] (Locality 25) 720 yd S. 15° W. of Cwm Head church about 90 ft of green, fine-grained flags are exposed with occasional massive beds up to 2 ft thick. The beds are generally very broken and unfossiliferous, but fragments of brachiopods and crinoids are present. Very steep dips prevail, the strike varying between N. 5° E. and E. 30° N. Behind the house at Blakemoor [42598810] (Locality 26) interbedded green micaceous flags and shales, with some crinoid debris, are well exposed, dipping at 60° to the east-south-east. North-north-eastwards the Flags are not exposed but fragments of greenish brown, flaggy sandstone are locally abundant, commonly with crinoid columnals and locally with Sowerbyellids. The north-eastern limit of this outcrop of the Flags is formed about 700 yd east-north-east of Cwm Head church by a north-westerly fault which throws down to the north-east, bringing Longville Flags to outcrop. D.C.G.

Little Stretton to Wall Bank. There are several areas underlain by the Flags in the faulted area between the Marshbrook valley and The Hough [451912], but exposures are poor. Debris of yellow-brown crinoidal sandstone is present in the soil [444908] about 900 yd W. 30° S. of The Hough. The Flags form the lower part of the steep wooded slope which runs from the eastern side of the main road [442909] about 1200 yd south of Little Stretton to a point about 250 yd north-west of The Hough. 10 ft of green fine-grained, banded sandstone is exposed in an old quarry [445911] 650 yd W. 4° S. of The Hough. The Flags were seen in 1953 in an old quarry 150 yd W. of The Hough, but are now obscured. Specimens of *Balacrinus basalis* from 'Marshbrook' are preserved in the Museum collections.

At The Hough the Flags are cut out by faulting and re-appear in the stream [461914] 1150 yd E. 13° N. of The Hough, where yellow flaggy sandstone dips at 34° to the south-west. From about 300 yd east of Dryhill [461923] the escarpment formed by the Flags can be followed to Soudley, and west and north of Haywood Common to Wall Bank. Exposures are few and small.

Cwms to Chatwall. A small outlier of Chatwall Flags, in the axial region of the syncline north-east of Cwms Farm (p. 265), caps the hill [479947] 900 yd N. 35° E. of the farm. Otherwise the Flags crop out in this area only east of the Cwms–Hoar Edge Fault. They occupy the lower part of the prominent escarpment which extends north-north-eastwards from the Sharpstones Thrust, and are well exposed in the lane [48589512] (Localities 28, 29, 30) 960 yd E. 17° S. of the summit of Caer Caradoc Hill where they dip at 65° to the south-east. The Flags have been quarried at Netchley, just west of Folly Bank, but no rock is now visible there. At Enchmarsh crinoidal flaggy sandstone from near the base of the formation is exposed in the lane [501967] leading to Hoar Edge, dipping at 55° to the south-east. Flags with many columnals of *Balacrinus* [*Glyptocrinus*] *sp.* are exposed in the lane [51289752] (Locality 33) 200 yd W., and in an old quarry [51379758] (Locality 32) 120 yd W. 34° N. of the road junction at Chatwall Hall. G.H.M.

Chatwall Sandstone

Sibdon Carwood to Whittingslow. In its southern outcrop the Chatwall Sandstone forms a prominent ridge between the depressions occupied by the Harnage Shales to the west and younger Ordovician and Silurian rocks to the east. South of the River Onny the dip slopes in the Sandstone are fairly extensive. It is obscured by drift as far north as the large disused Longlane Quarry [413842] (Localities 34, 35, 36) which was described briefly by Murchison (1839, p. 218). At its southern end, now partly obscured by tipped rubbish, 20 ft of massive, purplish, fine-grained sandstone are visible, weathering flaggy towards the top of the section. Abundant brachiopods occur in a band about 8 ft from the base. In its northern part some 52 ft of greenish and purple-banded sandstone are exposed. The lowest beds, seen in the north-western corner of the quarry, are mainly brown massive sandstones for the first 15 ft or so, above which the characteristic purple banding becomes conspicuous. Current bedding is common. Thin lenticular bands of shelly limestone occur at several horizons but are especially common, thicker, and more persistent in the higher beds. In the topmost 6 ft there are three persistent limestones between 4 and 9 inches thick, two of them apparently coalescing southwards to form a band of 1 ft 10 in. The limestones are commonly decalcified at the outcrop. Sowerbyellids and, to a less extent, Dalmanellids are the predominant forms in the fauna and both occur in places in the sandstone. Trilobites and gastropods are also present.

The highly inclined lowest beds described and figured by Murchison (*loc. cit.*) are not now visible. At the northern end of the quarry the beds dip at 26° to E. 20° S.; near the middle they dip at about 20° to E. 20° N. D.C.G., B.A.H.

About 180 yd east-north-east of Longlane Quarry, 1570 yd E. 18° S. of the summit of Wart Hill, is a smaller disused quarry [414843] in purple-banded dark green sandstone, of which some 24 ft are exposed, dipping eastwards at about 20°. The top 5 ft of rock are flaggy, apparently not merely as a result of weathering, but the lower beds are massive, commonly current-bedded, and locally slumped. An extensively decalcified limestone band, packed with the shells of *Sowerbyella sp.* occurs about 7 ft below the top, reaching a thickness of 2 ft locally. Several thin shell bands occur in the underlying 2 ft of sandstone. The higher beds of this quarry are also exposed in two smaller quarries some 80 yd to the south-south-east.

Purple sandstone debris is seen in several overgrown old workings [412844] in the copse some 1350 yd E. 16° S. of Wart Hill. A marked ridge extends north-north-westwards from Longlane Quarry past this copse and then turns northwards to reach the wood [412848] 600 yd west of Cheney Longville Castle. It is presumably formed by the lower beds of Longlane Quarry. A west-north-westerly fault, by which the Sandstone outcrop is displaced towards the west by about 350 yd, runs through the wood. For about 1 mile northwards the outcrop of the Sandstone is broken by the fault F3, the trend of which is nearly parallel to the strike of the beds. The fault has a westerly downthrow which causes a certain repetition of beds on either side of it and a consequent widening of the outcrop. At the southern end of the faulted part of the Chatwall Sandstone outcrop this effect is however counter-balanced by the increased dip of the lower beds as compared with the same beds farther south.

The sandstone is exposed in old quarries in the valley west of Cheney Longville. On the northern side of the valley [41208512] (Locality 38) 1280 yd E. 18° N. of Wart Hill, about 40 ft of massive sandstone are visible, generally purple in colour, but locally green, and commonly banded. Current bedding is very common and the rock is also slumped in places. Two thin bands of shelly limestone occur near the base, and several thicker bands are concentrated between 11 and 16 ft up and between 21 and 27 ft up. Strong joints dip very steeply to the N.E. and the beds dip at 17° to E. 40° S. Similar rock with a similar dip is exposed in two smaller old quarries in the plantation, 1350 yd E. 21° N. and 1410 yd E. 24° N, of Wart Hill, and in a third quarry [414853] at the road side 1620 yd E. 23° N. of that hill. The Sandstone is poorly exposed in the stream about 100 yd east of the big quarry. In contrast, the beds exposed in the old quarry [41018524] (Locality 40), farther west in the plantation, some 1150 yd E. 27° N. of Wart Hill, dip at 68° to E. 35° S. About 45 ft of purple and greenish brown sandstone are visible here, becoming markedly flaggy near the base in the transition to the underlying Chatwall Flags (p. 124).

In the south-eastern part of Longville Common the Sandstone has been worked in a group of old shallow workings [413855] on the eastern side of F3, some 1600 yd E. 30° N. of Wart Hill. Crushed brown and purple sandstone, considered to lie just west of the fault, is exposed on the common [413857] 1620 yd E. 38° N. of Wart Hill. In the western part of the outcrop the beds are vertical with a strike of N. 20° E., but a short distance to the east they dip at about 60° towards E. 40° S. Small exposures and old workings in the Sandstone occur in places between here and the old railway line.

The steeply dipping sandstone west of F3 is well exposed in an old railway cutting [414860] which extends for 130 yd to the S.E. from a place 1100 yd E. 8° N. of Lower Carwood. About 350 ft of dark green and purple sandstone are exposed, mainly massive, but flaggy in places, particularly near the base. A 3-in shelly limestone near the top and a thin band near the base are the only examples of this rock observed in this section. On a lithological basis the whole section is included in the Chatwall Sandstone although crinoid columnals, typical of the Chatwall Flags, are locally abundant in flaggy and massive beds in the basal 100 ft or so. The beds dip at between 80° and 90° throughout, to the east-south-east in the lower beds and to the west-north-west over most of the section. The dip changes within 5 yd, at the south-eastern end of the exposure, from vertical to 28° to the east-south-east, and it is here, where there is a short gap in the exposure, that the fault F3 is thought to cut the beds.

At least 150 ft, and probably as much as 200 ft, of sandstone crop out south-east of F3. About 90 ft in total are exposed in old quarries and cliffs [41558578] (Locality 37) immediately south-west of the railway, between 1250 and 1480 yd east of Lower Carwood. The lower rock is massive, current-bedded, purple and greenish brown sandstone with very occasional shell bands. The upper beds are similar, except that beds of shelly limestone, frequently rotted, are common in the topmost 20 ft or so, a limestone band, 2 ft 6 in thick, capping the exposed rock. *Sowerbyella sp.* is the dominant fossil; other brachiopods are locally common and Trinucleid trilobites also occur. These highest beds are closely comparable with the highest beds seen at Longlane Quarry (p. 125). The dip is about 8° towards the east or a few degrees more southerly.

Small exposures of sandstone occur in the river between 1330 and 1400 yd east of Lower Carwood. Between Glenburrell Farm and New House [418858] the sandstone has been worked in several quarries on the steep slopes immediately north-east of the road. In small quarries between 100 and 150 yd south-east of Glenburrell green and purple sandstone, with some bands of shelly limestone, is exposed. The rock, which is rather broken in places, dips at about 70° towards the east-south-east. In the principal quarry [41548593] (Locality 39) between 300 and 400 yd south-east of Glenburrell about 50 ft of beds are exposed, dipping very gently to the south-east. At least the lowest 20 ft of rock consist of dark green massive sandstone with many thin purple bands. It is current-bedded in detail and includes one or two thin bands of shelly limestone. The higher beds are rather less massive and less prominently banded, with more abundant bands of shelly limestone. Green is still the prevalent colour but in places the rock is uniformly purple. Use has been made in quarrying of prominent steep joints with a south-easterly trend, which are commonly marked by horizontal slickensides. Immediately north-west of New House [41738584] (Locality 41) 600 yd E. 37° S. of Glenburrell, some 18 ft of massive brown and green sandstone, dipping at 17° to the east-south-east, are exposed in an old quarry. A band of rotted limestone, packed with specimens of *Sowerbyella sp.*, occurs near the top of the section, and the top of the Chatwall Sandstone must lie at no great distance above.

Massive and flaggy, purple and green sandstone from near the base of the group is exposed in a small stream [416864], some 860 yd W. 25° N. of Upper Barn, dipping to the south-east at between 60 and 65 degrees. Higher beds form a prominent narrow ridge on the crest of Briar Edge, the constituent green, brown, and purple sandstones being exposed at intervals. Near the southern end they dip to the east-south-east at angles of between 70 and 90 degrees. In an old quarry [420870] 1080 yd N. 18° W. of Upper Barn, about 60 ft of massive and flaggy sandstone are exposed, with an inverted dip of 80° to W. 30° N. Vertical sandstone with the same strike is exposed to a thickness of about 30 ft in the old quarry [42228734] (Locality 42) at Woolston, 250 yd W.

33° N. of the road junction. A 15-in band of grey shelly limestone occurs near the top of the section. Apart from this the topmost 10 ft or so consist of fine-grained greenish grey sandstone with scattered shells of *Sowerbyella sp.* The underlying 20 ft consist of greenish brown sandstone with many purple bands and occasional thin limestone bands packed with Sowerbyellids.

North of the road at the Woolston quarry the Sandstone forms a distinct feature to within 200 yd of Blakemoor. Sandstone debris is abundant all along this feature, but the rock crops out only on a prominent knoll [425878] near its northern end and just west of the road. Small exposures of a massive brown sandstone occur here, the rock apparently dipping at a low angle to the north-east. This anomalous dip is uncertain, but the close fractures dipping steeply to the south-east appear to be joints rather than bedding planes. If the dip is correctly interpreted it is almost certainly a purely local contortion. The beds are thrown up on the northern side of an east-west fault following the depression south of Blakemoor, and the outcrops are dextrally displaced by about 120 yd. Much sandstone debris occurs in the field some 250 yd north-east of Blakemoor, and at the springs [429884] about 390 yd N.E. of that farm. The top of the Sandstone is indicated at the latter locality by an abundance of debris of the overlying limestone some 80 yd below the road. Pale brown and greenish brown, fine-grained sandstone is well exposed in the old quarry [430886] 660 yd N. 36° E. of Blakemoor. Specimens of *Sowerbyella sp.* are common and locally form thin bands of shelly limestone. The beds dip at 52° to E. 25° S., but on the top of the eastern face of the quarry they are more highly inclined. Similar beds crop out in the overgrown old quarry on the opposite side of the road, some 50 yd to the north. The most northerly outcrop of the Sandstone hereabouts is at the roadside [431887] 810 yd N. 35° E. of Blakemoor, and the distribution of the Cheney Longville Flags suggests that the rocks are cut at about 20 yd to the north-east, by a north-westerly fault with a considerable downthrow to the north-east.

The abundance of debris of green-brown sandstone, with many specimens of *Sowerbyella sp.* in places, at an old shallow working [423882] 370 yd W. 16° N. of Blakemoor, suggests that there may be a narrow strip of Chatwall Sandstone faulted down between Longmyndian rocks to the west and Chatwall Flags to the east. D.C.G.

Little Stretton to Wall Bank. The following section is seen in the railway cutting [44039050] 775 yd north of Marshbrook Station:

	ft
alternata Limestone (Localities 57, 58 and 59)	44
Chatwall Sandstone (Localities 45 and 46)	
Massive brown sandstone with 1 to 3-in bands of shelly limestone; flaggy towards top	10
Obscured	17
Massive sandstone, grey and purplish grey, medium-grained, with occasional fossil bands. Joints dip 40° to E. 40° N.	28

The beds dip at 57° to S. 35° W. Traces of the Sandstone are exposed in an old quarry [442903] on the east side of the main road, 500 yd N. 9° W. of Marshbrook Station, and small exposures of dark brown micaceous sandstone are seen in the track [445905] 200 yd N.N.W. of Oakwood. B.A.H.

Another outcrop of the Sandstone in this faulted area caps the escarpment west of The Hough. Isolated small outcrops of rock are present [445911] about 600 yd W. 8° S. of The Hough in which the dip is to the south-east at 33 to 36 degrees. To the north-west of Chelmick the Sandstone forms a small but prominent feature. It was formerly worked in a quarry [47119174] (Locality 27), now much overgrown, in the valley 870 yd S. 34° W. of Hope Bowdler church. The beds here belong to the lower part of the Sandstone and the upper part of the Chatwall Flags.

The best exposure in this area occurs at Soudley Quarry [47729182] (Plate 9A), 640 yd S. 24° E. of Hope Bowdler church. The section is as follows:

	ft	in
Head, Cheney Longville Flags, *alternata* Limestone	21	8
Chatwall Sandstone		
Fine-grained sandstone, current-bedded, purple, olive, and brown; soft; banding conspicuous; posts up to 2 ft 7 in thick; ½-in crinoidal parting 3 in up; (Locality 44)	6	0
Fine-grained sandstone, soft, purple, slightly, micaceous ..	4	0
Not exposed—probably sandstone	–	6
Fine-grained sandstone, purple and greenish brown, banded; bands with rounded flattened pellets of purple silty mudstone up to 1 in across all beds well banded but upper posts up to 4 ft 4 in thick; (Locality 43)	6	4
Soft band with small crinoid columnals and shells	–	–½
Fine-grained sandstone, purple and greenish brown, banded	2	9

The rocks dip south-south-east at 12°; the current bedding dips at about 15° to the south-east. Vertical joints strike S. 35° E. and E. 15° N.

The sandstone is softer than that at Chatwall or the flaggy sandstones of the Chatwall Flags and probably weathers more to a sand. It forms a readily worked freestone much in evidence in neighbouring walls and buildings.

In the ground near Common Farm the Sandstone can only with difficulty be separately mapped.

Cwms to Chatwall. East of the Cwms–Hoar Edge Fault the escarpment capped by the Chatwall Sandstone can be followed north-north-eastwards from the Sharpstones Thrust to Chatwall and beyond. Olive sandstone with onion-weathering is seen in a small exposure [485947] 1400 yd W. 34° N. of Cwms Farm. Purple, closely jointed sandstone with onion-weathering has been worked in a small old quarry [487951] 1860 yd N.E. of Cwms Farm. About 20 ft of beds with a reversed dip of 85° to the west-north-west have been worked, with some shelly bands near the top and bottom. Beds which are vertical or dip steeply to the south-east appear in the old quarry [48989562] (Localities 47 and 76) 1860 yd W. 15° N. of Cardington church, both Chatwall Sandstone and *alternata* Limestone being present.

The Chatwall Sandstone is exposed in the lane [497961] leading to Willstone, 80 yd from the cross roads at Folly Bank, and in a small quarry [49789622] (Locality 48) 50 yd east of the cross roads. Green, purplish, and yellow sandstone, with some pebbly bands is exposed at many places within the hamlet of Enchmarsh [5096] (e.g. [50159652] Locality 49). The Sandstone also crops out at a number of places on the road from Enchmarsh to Chatwall Hall, the pebble beds within it being well exposed. Here, as at Enchmarsh, the dip is south-eastwards at between 40 and 58 degrees. G.H.M.

ALTERNATA LIMESTONE

Cheney Longville to Whittingslow. Because of lack of exposure the *alternata* Limestone is not distinguished from the Cheney Longville Flags south of the latitude of Cheney Longville. Soft green flags with moulds of *Heterorthis alternata* form a small outcrop [414850] in the stream west of the village, 1510 yd E. 12° N. of the summit of Wart Hill. The rock is more extensively and typically exposed on the northern side of the road [41588524] (Locality 50) 1810 yd W. 13° S. of Wistanstow church. At one outcrop here 2 ft of grey flaggy limestone rest on 5 ft of green flags and shale, the index fossil being common throughout. At another, a thin rotted shelly limestone is developed within a group of grey flags. Green flags and grey bedded limestone with many specimens of *H. alternata*, and locally some Sowerbyellids, crop out for at least 70 yd in the old railway cutting [418857] some 1550 yd west of Wistanstow church, and under both banks of the River Onny [41828570] (Locality 52) a few yards to the north. On the left bank (Locality 53), 48 yd downstream from Locality 52, a 2-in band of grey limestone packed with specimens of *H. alternata* is overlain by about 2 ft of green shaly flags, with occasional partly decalcified bands containing that brachiopod and some Trinucleid trilobites. The beds dip at 12° to the east-south-east.

Northwards from the river the outcrop of the Limestone can be traced by sparse soil debris. A small exposure of green shale with many specimens of *H. alternata* occurs in the stream [418864] 670 yd W. 33° N. of Upper Barn. Green micaceous flags, dipping at a low angle to the east-north-east, which crop out between 50 and 95 yd upstream (Locality 54), must lie within the *alternata* Limestone group, since debris containing the index fossil occurs at a higher level to the north-east. In the roadside [42258733] (Locality 55) immediately south-east of the entrance to the Woolston quarry, interbedded green flags and grey limestone with abundant specimens of *H. alternata* form an outcrop between 25 and 30 yd wide, the beds being almost vertical, as in the quarry (p. 127). Debris of limestone containing *H. alternata* occurs near the spring [429884] 390 yd E. 43° N. of Blakemoor and in the soil [431887] 800 yd N. 38° E. of that farm. D.C.G.

Marshbrook. North-eastwards the next exposure of the *alternata* Limestone is in the railway cutting [44039050] 775 yd north of Marshbrook Station. Overlying the Chatwall Sandstone (Section, p. 128) are 44 ft of buff siltstone, shaly and micaceous, with shelly limestone in bands up to 1 ft thick. An 8-in limestone band marks the base of the formation. (Localities 57, 58 and 59).

Hope Bowdler. The *alternata* Limestone crops out in the lane [46619154] (Locality 61) 1400 yd W. 44° S. of Hope Bowdler church, where it was also seen in an excavation for a small reservoir. At Soudley Quarry [47729182], 640 yd S. 24° E. of Hope Bowdler church, the Limestone (Localities 66–71) overlies the sandstone for which the quarry was opened. It consists of only 3 ft 3 in of green mudstone, generally silty and shaly, with three bands of shelly limestone, 4 to 5 inches thick, and is overlain by the Cheney Longville Flags (p. 129).

Enchmarsh. The *alternata* Limestone occurs, with the Chatwall Sandstone, in an old quarry [48989562] (Localities 47 and 76) 1860 yd W. 15° N. of Cardington church, the beds dipping very steeply to the south-east. It is poorly exposed or turned up in the plough at many points south-west of Folly Bank, the dip at a place 770 yd W. 34° S. of the cross roads at Folly Bank being 66° to the south-east. It is also visible in a poor opening [499963] 190 yd E. 17° N. of the cross roads. The most northerly exposures of the Limestone within the Church Stretton district are at Chatwall Hall, where they occur in the road at the Hall itself and at the road junction [51379741] (Locality 78) 100 yd to the south-west. The stratigraphical relations of the Limestone at Chatwall have been discussed above (p. 111–2). G.H.M.

Cheney Longville Flags

Cheney Longville and Onny Valley. Largely obscured by boulder clay to the south, the Longville Flags are well exposed in the western part of Cheney Longville, from which village they take their name. Green micaceous flags near the base of the group crop out in the stream [415850] 370 yd W. 11° N. of the motte and bailey, dipping at 24° to S. 35° E. The beds exposed downstream and in the village dip to between east-south-east and east-north-east at angles ranging from 22° to 40°. Flags with a 6-in band of grey shelly limestone are seen in the stream bank [417849] 150 yd W. 8° S. of the motte and bailey, just north of the road. Green micaceous flags, weathered to a khaki colour, are seen in the road, and in both of the streams just south of it, south and south-east of the motte and bailey [419849]. Interbedded with these are rubbly silty beds, shales, and shelly limestone bands. Brachiopods, trilobites, and *Tentaculites sp.* occur sporadically throughout. On the eastern side of the road [41908493] (Locality 79) immediately east of the motte and bailey is a very fine exposure comprising more than 40 ft of beds. The section is here given in detail as typical of the upper part of the Cheney Longville Flags:

		ft	in
Khaki shale, rubbly and blocky	c.	4	0
Khaki sandstone	c.	1	0
Khaki shale, blocky	c.	3	0
Khaki flags, current-bedded	c.	2	0
Shelly limestone, decalcified		–	4
Khaki rubbly flags and blocky shale	c.	5	0
Grey and rust-coloured clay		–	3
Khaki flags, rubbly and blocky		1	3
Green flags and shale, weathering khaki; current-bedded and slumped locally; some grey and brown clay layers	c.	4	0
Green micaceous sandstone, finely laminated; in 2 main posts c. 8 in thick; current-bedded; shell band locally at base of upper post; transgressive base	c.	2	6
Khaki siltstone, blocky and laminated; some thin shale bands; transgressive base	c.	1	6
Green flags, current-bedded; siltstone wedge near middle	c.	1	0
Khaki shale, rubbly, micaceous; occasional flaggy bands	c.	3	0
Green micaceous sandstone	c.	–	6
Khaki siltstone, rubbly; some shales and flaggy bands	c.	6	0
Khaki flags, micaceous, finely laminated; shell layer at base	c.	1	0
Khaki siltstone, rubbly		–	6
Khaki flags, finely laminated, current-bedded		1	8
Grey shelly limestone; decalcified in top 2 in		–	7
Khaki flags, finely laminated; occasional rubbly bands	c.	3	0

The Flags form a long dip slope north-east of the road, but are not exposed. South of the old railway, the Flags are exposed in an old river cliff [42078546] (Locality 51) 1240 yd W. 7° S. of Wistanstow church. Approximately the same beds are exposed, and are easily examined, beside a farm track immediately to the south-west. They are mainly green flags, current-bedded in places, but with massive sandstone bands, up to 1 ft thick, in the upper half. Thin shelly crystalline limestones also occur. Altogether about 55 ft of beds are exposed, dipping at 20° to the east-south-east. In the river itself green flags are exposed in places (e.g. [42248542] Locality 80), but only in approximately the upper half of the group. At one locality the dip is only 7° to the south-east. Khaki flags and shales have been worked in two old quarries in the plantation 1270 yd W. 4° N. and 1180 yd W. of Wistanstow church. About 20 ft of beds, dipping at 17° to east-south-east, are seen in the more westerly quarry [420857].

Woolston and Bushmoor. Between the River Onny and Woolston the Cheney Longville Flags have a wide outcrop on what is fundamentally the dip slope of the underlying Chatwall Sandstone. Green and khaki flags, with some rotted limestone, crop out at Upper Barn [423860] and at the moat just north-west of it, dipping at 25° to east-south-east. A few exposures of flags, with some shell bands, and much rock debris, occur in the fields and road cuttings between Upper Barn and Woolston. The beds dip generally to the south-east at about 15 degrees. Green flags and sandstone are continuously exposed at the side of the road [423873] to Brokenstones between 90 and 210 yd north-west of the road junction at Woolston. The dip is steep and was recorded as 70° to W. 20° N. near the south-eastern end of the outcrops. The flags in the yard [424872] immediately west of the road junction are less highly, but more variably inclined, dips of 35° to S. 25° E. and 22° to E. 10° N. having been recorded. The fault F3 is considered to pass between the steep beds and those of more gentle inclination.

At the junction of the Whittingslow road and the old lane to Bushmoor [42508729] (Localities 81 and 82) khaki micaceous siltstone, variously flaggy and blocky, and locally calcareous, is well exposed. Brachiopods and large lamellibranchs are common here and the beds dip at only about 8° towards the north-east. Similar beds, with some thin limestone bands, crop out in places in the roadside banks for about 500 yd to the north and in the lane and streams between Woolston and Bushmoor. They dip generally at very low angles towards the north.

In the stream [43328783] (Locality 83) 1420 yd E. 38° S. of Cwm Head church is a small outcrop of greenish grey flaggy and rubbly siltstone, with bands of grey flaggy and nodular limestone, commonly decalcified. Large corrugated forms of *Kjerulfina sp.* are common and the abundant fauna includes also large Orthids, trilobites, and pentagonal crinoid columnals. The beds dip at 30° to S. 40° E., and, 30 yd downstream, at 43° to S. 15° E., in what is apparently a purely local flexure. Near the top of the steep northern side of this little valley are several old workings in flags, shelly limestone, and brown flaggy grit. In one of these [435879], 1020 yd E. 13° S. of Blakemoor, in a garden of a cottage, are two thin bands of limestone packed with specimens of *Sowerbyella sp.*, Dalmanellids, and *Tentaculites sp.* The flaggy siltstones are locally current-bedded within the thin posts, and the individual laminae are themselves undulatory in places. The general dip in this quarry is 8° to E. 40° N. Similar beds, dipping at low angles in a generally easterly direction, are exposed in a number of small sections (e.g. [43718801] Locality 86) in the Roman Road, between Bushmoor and a place [437883] 480 yd N.N.E. of the quarry. These beds are all high in the Cheney Longville Flags. Beds near the base of the group are exposed in an old quarry [43028834] (Locality 85) 850 yd E. 27° S. of Cwm Head church. The rocks are green micaceous siltstones, mainly flaggy below and blocky above, with trilobites and large coarse-ribbed brachiopods prominent in the abundant fauna. Here, just north-west of F3, the beds dip at 65° to E. 30° S.

Whittingslow to Marshbrook. In a well drilled in 1957 to a depth of 150 ft at a place [431887] 840 yd E. 2° N. of Cwm Head church, no trace of either the Chatwall Sandstone or the *alternata* Limestone was encountered although both crop out immediately to the south-west. Shales predominate in the driller's record, with two hard bands, 4 ft thick, which may represent limestone. The occurrence of green flags and grit with *Tentaculites sp.* in a ditch [429890] 700 yd E. 26° N. of Cwm Head church, and of green flags of Chatwall type a short distance to the south-west, helps to define the north-westerly course of the fault which must lie between the well and the Chatwall Sandstone outcrops (p. 271).

Green and brown sandstones and flags, with some shales and shelly limestones, are commonly exposed at the roadside and in the stream for about half a mile north-east from a place [430894] 950 yd E. 40° N. of Cwm Head church. Dean (1964, pp. 278–9) considered that strata belonging to the *alternata* Limestone occur at two places within this section. Due to the proximity of the faults F1 and F2 the dips in these outcrops are very varied. Large lamellibranchs are common at one locality, as at Woolston.

Only small exposures of weathered flags occur in the immediate vicinity of the farms at Whittingslow [432890], and on the steeper part of the slope to the east and south. In the lane [43578922] (Locality 87) north-east of the farms, 1450 yd E. 23° N. of Cwm Head church, there is a 50-yd outcrop of khaki micaceous flags with rotted shelly limestones. *Kjaerina typa* occurs here, characteristic of the Upper Longvillian Stage (Bancroft 1945, p. 183). The beds dip east at 47° and the outcrop is thought to lie just west of the line of F3. Brachiopods, trilobites, and *Tentaculites sp.* are very common in green micaceous flags extensively exposed in the stream [439888] between 1650 yd (Locality 84) and 1770 yd east of Cwm Head church. The direction of dip varies between E. 25° N. and S. 15° W. but the overall effect is of a low eastward dip. D.C.G.

An exposure [442888] 1420 yd W. 30° S. of Acton Scott church shows 2 ft of buff flaggy micaceous siltstone, dipping at 7° to E.S.E. At 380 yd N. of this exposure a stream section [44248914] (Locality 93) shows 56 ft of olive-green flaggy siltstone, with occasional bands of grey crystalline limestone, and with thin bands of coarse grit near the top of the section. These beds have a northerly dip of 5°.

The well-known quarry [44468901] in Marsh Wood, 1080 yd W. 28° S. of Acton Scott church, which was partly described by Bancroft (1945, pp. 195–6) and Dean (1958, pp. 209–10), is now largely overgrown. It is the type section for the Marshbrookian Stage (Upper Cheney Longville Flags) and has yielded extensive faunas. The following section can now be made out in the quarry and in the track to the south-east:

	ft
Track section and south end of quarry (Localities 89 and 90)	
Buff to olive-green siltstones and mudstones. Thin calcareous bands in the top few feet, with abundant fauna	30
Buff to olive-green micaceous flaggy siltstone, many 'gingerbread' bands (intermittent section)	c. 34
Central part of quarry	
About 10 ft of beds, equivalent to those at south end	
North end of quarry (Locality 91)	
Gap	c. 10
Buff to olive-green flaggy micaceous siltstone, with bands of silty mudstone and of fine-grained sandstone. Small and large scale current bedding. Abundant fauna	24

The highest beds seen in the track section are transitional to the Acton Scott Group (Dean 1958, pp. 209–10).

A stream section [447889] 770 yd N. 35° W. of Henley shows about 25 ft of grey to buff, flaggy, current-bedded sandstones and siltstones, overlain by gravel. The dip is 7° to S. 40° E. There is a poor exposure of the Flags in the railway cutting [443895] 350 yd south of Marshbrook Station.

In the Marshbrook area, and north-eastwards to near Ragdon, the outcrop of the Flags is divided into two areas, separated by a narrow outcrop of the Acton Scott Group which extends from F3 to about a mile north-north-east of Marshbrook, and is bounded on its southern side by a fault running about 200 yd south of Marshbrook Station through Swiss Cottage [450900] to Rag Batch [460911], the valley between Ragdon [458915] and Chelmick [468914]. The outcrops so far described lie to the south-east of this fault. B.A.H.

Grey shales and green flags, with nodular limestone locally, are exposed in the stream south of New House [4389], downstream from the outcrop of the *Pentamerus* Beds (p. 165). The most westerly outcrops are very small and are of shale, highly disturbed because of its proximity to F2. Flags which crop out about 220 yd downstream [43738973] (Locality 114) from the base of the *Pentamerus* Beds, dipping at 5° to the south-east, have yielded an Acton Scott Group fauna (Dean 1964, p. 281). Their relations with the surrounding rocks are obscure, and the area of outcrop is too small to show on the one-inch map. Between 85 and 120 yd downstream [43788974] (Locality 56) is an extensive exposure of Cheney Longville Flags, comprising green flags with two bands of shelly limestone near the base and shelly pockets and bands locally elsewhere. The lower beds dip at 35° to E. 40° S. but the angle increases downstream to 90° at the south-eastern end of the exposure. The fault F3 cuts across the stream near here. 100 yd farther downstream, silty flags with limestone bands which dip at 22° to the south, are exposed in the left bank [43928972] (Locality 88), 1580 yd W. 11° N. of Acton Scott church.

Beds low in the Marshbrookian Stage (Bancroft 1929, p. 34) are present at the roadside [43989002] (Locality 95) 370 yd E. 18° N. of New House. About 16 ft of buff-weathered flags and sandstones are exposed, dipping at low angles to between S. 30° W. and S. 40° E. Near the middle of the section an impersistent thin shelly limestone is developed at the base of a 1-ft band of massive sandstone, and a further 20 in below is a 3-in band of an unusual decalcified shelly grit ('gingerbread'). A similar grit was seen in old quarries west of Bushmoor. D.C.G.

On the south side of the road [44068976] (Locality 94) 180 yd W.S.W. of Marshbrook Station some 35 ft of buff, micaceous, flaggy and rubbly siltstones are exposed, with some massive bands of fine sandstone, thin shelly limestones, and a thin band of coarse-grained brown sandstone with many shells at its base. These beds, which dip at 30° to between S. and S. 30° E., have yielded an extensive fauna indicating a horizon in the middle part of the Flags. They are lithologically similar to the beds described in the preceding paragraph. Higher beds, also with a southerly dip, are exposed on the hillside 70 yd to the north and in a track section [44088990] (Locality 92) a further 50 yd to the north. They appear to be separated from the roadside exposure by a minor fault along the valley west-south-west of Marshbrook Station, throwing down to the north. D.C.G., B.A.H.

Marshbrook to Heywood Common. The junction of the Flags with the Acton Scott Group is sinistrally displaced for 200 to 250 yd by a north-north-westerly fault along the Marshbrook valley. The Flags cap the wooded slope which runs from the main road [442908] about 1200 yd south-south-west of Little Stretton Halt to a place [449913] about 250 yd north-west of The Hough. In this area flaggy sandstones crop out [445909] at 670 yd W. 21° S. of The Hough, and similar beds with calcareous bands are exposed in a stream section 150 yd to the north-east. The Flags also crop out in two small areas within the fault complex south-west of The Hough. They are present in an old quarry [442905] 525 yd W.N.W. of Oakwood, where greyish brown, micaceous, flaggy sandstone is exposed.

Traces of buff siltstone were seen 500 yd W. 39° S. of Oakwood, this rock carrying a fauna indicating a horizon high in the Flags. An old quarry [443902] 350 yd W. 6° S. of Oakwood shows 10 ft of alternating beds of buff micaceous flaggy and shaly siltstone, dipping at 25° to S. 25° E. The Flags form the dip slope north-east of Oakwood and between The Hough and Ragdon. They are well exposed in the stream (e.g. [45079094] Locality 98) 240 to 400 yd south of The Hough where they dip at 25 to 30 degrees to the south-east, and are also present in several small outcrops around Ragdon. To the east of Ragdon the outcrop ends against a fault running southwards from near Dryhill.

The Flags of the southern outcrop form most of the steep wooded slope in the angle between the Marshbrook–Craven Arms and the Marshbrook–Acton Scott roads, but exposure is very poor. An obscure exposure is seen in a gutter [44848995] (Locality 96) 785 yd W. 44° N. of Acton Scott church, with a fauna indicating a horizon high in the Flags.

An intermittent stream section [454904] in buff, flaggy, micaceous siltstone occurs 1000 yd north of Acton Scott church. The beds are folded, due to the proximity of the Swiss Cottage–Rag Batch fault to the north. 5 ft of micaceous siltstone and fine sandstone, with a 9-in limestone at the base, crop out in a track section [456902] 910 yd N. 15° E. of Acton Scott church, and higher beds, in the uppermost part of the Flags, are exposed in a hedge bank [45579010] (Locality 97) 110 yd to the south.

Disturbed flags with a dip of 50° to E. 35° S. occur in a stream [45669071] (Locality 99) 770 yd E. 37° S. of The Hough. Similar beds, anticlinally folded with dips of 12° to E. 35° S. and 50° to N. 30° W., are exposed in a cart track 300 yd to the north-east. Both exposures are affected by the adjacent Swiss Cottage–Rag Batch fault. At 120 yd to the east of the last exposure, on the east side of Rag Batch, brown flaggy sandstone with shale partings crops out in a cart track [45979089] (Locality 60), and

150 yd to the north-east along the track [461910] is a further exposure of similar beds. The faunas from these two exposures indicate a horizon near the base of the Flags. Immediately to the north the Chatwall Sandstone, which forms a marked scarp to the east, ends against a fault along Rag Batch. A stream section [461905], 1380 yd N. 37° E. of Acton Scott church, shows about 15 ft of buff fine-grained sandstone with siltstone bands. There is a 6-in lenticular band of rotted limestone about 7 ft from the base. B.A.H.

In the lanes and farmyards at Chelmick there are many outcrops, mainly of flagstone and flaggy mudstone with a south-easterly dip of from 10 to 20 degrees. A long dip slope extends below the hamlet towards Birtley. Fossiliferous horizons were found in the track [468913] 100 yd south of Chelmick, and in the stream [46759101] (Locality 100) 400 yd south of Chelmick, 1760 yd S. 29° W. of Hope Bowdler church.

At Soudley Quarry [47729182], 640 yd S. 24° E. of Hope Bowdler church, 15 ft 5 in of the Flags are exposed (Localities 72–74) above the *alternata* Limestone (p. 129). The beds comprise alternating and lenticular bands of current-bedded, greenish yellow sandstone and siltstone and silty mudstone, with a 4-inch shell bed 7 ft 6 in. from the base. Exposures of the Flags are common in the valley below the fish ponds at Soudley. 12 ft of flags and silty shales are seen in the stream [47789159] (Localities 62 and 63) 900 yd S. 16° E. of Hope Bowdler church, and a good section is exposed 100 yd farther downstream (Locality 64). On Heywood Common the Flags make a long dip slope facing south-east. The dip of the strata is, however, greater than that of the hill slope, measured dips ranging from 7 to 20 degrees. The rocks are exposed in a number of ditch sections on the Common as well as in streams and other natural exposures. About 70 ft of flaggy sandstone and shales were seen in a newly cleaned ditch [484923] 600 yd north of Hollies, and other similar exposures occur at a barn [48139177] (Locality 65) 350 yd west of Hollies and at Hollies itself [48459176] (Locality 101). The latter exposure is celebrated for the controversy about the age of the fossils obtained from there (Salter and Aveline 1854; Cobbold 1900, p. 37). The house, built on Caradoc rocks, adjoined a lime kiln built of the same material but in which *Pentamerus* Beds from a little downstream were burnt. The calcined remains of both rocks furnished a 'mixed' fauna reputed to come from Hollies.

Fossiliferous yellow silty mudstone is exposed [48939199] (Locality 102) 600 yd S. 26° W. of Common Farm, with flagstones 80 ft farther upstream (Locality 103). A small exposure [49639262] (Locality 75) 500 yd E. 15° N. of the farm, showing greenish yellow, fossiliferous, fine-grained sandstone, is probably in the lower part of the Flags.

Willstone to Chatwall. North of the Sharpstones Thrust the Cheney Longville Flags form a long dip slope between Willstone and Chatwall. The beds west of Willstone are greatly disturbed and evidently overturned, with dips of 45 to 62 degrees to the north-west in the lane [489952] 300 to 400 yd W.S.W. of the hamlet. A strike fault extending through Willstone is probably connected with these disturbances and bounds the outcrop on its eastern side.

Yellow flaggy siltstone is exposed in the lane [50019571] (Locality 104) 900 yd W. 43° N. of Cardington church. Near Enchmarsh there are exposures in old lanes below the village. Fossiliferous horizons occur in weathered rock [510963] 430 yd E. 11° N. of the cross roads at Folly Bank and at an outcrop [50329639] (Locality 77) 670 yd E. 17° N. of the cross roads. Fossiliferous flags are exposed in a ditch [51579738] (Locality 105) 190 yd S. 40° E. of the road junction at Chatwall Hall. G.H.M.

Acton Scott Group

River Onny to Woolston. The Acton Scott Group is not exposed south of the River Onny, but in the river the best exposure is in the right bank [42388538] (Localities 106 and 107. Localities 108–110 are a short way upstream) at a place 940 yd W. 16° S. of Wistanstow church. The rock here is a grey blocky mudstone, with two thin bands of clay, dipping at 22° to the east-south-east. About 6 ft of khaki-weathered blocky mudstone are exposed on the northern roadside bank [42418547] (Locality 111) 860 yd W. 9° S. of the church, and elsewhere within 40 yd to east and west. For about 750 yd to the north rubbly grey shales are seen in places in the stream which passes under the road just east of these outcrops. They include, at the bend 100 yd above the road [425856], a thin band of partly decalcified shelly limestone. Small brachiopods are common in the shales, the calcite of the shells being often preserved. In a small old quarry [425860] high on the left bank, 230 yd E. of Upper Barn, about 8 ft of beds are exposed beneath the drift. In the stream about 100 yd to the south the beds dip at 15° to the south-east. Many small exposures occur near the top of the steep slope of the right bank [42478581] (Locality 112), 270 yd S. 42° E. of Upper Barn. The rock is the usual khaki-weathered micaceous mudstone, poorly bedded, with abundant small brachiopods, as well as orthoconic cephalopods and crinoid columnals.

Brown-stained, khaki, blocky mudstone, from near the base of the group, is exposed in a roadside bank [42568648] (Locality 113) between 1150 and 1200 yd N. 36° W. of Wistanstow church. The lowest rock exposed here is a crystalline limestone of which 6 in is visible. These beds dip eastwards at about 15°. In the stream [42848674] 1310 yd N. 18° W. of the church, green micaceous, knobbly shales and flags are exposed, dipping at about 10° to the south-east. The characteristic fauna of small brachiopods including *Sericoidea?* occurs here, as elsewhere in the Acton Scott Group to the south. An isolated outcrop of flags of the Group occurs in the stream [43738973] (locality 114) 1900 yd E. 37° N. of Cwm Head church (Dean 1964, p. 281). The structural relations of this outcrop with the surrounding rocks are obscure, and its area is too small to be shown on the one-inch map. D.C.G.

Marshbrook. The Acton Scott Group has been recorded by Bancroft (1928, p. 190; 1949, p. 307) from the small wood 250 yd south-south-west of Marshbrook Station, but these exposures are now obscured. At the overflow from Chuney Pool [44608992] (Locality 115), 400 yd S. of Oakwood, 20 ft of roughly bedded, olive-green, micaceous siltstones are seen, folded into a broad anticline with an east-north-easterly axis. These beds have an abundant fauna. There are several small exposures of shaly siltstones and mudstones in the stream between places 400 yd E. and 400 yd S.E. of Oakwood. The dip is 25° to 30° to the north-north-west. Farther north in the same stream [45059053] (Locality 119), 600–730 yd south of The Hough, there is a good exposure of the lowest part of the Acton Scott Group. These beds are buff to olive-green, shaly, silty, micaceous mudstones, frequently orange-stained, with occasional calcareous bands up to 6 in thick. Lithologically they resemble the Harnage Shales of the Hope Bowdler area. They dip at 12° to S. 35° E., a return to the regional direction of dip.

Bushmoor, Acton Scott, Ticklerton. The Group is drift-covered and is not exposed west of the Craven Arms–Church Stretton railway. A poor exposure of dark green siltstone and fine-grained sandstone occurs in the west bank [44808850] (Locality 116) of the Quinny Brook, 380 yd W. 33° N. of Henley.

Between the railway and Acton Scott the Group forms a strong feature, due to the development of the Acton Scott Limestone which caps the high ground. Its outcrop is broken up by several small faults west of the village. The beds above and below the Limestone are virtually unexposed, but there are several good sections in the Limestone itself, and its outcrop is easily traced by debris in the fields. Fifteen feet of Limestone, here very calcareous siltstone or fine-grained sandstone, are exposed in an old quarry [44968955] (Locality 117) 470 yd W. 15° N. of Acton Scott church, dipping 5 to 10 degrees to S. 25° E. Similar beds are exposed in an old quarry by the roadside 80 yd E.S.E. of the church, and also at 150 yd to the north of the church. Fifteen feet of greyish blue sandy limestone, weathering yellowish brown, are exposed in the rock garden of Acton Scott Hall [45488928] (Locality 118), 200 yd S. 35° E. of the church. The most easterly exposure of the Limestone is in a stream section

560 yd E. 20° N. of the church. East of this the Limestone dies out, and the feature formed by the Acton Scott Group diminishes.

Twenty four feet of steel-grey, micaceous, silty mudstone, with bands of grey limestone, are exposed in the south bank of the stream [46459013] (Localities 120, 121), by the road bridge 1370 yd E. 35° N. of Acton Scott church. The fauna indicates a horizon in the lower part of the Acton Scott Group. There are occasional exposures of similar beds in the stream for 200 yd to the west, and there is a small exposure [46459025] (Locality 122) in the tributary stream 120 yd north of the bridge. This is probably the locality given by Bancroft (1945, p. 217) as type locality for *Onniella grandis* Bancroft, "a few feet above the base of the Acton Scott Beds".

Ten feet of buff-weathering, micaceous, flaggy siltstone, with sandstone bands, are exposed in a sunken track [466901] 700 yd N. of Hattongrove. These beds are higher in the sequence than those at the bridge. Fine-grained pale yellow sandstone with a southerly dip of 12° was seen in the stream 170 yd to the south. Eight feet of beds similar to those in the track section, and probably at about the same horizon, are exposed in a farmyard on the north side of the road at Hatton [469904], 1020 yd N. 19° E. of Hattongrove, and 2 ft of fine-grained, micaceous yellow sandstone crop out in a farm entrance on the south side of the road 140 yd to the west-south-west, dipping at 8° to the south-east. These siltstones and sandstones in the higher part of the Acton Scott Group form the small ridge on which the village of Hatton stands.

B.A.H.

Grey earthy mudstone with shelly bands is exposed in the stream [47929104] (Locality 123) 975 yd S. 36° W. of Hollies, and similar beds occur in the stream [481912] 740 yd W. 35° S. of Hollies. The general dip of these beds, which are overstepped by the Silurian some 200 yd north-east of the latter exposure, is between 15 and 20 degrees to the south-east.

Cardington. In the area north of Cardington Hill the Acton Scott Group, while retaining its predominantly argillaceous character, includes two thick bands of sandstone which form extensive conspicuous outcrops.

The basal beds, estimated to be about 300 ft thick, are so badly exposed that very little is known of them. Assumed to be shales, they occupy the low ground at the foot of the dip slope of the Cheney Longville Flags, extending some distance up the face of the escarpment formed by the overlying sandstone.

The lower sandstone, probably about 50 ft thick, crops out in the farmyard at Willstone [492952], where about 14 ft of yellow, sandy and silty flagstones are exposed, dipping southwards at 30°. Immediately to the west the Group is faulted against the Cheney Longville Flags. Eastwards from Willstone the sandstone can be followed by the feature it forms, and yellow sandstone is exposed at the front gate of Cardington vicarage [50499520] (Locality 124), 135 yd W. 24° N. of Cardington church. The feature runs north-eastwards from the village to about 400 yd from the church where it is displaced to the east for some 300 yd by an east-west fault. The feature, on which the mapping of the sandstone to the north is dependent, cannot be followed beyond a locality some 300 yd north of Plaish Park [521971].

The shales which succeed the sandstone occupy a 'slack' between its outcrop and that of the sandstone above, but are exposed only in the sunken lane [514959] 550 yd S. 39° E. of Leyhill where, beneath 4 ft of drift, there are 2 ft of yellow, micaceous, shaly mudstone with calcareous bands.

The shales are thought to be some 60 ft thick and are succeeded by another sandstone, about 75 ft thick, which has a wide outcrop by reason of its long dip-slope. South of Willstone and east of Cardington the outcrops are followed by the features they form, but the solid rock is concealed by drift deposits. Abundant fragments of fossiliferous yellow flagstones were found in the fields [51709630] (Locality 125) 1700 yd N. 43° E. of Cardington church. The feature continues to Plaish Park where a temporary pit [51989676] (Locality 127) furnished fine-grained, fawn-coloured, shelly flagstones. Yellow flags and shales were seen in another temporary pit [52889646] (Locality 126) 125 yd west of Plaish Hall. Just north of the limit of the Church Stretton district, about a mile north-east of Plaish Park [537977], the sandstone is overstepped by the Silurian. About half a mile from that house its feature is cut by a prominent cross-strike valley.

The highest beds in this area are shales which occur between Gretton and Plaish. They are drift covered, the only exposure being in the brook [512945] 30 yd south of the Mill, Cardington. At this locality, earthy grey mudstones have yielded the trilobite *Tretaspis ceriodes favus* Dean (Dean 1963, pp. 8–9) which suggests a high horizon in the Acton Scott Group. G.H.M.

Onny Shales

Wistanstow. The principal outcrop of the Onny Shales in this area is in the well known cliff section beside the River Onny [42588536] (Locality 128. Localities 129–131 are a short way upstream) some 720 yd W. 22° S. of Wistanstow church. The topmost 30 ft or so of the group are exposed here and consist of green micaceous shale, weathered to a blue-grey colour in places and locally exhibiting onion-weathering. A band of grey crystalline limestone, 1 inch thick, occurs within a few inches of the unconformity, discovered by Salter and Aveline (1854), which separates these Ordovician rocks from the overlying Silurian (p. 169). The dip of the Onny Shales is difficult to determine but appears to be about 22° to E. 30° S. North of the Onny the beds are largely obscured by superficial deposits. Weathered grey shale is exposed to a thickness of 2 ft in the stream [433865] 985 yd N. of Wistanstow church, and poorly exposed, rusty-weathered grey shale occurs below the drift in the railway cutting [43888666] (Locality 132) between 1330 and 1530 yd N. 30° E. of the church. D.C.G.

Felhampton to Hatton. Between the railway exposure and Marsh Mill [448881], 400 yd S.W. of Henley, the outcrop is obscured by drift. 50 yd N. of Marsh Mill 6 ft of buff, shaly, micaceous siltstone are exposed below the weir. This exposure is probably within the Onny Shales, although no fossils were found to confirm this belief. Similar beds, also unfossiliferous, are present in the south side of the road [451884] 125 yd N.N.W. of Henley, dipping at 15° to S. 35° E. These may be the beds referred to by Salter and Aveline (1854, p. 68) as '*Trinucleus* shales' near Henley. Exposures of definite Onny Shales are seen in the disused railway cutting [45268851] (Locality 133) 250 yd N. 40° E. of Henley, and intermittently for 300 yd to the east, and an exposure of buff, flaggy, micaceous siltstone occurs by the weir [46228942] (Locality 134) at Hatton Pool, 900 yd E. of Acton Scott church. The Onny Shales are overstepped by the Silurian to the south of Hatton. B.A.H.

References

BANCROFT, B. B. 1928. The Harknessellinae. *Mem. Proc. Manch. Lit. Phil. Soc.*, **72,** 173–96.

—— 1929. Some new genera and species of Strophomenacea from the Upper Ordovician of Shropshire. *Mem. Proc. Manch. Lit. Phil. Soc.*, **73,** 33–65.

—— 1933. *Correlation-tables of the stages Costonian–Onnian in England and Wales.* (Privately printed). Blakeney, Gloucestershire.

—— 1945. The Brachiopod Zonal Indices of the Stages Costonian to Onnian in Britain. *J. Paleont.*, **19,** 181–252.

—— 1949. Upper Ordovician trilobites of zonal value in south-east Shropshire (ed. A. Lamont). *Proc. Roy. Soc.*, (B), **136.** 291–315.

BULMAN, O. M. B. 1948. Some Shropshire Ordovician Graptolites. *Geol. Mag.*, **85**, 222–8.

COBBOLD, E. S. 1900. in *Church Stretton*, **1**, ed. C. W. Campbell-Hyslop. Shrewsbury.

DEAN, W. T. 1958. The Faunal Succession in the Caradoc Series of South Shropshire. *Bull. Brit. Mus. (Nat. Hist.) Geol.*, **3**, 191–231.

—— 1960a. The Ordovician Rocks of the Chatwall District, Shropshire. *Geol. Mag.*, **97**, 163–71.

—— 1960b. The Ordovician Trilobite Faunas of South Shropshire. I. *Bull. Brit. Mus. (Nat. Hist.) Geol.*, **4**, 71–143.

—— 1963. The Ordovician Trilobite Faunas of South Shropshire. IV. *Bull. Brit. Mus. (Nat. Hist.) Geol.*, **9**, 1–18.

—— 1964. The Geology of the Ordovician and adjacent strata in the southern Caradoc District of Shropshire. *Bull. Brit. Mus. (Nat. Hist.) Geol.*, **9**, 257–96.

—— and DINELEY, D. L. 1961. The Ordovician and associated Pre-Cambrian rocks of the Pontesford district, Shropshire. *Geol Mag.*, **98**, 367–76, pl. 20.

ELLES, GERTRUDE L. 1922. The Bala Country: its Structure and Rock-Succession. *Quart. J. Geol. Soc.*, **78**, 132–75.

LAPWORTH, C. 1879. On the Tripartite Classification of the Lower Palaeozoic Rocks. *Geol. Mag.* (2), **6**, 1–15.

—— 1916. [Vertical Sections of the Lower Palaeozoic Rocks of the Caradoc–Wrekin country.] *Sum. Prog. Geol. Surv.* for 1915, 38.

—— and WATTS, W. W. 1894. The Geology of South Shropshire. *Proc. Geol. Assoc.*, **13**, 297–355.

LA TOUCHE, J. D. 1884. *A Handbook of the Geology of Shropshire.* London and Shrewsbury.

MURCHISON, R. I. 1839. *The Silurian System*, Part I. London.

POCOCK, R. W., WHITEHEAD, T. H., WEDD, C. B. and ROBERTSON, T. with contributions by WRAY, D. A., STUBBLEFIELD, C. J., CANTRILL, T. C., and DAVIES, W. M. 1938. Shrewsbury District. *Mem. Geol. Surv.*

POCOCK, R. W. and WHITEHEAD, T. H. 1948. The Welsh Borderland. 2nd Edit. (1961 reprint). *British Regional Geology*, *Geol. Surv.*

RAMSAY, A. C. 1853. On the Physical Structure and Succession of some of the Lower Palaeozoic Rocks of North Wales and part of Shropshire. *Quart. J. Geol. Soc.*, **9**, 161–79.

—— and AVELINE, W. T. 1848. Sketch of the Structure of parts of North and South Wales. *Quart. J. Geol. Soc.*, **4**, 294–9.

SALTER, J. W. and AVELINE, W. T. 1854. On the 'Caradoc Sandstone' of Shropshire. *Quart. J. Geol. Soc.*, **10**, 62–75.

SEDGWICK, A. 1853. On a proposed Separation of the so-called Caradoc Sandstone into two distinct Groups; viz. (1) May Hill Sandstone; (2) Caradoc Sandstone. *Quart. J. Geol. Soc.*, **9**, 215–30.

STRACHAN, I., TEMPLE, J., and WILLIAMS, A. 1948. The age of the Neptunian Dyke at Hazler Hill. *Geol. Mag.*, **85**, 276–8.

STUBBLEFIELD, C. J. 1930. [Palaeozoic Rocks in Shropshire.] *Sum. Prog. Geol. Surv.* for 1929, pt. I, 87–8.

WHITTARD, W. F. 1952. A Geology of South Shropshire. *Proc. Geol. Assoc.*, **63**, 143–97.

—— with contributions by BALL, H. W., BLYTH, F. G. H., DINELEY, D. L., JAMES, J. H., MITCHELL, G. H., POCOCK, R. W. and STUBBLEFIELD, C. J. 1953. Report of Summer Field Meeting in South Shropshire, 1952. *Proc. Geol. Assoc.*, **64**, 232–50.

—— 1960. *Lexique Stratigraphique International*, **1**, fasc. 3 a IV, (Ordovician of Great Britain). Paris.

—— 1966. The Ordovician Trilobites of the Shelve Inlier, West Shropshire, Pt. 8. *Palaeont. Soc.*, 265–306.

Chapter 6

SILURIAN

GENERAL ACCOUNT

The Silurian rocks of the Church Stretton district occur in two chief areas of outcrop separated by the Church Stretton Fault Complex and the Eastern Uriconian axis. The rocks of the two areas (described as the Western Outcrop and the Main Outcrop) differ considerably in facies at some horizons, and separate classifications have been used in their description.

In the Western Outcrop, on the western side of the Church Stretton Fault Complex, Ludlow Series rocks extend northwards from the extreme south-west corner of the district to the Edgton area. Beds of the Wenlock and Llandovery series, with a general southerly or south-easterly dip, occupy the ground between Edgton and the southern end of the Long Mynd and occur on the western side of the Long Mynd as far north as the Wentnor area. Between the Church Stretton Fault Complex and the eastern margin of the Long Mynd, Llandovery and Wenlock beds, with a general south-easterly dip, extend from Plowden to the All Stretton area.

On the eastern side of the Church Stretton Fault Complex Silurian rocks extend in a continuous outcrop (the Main Outcrop) from Clungunford northwards to the Presthope area. The Wenlock and Ludlow series are fully developed along the whole length of the outcrop, but south of Cheney Longville the Llandovery Series is absent, cut out by the overlap of the Wenlock Shales. Around Clungunford the beds have an east-south-easterly or south-easterly dip of about 10°. North of a faulted area around Stokesay and Dinchope the 10° dip is maintained in a south-easterly direction.

The classification of the Silurian rocks of the Welsh Borders was initiated in the classic works of Murchison. In 1833 (p. 475) he divided the upper part of the rocks underlying the 'Old Red Sandstone' of the Welsh Borders into I. Upper Ludlow Rock, II. Wenlock Limestone (with which he included the limestone at Aymestrey) and III. Lower Ludlow Rock, the three divisions making up his 'Ludlow formation' (p. 477). The confusion of the Wenlock with the Aymestry Limestone was corrected in 1834 when he separated Ludlow rocks from Wenlock rocks in the following classification:

Ludlow rocks	Upper Ludlow rock Aymestry and Sedgeley limestone Lower Ludlow rock
Wenlock and Dudley rocks	Wenlock and Dudley limestone Wenlock and Dudley shale
Horderley and May Hill rocks	Flags Sandstones, grits and limestones

In 1839 he introduced the term 'Silurian System' and divided it (in descending order) into the Ludlow, Wenlock, Caradoc and Llandeilo formations. Salter and Aveline (1854) showed that there was a marked unconformity in Shropshire within the Caradoc formation of Murchison, with their '*Pentamerus* beds' (now Upper Llandovery Series) resting unconformably on 'Llandeilo and Bala rocks' (now the Caradoc Series of the Ordovician). Thus by 1854 the main divisions of the present day lithological grouping of the Silurian had been recognized. The further development of classification within the main subdivisions is considered under the sections dealing with the Upper Llandovery Series (p. 144), Wenlock Series (p. 148) and Ludlow Series (p. 155).

In the present account the classification adopted for the Silurian of the Main Outcrop is adapted from that used by Pocock and others (1938) for the Shrewsbury district to the north. For the Western Outcrop, differences in facies have necessitated some variation in this classification, particularly in the Wenlock and Ludlow Series. The upper limit of the Silurian System has, in conformity with most recent authors, been taken at the top of the Ludlow Series. Along the Main Outcrop this horizon is equated with the base of the Ludlow Bone Bed. In the Western Outcrop the Ludlow–Downton Series junction is not seen, and it is not known if the Ludlow Bone Bed is present. The detailed classification adopted is as follows:

	Approx. thickness in ft
Western Outcrop (West of Church Stretton Fault Complex)	
Ludlow Series (undivided)	
Grey and olive-green flaggy and shaly siltstones and calcareous siltstones	at least 2000
Wenlock Series	
Edgton Limestone:[1] grey, thinly bedded, fine-grained argillaceous nodular limestone interbedded with bands of grey, shaly mudstone, locally calcareous ..	125 to 200
Wenlock Shales: grey siltstones and silty mudstones, in places calcareous, with thin nodular limestone beds in the upper part, and calcareous nodules towards the base. The basal beds are commonly purple and greyish green in colour	1000
Llandovery Series (Upper)	
Hughley (Purple) Shales: purple shaly mudstones with subsidiary green bands	300 to 550
Pentamerus Beds (Shales and Limestones): grey, shaly siltstones and silty mudstones with thin bands of shelly limestone and sandstone	0 to 300
Pentamerus Beds (Grits and Conglomerates): purple and greenish grey coarse pebbly sandstones and conglomerates	0 to 180
Main Outcrop (East of Church Stretton Fault Complex)	
Ludlow Series	
Upper Ludlow Shales: olive-green and buff flaggy siltstones, commonly calcareous, with occasional thin limestone bands	100 to 400

[1]New name, see p. 150.

	Approx. thickness in ft
Ludlow Series—*continued*	
Aymestry Group: grey nodular crystalline limestone with siltstone bands; less calcareous in the north ..	80 to 210
Lower Ludlow Shales: olive-green and grey mudstones and siltstones with some thin limestone bands ..	600 to 850
Wenlock Series	
Wenlock Limestone: grey nodular crystalline limestone, flaggy and crinoidal in the north	0 to 50
Wenlock Limestone Reef Facies: irregular masses of unbedded reef limestone included in bedded nodular limestone with siltstone bands	0 to 90
Tickwood Beds: alternating bands of grey fine-grained argillaceous limestone and grey calcareous siltstone	50 to 180
Wenlock Shales (Buildwas and Coalbrookdale beds): grey siltstones and silty mudstones, locally calcareous, with thin nodular limestone beds in the upper part, and calcareous nodules towards the base	1000
Llandovery Series (Upper)	
Hughley (Purple) Shales: purple shaly mudstones with subsidiary green bands	0 to 250
Pentamerus Beds: grey, shaly siltstones and silty mudstones with thin beds of shelly limestone and sandstone	0 to 225
Kenley Grit: yellowish brown, coarse-grained sandstones, grits and conglomerates	0 to 75

The two major palaeogeographical features which influenced sedimentation in the Church Stretton district in Ordovician times, the relatively stable shelf area to the east and the intermittently subsiding basin to the west, continued to affect sedimentation at times during the Silurian Period. The boundary between the two areas lay in the general region of the Church Stretton Fault Complex, but it must be considered as a diffuse zone rather than one sharply defined along the line of the Fault Complex. The basal deposits of the Silurian, the Upper Llandovery Series, do not exhibit a contrast in facies between the shelf and the basin. They were laid down as the earliest deposits of a transgressive sea, and rest with strong unconformity on Ordovician and older rocks. In some areas, as for example around the Long Mynd, the basal deposits exhibit a marginal littoral facies with shoreline features such as pebble banks and sea stacks. Elsewhere the basal beds are of finer grain, comprising siltstones and mudstones. By the close of Upper Llandovery times the sediments were of uniformly fine grain (Hughley Shales) over the whole district.

The first obvious difference in facies between the sediments of the shelf and the basin is in the upper part of the Wenlock Series. There is a marked contrast between the shallow water reef-limestone facies of the Wenlock Limestone of the Much Wenlock–Presthope area and the graptolitic shales of equivalent horizon in the Long Mountain area (Das Gupta 1932) twenty miles to the west. Whittard (1952, pp. 170–1) and Dineley (1960, pp. 101–2) considered that the change from a shelly to a graptolitic facies took place across the Church Stretton Fault Complex, with the development of an intermediate facies at Brokenstones, within the Complex. However, as Dean (1964, pp. 285–6) has pointed out, and as has been confirmed in the present study, the westward change of facies is much more gradual, at least in the southern part of the district. The Wenlock Series sediments of the Clungunford–Broome area (east of the Fault Complex) and of the Edgton area (west of the Fault Complex) are essentially similar, consisting of a siltstone series with a mixed graptolite and trilobite fauna with the development of nodular limestones (Tickwood Beds, Edgton Limestone) in the upper part. In effect, this sequence can be taken as a facies intermediate between the shelly facies of Much Wenlock and the graptolitic facies of the Long Mountain.

The distribution of sedimentary facies in the Ludlow Series of Wales and the Welsh Borders was described by Holland and Lawson (1963). They concluded that, although a distinction between shelf and basin areas is valid, the boundary between them is not a simple Caledonoid line such as the Church Stretton Fault Complex, but has a curved form convex to the east. Moreover the position of the boundary fluctuated considerably during Ludlow times. In Upper Ludlow times there was no distinction in facies between basin and shelf, although the thickness of the sediments (calcareous shelly siltstones) continued to increase towards the west. This elimination of the distinction between the shelf and basin areas in the uppermost Silurian heralded the great palaeogeographical changes which were to take place during the Old Red Sandstone Period. B.A.H.

UPPER LLANDOVERY SERIES

Salter and Aveline (1854) proposed a three-fold subdivision of the Upper Llandovery Series (their '*Pentamerus* beds') of the Main Outcrop. These subdivisions, with variations in the names, were adopted by Whittard (1925, 1928) and Pocock and others (1938) as shown in the following table:

Salter and Aveline (1854)	Whittard (1925, 1928)	Pocock and others (1938)
Purple Shales	Purple Shales	Hughley Shales
Pentamerus limestone	*Pentamerus* Beds	*Pentamerus* Beds
Coarse grits of Church Preen, Kinley (sic) etc.	Arenaceous Beds	Kenley Grit

The terminology adopted by Pocock and others (Shrewsbury Memoir, 1938) has been followed in the present survey. All these subdivisions are present at the northern margin of the district, but as the outcrop of the Upper Llandovery Series is followed southwards there is successive overlap of each subdivision by that above until, south of Cheney Longville, the whole of the Upper Llandovery Series is cut out, and the Wenlock Shales rest on Ordovician rocks.

The stratigraphy of the Upper Llandovery Series of the Western Outcrop has been studied in detail by Whittard (1932). He concluded, on palaeontological grounds (1932, pp. 861–2), that the basal grits and conglomerates of this outcrop should be regarded as a basal facies of the *Pentamerus* Beds, and he classified the strata as shown below:

Whittard (1932)		Geological Survey	
Purple Shales		Hughley Shales	
Pentamerus Beds	Mudstone Phase	*Pentamerus* Beds	Shales and Limestones
	Arenaceous Phase		Grits and Conglomerates

Whittard's subdivisions have been followed in the present survey, with alterations in terminology (see above). The Hughley Shales are named from the equivalent beds of the Main Outcrop, while the two facies of the *Pentamerus* Beds have been given more accurate lithological descriptions.

The subdivisions of the Upper Llandovery are easily distinguished in the field, and there are only thin transitional strata between them. In the Western Outcrop there is some difficulty in distinguishing the Hughley Shales from the basal Wenlock Shales, which are commonly purple in colour (p. 151). The palaeontology of the Upper Llandovery Series of Shropshire has been described in detail by Whittard (1928; 1932; 1938) and his co-workers Pitcher (1939), Harper (1940) and Barker (Whittard and Barker 1950), and no additional evidence has been revealed in the present survey.

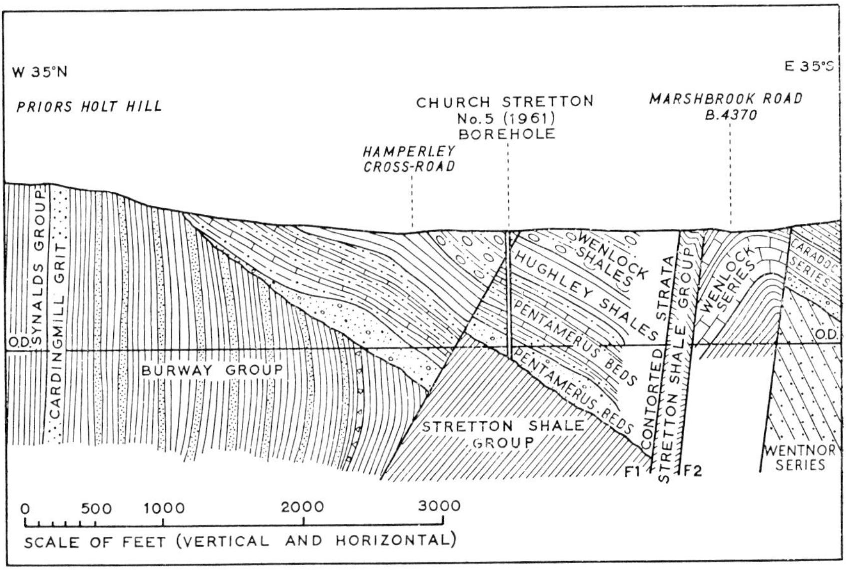

FIG. 13. Horizontal cross section at Hamperley showing the relationships between the Pre-Cambrian and Palaeozoic rocks

WESTERN OUTCROP (WEST OF CHURCH STRETTON FAULT COMPLEX)

PENTAMERUS BEDS (GRITS AND CONGLOMERATES)

On the western side of the Long Mynd, the Grits and Conglomerates are known only from three isolated exposures, and were not present in the Geological Survey Church Stretton No. 1 [37238978][1] or No. 2 [39409302]. boreholes. In contrast, the beds are up to 180 ft thick along the southern and eastern margins of the Long Mynd where there is a discontinuous series of outcrops from Plowden [383876] to Little Stretton [444921] which show many of the features of an Upper Llandovery shoreline. These features, which include sea stacks, beach deposits and pebble bars, have been described in detail by Whittard (1932, pp. 890–5). The beds dip away from the Long Mynd at angles up to 30°. This is, in part, an original depositional dip. A small outlier of *Pentamerus* Beds, including the Grits and Conglomerates, occurs immediately to the east of the Church Stretton Fault (F1) in the vicinity,

[1]Such numbers are eight-figure National Grid References within the 100-km square SO (32).

and to the north-east, of New House, Marshbrook [436899]. The basal beds, which have a north-westerly dip, are estimated by Whittard (1932, p. 863) to be at least 55 ft thick, and appear to rest unconformably on Ordovician rocks. The presence of 158 ft of sandstones and conglomerates resting unconformably on the Pre-Cambrian in the Geological Survey Church Stretton No. 5 Borehole [42178912] shows that the marginal facies extends locally for at least 800 yd from the present margin of the Long Mynd (Fig. 13).

The Grits and Conglomerates comprise purple and purplish brown, greenish grey and grey, medium to coarse-grained sandstones, pebbly sandstones and conglomerates. The conglomerates are generally poorly sorted, with subrounded pebbles of quartz and quartzite and subangular pebbles and fragments of local Longmyndian rocks in a matrix of coarse-grained sandstone. Longmyndian pebbles and fragments are most abundant in the lower beds. There is a tendency for the higher beds to be less conglomeratic than those near the base of the sequence. This can be demonstrated in some of the surface exposures, but was particularly noticeable in the Church Stretton No. 5 Borehole.

PENTAMERUS BEDS (SHALES AND LIMESTONES)

The outcrop of the *Pentamerus* Beds (Shales and Limestones) closely follows the margin of the Long Mynd from the Long Mynd Scarp Fault at Plowden to Little Stretton and probably thence northwards to Church Stretton. Locally the Grits and Conglomerates are absent, and the Shales and Limestones rest directly on Longmyndian rocks. Although the ground is largely drift-covered it has been possible to show that the outcrop of the *Pentamerus* Beds, together with the overlying Hughley and Wenlock shales, is affected by several faults. The structural interpretation of this ground put forward in this memoir (p. 264) differs in some respects from that proposed by Whittard (1932, pp. 882–4). The Shales and Limestones are also present in the All Stretton area (p. 202) and the outlier at New House (p. 166), and they form a small outcrop [421883] of uncertain extent north of Brokenstones.

The *Pentamerus* Beds (Shales and Limestones) consist of grey shaly siltstones and silty mudstones with many sandy laminae and small sandy lenses. Pale grey limestone bands and fine to medium-grained calcareous sandstone bands, usually less than 3 in thick but in places occasionally as much as 1 ft in thickness, are common. Most of the limestones are very sandy, but rare argillaceous bands occur. *Pentamerus oblongus* J. de C. Sowerby is abundant in most of the limestones and locally in the siltstones.

Whittard (1932, p. 868) stated that there is a gradual change in lithology from the sandy shales and sandstones of the southern part of the outcrop to soft mudstones around Hamperley and Minton. This has not been entirely borne out by the present survey, as the predominant rock type in the Church Stretton No. 5 Borehole at Hamperley [42178912] was siltstone with sandy laminae and lenses, and a similar lithological type predominated in the No. 3 Borehole in the All Stretton area (p. 202), four miles to the N.N.E.

The thickness of the Shales and Limestones is difficult to determine owing to faulting and an extensive drift cover, but they were estimated to be at least 300 ft thick in the Church Stretton No. 5 Borehole. B.A.H.

HUGHLEY SHALES

The Hughley Shales probably occupy most of the low ground between Wentnor and the Long Mynd, and southwards to Plowden. The exact limits of the outcrop cannot be determined as there is a thick drift cover, but it must extend at least as far north as the Church Stretton No. 2 Borehole [39409302] on Wentnor Prolley Moor. This borehole is about 1000 yd to the north of the position suggested by Whittard (1932, p. 860, fig. 1) for the base of the Llandovery. At the southern end of the Long Mynd the outcrop of the Shales is displaced southwards by about 300 yd by the Long Mynd Scarp Fault to the southern side of the River Onny. From this locality, the Shales have a continuous outcrop, largely obscured by drift, through Churchmoor Rough, Hamperley and Minton, to the All Stretton area. The general dip of the Shales is to the south-east or east at about 25°, but is more complicated in detail. The faulting in this area is described in Chapter 9 (p. 264). B.A.H., J.E.W.

The Shales show little variation in lithology. They comprise purple and maroon shaly mudstones, occasionally silty, with many green bands and patches. Thin calcareous laminae are common, and thin bands of shelly and argillaceous limestone, rarely more than 2 in thick, occur locally, notably in the Church Stretton No. 1 Borehole [37238978] (Montgomery Sheet 165). The thickness of the Shales appears to be about 300 ft over most of the outcrop, increasing to about 550 ft in the No. 1 Borehole. Although the Shales rest directly on the Pre-Cambrian in both the No. 1 and No. 2 boreholes, the *Pentamerus* Beds being absent, there is no development of a sandy facies at their base.

MAIN OUTCROP (EAST OF CHURCH STRETTON FAULT COMPLEX)

KENLEY GRIT

The Kenley Grit, the basal member of the Upper Llandovery Series of the Main Outcrop, occurs only in the northern part of the district. It extends southwards from the margin of the one-inch sheet [540977] north of Holt, to a place [526952] about two-thirds of a mile east of Gretton where it is overlapped by the *Pentamerus* Beds. The Grit, which comprises yellowish brown coarse-grained sandstones, grits and conglomerates, is about 75 ft thick at the northern limit of the district, but decreases rapidly in thickness towards the south.

PENTAMERUS BEDS

The *Pentamerus* Beds have a continuous outcrop from Wistanstow to the northern edge of the district. They have a general south-easterly dip of up to 20°. Between the southern limit of the Kenley Grit [526952] and Wistanstow they rest unconformably on rocks ranging from the Uriconian to the Acton Scott Group. Southwards from Wistanstow they are cut out by the overlap of the Hughley Shales. Most of the outcrop is drift covered, and exposures are few. Lithologically the *Pentamerus* Beds closely resemble the Shales and Limestones facies of the Western Outcrop, consisting essentially of grey siltstones and silty mudstones with shelly limestone bands. Their thickness is difficult to determine, but appears to be about 225 ft at the northern margin of the district, decreasing gradually towards the south.

HUGHLEY SHALES

The outcrop of the Hughley Shales extends from Cheney Longville to the northern limit of the district [556976], near Brook House. They have a general south-easterly dip of up to 20°. At Wistanstow they overlap the *Pentamerus* Beds and to the south rest unconformably on Ordovician rocks, and they are themselves cut out by the transgressive base of the Wenlock Shales near Cheney Longville. Lithologically the Shales closely resemble the Hughley Shales of the Western Outcrop. They are about 200 to 250 ft thick along most of the outcrop, but decrease in thickness south of the disused railway [462887] near Henley.

B.A.H.

WENLOCK SERIES

The subdivision of the Wenlock Series of the Main Outcrop adopted in this memoir is essentially lithological, and differs in several respects from that adopted by Pocock and others (1938) for the Shrewsbury District to the north. The two classifications are as follows:

Pocock and others (1938)		Present memoir
Wenlock Limestone		Wenlock Limestone Wenlock Limestone Reef Facies
Wenlock Shales	Tickwood Beds	Tickwood Beds
	Coalbrookdale Beds Buildwas Beds	Wenlock Shales

The Wenlock Shales outcrop is largely drift covered, and it has not proved possible to trace separately the Coalbrookdale and Buildwas Beds.

The Tickwood Beds, as originally described by Davidson and Maw (1881, pp. 102–3), comprise a transition zone between the Wenlock Shales and Wenlock Limestone. They estimated the thickness of the Beds to be 300 to 500 ft, but Pocock and others (1938, p. 112) restricted the subdivision to include only the 80 to 100 ft of shales and nodular limestones below their Wenlock Limestone. This lithological unit has been traced southwards on the Church Stretton Sheet.

Downie (1963) studied the 'hystrichospheres' (acritarchs) and spores of the Wenlock Shales in the Wenlock area and grouped the assemblages into three main types, characteristic of the lower, middle and upper parts of the Shales respectively.

The Wenlock Limestone as interpreted by Pocock and others (1938) can be subdivided into two lithological units within the northern part of the district. These comprise nodular limestones and siltstones (of Tickwood Beds lithology) with sporadic masses of reef-limestone (Wenlock Limestone Reef Facies), overlain by flaggy crystalline crinoidal limestone (Wenlock Limestone). Traced southwards the reef limestones die out and the crinoidal limestones pass laterally into bedded crystalline limestones which rest directly on the Tickwood Beds. Still farther south the bedded crystalline limestones pass laterally into bedded nodular limestones and eventually die out south of Craven Arms, resulting in strata of Tickwood Beds lithology being directly overlain by the Lower Ludlow Shales (Fig. 14). There is no evidence of unconformity at the base of the Lower Ludlow Shales, and it is probable that the strata of Tickwood Beds lithology south of Craven Arms are equivalent to the Wenlock Limestone, Wenlock Limestone Reef Facies and Tickwood Beds (or part of

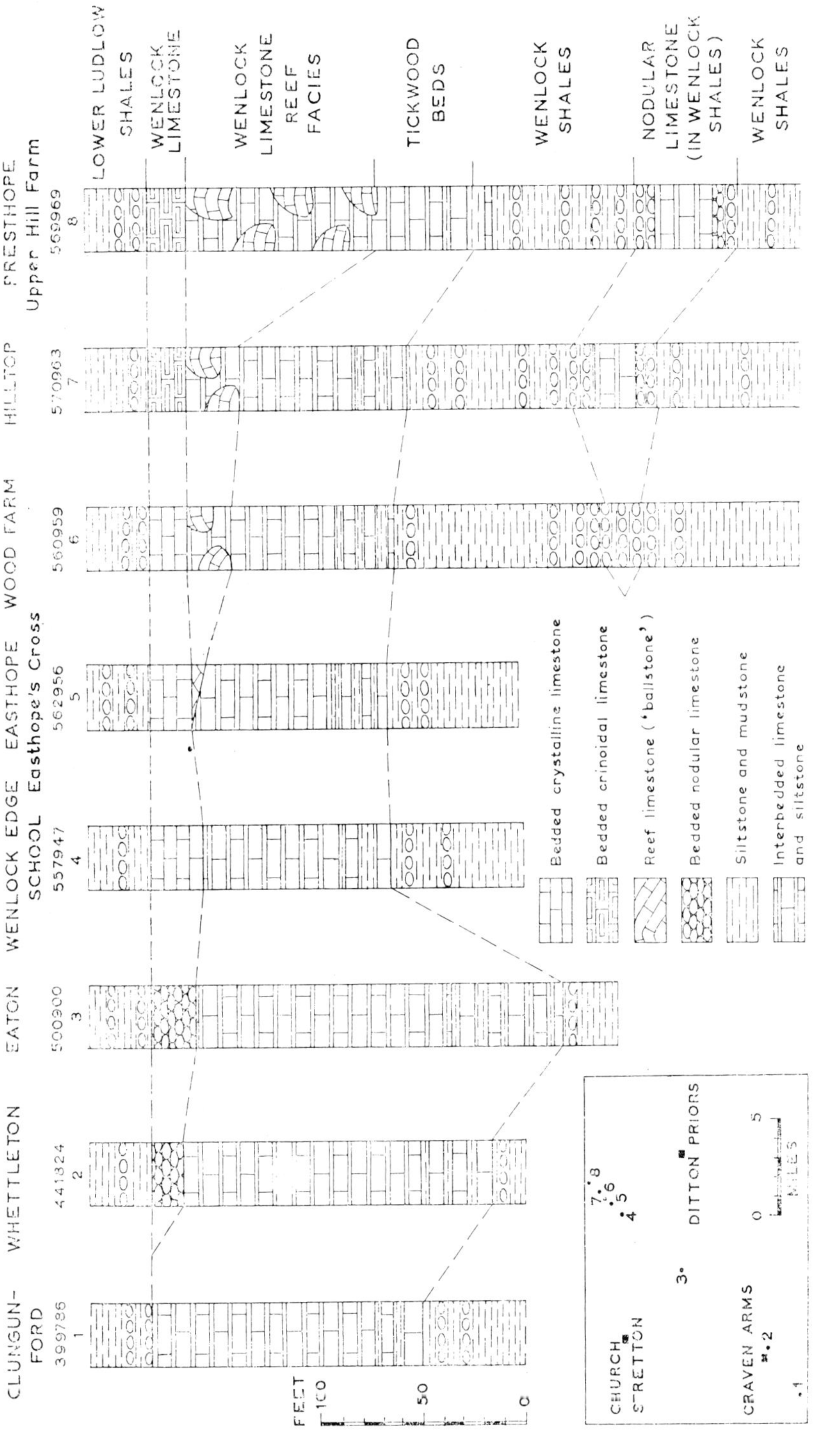

FIG. 14. Lateral variation in the Wenlock Limestone, Reef Facies and Tickwood Beds of the Main Outcrop

them) at the northern edge of the district. There is insufficient palaeontological evidence to determine whether the Tickwood Beds lithology is in fact diachronous, and consequently a purely lithological classification has been used in the preparation of the map. B.A.H., R.H.H.

In the Western Outcrop of the Silurian, the Wenlock Shales pass upwards into a group of beds in which layers of nodular limestone are abundant. These beds are here given the local name of 'Edgton Limestone' from the village of Edgton [386858] where they are well exposed. The Limestone, which resembles the Tickwood Beds of the Main Outcrop, is succeeded by siltstones and silty mudstones which closely resemble the Wenlock Shales below. No clear cut lithological boundary can be drawn within these higher siltstones and silty mudstones, which extend upwards into rocks of known Ludlow age. Since the Edgton Limestone is lithologically similar to the Tickwood Beds, and as no limestone is present similar to the Wenlock Limestone of Wenlock Edge, the top of the Edgton Limestone has been taken as the top of the Wenlock Series on lithological grounds. The fauna collected from the beds immediately above the Edgton Limestone is not sufficiently diagnostic to determine whether any part of the these beds is of Wenlock age, and they have been included in the Ludlow Series. D.C.G.

The Cwm Head Limestone of Dean (1964, p. 285) which occurs within a small area of Silurian rocks at Cwm Head [423887], has not been mapped separately in the present survey. Dean pointed out that it may be the lateral equivalent of the "calcareous beds of the west of F1 near Horderley" (i.e. the Edgton Limestone). Whittard (1952, p. 171) originally described the Wenlock Series of the Cwm Head area as being intermediate in facies between the true graptolitic and shelly facies, and brought into juxtaposition with the graptolitic facies of the Western Outcrop by the Church Stretton Fault (F1). However, Dean (1964, p. 285) stated that the lower Wenlock beds on the west side of F1 contain both trilobites and graptolites, and concluded that the apparent change in facies between the Cwm Head area and the Wenlock beds on the western side of F1 is due to a difference in age of the strata. He suggested that the succession at Cwm Head and on the western side of F1 is probably the same, but that this is difficult to prove owing to the paucity of exposure. B.A.H.

PALAEONTOLOGY (MAIN OUTCROP)

In the Wenlock Shales of the area under survey, the Buildwas and Coalbrookdale Beds have not been differentiated. The Wenlock Shales have a rich brachiopod fauna, typically with abundant *Atrypa reticularis* (Linné), *Brachyprion* cf. *arenaceus* (Davidson), '*Chonetes*' *minimus* (J. de C. Sowerby), *Dicoelosia biloba* (Linné), *Glassia obovata* (J. de C. Sowerby), *Leptaena rhomboidalis* (Wilckens), *Leangella segmentum* (Lindström) and *Lingula symondsii* Davidson. The most typical trilobite is *Dalmanites caudatus* (Brunnich), but species of *Proteus* and *Otarion* are not uncommon.

There is no distinct faunal change in the Tickwood Beds or in the succeeding Reef Facies and Wenlock Limestone, except that the Reef Facies has a rich coral fauna including *Favosites gothlandicus* Lamarck forma *forbesi* Milne Edwards and Haime, and *Heliolites interstinctus* (Linné).

Gothograptus nassa (Holm) has been identified from only one locality (9)[1] in the Wenlock Shales. More typical graptolites are *Monograptus dubius* (Suess), *M. flemingii* (Salter), *M. priodon* (Bronn) and *M. vomerinus* (Nicholson), only the first named species also occurring in the Lower Ludlow Shales.

J.D.D.S.

WESTERN OUTCROP (WEST OF CHURCH STRETTON FAULT COMPLEX)

WENLOCK SHALES

To the west of the Long Mynd, the Wenlock Shales outcrop is obscured by drift north of Myndmill Farm [382887], and its northern limit has been inferred from evidence from the Church Stretton No. 1 Borehole (p. 170). Southwards the outcrop extends along the margin of the district to Plowden Woods, where it is affected by faulting and swings eastwards through the woods to the River Onny. The dip also changes in direction from westerly at Plowden to southerly or south-south-easterly in Plowden Woods. A group of limestone bands towards the top of the Shales gives rise to the steep scarp along the southern margin of Plowden Woods to the west of Ridgway Hill. A small area of probable Wenlock Shales is present immediately north-west of Edgton, and another area (Fig 16) of supposed Wenlock Shales occurs within the Church Stretton Fault Complex to the south of Horderley. On the north side of the River Onny the main part of the outcrop extends north-eastwards to Little Stretton. To the south-east it is bounded by the Church Stretton Fault (F1) between the Onny Valley and Cwm Head, and by subsidiary faults from Cwm Head to Little Stretton. The Shales appear to have a general south-easterly dip, but much of the outcrop is obscured by drift. An outlier of the Shales at Minton [431907] is largely hidden by drift, and its extent is uncertain. At Cwm Head [424886] an area of the Shales occurs within the Church Stretton Fault Complex (p. 174). The Shales are also present in the All Stretton area (p. 202).

The Shales comprise about 1000 ft of flaggy and shaly grey siltstones and silty mudstones, locally calcareous, with thin nodular argillaceous limestone bands in the upper part, and calcareous nodules and bullions towards the base. The limestones in the upper part are locally sufficiently well developed to form a strong feature as, for example, in Plowden Woods. There is a gradual upward transition into the Edgton Limestone. In the Plowden–Horderley area the basal beds of the Shales are reddish brown and purple silty mudstones, with local greenish coloration. These 'Purple Wenlock Shales', which are up to 40 ft thick, do not represent a transition from the Hughley Shales below since, as Whittard (1932, p. 869) has pointed out, the upper part of the Upper Llandovery Series is absent.[2] Because of the extensive drift cover it is not known if these purple beds extend to the north of the River Onny.

---

[1]Such numbers refer to Silurian fossil localities listed in Appendix 2.

[2]Recent work on the fauna from Geological Survey Boreholes Nos. 1 and 5 by Dr. L. R. M. Cocks and Dr. R. B. Rickards, of the British Museum (Natural History), has shown that the Llandovery Series is complete, at least to the *griestonensis* Zone, with room for the development of the *crenulata* Zone above. This is overlain without angular discordance by the *riccartonensis* Zone of the Wenlock Series, which is complete to the *rigidus* Zone at the top of Borehole No. 1. Thus, contrary to Whittard's suggestion, the Llandovery sequence is almost complete, but the two lower zones of the Wenlock, those of *centrifugus* and *murchisoni*, are missing, at least in the south Long Mynd area.

EDGTON LIMESTONE

The main outcrop of the Edgton Limestone extends eastwards from the western margin of the district [380860] half a mile west-north-west of Edgton, to Horderley [408871] where it terminates against F1. The Limestone caps the prominent escarpment of Ridgway Hill [398866]. Although the Limestone has a regional dip to the south or south-south-east, there is much local variation, and at the eastern end of the outcrop, close to F1, the beds are highly disturbed.

There are two other small outcrops of the Limestone, separated from the main outcrop by two strike faults. The first extends eastwards from the western margin of the district to the village of Edgton, where it ends against a N.–S. Fault. The other occurs on the northern side of the small area of Ludlow rocks in Plowden Woods (p. 174), where it caps a steep north-west-facing scarp. It is also cut off to the east by a N.–S. fault.

The Limestone consists of grey, thinly bedded, fine-grained argillaceous nodular limestone interbedded with bands of grey shaly mudstone, locally calcareous. The individual bands of limestone are usually about 2 to 3 inches thick, and typically make up about 50 per cent of the rock. There appears to be a transition from the Wenlock Shales to the Edgton Limestone by a gradual increase in the proportion of limestone bands. The junction between the Limestone and the Ludlow Series is more clearly defined, because of a change to grey calcareous flaggy shales with a few bands and scattered nodules of limestone. The thickness of the Limestone is difficult to estimate owing to variations in the dip. In its main outcrop it appears to vary from 125 ft to 200 ft in thickness, being at a maximum at the eastern end of the outcrop near Horderley. In the most northerly of the three areas of outcrop it may be as little as 50 ft thick. The Wenlock Limestone of the All Stretton area is described on p. 203.

MAIN OUTCROP (EAST OF CHURCH STRETTON FAULT COMPLEX)

WENLOCK SHALES

The Wenlock Shales have a broad continuous outcrop which extends from the southern margin of the district, at Clungunford [395787], north-eastwards through Craven Arms [432828] and Rushbury [514919] to the northern margin of the district [568976] west of Presthope. North of Strefford [444857] the Shales occupy the broad drift-covered valley of Ape Dale, and the lower slopes of Wenlock Edge. South of Strefford, where the Shales occupy only low ground, they are almost completely obscured by drift. At the southern margin of the district, the outcrop of the Shales is limited to the west by the Church Stretton Fault (F1). North of New House [384798], near Coston, the Shales rest on successively higher divisions of the Caradoc Series, and north of Cheney Longville they overlie the Hughley Shales. Along the south-eastern limit of the outcrop there is everywhere an upward transition into the Tickwood Beds. The Shales have an east-south-easterly dip of some 15° from Clungunford to Broome [400810]. North of Broome the dip direction swings to south-easterly, at about 10° to 15°.

Lithologically the Wenlock Shales of the Main Outcrop resemble those of the Western Outcrop (p. 151). The basal part of the Shales contains numerous bands of calcareous nodules and limestone bullions up to 9 inches in diameter. These beds, probably equivalent to the Buildwas Beds of Davidson and Maw

(1881, p. 104), are also typified by the presence of a large number of small fossils, particularly brachiopods. There appears to be no development of basal Purple Wenlock Shales, although lack of exposures makes this difficult to confirm. Whittard (1928, p. 752) considered that the change in lithology from the Hughley (Purple) Shales to the Wenlock Shales is so sharp that an overlap junction is indicated. In the southern part of the outcrop the Shales clearly overlap the Hughley Shales on to the Ordovician. At the top of the Shales, the transition into the Tickwood Beds is not always continuous and at the northern end of the outcrop there is a development of Tickwood Beds facies within the upper part of the Shales (p. 177). The Shales are about 1000 ft thick throughout their outcrop.

TICKWOOD BEDS

The outcrop of the Tickwood Beds extends from the southern margin of the district, near Clungunford, to the northern margin, at Presthope. South-west of the River Onny, the Beds are overlain by Lower Ludlow Shales, and to the north-east by Wenlock Limestone or, north-east of Easthope Wood, by the Reef Facies. From Clungunford to Rowton [410805], the Beds form a low feature, largely covered by boulder clay. They have an east-south-easterly dip of about 15°. Between Rowton and Paddock [427821] the direction of dip changes to south-easterly, and the outcrop is affected by several faults, the most important of which, the Stokesay Fault, displaces it to the south-east by about 500 yd. On the eastern side of the River Onny, between the Stokesay and Bache faults, the Beds dip to the east at about 10°, and the feature formed by the Tickwood Beds and Wenlock Limestone is poorly developed. North-east of the Bache Fault, the direction of dip changes again to south-east at about 10°, and the Tickwood Beds and Wenlock Limestone cap the prominent scarp of Wenlock Edge.

The Tickwood Beds comprise alternating beds of grey fine-grained argillaceous limestone and grey calcareous siltstone. The limestone beds are generally tabular in form, although at the north-eastern end of the outcrop, in the Presthope area, they become more nodular. The limestone beds are usually 1 to 3 in thick, and in the main part of the Beds they make up about 50 per cent of the rock. Towards the base of the Beds the siltstones are progressively more dominant and there is a passage into the Wenlock Shales beneath. Between Stokesay [436818] and Easthope Wood [558952] there is a sharp upward transition into the Wenlock Limestone. Farther to the north-east, where the Beds are succeeded by the Reef Facies, the top of the Tickwood Beds is less clearly defined, as might be expected with the lateral, though spasmodic, development of reefs in strata of Tickwood Beds lithology. The Beds vary in thickness from 100 to 180 ft over most of their outcrop, but, mainly as a result of the facies change in the upper part due to the inclusion of reefs, decrease to about 50 ft in thickness north-east of Hilltop [567960] (Fig. 14).

WENLOCK LIMESTONE REEF FACIES

The outcrop of the Wenlock Limestone Reef Facies extends north-eastwards from Easthope Wood [558952] along the crest of Wenlock Edge to the northern limit of the district at Presthope. An up-faulted area of the Reef Facies lies to the east of Wenlock Edge, north of Easthope [567952].

The Wenlock Limestone Reef Facies consists of masses of reef limestone ('ballstone'), surrounded by bedded nodular limestone with siltstone bands. The surrounding strata, which are of similar lithology to the underlying Tickwood Beds, arch over and sometimes sag beneath the 'ballstones', probably as a result of differential compaction. The 'ballstones' are flat-bottomed lenticular masses which may be up to 80 ft in thickness (Murchison 1839, p. 211; Crosfield and Johnston 1914, p. 199). They are reef limestones, made up largely of the skeletons of reef-forming organisms, particularly corals and stromatoporoids, commonly in the position of growth and set in a matrix of bluish grey and green calcite-mudstone. They have been described in detail by Crosfield and Johnston (1914) who concluded (pp. 221–2) that they were formed under conditions similar to those behind a barrier reef or in a lagoon. They also concluded that, since the surrounding strata contain a large coral fauna identical to that in the 'ballstones' but not in the position of growth, the 'ballstones' underwent partial destruction by wave action before final burial. Butler (1939) has described reef masses from the Wenlock Limestone of the Dudley area. Hill (*in* Hill and others 1936, pp. 130, 131) described the ballstones of Wenlock Edge as having been formed "at a locus of intense activity of coral, stromatoporoid and algal growth". They were compared with the reef limestone of the Silurian of the Dudley area where Butler and Oakley (*in op. cit.*, p. 136) suggested that the fine matrix of the ballstone originated "as a chemical precipitate resulting from the decomposition of soft algae such as envelop many present-day reefs". These Silurian deposits were also compared with Jurassic coral reefs in the Oxford area. Hazel M. Johnson (1966) has lately described various species of algae from the Wenlock Limestones of Wenlock Edge and elsewhere.

The Wenlock Limestone Reef Facies reaches a thickness of about 90 ft at Presthope. From Hilltop southwards it thins rapidly, and dies out within half a mile.

WENLOCK LIMESTONE

The Wenlock Limestone has not been traced to the south of Stokesay. Its outcrop parallels that of the Tickwood Beds from Stokesay to Easthope Wood, and that of the Reef Facies farther to the north-east. The Limestone forms the higher part of the dip slope south-east of Wenlock Edge, with its base lying along the crest line of the Edge except where the Reef Facies is developed. An up-faulted area of the Limestone and the Reef Facies lies to the east of Wenlock Edge, north of Easthope.

At its most southerly occurrence, in the River Onny [43868221], the Limestone is a roughly bedded nodular finely crystalline argillaceous limestone with thin siltstone bands and partings. This lithology is maintained north-eastwards to the vicinity of Wenlock Edge School [557947] where there is a lateral change to a rather more evenly bedded, sometimes flaggy, nodular crystalline limestone. A further lateral change to a flaggy crinoidal crystalline limestone takes place between the north end of Easthope Wood [562956] and Hilltop [570963] (Fig. 14). The Limestone is 20 to 25 ft thick in the northern part of the outcrop, thickening to about 30 ft, and locally to 50 ft, south of Easthope before dying out at Stokesay.

A. View of Hope Dale between the Wenlock and Aymestry escarpments (A9493)

B. Callow Hill, part of the Aymestry escarpment (A9495)

Geology of Church Stretton
(*Mem. Geol. Surv.*)

PLATE 9

A. Colour banding in Chatwall (Soudley) Sandstone, Soudley Quarry (A8888)

B. Quarry face in bedded Wenlock Limestone (A9540)

C. Reef development in Wenlock Limestone (A9543)

Ludlow Series

Early work on the Ludlow Series rocks of the Welsh Borders was largely concentrated on the type area around Ludlow, to the south of the present district. Recently, various members of the Ludlow Research Group have carried out detailed studies of most of the Ludlow Series outcrops of Wales and the Welsh Borders, but comparatively little work has been published concerning the Church Stretton district.

In 1906, Elles and Slater divided the higher part of the Silurian (including the Temeside Group, now classed as Old Red Sandstone) of the Ludlow district as follows:

III. Temeside Group	F.	Temeside or *Eurypterus*-Shales
	E.	Downton-Castle or Yellow Sandstones
II. Upper Ludlow Group	D.	Upper-Whitcliffe or *Chonetes*-Flags
	C.	Lower-Whitcliffe or *Rhynchonella*-Flags
I. Aymestry Group	B.	Mocktree or *Dayia*-Shales
	A.	Aymestry or *Conchidium*-Limestones

Although their work was centred on the Ludlow district, they described a number of sections around Onibury and Norton. Their classification, with minor modifications, was adopted by Frances Alexander (1936) in her account of the Aymestry Limestone of the area between Ludlow and Craven Arms. She concluded that, although the Aymestry Limestone lay mainly within the Zone of *Monograptus leintwardinensis*, its base was diachronous and extended downwards into the *M. tumescens* Zone in the western part of the outcrop. She also considered that the overlying *Dayia* Shales rested unconformably on the Limestone.

The Aymestry Group and Upper Ludlow Group of the area north-east of Norton Camp was described briefly by Robertson (1927) in a paper dealing mainly with the Wenlock district. His estimates of 80 to 100 ft for the thickness of the Aymestry Group and 100 ft for the Upper Ludlow Group for this area have proved to be too small, especially in the southern part of the outcrop (p. 161 and p. 163). Shirley (1952) also gave a brief description of the Ludlow rocks north-east of Craven Arms, and related them to the succession in the 'basin' area of Builth Wells.

The Ludlow rocks of the area between Shelderton and Downton on the Rock, immediately to the south of the present district, have been described in detail by Whitaker (1962). In this area he described several submarine channels or canyons within the Lower Leintwardine Beds (see Table, p. 156), which he considered were formed at the margin of the 'shelf'. There is no trace of similar channels in the Brandhill and View Edge area immediately to the north.

In 1959, Holland, Lawson and Walmsley put forward a synopsis of a revised classification of the Ludlovian succession at Ludlow. This classification was amplified in their 1963 paper on the Silurian rocks of the Ludlow district and was correlated with the successions in other areas of Ludlow rocks of both shelf and basin facies in Wales and the Welsh Borders. They stated (1963, p. 98) that "The divisions [of this classification] in the Ludlow district have been mapped mainly on the basis of their faunal assemblages but lithological characteristics have also proved very helpful in their identification. These units are therefore biostratigraphical and to a lesser extent rock-stratigraphical".

Table 7 compares the Ludlow sequence established by Elles and Slater and Alexander with that of Holland, Lawson and Walmsley and with the subdivisions used by the Geological Survey in the View Edge area, south-west of the River Onny. Fig. 15 compares the stratigraphical relationships of the Ludlow rocks of the Norton Camp, Diddlebury and Bourton areas with those of the View Edge area and with the classification of Holland, Lawson and Walmsley.

B.A.H.

TABLE 7

Compiled from Elles and Slater 1906 and Alexander 1936	Holland, Lawson and Walmsley 1959, 1963		Present classification (area S.W. of River Onny)
Upper Whitcliffe or *Chonetes* Flags	Whitcliffian	Upper Whitcliffe Beds	Upper Ludlow Shales
Lower Whitcliffe or *Rhynchonella* Flags		Lower Whitcliffe Beds	
Mocktree or *Dayia* Shales	Leint-wardinian	Upper Leintwardine Beds	
		Lower Leintwardine Beds	Aymestry Group
Aymestry or *Conchidium* Limestones	Bringe-woodian	Upper Bringewood Beds	
		Lower Bringewood Beds	Lower Ludlow Shales
Lower Ludlow Shales	Eltonian	Upper Elton Beds	
		Middle Elton Beds	
		Lower Elton Beds	

The classification adopted in the present survey for the Ludlow Series of the Main Outcrop is essentially that used by Pocock and others (1938) for the Shrewsbury district to the north. It is basically a rock-stratigraphical classification, separating the calcareous beds of the Aymestry Group from the Upper and Lower Ludlow Shales, but it has been possible to relate these lithological divisions to the biostratigraphical subdivisions of Holland, Lawson and Walmsley (1963). The term Aymestry Group has been used (as in Pocock and others 1938, p. 117) in preference to Aymestry Limestone because, in the northern part of the district, the Group comprises beds which are calcareous siltstones and not limestones; consequently the term Aymestry Limestone if applied to these beds would be misleading. The top and bottom limits of the Ludlow Series have been taken at the base of the Ludlow Bone Bed and the top of the Wenlock Limestone (or Tickwood Beds where the Limestone is absent) respectively. The base of the Aymestry Group corresponds approximately with the base of the Upper Bringewood Beds over most of the district, although in the northernmost part of the outcrop it is possible that part of the

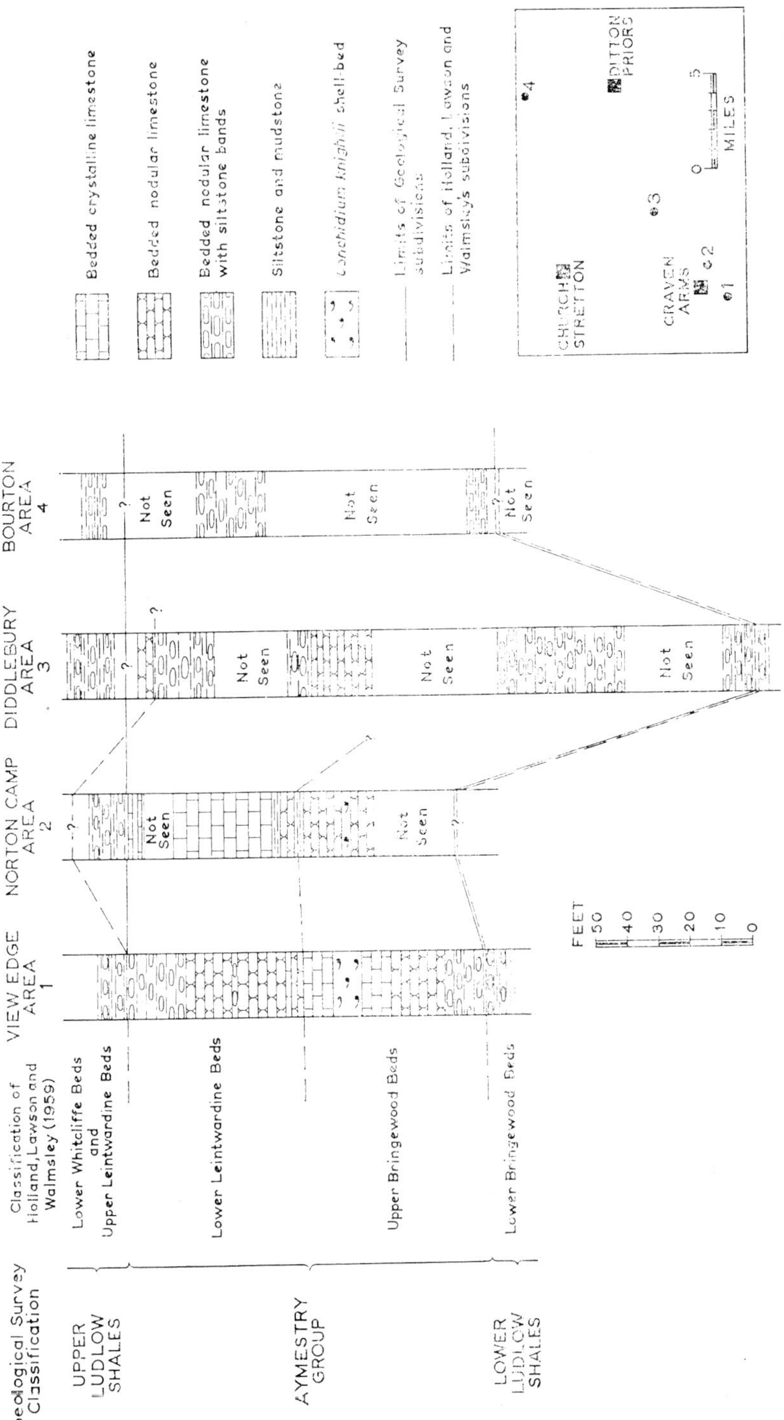

FIG. 15. Composite sections in the Aymestry Group of the Main Outcrop showing the lateral variation within the Group

Lower Bringewood Beds has been included within the Group. The Lower Bringewood Beds are essentially a transition facies between the shales of the Lower Ludlow and the limestones and calcareous siltstones of the Aymestry Group, and are here considered to be more closely related lithologically to the Lower Ludlow Shales. Elles and Slater (1906) and Alexander (1936) both appear to have included these beds within the Aymestry Limestone. The upper limit of the Aymestry Group shows considerable variation in its vertical position along the outcrop when compared with the biostratigraphical classification. At Brandhill [420790] and View Edge the Lower Leintwardine Beds are mainly limestones and have been included within the Group in their entirety. Farther east, on the eastern side of the Onny valley, around Park Farm [447802] and Norton Camp, only the lower part of the Lower Leintwardine Beds is calcareous and the top of the Aymestry Group has been taken within these Beds. As the Group is traced north-eastwards from Norton Camp it is less easy to define its upper limit as there is a more gradual transition upwards into the Upper Ludlow Shales. At Siefton Batch [476849] the whole of the Lower Leintwardine Beds are included within the Group, and to the north-east calcareous beds with abundant *Shaleria ornatella* (Davidson), probably equivalent to the Upper Leintwardine Beds of the Ludlow area are also included within it (Fig. 15).

The Ludlow Series rocks of the Western Outcrop have not been subdivided in the present survey since the comparatively small area mapped did not yield sufficient evidence for a satisfactory subdivision and correlation with the beds of the Main Outcrop. Ludlow Series rocks are also present in the All Stretton area, and there it has been possible to separate a calcareous division (Aymestry Group) from the Upper and Lower Ludlow Shales (p. 203).

B.A.H., D.C.G.

PALAEONTOLOGY

The lowest beds of the Lower Ludlow Shales have a fauna which is predominantly Wenlockian. Typical brachiopods which are not found above the lowest beds of the Ludlow Series are *Cyrtia exporrecta* (Wahlenberg), *Dicoelosia biloba*, *Plectodonta transversalis* (Davidson) and *Skenidioides lewisii* (Davidson). *Atrypa reticularis*, *Chonetoidea grayi* (Davidson) and *Leptaena rhomboidalis* are common members of this fauna, as is the trilobite *Dalmanites myops* (König). Species of *Monograptus* typical of the basal Ludlow are *M.* cf. *colonus* (Barrande), *M.* cf. *dubius* and *M.* cf. *priodon.*

Higher beds of the Lower Ludlow Shales are characterized by the commonness of Strophomenid brachiopods, particularly *Amphistrophia funiculata* (McCoy), *Brachyprion arenaceus*, '*Chonetes lepisma*' (J. de C. Sowerby), '*Fardenia*' *pecten* (Linné), *Howellella elegans* (Muir-Wood), *Isorthis orbicularis* (J. de C. Sowerby), *Leptostrophia filosa* (J. de C. Sowerby) and *Strophonella euglypha* (Hisinger) and the lamellibranchs *Paracyclas perovalis* (Salter) and *Cardiola interrupta* (J. de C. Sowerby) are common, and the most characteristic graptolites are *Monograptus chimaera* (Barrande), *M. dubius*, *M. tumescens* Wood and *M. varians* Wood.

In the southern part of the outcrop a prominent fossil of the limestones of the Aymestry Group is *Conchidium knightii* (J. Sowerby), and this species occurs prolifically in bands which are well seen in the View Edge quarries

[425808] and at Norton Camp [446820]. In the present survey the species has been found as far north as Callow Hill (Locality 164) about two and a half miles north-east of Craven Arms. *Atrypa reticularis*, '*Camarotoechia*' *nucula* (J. de C. Sowerby), *Isorthis orbicularis* and *Sphaerirhynchia wilsoni* (J. Sowerby) are prolific brachiopods in the limestones, and such species as *Dayia navicula* (J. de C. Sowerby), *Leptaena rhomboidalis*, *Howellella elegans* and *Strophonella euglypha* also occur abundantly. *Dalmanites* does not typically occur in the Group, although it has been found in association with a band rich in *Conchidium knightii* at Norton Camp. *Monograptus leintwardinensis* Lapworth has been found locally in the Aymestry Group, and also in the topmost beds of the Lower Ludlow (FGD 3697). At several exposures in and to the north-east of Siefton Batch *Protochonetes ludloviensis* Muir-Wood, *Salopina lunata* (J. de C. Sowerby) and *Shaleria ornatella* occur, but these species are not found in the limestones to the south and west. These three species are more characteristic of the Upper Ludlow Shales, and demonstrate the diachronism of the limestone facies.

The name *Dayia*-shales was applied by Elles and Slater (1906) to their upper division of the Aymestry Group. They contain typically *Dayia navicula*, and other common fossils are *A. reticularis*, '*C.*' *nucula*, '*Chonetes lepisma*', *I. orbicularis*, *H. elegans*, *L. filosa* and *Sphaerirhynchia wilsoni*. The higher beds of the *Dayia* Shales contain *Shaleria ornatella*, which commonly occurs in bands where the rock surface is covered by this characteristic brachiopod. The trilobite *Calymene neointermedia* R. and E. Richter has been found associated with *Shaleria ornatella* at localities 231 and 265. Beds containing *S. ornatella* have yielded *M. leintwardinensis* and *M. leintwardinensis* cf. *incipiens* Wood. South and west of Craven Arms beds yielding *S. ornatella* are typically non-calcareous siltstones, but they become progressively calcareous when traced towards the north-east along the Aymestry Group ridge. These beds are commonly characterized by a large form of '*C.*' *nucula*.

The succeeding beds of the Upper Ludlow Shales are the *Rhynchonella*-Flags of Elles and Slater (1906), and are characterized by the abundance of '*C.*' *nucula*. *Protochonetes ludloviensis* and *Salopina lunata* occur in these beds although these species are more typical of the succeeding siltstones—the *Chonetes*-Flags of Elles and Slater (1906). The *Chonetes*-Flags fauna is rich in lamellibranchs of which *Fuchsella amygdalina* (J. de C. Sowerby) and *Pteritonella retroflexa* (Wahlenberg) are typical. *Orbiculoidea rugata* (J. de C. Sowerby) occurs commonly, and *Campylites* [*Serpulites*] *longissimus* (J. de C. Sowerby) is characteristic. The trilobite *Homalonotus knightii* König has been found in the highest beds of the Upper Ludlow Shales immediately underlying the Ludlow Bone Bed, which is taken as the basal member of the Old Red Sandstone. Thelodont denticles collected 3 ft below the Ludlow Bone Bed at Burley (Locality 240), one mile west-south-west of Culmington, constitute the only trace of vertebrates found in the Silurian rocks of the district.

J.D.D.S.

WESTERN OUTCROP (WEST OF CHURCH STRETTON FAULT COMPLEX)

LUDLOW SERIES (UNDIVIDED)

The outcrop of the Ludlow Series extends from the south-west corner of the map northwards to the Edgton Limestone outcrop of Edgton and Ridgway

Hill. To the west it extends to the margin of the sheet, and eastwards it is bounded by the Church Stretton Fault (F1). A small faulted area of the Series occurs in the western part of Plowden Woods [381865], 1000 yd north-west of Edgton. In the southern part of the outcrop, as far north as Grist House [394850], the Ludlow rocks are folded into the Oaker Syncline (p. 262). In the northern part of the Ludlow Series outcrop, the beds have a general dip of about 30° to the south-east or south-south-east, with local minor complications.

The Ludlow Series comprises hard grey and olive-green, locally blue-hearted, siltstones and calcareous siltstones. Bands of argillaceous or silty limestone nodules occur sporadically, and slumped beds are present in a few localities. The beds are estimated to be least 2000 ft thick.

The Ludlow strata of the Western Outcrop have not been subdivided in the present survey (see p. 182). South and west of Hopesay the strata appear to be entirely of Upper Ludlow age. At Hopesay (p. 182) (Dean 1964, p. 283) calcareous beds of uncertain thickness, which may be equivalent to the Aymestry Group of the Main Outcrop, appear to be faulted against shales that contain undoubted Lower Ludlow graptolites. Lower Ludlow graptolites also occur in beds in the northern part of the outcrop, adjacent to the Edgton Limestone. Thus at Edgton Cross (Locality 317, p. 183) *Monograptus colonus compactus* Wood and *Spinograptus spinosus* (Wood) occur with a fauna typical of the Lower Ludlow, including *Dalmanites myops*. Similarly, Locality 292 (p. 183) has yielded numerous graptolites including *Monograptus bohemicus* (Barrande), *M. chimaera* and *M. chimaera* cf. *salweyi* (Lapworth), a fauna which, together with *Plectodonta transversalis* (Davidson), indicates Lower Ludlow age.

J.D.D.S., B.A.H.

MAIN OUTCROP (EAST OF CHURCH STRETTON FAULT COMPLEX)

LOWER LUDLOW SHALES

The outcrop of the Lower Ludlow Shales extends from the southern margin of the district, half a mile east of Clungunford, to the northern margin just east of Presthope. Between Clungunford and Clapping Wicket [421812], and also to the north of Wilderhope [545929], the lower part of the Shales sequence is obscured by drift, and the only available sections are in the higher beds on the slopes of the Aymestry Group scarp. From Clungunford to the Stokesay Fault the Shales have a general south-easterly dip of 5° to 15°, but between the Stokesay and Dinchope faults this is replaced by an easterly dip of about 10°. A south easterly dip of 10° is maintained over the rest of the outcrop from Lower Dinchope [451843] to the northern limit of the district. North-east of Lower Dinchope the Shales floor the long strike valley of Hope Dale between the ridges formed by the Wenlock Limestone and Aymestry Group.

The Shales are about 850 ft thick between Clungunford and Westhope [475867], but north-east of Westhope they thin to about 600 ft at Stanway Manor [527912]. This thickness appears to be maintained north-eastwards to the margin of the district, although Pocock and others (1938, p. 117) recorded a thickness of only 400 ft in the Shrewsbury district immediately to the north.

The Lower Ludlow Shales show little lateral variation in lithology within the Church Stretton district. The lowest 100 or 150 ft of beds comprise olive-green and grey siltstones, with layers of fine-grained limestone nodules particularly in the lower part where there is a sharp upward transition from the

Wenlock Limestone (or Tickwood Beds). These beds commonly have a speckled appearance due to an abundance of small shell fragments. They are equivalent to the Lower Elton Beds of Holland, Lawson and Walmsley (1959). The main part of the Shales, about 300 to 500 ft thick, comprises olive-green and grey flaggy mudstones and siltstones with occasional thin limestone bands, especially in the upper part. The fauna of these beds is predominantly graptolitic (p. 158), and they are equivalent to the Middle and Upper Elton Beds of Holland, Lawson and Walmsley (1959). The topmost 150 or 200 ft of the Shales, approximately equivalent to the Lower Bringewood Beds of Holland and others (1959), are transitional to the Aymestry Group and have been included in that Group (at least in part) by previous authors (see p. 156). They consist of flaggy calcareous siltstones with fine-grained limestone nodules, the limestone nodules becoming more abundant upwards and coalescing into nodular limestone bands as the base of the Aymestry Group is approached.

The Lower Ludlow Shales are also present in the All Stretton area (p. 203).

AYMESTRY GROUP

The Aymestry Group outcrop extends from the southern edge of the Church Stretton district at Saddle Hill [413785], about a mile east of Clungunford, to the northern limit near Henmoor Hill [596971]. In the area between Saddle Hill and the Stokesay Fault the beds have a general south-easterly dip of up to 10°, although this is locally affected by the minor faults between View Edge [425806] and the Stokesay Fault. Between the Stokesay and Dinchope faults the dip has changed direction to easterly or east-south-easterly at about 10°, although it is again affected by faults between Greenway Cross [460828] and Ireland [445839]. A south-easterly dip of about 10° is resumed on Callow Hill [461850] and is maintained to the northern limit of the district.

The Group is about 115 ft thick at View Edge, thinning to about 80 ft at Park Farm [447802]. It increases in thickness towards the north, being about 100 ft at Norton Camp [447820], 145 ft at Callow Hill and 200 ft near Diddlebury. This thickness is maintained to the Easthope area, but the Group thins to about 100 ft at the margin of the district. Robertson (1927, p. 96) had previously estimated the thickness of the Group to be 80 to 100 ft from the Wenlock area to within a mile or so of Norton Camp. Estimation of thickness is difficult owing to the paucity of exposure in many areas and the transitional nature of the base and top of the Group.

Unlike the Lower Ludlow Shales, the Aymestry Group exhibits considerable lateral variation when traced along the outcrop (Fig. 15). South-west of the River Onny, in the area around Brandhill and View Edge, grey nodular limestones (with siltstone bands in the lower part) form the basal 30 ft of the Group. These pass upwards into massive blue-hearted crystalline limestones, also about 30 ft thick. In the middle of these massive limestones there is a band, up to 12 ft thick, composed almost entirely of the shells of the brachiopod *Conchidium knightii* (Plate 10A). The same fossil occurs sparsely in the limestones above. Only one band of *C. knightii* has been found in this area, and it has been used as a marker horizon to elucidate the structure around View Edge. The massive limestones lie within the Upper Bringewood Beds. There is a rapid passage upwards into grey nodular limestones (Lower Leintwardine Beds) which are about 55 ft thick. Their upper limit is not well defined and there is a gradual passage into the Upper Ludlow Shales.

Across the River Onny, at Norton Camp, the lower part of the Group (Upper Bringewood Beds) is poorly exposed. It comprises about 50 ft of nodular grey limestone with thin siltstone bands, with a 3-inch band containing *Conchidium knightii* apparently at a similar horizon to the shell bank of View Edge (Fig. 15). The Lower Leintwardine Beds include about 8 ft of nodular limestone at the base, overlain by massive, bedded, grey crystalline limestone at least 30 ft thick. On the assumption that *C. knightii* marks a time horizon the massive limestone facies occurs later at Norton Camp than at View Edge. The upper part of the Lower Leintwardine Beds consists of siltstones with limestone nodules which have been included, in the present account, with the Upper Ludlow Shales because of their basically non-calcareous nature.

North-eastwards from Norton Camp the massive limestone facies dies out and the whole of the Aymestry Group consists of grey nodular limestones with siltstone bands and partings, with a transitional base and top. Within the Group there are several ill-defined calcareous subgroups of beds separated by less calcareous beds. The calcareous subgroups which comprise nodular limestones with thin siltstone partings as opposed to the alternations of limestone and siltstone of the less calcareous beds, may be lenticular in form although it has been possible to trace some of them individually for more than a mile. As the Aymestry Group is traced to the north-east it gradually becomes less calcareous in aspect, and at the northern margin of the district it consists chiefly of alternations of argillaceous nodular limestone and calcareous siltstone. The decrease in calcareous content is accompanied by a decrease in the height of the ridge formed by the Group. B.A.H.

One of the most important conclusions arising from the mapping and analysis of the Aymestry Group escarpment from the Dinchope Fault northwards is that the escarpment itself and the valleys and 'deans' either breaching the escarpment or situated on the main dip slope, are controlled by jointing. Directions of dominant joints have been noted as follows: (i) a few degrees east of north, (ii) a few degrees north of north-west and (iii) a few degrees north of west. A system of less well-defined joints appears to lie in a north-east direction. That part of the escarpment called Middlehope Hill [501879–508888] is a good example of a joint feature. The main face of the escarpment is a little under 100 ft high and strikes north-north-east in a straight line for nearly three-quarters of a mile. In that distance the boundary marking the base of the Aymestry Group rises in a northerly direction from the foot of the escarpment to near its crest. Also, in nearly all the valleys which breach the escarpment it can be demonstrated that the stratigraphical boundaries cross the main scarp face, which is controlled in detail by the joints.

The 'deans', situated on the dip slope of the Aymestry Group and Upper Ludlow Shales, seem to be similarly controlled by jointing. These mainly dry valleys have steep sides, and their courses are either straight or consist of a series of straight sections joined by sharp angular bends, here interpreted as an abrupt change from one joint direction to another. Aston Deans [511875] is an example of a straight valley controlled by one joint direction a little east of north, while the valley running north of Hungerford [537893] to Blackwood [527906] is an example of a valley controlled by more than one joint direction.

Lyell (1841) noted that although there were many valleys through the Aymestry Limestone [Group] scarp between Wenlock and the Onny valley, there were no corresponding breaks in the Wenlock Limestone scarp (Wenlock Edge). He concluded that this was largely due to the presence of thick massive limestone in the Wenlock Limestone and its absence in the Aymestry beds, since it was unlikely that the amount of fracturing and fissuring would differ greatly in the two limestones. He also pointed out that the reverse situation obtained between Shelderton and Aymestrey, where massive limestone is developed in the Aymestry Limestone and not in the Wenlock Limestone.

The main dip slope formed by the Aymestry Group and Upper Ludlow Shales is a strikingly regular feature forming the north-western side of Corvedale, broken only by the occasional valleys and deans referred to above. The actual dip of the rocks on the dip slope is greater than the angle of the slope and detailed feature mapping has made it possible to trace the outcrops of many thin groups of beds whose relative hardness is attributed to a higher proportion of limestone. One such bed has been mapped as the topmost bed of the Aymestry Group north-east of Siefton Batch. The beds immediately underlying this mapped bed are characterized by *Shaleria ornatella* and are probably equivalent to the Upper Leintwardine Beds of the Ludlow district.

R.H.H.

UPPER LUDLOW SHALES

The outcrop of the Upper Ludlow Shales extends from the southern margin of the district at Brandhill [420790] to the northern margin at Callaughton [620976]. The Shales have an easterly dip of 7° in the south, which swings to south-easterly in the region of the Stokesay Fault. The easterly dip is resumed between the Stokesay and Dinchope faults, but north-east of the Dinchope Fault the Shales have a south-easterly dip of 8° to 12°. The Shales form the main part of the dip slope of the Aymestry Group ridge and consequently their thickness is difficult to estimate with accuracy. Robertson (1927, p. 96), in a brief account of the Shales, estimated that they maintained a thickness of about 100 ft from Wenlock as far south as Norton Camp. However, the present survey has shown that they are about 400 ft thick at Brandhill, thinning north-eastwards to 200 ft at Corfton Bache [493854] and 100 ft at Bourton [595963].

Unlike the Aymestry Group, the Upper Ludlow Shales maintain a fairly constant lithology along the outcrop. They comprise olive-green and buff flaggy siltstones, commonly calcareous, with occasional thin platy and nodular limestone and silty limestone bands. The limestone bands are more common towards the base of the Shales. Towards the top of the Shales the siltstones become coarser and locally there are thin beds of fine-grained sandstone. The Concretion Band of Elles and Slater (1906, p. 198) (Plate 10B), now considered to be due to slumping (Whittard *in* Whittard and others 1953, p. 232), has been traced as far north as Bache [470819] and slumped beds also occur at Munslow Aston (p. 200).

B.A.H.

DETAILS

Upper Llandovery Series

WESTERN OUTCROP (WEST OF CHURCH STRETTON FAULT COMPLEX)

PENTAMERUS BEDS (GRITS AND CONGLOMERATES)

Western Escarpment of the Long Mynd. In three isolated outcrops between Myndtown and Plowden, each extending to no more than a few square yards, pebbly grit and conglomerate rest on the Pre-Cambrian rocks on the lower part of the steep western face of the Long Mynd. Of similar lithology to the basal Llandovery rocks to be described in the following paragraphs, the rocks in these outcrops are considered to be remnants of a spread of littoral Llandovery rocks banked against this face of the hill. Just south-east of the road [389888], 790 yd S. 6° W. of Myndtown church, the Portway Group shales are overlain with marked disconformity by a 4-in bed of quartz-conglomerate with fragments of weathered Longmyndian shale, followed upwards by green sandstone, which appears to dip gently to N. 40° W. Malachite veins are common in both the Pre-Cambrian and Llandovery rocks at this locality. A similar exposure is seen about 40 yd east of the road [388884], 1270 yd S. 10° W. of the church. The green pebbly grit, exposed to about 8 ft some 350 yd E. 19° N. of the church [394897], is probably of Llandovery age but could belong to the Wentnor Series.

South-eastern margin of the Long Mynd. The outcrops of basal *Pentamerus* Beds in Park Plantation [384875] have been described in considerable detail by Whittard (1932, pp. 867–8, 893–4, and 1952, p. 165) and no significant new information has been obtained during the recent survey. At the southern end of an old quarry [383876] 470 yd N.N.W. of the Mill, Plowden, the grit, dipping at about 30° to S. 25° W., rests directly on very steeply dipping shales. It is green and pebbly at its base, purple and finer grained above. Some 50 yd farther north the grit is seen to abut against an approximately vertical face of shale. No Llandovery grit is seen south of the Longmyndian inlier near the south-western corner of the plantation, the solid rock being obscured by drift.

To the east the Llandovery grit is next seen [390874] 630 yd E. 20° N. of Plowden Mill. The exposed rock is about 15 ft thick, green and purple, well bedded, and with small rounded quartz pebbles in places. Dipping at about 35° to about S. 15° E. it extends for nearly 200 yd to the east. Several spurs and embayments can be traced in the immediately underlying Longmyndian rocks.

Grits and conglomerates are extensively exposed between places about 530 yd W. 18° S. [393875], and about 250 yd west [395877], of Hillend Farm. In a path 445 yd W. 10° S. of the farm a gap of only 2 ft separates the Silurian outcrop from the Longmyndian. The Llandovery grit, which includes fragments of purple shale, forms beds of the order of 1 ft thick which dip at from 25 to 40 degrees towards S. 25° E.

A small exposure of purple pebbly grit, dipping at 30° to S. 35° E., occurs about 250 yd N. [398879] of Hillend Farm, and may be faulted against argillaceous *Pentamerus* Beds lying nearby to the north and north-east. About 425 yd N. 15° E. of the farm [399880] the grit is seen to end against a bank of Longmyndian rocks facing south-westwards, and it is not seen again, apart from small outcrops in Mount Gutter [403883] about 850 yd N. 37° E. of the farm, until the valley, Woolers Batch, immediately west of Churchmoor Hall, is reached. The abundance of grit debris indicates that the outcrop is probably continuous between these two valleys but its width is uncertain. Conglomerate debris is particularly abundant round the Longmyndian inlier just west of Woolers Batch. In the stream fine-grained conglomerates and pebbly grits crop out for about 190 yd downstream from a place [410887] 1490 yd W. of Cwm Head church. They dip at some 23° to E. 35° S. and have a calculated thickness of about 140 ft. Fragments of Longmyndian shale and sandstone and poorly rounded quartz pebbles occur, and there is a tendency for the beds to become finer-grained upwards.

The presence of grit debris, and a small outcrop [413893] 750 yd N. 12° E. of Churchmoor Hall, indicate the probable continuity of the outcrop from Woolers Batch to beyond the next valley, apart from a break round a narrow Longmyndian promontory on the right bank of the latter [414893]. The next outcrop extends from near a place [420899] about 780 yd north of Hamperley cross roads, where abundant debris of purple pebbly grit is so distributed as to show that the rock occupies a small embayment in the Longmyndian, the 'landward' margin being inclined steeply towards the south-east. On the spur [422901] on the right bank of Minton Batch purple pebbly grit is widely exposed, in close association with the Longmyndian. It dips at 25 to 35 degrees to about S. 30° E., but probably to the north-east on the northern side of the spur. Purple gritty sandstone is exposed across the valley on a wooded bluff [424901] some 50 yd north-west of the road bridge. The dip here is obscure, but about 100 yd to the north-north-west it is seen to be about 20° to south-south-east. In the latter outcrop the conglomerate consists of rounded pebbles of colourless and pale red quartz with some green siltstone and sandstone, in a matrix of sub-rounded quartz grains. D.C.G.

In the Geological Survey Church Stretton No. 5 Borehole [42178912], 490 yd N. 19° W. of Cwm Head church, 158 ft 3 in of the Grits and Conglomerates were proved, resting unconformably on Stretton Shales (Fig. 13). B.A.H.

Minton to Little Stretton. Exposures of the basal grits and conglomerates occur to the north and south of Callow Hollow [431900–435914] on the lower parts of the steep south-easterly slopes of the Long Mynd. The rocks consist of rather massive greenish grey and purple sandstone and grit which is locally pebbly and conglomeratic. The dip, where seen, is to the east-south-east at angles from 48° to 67°. The plane of unconformity with the Longmyndian is not exposed, but exposures of the two groups can be seen only short distances apart, and the base of the Llandovery can be drawn with reasonable accuracy in the area. There are no exposures of conglomerate on the eastern side of Callow Hollow, but debris of coarse pebbly grit is found [437916] 1200 yd N. 30° E. of Minton Manor House. Between this locality and Little Stretton there is no evidence of conglomerate, but at the mouth of Ashes Hollow there are exposures of conglomerate on both sides of the valley [441919, 442920]. On the lower part of the spur to the north of Little Stretton, 390 yd W. 13° N. of Little Stretton Halt, the conglomerate is a massive purple rock with a coarse sandy matrix containing rounded pebbles of quartz, quartzite and rhyolite. Thirty yards to the north, at the top of this spur, there is a small excavation which shows a section in a 3-ft band of conglomerate composed almost wholly of flattish pebbles of grey shale which have a rough-bedded arrangement. This band overlies the purple sandy conglomerate and, unlike the latter, appears to consist of material derived from the adjacent Stretton Shales. It is thought that the conglomerate in the Little Stretton area is separated by a fault from the Stretton Shales to the north-west. J.E.W.

Marshbrook. Near New House farm [437899] and the stream about 150 yd south-west of it Upper Llandovery grits are widely exposed, apparently resting unconformably on rocks of the Caradoc Series, the only older rocks exposed nearby. The weathered pebbly grit in the wood [435897] on the right bank of the stream is very broken and seems to be faulted at its north-western limit against grey shelly mudstone. In the stream brown and green grits overlie conglomeratic purple grits, the dip being about 20° to W. 20° N. in the lower beds, steepening upstream to 90° to W. on the line of the fault in the wood. Pebbly gritty sandstone, commonly with brachiopod moulds has been worked on the crest of the bank between the stream and the farm, where it dips at about 45° to north-north-west. Debris and small outcrops of grit and conglomerate are found up to about 600 yd N. 30° E. of the farm, near the termination of the outcrop against a fault.

PENTAMERUS BEDS (SHALES AND LIMESTONES)

South-eastern margin of the Long Mynd. Grey shales with thin bands of limestone are exposed to about 12 ft at the eastern end [388872] of extensive old workings between the road and the old railway 360 yd E. 14° N. of the Mill, Plowden. *Pentamerus oblongus* is very common in thin limestone bands in the old railway cutting [390871] some 200 yd to the east-south-east and in a road cutting [391873], made in 1959, 650 yd E. 10° N. of the Mill. In the latter exposure the beds must be near the base of the argillaceous group and are somewhat sandy in places. From west to east the dip, which is about 25°, changes in direction from S. 10° W. to S. 30° E.

In the old quarry [396876] 250 yd W. 25° S. of Hillend Farm grey shales and blue-grey limestones, generally decalcified near the top of the outcrop, are exposed to 18 ft, dipping at 25° to S. 15° E. Three layers of cream-coloured clay also occur. The limestone bands are packed with specimens of *P. oblongus*, and other brachiopods are common in both the limestones and the shales. There are small exposures of weathered shale west and north of the quarry and in the tracks about 350 yd north of the farm. A larger outcrop, with thin bands of shelly limestone, occurs in the stream [398879] 285 yd N. 11° E. of the farm. Small exposures are seen in a stream [408882] 1250 yd E. 30° N. of the farm, where the dip is steep to the south, and in the stream [411885] about 70 yd W. 33° S. of Churchmoor Hall, where brachiopods occur in somewhat cleaved sandy shale.

In a disturbed section in the stream [41308876] (Locality 1) 220 yd N. 39° E. of Churchmoor Hall interbedded grey and pale purple shaly and blocky mudstones include a 1-ft band of grey crystalline limestone resting on 4 in of blue-grey micaceous sandstone. *P. oblongus* occurs in the shales and is abundant in the limestone. Purple shales, taken to be the basal beds of the Hughley Shales, are seen upstream at a slightly higher horizon. The highest *Pentamerus* Beds, like those just described, are seen again about 150 yd upstream and in the stream [418893] about 240 yd W. 9° N. of Hamperley cross roads.

The argillaceous *Pentamerus* Beds are concealed by drift or cut out by faulting between Hamperley and Minton. Their presence is indicated by abundant debris of grey sandy shale and sandstone beside a spring [429905] 300 yd W. 31° S. of the Castle Mound at Minton. D.C.G.

The Church Stretton No. 5 Borehole [42178912], 490 yd N. 19° W. of Cwm Head church, proved 425 ft 9 in of grey siltstones with thin bands of sandstone and shelly limestone (Fig. 13). The true thickness of the beds appears to be atleast 300 ft, although an accurate estimate is not possible as they are affected by several faults of indeterminate throw. B.A.H.

Brokenstones and Marshbrook area. A small stream exposure [422883] 425 yd S. 28° W. of Cwm Head church has yielded to the late Professor Whittard (private communication) fossils diagnostic of the *Pentamerus* Beds. Similar shales are seen a further 25 yd from the church but the extent of the small outcrop of this formation is unknown.

In a small stream [43428983] (Locality 2) 200 yd W. 24° S. of New House there is a small outcrop of green soft mudstone with some thin bands of fossiliferous limestone. Visible faults in the main stream delimit the argillaceous *Pentamerus* Beds to north-west and south-east, but nothing is known of their extent in directions at right angles to the stream.

HUGHLEY SHALES

West of the Long Mynd. Exposures of Hughley Shales between the Long Mynd and the western limit of the district are restricted to the more deeply eroded valleys and the drift-free ground immediately adjacent to the Long Mynd. The most southerly exposure is in the ditch on the east side of the road [383877] 1070 yd S. of Myndmill Farm, where purple shales dip at about 20° to W. 20° S. In the lane 250 yd to the north purple shales, dipping steeply to west-north-west, are overlain by grey shales assigned to the Wenlock Series.

The Long Mynd Scarp Fault is exposed, although considerably overgrown, in a small old working on the east side of the road [388886] 1080 yd S. 12° W. of Myndtown church. West of the fault, which is marked on the southern face of the working by a 1-ft band of grey and yellow clay 15 yd from the middle of the road, purple and khaki shales dip steeply towards the west. This locality is Whittard's 'New Quarry' (1932, p. 884).

Purple and green soft shale forms a prominent bank on the east side of the road [390889] about 720 yd S. 4° W. of Myndtown church and similar rock with small brachiopod fragments has been worked some 90 yd to the south-west. D.C.G.

The Geological Survey Church Stretton No. 1 Borehole (Montgomery Sheet 165) [37238978], 1975 yd W. 7° N. of Myndtown church, proved the unusually large thickness of 584 ft 6 in of Hughley Shales (approximate true thickness 550 ft), overlain by Wenlock Shales and resting unconformably on the Pre-Cambrian. B.A.H.

North of Myndtown the only exposure of Silurian rocks which has been found occurs in a deep ditch [393928], 1050 yd E. 11° N. of Wentnor church. The rocks here are gently dipping, purple, flaggy siltstones with a rather poor fauna which suggests that they belong to the Hughley Shales. About 200 yd to the north the Geological Survey Church Stretton No. 2 Borehole [39409302], 1940 yd N. 17° W. of the chapel of Asterton, proved 216 ft 9 in of Hughley Shales, overlain by boulder clay and resting unconformably on the Pre-Cambrian.

While it is thought that the greater part of the boulder clay-covered hollow to the north and west of Asterton is underlain by Llandovery rocks, the limits of these deposits as shown on the map must be regarded as conjectural, there being no direct evidence, other than the above, on which to base the lines. J.E.W., B.A.H.

South-eastern margin of the Long Mynd. In frequent exposures on the steep right bank of the Onny, up to 300 yd west of the Mill [385871] at Plowden, the purple shales show very variable dips. At one place the inclination is 15° to south-south-east but near the Mill a figure of 37° to west-south-west is recorded. South-west of the Mill the upper limit of the formation is closely defined by the proximity of debris of purple shale to an outcrop of Wenlock Shales. In the right bank of the Onny [389870] 530 yd E. 18° S. of the Mill purple shales dip at 25° to about E. 35° S. Purple shales near the water level on the right bank [393871] 810 yd N. 31° W. of the Ridgway road junction (mid-way between Edgton and Horderley) seem to have a moderate southward dip. Greenish grey rocks, almost certainly of Wenlock age, occur higher on the bank, dipping steeply towards the south-south-west (p. 172). The purple and the grey rocks appear to be separated by a fault trending north-east and with a south-easterly hade. Whittard (1932, p. 869) considered the section to be unfaulted, the purple beds being typical of the basal purple shales of the Wenlock Series. No evidence diagnostic of their age has been obtained, but the appearance of faulting and consideration of the thickness of the Hughley Shales together suggest that they belong to that formation.

Purple shales inclined gently to the south are exposed in an old railway cutting [395873] about 1000 yd N. 7° W. of the Ridgway road junction. They are described by Evans (1957, pp. 13–4) as Purple Wenlock Beds, although his description does not in this respect accord with his map (pl. ii). The occurrence of purple shales at the top of the bank immediately south of the cutting suggests that the beds in the cutting itself belong to the Hughley Shales.

At Hillend Farm [398877] purple shales dip at 5° to the south-south-east. Upstream from a place [408877] 1050 yd E. 2° N. of the farm, where highly contorted purple shale dips steeply to the south-east and is locally overturned to W. 40° N., there are intermittent exposures of purple and green shale dipping at from 20° to 45° towards the south or south-south-east. More continuous exposure of purple blocky shale extends from 400 to 800 yd upstream, the dip being generally about 30° or 35° to points between east and south-east. The beds are locally crushed, for example at the upstream end of these outcrops [413882] where they are cut by an east-north-east fault. Intermittent exposures continue for a further 800 yd upstream and on the steeper banks within the valley, the dip being usually rather low to the south-east.

Purple shale contains a thin band of grey limestone in the stream [413887] 190 yd E. 25° N. of Churchmoor Hall. The base of the Hughley Shales is exposed 90 yd upstream (p. 166).

Hughley Shales, 168 ft in thickness and comprising maroon mudstones with green patches, were proved in the Church Stretton No. 5 Borehole [42178912], 490 yd N. 19° W. of Cwm Head church. The upper part of the Shales is cut out by faulting (Fig. 13). At Hamperley [420892] purple blocky shales crop out in the stream to south and west and in a roadside ditch about 160 yd W. 40° S. of the cross roads. A typical dip is 28° to E. 10° S. About 5 ft of purple shales are exposed at the sides of the road [42348980] (Locality 3) 700 yd N. 30° E. of the cross roads. A small exposure of weathered purple shale, at the roadside [426903] 1340 yd N. 30° E. of the cross roads, is significant in showing the proximity of the Hughley Shales at this locality to the arenaceous *Pentamerus* Beds and to the Longmyndian. The greater part of the argillaceous *Pentamerus* Beds is considered to be cut out by a south-south-westerly prolongation of the Long Mynd boundary fault of the Little Stretton area.

A few small exposures of purple and green blocky shale are seen below the drift in the stream about a quarter and half a mile west-north-west of New House farm, Marshbrook. North of the stream the purple colour seen in the deeper exposures of boulder clay is considered to be evidence of the presence of Hughley Shales at no great depth.
D.C.G.

In a farm track [434906] 300 yd E. 20° S. of Minton Manor House there are small exposures of weathered reddish brown siltstones, probably Hughley Shales. Deeply weathered Hughley Shales are seen in a deep ditch [440916] 675 yd W. 24° S. of Little Stretton Halt. Because of the dearth of exposures the boundaries shown on the map between Minton and Little Stretton must be regarded as conjectural. The repeated succession shown on the map, due to faulting, is inferred from more definite evidence in the ground to the south and from information kindly provided by the late Professor W. F. Whittard (personal communication) who dug a number of trial excavations during his investigations in this area (Whittard 1932). A borehole sunk in 1966 near Brockhurst Castle [44769236] on behalf of Birmingham University Geology Department (information kindly provided by Professor F. W. Shotton), found Hughley Shales at a depth of 117 ft beneath boulder clay and gravel. This is the only evidence of Silurian rocks west of F1 between Little Stretton and Church Stretton, but it is possible that the *Pentamerus* Beds and Hughley Shales are present beneath the drift on the east side of Brockhurst Hill, and connect the known outcrops of the Little Stretton and All Stretton areas.
J.E.W.

MAIN OUTCROP (EAST OF THE CHURCH STRETTON FAULT COMPLEX)

KENLEY GRIT

Holt. The most southerly exposures of the Kenley Grit are in a group of old quarries [531961] immediately north of Sheaves Farm, where grit and conglomerate debris is seen. South of the farm the feature formed by the grit dies away, and the rock is probably overlapped by the *Pentamerus* Beds about half a mile south-south-west of the farm.

Conglomerate is exposed in the stream [536965] 580 yd E. of Plaish Hall, and similar, though finer-grained, rock is exposed about 100 yd upstream. From 200 to 400 yd north of Holt there are many overgrown old quarries [535968] on the dip slope of the Grit.
G.H.M.

PENTAMERUS BEDS (SHALES AND LIMESTONES)

Wistanstow to Hatton. In this area the *Pentamerus* Beds (Shales and Limestones) are largely drift-covered. It is thought that they are overlapped by the Hughley Shales about 250 yd north-west of Wistanstow church. North of the disused railway north of Henley Common their outcrop is marked by a slight feature which has been traced north-eastwards towards Ticklerton. One foot of grey limestone with *Pentamerus oblongus* and traces of brown shale are seen in a ditch [472901] 1050 yd E. 43° N. of Hattongrove. D.C.G., B.A.H.

Ticklerton to Holt. Grey, locally silty, laminated mudstones with beds of limestone packed with *P. oblongus* and grey, flaggy, micaceous sandstone are exposed 150 to 400 yd upstream from the road bridge at Ticklerton. At a place 250 yd [482909] from the bridge the beds dip to the south-east at 20°. Poor exposures of brown shaly mudstone are seen in the stream [495919] 100 to 150 yd south of Hargrove Farm, and fossiliferous bands were temporarily exposed in land drainage excavations [503925] 800 yd E. 5° N. of Peartree. Shelly bands were also located in the stream [505929] 740 yd E. 13° S. of Gutter Farm and again 60 yd downstream.

The *Pentamerus* Beds are well exposed in the brook [511935] 1050 to 1220 yd downstream from the Mill, near Hill End ('Gilberries (Heath) Brook' of Whittard 1928, p. 741), where the dip apart from certain minor rolls, perhaps superficial, is to the south-east at 10° to 19°. The beds consist of grey mudstones and siltstones interbedded with which are hard shelly limestone bands with *Pentamerus oblongus* weathering to form the characteristic 'Government Rock'. Fossiliferous beds are well exposed 1200 and 1220 yd downstream from the Mill.

The outcrop of the *Pentamerus* Beds between Gilberries [515938] and the northern margin of the district is largely a calculated one as plotted on the map since the ground is drift-covered and there are no exposures of solid rock. G.H.M.

HUGHLEY SHALES

Wistanstow. Near the River Onny the Hughley Shales are exposed only in the well-known outcrop on the left bank [42578534] (Locality 4) 720 yd W. 22° S. of Wistanstow church. They are green shales, commonly purple and purplish green in the upper beds, with many beds of hard, micaceous, calcareous flags up to 8 inches thick. The underlying Ordovician rocks are overstepped and the successive Silurian beds overlap towards the north-west, apart from an apparently persistent basal bed of grey shelly limestone, half an inch thick. Some 35 ft of Hughley Shales are exposed, dipping at about 18° to the south-east.

Descriptions and sketches of this important section have been provided by Salter and Aveline (1854, pp. 69–70), Cobbold (1900, pp. 38–9), and Whittard (1928, pp. 747–52). The lowest Silurian beds, apparently the green shales and flags, were assigned by Salter and Aveline to the *Pentamerus* Beds, but in Whittard's opinion the elements in the fauna which belong to that formation have been derived from it. The state of the exposure is evidently variable, and it is not clear whether the thin shelly limestone seen at the base of the Silurian in 1959 is a *Pentamerus*-limestone as seen by Salter and Aveline, invisible when Whittard examined the section in 1925, or the thin conglomeratic limestone visible in 1925 only below normal water level, immediately above the surface of unconformity (Whittard 1928, pp. 749–51). Although it has a few small fragments of green mudstone it could scarcely be described as conglomeratic, being predominantly composed of abundant brachiopod fragments in a matrix of crystalline calcite. It is similar to the limestones of the *Pentamerus* Beds in every respect except that *P. oblongus* is not identified in it. D.C.G.

Felhampton to Red Barn. The outcrop of the Hughley Shales is largely obscured by drift, and exposures of the solid rock are poor and infrequent. Between the valley of the Quinny Brook and the disused Wenlock branch railway the outcrop was partly traced by debris thrown out of excavations for electricity pylons. One foot of purple and grey shale with thin siltstone bands is exposed in the stream [465891] 350 yd south of Hattongrove. Similar beds, with a south-easterly dip of about 20°, are seen in a small stream [472893] 800 yd E. 10° S. of Hattongrove, and further exposures occur in the stream [477901–481901] 1450 yd E. 27° N. of Hattongrove and for 500 yd to the east. B.A.H.

Traces of reddish purple shaly mudstone are seen in the roadside [484907] at Ticklerton, 80 yd N.E. of the bridge. Brown fossiliferous mudstones, dipping at 12° to E. 30° S., are exposed in the stream [490919] 350 yd W. of Claybrook, and poor exposures at or near the junction of the Hughley Shales with the underlying *Pentamerus* Beds are seen in the stream [495919] 150 to 200 yd S. of Hargrove Farm. Greenish grey and purple, micaceous, earthy mudstone occurs in the banks of the brook [511933] 850 yd N. 19° E. of the inn at Wall under Heywood. G.H.M.

North-eastwards from Wall under Heywood the Shales are almost entirely obscured by boulder clay. Disturbed reddish brown, purple and grey mudstone is seen in the stream [553972] 180 yd W. 10° S. of Brook House, and 6 ft of purple and grey shaly mudstone with thin ribs of flaggy laminated calcareous siltstone are exposed farther down the same stream [559974] 550 yd E. 25° N. of Brook House. B.A.H.

Wenlock Series

Western outcrop (west of Church Stretton Fault Complex)

Wenlock Shales

Plowden to Horderley. North of Myndmill Farm [382887] the Wenlock Shales are not exposed. The position of the boundary with the Hughley Shales is inferred from evidence yielded by the Church Stretton No. 1 Borehole (Sheet 165) [37238978], 1975 yd W. 7° N. of Myndtown church. In this borehole 259 ft of Wenlock Shales and ? Purple Wenlock Shales were proved. They are overlain by 25 ft 6 in of drift and overlie the Hughley Shales. The Wenlock Shales comprise grey siltstones and silty mudstones with calcareous nodules towards the base, and an abundant shelly and graptolitic fauna. The basal 26 ft 4 in, the ? Purple Wenlock Shales, consist of reddish brown silty mudstones with local green coloration. They have been separated from the Hughley Shales on lithological grounds since there is no conclusive faunal evidence. B.A.H.

Between Myndmill Farm [383887] and Plowden there are many small outcrops in the steep bank west of the Onny and the wooded cuttings of the old railway. The rock is mainly grey, calcareous, rather silty mudstone, with in places highly calcareous nodules and large flat bullions. The beds normally dip at about 25° to the west-south-west, but steeper and more westerly dips are seen southwards from a lane [382879] 820 yds S. of Myndmill Farm. About 16 ft of grey calcareous silty mudstone are exposed in the lane about 80 yd east of the railway. There is a highly calcareous band at the top, and the rock is typical of the local Wenlock Shales in being poorly stratified and irregularly fractured, and bearing few fossils apart from reddened polyzoa. Some 120 yd farther east the downward transition to the purple Hughley Shales is seen (p. 166). Grey calcareous flags and shales with some calcareous bullions are well exposed in a stream course [379878] 1040 yd S. 18° W. of Myndmill Farm, just north of the road from Plowden to Eyton (Montgomery Sheet 165).

In the old railway cutting [381878] 1000 yd S. 7° W. of Myndmill Farm purple and greenish grey shale, weakly calcareous but containing some very calcareous fine-grained nodules, is overlain by about 20 ft of grey calcareous shale with bands of nodular limestone. Whittard (1932, p. 869) included all the lower, purple, beds in the Wenlock Series, but Evans (1957, pp. 11–3, fig. 2, 3) included only their topmost 5 ft, which he considered to be Whittard's Purple Wenlock Beds. However, he recorded no positive evidence of the Llandovery age of the subjacent beds, and for this reason, and considering also the outcrops in the lane (preceding paragraph), the outcrop of the base of the Wenlock Series is taken to be some distance east of the railway. The grey calcareous shale of the cutting is well exposed at the roadside immediately to the west.

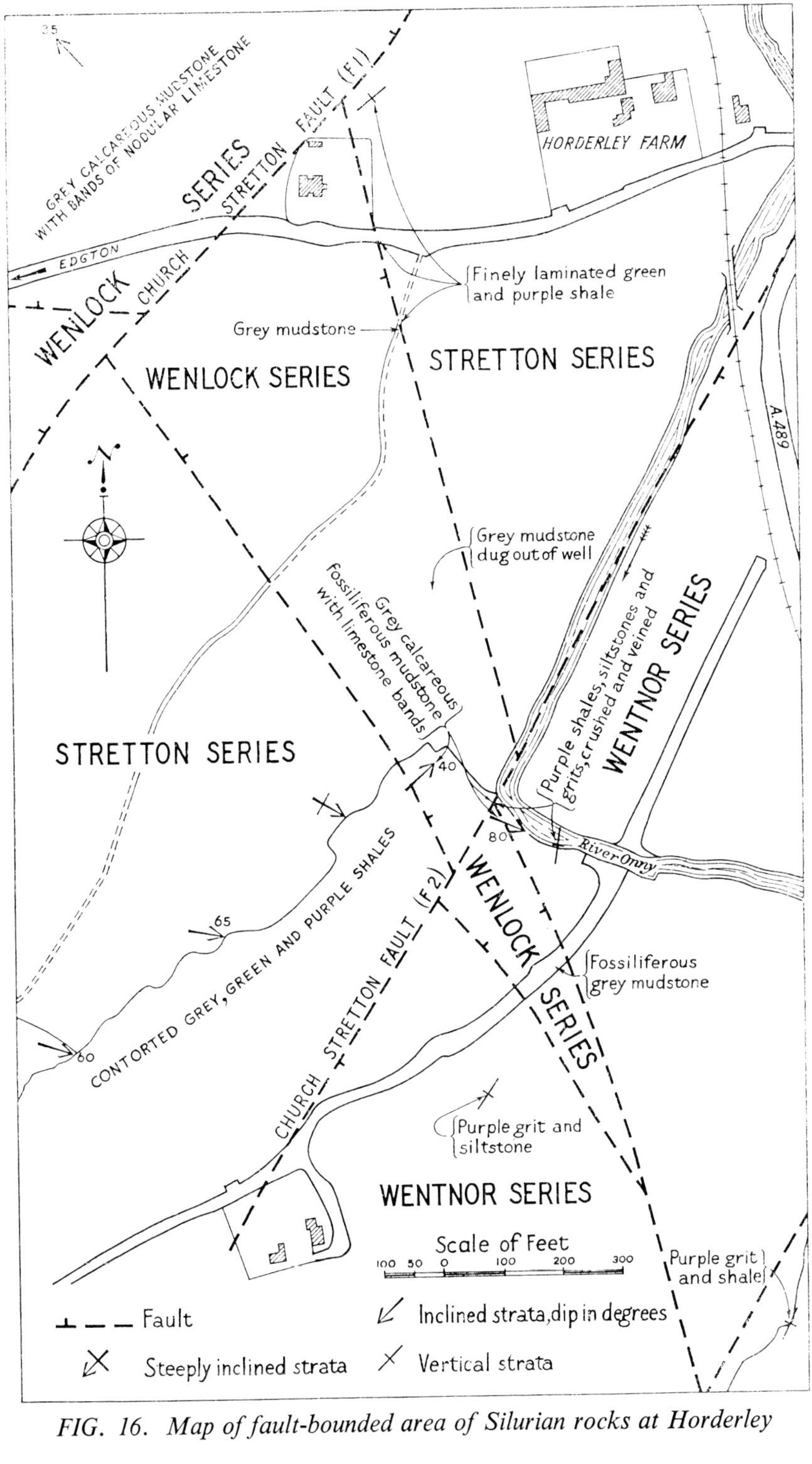

FIG. 16. Map of fault-bounded area of Silurian rocks at Horderley

Beside the private road [381871] to Plowden Hall (Montgomery Sheet 165), and on the face of the escarpment (Localities 5 and 6) south of Plowden Mill, are many small outcrops of grey calcareous shale, commonly with much nodular limestone. Beside the road they are close to the Long Mynd Scarp Fault and are somewhat disturbed, but elsewhere the dip is between 20° and 35° to the south-south-east. The local abundance of limestone nodules suggests a transition to the limestones above, and an early limestone within the Wenlock Shales, of restricted lateral extent, probably forms the ridge about 800 yd south of the Mill.

The existence of the inlier of Wenlock Shales some 200 yd north-west of Edgton church is inferred largely from the distribution of the overlying limestones.

Several small outcrops of grey, calcareous, flaggy and shaly mudstone occur near the top of the escarpment [395865] west-north-west of the Ridgway road junction (747 ft spot height), dipping at upwards of 40° to south and south-east, and near the top of the steep northern face of Ridgway Hill. On a footpath [40548714] (Locality 9) 1250 yd E. 40° N. of the Ridgway road junction the upward transition from shale to shale with abundant limestone bands, is seen in beds dipping at 20° to E. 10° S. One of the few exposures between the scarp and the Onny is in the stream [399872] about 900 yd N. 20° E. of the road junction where grey shales dip southwards at about 35°. The most westerly outcrop in the Onny is the apparently faulted exposure on the right bank [393871] already discussed (p. 167). The upper beds here are grey and greenish grey, massive and flaggy, calcareous siltstone, with some nodules and much calcite veining, and dip at varying high angles to the south-south-west. Notably flaggy beds, dipping at 35° to S. 36° E., occur on the left bank [402874] 1180 yd N. 27° E. of the Ridgway road junction, and flaggy shales crop out for about 200 yd from 150 yd downstream. Grey blocky calcareous shale, with some thin bands of nodular limestone, is exposed to over 50 ft in old river cliffs on the right bank [40808708] (Locality 10) 1450 yd E. 31° N. of the road junction. These beds dip at from 17° to 25° towards S. 25° E.

Grey calcareous shales with some limestone bands, which lie south-east of the Church Stretton Fault (F1) from the point at which it crosses the Edgton–Horderley road, are doubtfully assigned to the Wenlock Shales (Fig. 16). They are well exposed between a tributary and the right bank of the Onny [40758651] (Locality 7), 1220 yd E. 5° N. of the Ridgway road junction, dipping at 40° to the north-east, and on the right bank of the Onny [408864] about 50 yd to the south-east. At the latter locality they are faulted against Longmyndian grits (p. 90) along two lines, one trending south-south-east and the other north-north-east, F2 of the Church Stretton Fault Complex. The eastern limits of these Silurian rocks are usually closely defined and it is clear that the outcrop cannot extend very far to the west or south.

Horderley to Minton. Wenlock Shales are commonly exposed in the steep banks on the north side of the road between Hillend Farm [398877] and The Bank, some 500 yd north of Horderley. At 480 yd E. 32° S. of Hillend Farm grey shales with limestone nodules, dipping at 35° to the south-east, form a prominent roadside bluff [402875]. In a smaller exposure 50 yd farther east similar beds dip at nearly 90° to the east-south-east. Irregularly fracturing grey mudstone with occasional large calcareous bullions is extensively exposed at The Bank [408873], 300 yd to the north-west of the junction with the Marshbrook road, dipping at 25° to the south-east. Similar dips are seen in the occasional outcrops (e.g. Locality 8) of grey shale with calcareous nodules in the streams to the north-west and north-east. In the more easterly outcrops, however, up to about 200 yd north-west of the Church Stretton Fault (F1), the beds are locally inclined at up to 70°.

Several faults are exposed in the tributary valley [414877] about 1740 yd E. 4° N. of Hillend Farm. Greenish grey shales with occasional flat calcareous nodules crop out 10 yd west of the main stream, and immediately to the east is a band, 2 ft 3 in thick, of greenish grey slickensided clay, followed by at least 2 ft of purple slickensided clay. These clays are considered to be respectively the crushed representatives of the Wenlock Shales and the Wentnor Series grits, a less crushed form of the latter being exposed about 11 ft farther to the east. The boundary between them is thus the Church Stretton Fault, F1, which is here reversed, the plane hading to E. 45° S. at about 15°. Between F1 and a fault of uncertain trend, seen about 20 yd to the west, the Wenlock Shales lie parallel to F1 and are much crushed and slickensided. West of the second fault the shales dip at about 25° to W. 20° N. and are much less broken. Occasional thin layers of cream-coloured clay occur in the shales throughout these outcrops.

In the tributary valley about 330 yd to the north-east the Wenlock Shales up to about 80 yd from F1 have a nearly vertical dip towards W. 45° N. In the 'Sand Pit' [417882] 870 yd W. 43° S. of Cwm Head church, 35 yd north-west of the fault, olive-green weathered shales are nearly vertical on a strike of N. 40° E., parallel to the fault and to the valley. South-eastward dips of from 25° to 80° predominate in weakly calcareous grey shales, commonly crushed, in the tributaries which join the main stream 725 yd S. 40° W. of Cwm Head church. A bed of pale grey clay, about 11 inches thick, occurs 220 yd from the main stream [41718832] (Locality 11). This clay has been examined by X-ray by Mr. B. R. Young, who reports it to be "composed almost entirely of a mixed-layer illite-smectite" apparently containing "about 80 per cent illite and 20 per cent smectite layers. These layers are randomly distributed throughout the individual crystals of which the clay is formed". The topmost and lowest ½ in of the bed are of a pale cream colour, similar in appearance to the thin bands of clay seen in the tributary 700 yd to the south-west (preceding paragraph).

The Wenlock Shales form a broad ridge of subdued relief between the steep-sided valley along the line of the Church Stretton Fault (F1) and the road which runs south-westwards from Hamperley to the Onny. The Shales, which are deeply weathered and overlain generally by a thin spread of boulder clay, are exposed only occasionally, as in the road about 800 yd S. 31° W. of Hamperley.

Beds near the base of the Wenlock Shales are exposed in the stream [420891] 130 yd S. 9° E. of Hamperley cross roads. The grey calcareous shale includes two bands of limestone, up to 7 inches thick, and calcareous bullions are common in the next exposure, about 120 yd downstream. The higher beds, commonly exposed downstream, are fairly uniform grey calcareous mudstones with irregular fracture, commonly crushed and closely jointed. The general dip is from 30° to 45° towards the east-south-east. Small brachiopods, Monograptids, and orthoconic nautiloids are the most common fossils. In the streams (e.g. Locality 12) and the old quarry [42248877] (Locality 15) near Pillocksgreen the shales are highly contorted and often veined by calcite. The rock in the quarry, some 20 yd from the Church Stretton Fault, F1, is an olive-green massive or shaly mudstone with some thin bands of nodular limestone. Some of the many fractures are filled with yellow clay and the beds dip generally from 70 to 90 degrees towards S. 30° E. or N. 30° W., slightly oblique to the fault.

Between Pillocksgreen and the stream from Minton Batch the Wenlock Shales are exposed only in the valley [429891] half a mile east of Hamperley, where grey calcareous shale, locally crushed, crops out for about 100 yd. The rock dips steeply towards the Church Stretton Fault, which must pass at a very short distance to the south-east. Old trial shafts at White Birches [429894], shown on the six-inch map, were thought by Cobbold (1900, p. 33) to have been sunk for coal. No record of these shafts is known to the landowner, nor could any evidence of their presence be found on the ground.

Very broken grey shale, closely veined with calcite, which marks the position of the Church Stretton Fault in the stream [434899] 200 yd west of New House farm, may belong to the Wenlock Shales or the *Pentamerus* Beds. Undoubted Wenlock Shales are seen in a track 150 yd to the west-north-west and their debris is abundant about 170 yd to the north-north-west. Olive-green shale with limestone nodules and some crinoid fragments has been worked [437904] 520 yd N. of New House farm, where it dips at 40° to S. 30° E. D.C.G.

Closely jointed grey siltstones, seen in a sunken lane [43189078] (Locality 16) 130 to 170 yd N. 11° E. of the Castle Mound, Minton, occur in a fault-bounded outlier of Wenlock Shales, which extends at least as far south as a temporary exposure east of the lane [431904] about 825 yd W. 40° N. of New House farm. D.C.G., J.E.W.

Cwm Head. Wenlock Shales occupy a prominent rounded ridge at Cwm Head, succeeded to the south-west by Ludlow Shales, forming low ground, and concealed to the north-east by boulder clay. The fault-bounded outcrop lies within the Church Stretton Fault Complex. Old workings on the east side of the ridge, 250 yd S. 11° E. and 280 yd S. of Cwm Head church, are very overgrown but contain debris of shale and nodules of limestone. At Cwm Head house [42438854] (Locality 13) 210 yd S. 39° E. of Cwm Head church, and in the old quarry (Locality 14) 180 yd to the north, the rock is a khaki blocky mudstone with many thin bands of grey fine-grained limestone ('Cwm Head Limestone' of Dean 1964, p. 285), apparently lithologically intermediate, as Whittard has pointed out (1952, pp. 170–1 and 1953, p. 247), between the basin facies of the west and the shelf facies of the east (but see p. 143). The dip of the beds near the house varies between 30° to S. 10° W. and a lower angle to the east-south-east, and in the quarry to the north it is very steep to W. 30° S.

EDGTON IMESTONE

Grey shales alternating with thin bands of nodular limestone cap the wooded escarpment some 1200 yd north-west of Edgton church. The only good exposure in this outcrop is within the Montgomery (165) Sheet, in an old quarry [378863] 1150 yd W. 35° N. of the church, where about 24 ft of bedded nodular limestone are seen, with thin partings of grey, calcareous, flaggy siltstone. The total thickness of the limestone group is estimated to be at least 50 ft, the upper limit being inferred from topographic evidence only.

The main outcrop of the Edgton Limestone, separated by strike faults from the other two outcrops, is well exposed in a stream [38468615] (Locality 48) 510 yd N. 23° W. of Edgton church. Grey calcareous flags and shale with much interbedded nodular limestone crop out for about 140 yd, dipping at about 30° in a variable southerly direction. Close joints, steeply inclined towards N. 30° W., could easily be mistaken for bedding in flat-lying exposures, a potential source of error observed at other outcrops of the Limestone. A specimen of *Halysites sp.* was obtained from the Limestone beside the footpath [38678620] (Locality 47) some 550 yd north of the church. At both these localities the Limestone is succeeded by shales with very little limestone.

The lithological character of the group is very well displayed in the old quarry [39648651] (Locality 49) 100 yd N. of the Ridgway road junction, where about 30 ft of interbedded limestone and shale are exposed. The limestone is finely crystalline and occurs in flaggy nodular bands about 2 or 3 inches thick, separated by bands of flaggy mudstone, on the average rather thinner. The beds dip at about 55° to S.S.W., decreasing in the south-east corner, but increasing to 73° in the south-west corner. Some of this variation may be of superficial origin but indication of more deep-seated disturbance is afforded by the abundance of calcite veins, up to 2 inches thick, in many joint planes, and the local development of horizontal slickensiding. The abrupt termination, just west of the quarry, of the hog's back of Ridgway Hill is more probably due to a decrease in the dip of the beds than to faulting. In the lane about 170 yd west of the quarry the Limestone includes near the top a band rich in small brachiopods and simple corals.

Interbedded shales and limestones are widely exposed in the stream [399864–403864] (includes Locality 51) east of the Ridgway road junction. They dip at about

20° to the south but the angle increases downstream and the strike swings to east-north-east. Near the Church Stretton Fault the beds are vertical. A thin band of limestone containing many simple and compound corals occurs near the base of the exposed beds [401864], some 80 ft below the top of the group, 470 yd E. of the road junction. Small exposures of the Limestone occur on the crest of Ridgway Hill for about a quarter of a mile east of the quarry, at the roadside up to about 700 yd east of the road junction, and in the wood north of the road. Although much of the wide outcrop constitutes a dip slope it is clear that unless there is considerable folding or strike faulting, the thickness of the Limestone is increasing eastwards from the Ridgway quarry.

In the fault-bounded outcrop of the Limestone which extends south-westwards from the farm, known as Highfield [408871], some 1420 yd E. 29° N. of the Ridgway road junction, the beds generally dip at upwards of 65° to the north-west. This dip direction is unknown elsewhere in the Silurian ground between here and Edgton. In an old quarry [40448678] (Locality 50) 940 yd E. 25° N. of the Ridgway road junction about 12 ft of grey calcareous shale with many bands of fine-grained nodular limestone are exposed. The unusual dip of 8° to N. 25° E. is seen here. Small exposures with much nodular limestone are seen elsewhere in and near the coppice to the north-east. Highly disturbed shale and limestone, with a general vertical dip, are widely exposed beside a strong spring [408870] 1350 yd E. 17° N. of the Ridgway road junction, and some 45 yd north-west of the Church Stretton Fault (F1).

Shales and limestones crop out amongst the buildings of Highfield. They are highly disturbed, in marked contrast to the regularly dipping beds in the cliffs [408871] 40 yd north of the farm (p. 172). The springs 20 yd north-east of the farm mark approximately the northern limit of the disturbed beds. In the lane just south of the springs contorted shales with limestone bands are approximately vertical on a strike of E. 30° N. and show several small east–west faults. A few yards to the south-west, neither contorted nor fractured, they dip steeply to W. 25° N. Farther west, behind the house, are shaly beds dipping at a relatively low angle to S. 10° E.

About 1275 yd E. 22° N. of the Ridgway road junction [408868] small exposures of Longmyndian shale, and of grey Silurian shale taken to be part of the Edgton Limestone, define closely the course of the Church Stretton Fault (F1).

The Limestone forms the prominent escarpment between Highfield and the Castle Ring [401869], about 700 yd N.E. of the Ridgway road junction, separated by an east-north-east fault from the Ridgway Hill and Highfield outcrops just described. Limestone nodules are abundant in the soil on the crest, and there are one or two small outcrops of grey shale and limestone on the higher part of the scarp slope.

The Edgton Limestone in the third principal outcrop is well exposed at the roadside [386858] about 110 yd N.W. of Edgton church, and for about 70 yd to the south, as well as in the farmyard west of the church. It is cut off to the east against Ludlow Shales by a north-south fault passing through the church. At the roadside the rock is seen to consist of interbedded grey nodular limestone and olive-green flaggy shales. About 100 ft of rock are exposed dipping at 26° to 30° to S. 20° E. Grey calcareous siltstone with many nodules has been worked south of the road [382857] 500 yd W. 8° S. of the church and similar rock is seen at the roadside nearby. D.C.G.

MAIN OUTCROP (EAST OF CHURCH STRETTON FAULT COMPLEX)

WENLOCK SHALES

Clungunford to Craven Arms. Wenlock Shales occupy the low ground of the Clun valley from Clungunford to Broome, and extend across the col from Broome to the Onny valley at Craven Arms. This area is almost entirely drift covered, and exposures of the Shales are very infrequent.

Traces of weathered, yellowish brown siltstone were seen below boulder clay [38097869] (Locality 20) 995 yd S. 12° W. of New House, near Coston. The next exposure of the Shales to the north-east is at the base of a deep railway cutting one mile west of the Central Wales Junction, Craven Arms, where grey, shaly, silty mudstones with occasional limestone nodules are seen. The basal beds of the Shales are exposed on the southern side of the pool [414833] 190 yd N. 28° E. of Sibdon Carwood church, where 6 ft of grey, silty mudstone with bands of calcareous bullions up to 9 inches in diameter are seen. Grey shale has been recorded from a well [412824] near Long Meadowend and, below a thick cover of gravel, from several boreholes near Craven Arms. B.A.H.

Wistanstow area. South of the River Onny the Wenlock Shales are exposed only at a pond [425843] 1590 yd S. 30° W. of Wistanstow church, where weathered shales contain ostracods and trilobite fragments. Small exposures of grey calcareous shale occur in the river banks between 720 and 580 yd [42778509] (Locality 21), and again about 100 yd, above the railway bridge. In the north bank of the road [43028518] (Locality 23) 480 yd S. 25° W. of Wistanstow church 4 ft of khaki shale with ostracods are overlain by buff flaggy sandstone, but between the Onny and the Quinny Brook the solid rock is nearly everywhere concealed by drift. Khaki shales with ostracods are seen in the railway cutting 560 yd north of the Onny bridge. Shale with nodular flaggy limestone is exposed on the left bank of the Onny [43598423] (Locality 22) 1530 yd S. 16° E. of Wistanstow church, under the Wenlock Limestone escarpment. D.C.G.

Strefford to Presthope. In Apedale the Wenlock Shales are largely drift-covered, and most exposures are either in stream sections or in the higher part of the shales on the lower slopes of Wenlock Edge. A good section in the upper part of the shales is seen along a footpath for 150 yd east of a place [44688533] (Localities 24 to 27) 1550 yd W. 11° N. of the Tower, Callow Hill. About 185 ft of buff-weathering, pale greenish grey, flaggy, argillaceous siltstones are exposed. Grey, finely crystalline, argillaceous limestone bands occur except in the basal 30 ft, and these bands gradually become more frequent and more nodular towards the top of the section. The beds have yielded an extensive fauna. Traces of buff shaly siltstone are seen in old clay pits [459883] a quarter of a mile north-east of Henley Common and in a stream section [465881] 600 yd W. 20° N. of Wolverton.

There are several small sections in buff, silty mudstones in the roadside [483888] on the north side of Harton, and 70 ft of pale greyish green, shaly and flaggy, silty mudstones with occasional bands of grey, calcareous nodules are seen above the Eaton Brook [48108870] (Locality 28) 740 yd S. 19° W. of Harton Road Station, dipping to the south-east at 15°. About 20 ft of similar beds are exposed in a river cliff [48808908] (Locality 29) 600 yd E. 28° S. of the station.

Beds near the base of the Wenlock Shales are exposed in a stream section [482901] 820 yd N. 8° W. of Harton Road Station. This section shows about 10 ft of grey and pale olive-green, flaggy, weakly calcareous, silty mudstones with occasional bands of small calcareous nodules and a 6-inch band of calcareous bullions at the base. A 6-inch lenticular band of white clay occurs 2 ft from the top of the section, and a 4-inch band 7 ft from the top. The fauna consists mainly of small brachiopods. Several small outcrops of similar beds occur in the stream which runs southwards from Ticklerton towards Harton. B.A.H.

Old brick pits [486905] 400 yd E. 38° S. of the bridge at Ticklerton, in beds near the base of the Wenlock Shales, show 12 ft of calcareous grey mudstone with bands of fossiliferous limestone nodules. These beds dip at 24° to S. 30° E.

There are good sections in the upper part of the Shales at Eaton. The entrenched lane [50098999] (Locality 31) 100 yd E. 4° S. of the church and for 140 yd to the west-north-west shows shaly-weathering grey and olive mudstone, with bands of limestone nodules. Traces of calcareous and silty mudstone are seen farther up the track to the east [50348997] (Locality 30), and 420 yd E. 2° S. of the church there is a section (Locality 36) showing the transition to the Tickwood Beds above (p. 179).

An excellent section occurs in a stream [514915–517912] 410 yd S. of Rushbury church and for 450 yd towards E. 35° S. About 235 ft of beds are intermittently exposed, consisting of pale greenish grey, flaggy silty mudstones with very occasional calcareous bands and bands of fine-grained calcareous nodules. Similar beds, with an extensive fauna, were seen in temporary trenches (Localities 32–34) along the road between Rushbury village and the station, and also in several sections in the stream on the south-east side of the village. The general dip of the beds is about 10° to the south-south-east. An old quarry [535923] 1060 yd E. 19° S. of Coats, near Rushbury, shows 16 ft of beds near the top of the Shales. The basal 6 ft are blocky olive-green and buff mudstones with occasional nodular bands of fine-grained, grey limestone. In the top 10 ft the limestone bands become more frequent.

In the stream bed [525937] 1100 yd E. 6° S. of Gilberries, greenish grey shale with limestone bullions up to 8 inches in diameter is exposed, dipping at 24° to E. 35° S. Greenish grey shale was also seen [536938] 800 yd W. 9° N. of Longville in the Dale Station, and silty mudstones with sparse limestone nodules and nodular limestone bands, near the top of the Shales, occur in an old quarry [540928] 510 yd W. 7° S. of Wilderhope.

North of Easthope's Cross [562955], 650 yd N.W. of Easthope church, a development of mudstone and nodular limestone of Tickwood Beds facies occurs in the upper part of the Shales (Fig. 14). A section in the lane [563958] 870 yd N. 23° W. of the church shows buff-grey silty mudstone with slightly nodular weathering overlain by silty mudstone with limestone nodules and bands of nodular limestone. Higher beds, exposed 100 yd up the lane to the south, comprise grey silty mudstones with only occasional limestone nodules. Another section in the nodular limestone development is seen in the old railway cutting [570967] 1700 yd N. 13° E. of the church. This local development of Tickwood Beds facies is also seen in the old railway cutting [574970] 1450 yd W. 24° S. of Presthope Station. The section shows bedded silty mudstones with nodules of argillaceous limestone and nodular limestone bands. At the top there is a rapid transition into shaly mudstones with only scattered limestone nodules, while towards the base of the section there is a more gradual transition into less calcareous beds. The thickness of the calcareous group is about 25 to 30 ft. The nodular limestone development is seen again in a gully section 600 yd W. 10° S. of the station.

The basal beds of the Shales are seen in several sections in the vicinity of Brook House [554972]. In the stream 110 yd W. 17° S. of the house greyish buff and olive mudstones are exposed, and 15 yd downstream there are traces of mudstone with large flat nodules of calcareous silty mudstone. Reddish brown and purple mudstones of the Hughley Shales occur a few yards to the west of these localities. Bluish grey mudstone occurs in the stream [558974] 520 yd E. 28° N. of the house. These beds are faulted against Hughley Shales about 30 yd to the east. About 4 ft of greyish buff mudstone, with bands of silty limestone nodules and a thin bed of white clay, crop out in the stream [562973] 840 yd E. 8° N. of the house, and similar beds are seen 40 yd downstream. All these localities yield an abundant fauna of small brachiopods and other fossils typical of the lowest part of the Wenlock Shales (Buildwas Beds).

B.A.H., R.H.H., G.H.M.

TICKWOOD BEDS

Clungunford to Presthope. Several small sections in typical Tickwood Beds are seen in an old quarry and adjacent road banks [402790] 850 yd E. 18° N. of Clungunford church. The beds comprise alternating bands, 1 to 2 inches thick, of grey or khaki siltstone and grey, fine-grained, argillaceous limestone. About 10 ft of similar beds, dipping at 15° to the east-south-east, are exposed in an old quarry 300 yards to the north. For about a mile north of this section the Beds are unexposed.

Northwards from Rowton the feature formed by the Tickwood Beds becomes more marked, and it is displaced by several faults between Broome and Stokesay. About 25 ft of roughly alternate-bedded, buff siltstone and grey, fine-grained limestone are exposed in an old quarry [409812] 1200 yd S. 7° W. of the cross roads at Long Meadowend, and 20 ft in an old quarry [419818] 720 yd N. 17° W. of Clapping Wicket. Similar beds are seen in another quarry 875 yd farther to the north-east and at Paddock [428821].

A good section through the greater part of the Tickwood Beds is seen in the railway cutting [433821] 1000 to 1300 yd south of Craven Arms and Stokesay Station, where about 120 ft of beds are intermittently exposed. They consist of alternating bands of grey, fine-grained, argillaceous limestone and buff, shaly siltstone, dipping at 11° to E. 35° S. No traces of the Wenlock Limestone facies were seen in this cutting. About 50 ft of similar beds are exposed in the east bank of the River Onny [43868250 to 43878224] (Locality 35) 970 yd E. 39° S. of Craven Arms and Stokesay Station and for 270 yd to the south. The dip here has swung to E. 25° S., and in an old quarry 200 yd to the north it is 8° to the east.

South-east of Grove the Tickwood Beds and Wenlock Limestone outcrops are affected by the north-western end of the Bache fault belt. North-east of this area the strike of the Tickwood Beds becomes north-easterly, and the scarp (Wenlock Edge) formed by the Tickwood Beds and Wenlock Limestone becomes very prominent. The Tickwood Beds crop out on the upper part of the scarp slope, the dip slope being formed by the Wenlock Limestone.

An old quarry [448844] 970 yd E. 7° S. of Berrymill Cottage shows the following section:

		ft
Wenlock Limestone	Limestone, rubbly bedded, nodular and crystalline	c. 3
Tickwood Beds	Limestone, well bedded, nodular and fine-grained. Many siltstone beds, some wrapping round limestone nodules	c. 11
	Siltstone, calcareous. Occasional limestone nodules	$\frac{1}{2}$
	Limestone, as 11-ft above	c. 6

The boundary between the Tickwood Beds and Wenlock Limestone is rather transitional here, although there is a distinct change from fine-grained to crystalline limestone.

The series of quarries on the scarp edge west of Moorwood shows the top part of the Tickwood Beds and the lowest beds of the Wenlock Limestone. The most northerly quarry [454856], 220 yd W. 21° N. of Moorwood, shows 40 ft of Tickwood Beds, consisting of alternate bands of limestone and siltstone, with the limestone bands most frequent towards the top of the section. There is an unexposed interval of about 3 ft from the top of this section to the bottom of the next quarry to the south, in which is seen the following section:

		ft
Wenlock Limestone	Roughly bedded, nodular, greyish blue, crystalline limestone with thin siltstone partings	5
	Marked parting	
Tickwood Beds	Alternating bands of grey and buff calcareous siltstone and bluish grey, fine-grained, argillaceous limestone, rather roughly bedded and locally nodular	9

Further sections in both divisions of beds are seen for 150 yd to the south-west.

Some 35 ft of alternate-bedded, buff siltstone and grey, fine-grained, argillaceous limestone are seen in an old quarry and track [461865] 650 yd S. 20° E. of the moat at Alcaston. Exposures in the hollow [482880] 800 yd S. of Harton show the lower part of the Tickwood Beds.

The base and top of the Tickwood Beds are closely defined in track sections east of Eaton. In the track [50378997] (Locality 36) 420 yd E. 2° S. of Eaton church, 4 ft of grey and olive calcareous and silty mudstone with occasional limestone nodules (Wenlock Shales) are overlain by about 12 ft of interbedded mudstone and fine-grained limestone (Tickwood Beds). In the same track, at the top of the scarp 630 yd E. 9° N. of the church, there are further exposures of interbedded mudstone and fine-grained limestone. The junction with the overlying Wenlock Limestone must occur in the 7 ft of unexposed beds between this section and a quarry in the limestone on the crest of the scarp. The thickness of the Tickwood Beds at Eaton is estimated to be about 180 ft.

An obscure section in a track and old working [531920] about 900 yd W. 43° S. of Coats, exposes about 80 ft of alternating beds of buff, silty mudstone and grey, fine-textured, argillaceous limestone. Towards the base of the section the limestone bands are rather nodular, probably indicating the start of a transition into the Wenlock Shales below.

The Tickwood Beds are well exposed in a road cutting [539926] 650 yd south-south-east of Wilderhope, where there is an almost continuous section of 50 to 55 ft of alternating beds (usually 1 to 3 in thick) of buff mudstone and fine-grained limestone, dipping south-east at 9°. Twenty feet of similar beds are exposed in an old quarry [54919383] (Locality 37) 410 yd W. 8° N. of Pilgrim Cottage. Another old quarry 170 yd to the north, on the northern side of a small cross fault, shows about 40 ft of similar alternating beds of mudstone and limestone, although in the top 6 ft the bedding is less regular and the limestones thicker, more nodular, and more abundant. A road cutting [552944] 670 yd N. 6° W. of Pilgrim Cottage shows a gradual transition from the Tickwood Beds to the Wenlock Limestone. The lowest 6 ft comprise thinly bedded nodular limestone and shaly mudstone bands, probably at about the same horizon as the highest beds at the previous section. These beds are overlain by 20 ft of irregularly bedded nodular limestone with thin shaly films, and calcareous silty mudstone with limestone nodules, in turn overlain by 5 ft of bedded nodular limestone with rare shaly mudstone films. North of this section the Tickwood Beds are poorly exposed. At the western end [581974] of Presthope tunnel a section in the railway cutting mentioned by Robertson *in* Pocock and others (1938, p. 113) shows an atypical development of the Beds. There is a rapid downward transition from bedded nodular limestone with 'ballstones' (Wenlock Limestone Reef Facies) into bedded nodular limestone and silty mudstone (Tickwood Beds). Although the proportions of limestone and mudstone are approximately equal, they are not evenly bedded as in typical Tickwood Beds. About 50 ft of these beds are exposed, with the proportion of limestone decreasing towards the base of the succession. B.A.H.

WENLOCK LIMESTONE REEF FACIES

The most southerly occurrence of the Reef Facies or 'ballstones' on Wenlock Edge is in a section [561955] immediately below the road 680 yd W. 28° N. of Easthope church, where 15 ft of bedded nodular limestone with mudstone films and partings are seen. Some small 'ballstone' masses, up to 3 ft in diameter, are present. Larger 'ballstones', one 25 yd long and 12 ft high, are seen in another section 100 to 300 yd to the north-east.

North-eastwards for about half a mile from a small valley [569964] in the scarp face, 1400 yd N. 14° E. of Easthope church, the top of the scarp is formed by a discontinuous line of cliffs up to 40 ft high. The cliffs are largely formed by 'ballstone' masses. Locally, as at Major's Leap, 1470 yd N. 13° E. of the church, the 'ballstones' can be seen to rest on bedded nodular limestone with calcareous mudstone partings which sags beneath them.

An excellent section of the Reef Facies is seen at Lilleshall Quarry [57439682] (Locality 46), 1600 yd W. 34° S. of Presthope Station. The main face of the quarry (Plate 9c), immediately below the road, is chiefly in 'ballstone'. Bedded nodular limestone, with shale films and partings, dips off the 'ballstone' masses, and at the lowest point of the face 6 ft of grey nodular limestone interbedded with mudstone are seen. In the north-eastern extension of the quarry 15 ft of flaggy crystalline limestone with pink mottling (Wenlock Limestone) rest on about 20 ft of bedded nodular limestone with silty mudstone partings up to 1 inch thick. These beds, the upper part of the Reef Facies, rest on an uneven surface of 'ballstone' masses. A similar section is seen in the south-eastern part of the quarry, 1590 yd W. 38° S. of Presthope Station, where fairly high south-easterly dips may be due to the accentuation of the normal south-easterly dip of 10° by differential compaction off the flanks of 'ballstone' masses. Two subparallel faults, about 20 yd apart, with a north-north-easterly trend, cut through the quarry about 40 yd to the south-east of the main face. In the trough between the faults, at a locality 1650 yd W. 37° S. of Presthope Station, massive-bedded crystalline limestone with calcite veining (Wenlock Limestone) is overlain by 8 ft of Lower Ludlow Shales comprising buff calcareous silty mudstone with occasional bands of nodular limestone. The outcrop of Lower Ludlow Shales is too small in extent to be shown on the one-inch map. Traces of the faults are also seen at the extreme north-east end of the main quarry face.

The old quarry [582974] at Presthope (Robertson *in* Pocock and others 1938, p. 115), 540 yd W. 23° S. of the station, has a 30-ft face, largely in 'ballstone', and the old railway cutting 150 yd to the east-north-east shows a section in bedded nodular limestone with thin shale bands, and occasional 'ballstone' masses up to 2 yards across.

An area of Reef Facies and Wenlock Limestone extends from Easthope [566952] to a locality [578966] 400 yd south-east of Lilleshall Quarry, on the south-eastern side of two faults. In Easthope village, 230 yd N. 13° E. of the church, 15 ft of bedded limestone with some small 'ballstones' are exposed, and old quarries [572963] in similar beds are located 1350 yd north-north-east of the church. B.A.H., R.H.H.

WENLOCK LIMESTONE

Stokesay to Grove. The most south-westerly exposure of Wenlock Limestone is in the banks of the River Onny [43868221] (Locality 40) 300 yd W. 35° S. of Whettleton where grey and olive-green finely crystalline limestone with irregular buff siltstone bands and partings is seen. Southwards from this place it is not possible to trace the nodular crystalline limestone, and only the Tickwood Beds facies is represented. Northwards from Whettleton the Limestone is not exposed for about a mile, although nodules of limestone can locally be found in the soil.

Lower Dinchope to Harton. A good section is exposed in old quarries [451851] (including Localities 41 and 42) 600 to 700 yd south-west of Moorwood, where the following beds are seen, dipping at 10° to the south-east:

Nodular, grey-hearted, finely crystalline limestone, with many bands of siltstone. Some of the siltstone bands are regular, others wrap around limestone nodules. Large blue-hearted nodules with shell debris immediately above parting	13 ft
Marked parting—soft buff siltstone	2 in
Nodular, grey or bluish grey, crystalline limestone, with irregular bands of siltstone wrapping round the nodules	9 ft

The outcrop of the Limestone can easily be traced to the north by crystalline limestone nodules in the soil, but exposures are few.

An old quarry [48058762] (Locality 43) by the roadside 550 yd N. 43° W. of Burwood shows 25 ft of bedded, nodular limestone and argillaceous limestone, with thin films and bands of silty mudstone in the basal 15 ft. The basal beds are probably transitional into the Tickwood Beds since the limestone nodules of these beds are smooth-textured and platy in contrast with the irregular nodules of the Wenlock Limestone in the upper part of the section. About 12 ft of nodular, argillaceous limestone with thin silty bands are seen 70 yd to the south-west (Locality 44) and similar beds with some shell bands are exposed in old quarries 300 to 400 yd to the east-north-east.

B.A.H.

Middlehope, Eaton. Sections in the neighbourhood of Middlehope [497885] show the transitional nature of the junction between the Wenlock Limestone and Lower Ludlow Shales in this area. Old workings [493883] in the upper part of the Limestone, 2000 yd S. 20° W. of Eaton church, show 6 ft of bedded nodular limestone and silty limestone, the upper beds becoming more argillaceous. Bedded nodular limestone, with mudstone partings, is seen in the stream [498888], 1390 yd S. 7° W. of the church, and 40 to 120 yd downstream the lowest beds of the Lower Ludlow Shales are exposed, comprising interbedded buff shaly mudstone and smooth textured nodular limestone.

The lowest beds of the Limestone are exposed in an old quarry [50579007] (Locality 45) 630 yd E. 9° N. of Eaton church, where 6 ft of well-bedded nodular limestone with thin beds and films of silty shale are seen. The topmost strata of the Tickwood Beds are exposed in the track below the quarry (p. 179). At 150 yd to the east, the outcrop of the Limestone is displaced about 500 yd to the south by a N.–S. fault. On the eastern side of this fault, 5 ft of bedded nodular crystalline limestone are exposed in an old quarry [508897] 920 yd E. 18° S. of the church, and a further 6 ft in another old quarry 530 yd E. 30° N. of the last. This quarry was the source of lime for the Millichope Estate. R.H.H., B.A.H.

Roman Bank, Stanway. The dip-slope of the Wenlock Limestone about half a mile south of Rushbury Station is pock-marked with shallow workings in the Limestone, and burnt limestone nodules are frequent. The best section [518907] is 840 yd S. 15° E. of Rushbury Station where 16 ft of roughly bedded, bluish grey, nodular, crystalline limestone with thin irregular siltstone partings are exposed. Crinoids and brachiopods are abundant. The beds dip at 10° to S. 35° E. A long quarry [521911] on Roman Bank, 650 yd E. 40° S. of Rushbury Station, shows 15 ft of buff-weathering, blue-hearted, nodular, crystalline limestone, commonly crinoidal. A persistent limestone band about 4 ft from the top of the section shows a slight development of chert. About 10 ft of similar limestone, with some siltstone partings, are exposed in an old quarry [528914] 300 yd N. 37° E. of Stanway Manor. This section is near the top of the Wenlock Limestone. B.A.H.

Wilderhope to Presthope. Some 3 ft of grey, bedded, nodular crystalline limestone are seen in a road section [539924] 850 yd S.W. of Wilderhope, and there are numerous small sections in the ground between the road and the house. Between Wilderhope and Easthope exposures are few, though nodules of crystalline limestone are common in the soil.

A lateral change in the Wenlock Limestone to a flaggy crinoidal limestone occurs in the Easthope area (see p. 149). On the south-west side of Lilleshall Quarry a small working, 1850 yd N. 25° E. of Easthope church, shows 10 ft of flaggy weathering, massive-bedded, coarsely crystalline and crinoidal limestone. Similar limestone is seen in part of the main quarry, above the nodular limestone with 'ballstones' (p. 180). A pavement of crystalline limestone with a south-easterly dip of 12° is exposed at Presthope Farm [580971], 830 yd W. 39° S. of the station, and a similar pavement with large scale 'ripple marks' (amplitude 4 to 6 inches, wavelength 18 in) occurs at the northern end of the old quarry 150 yd to the north-north-east. The main part of this quarry shows about 10 ft of bedded nodular limestone with shale partings passing upward into 10 ft of flaggy, crinoidal, crystalline limestone.

In the faulted outcrop of Wenlock Limestone north of Easthope (p. 180), flaggy, crystalline limestone with an east-south-easterly dip of 20° is exposed [569958] 750 yd N. 20° E. of Easthope church. About 12 ft of bedded, crystalline, crinoidal limestone occur in an old quarry [573963] 1600 yd N. 30° E. of the church. B.A.H., R.H.H.

Ludlow Series

Western Outcrop (West of Church Stretton Fault Complex)

Ludlow Series (Undivided)

Coston to Barlow. Forty feet of hard greyish green siltstone with an easterly dip are exposed in the roadside [379789] 600 yd S. of Hope, near Coston. Several sections are seen in old workings (e.g. Locality 282) and a track [383803] a quarter of a mile west of Coston. The beds are olive-green and grey siltstone and calcareous siltstone with a north-westerly dip of 35° to 40° and occasional minor puckers. The fauna includes may specimens of '*Camarotoechia*' *nucula* and *Protochonetes ludloviensis*. An old quarry [38608100] (Locality 281) 450 yd N. 8° W. of The Hollies exposes 44 ft of hard, pale olive-green siltstone, locally weakly calcareous, dipping at 65° to the north-west. There are indications of a syncline with a north-north-easterly trend to the west of these exposures, since an old quarry [380810] 1050 yd N. 42° W. of Coston shows about 80 ft of hard, greyish green, strongly cleaved, silty mudstone dipping at 25° to E. 20° S. In the track which runs eastwards from the quarry there are several minor folds in similar strata.

To the north of the River Clun, the best exposure on the western side of the Oaker Syncline (Fig. 2) is the large quarry [378815] (Montgomery Sheet 165), 750 yd W. 10° S. of Oaker, where about 90 ft of massive, grey, calcareous siltstone with rare bands of rotted calcareous nodules are exposed, dipping at 21° to E. 30° S. This quarry is the type locality for *Salopina lunata*, and other fossils found are indicative of an Upper Ludlow horizon. There are several exposures (e.g. Locality 283) of similar beds on the eastern slopes of Burrow, and at Barlow [383840] hard, grey, micaceous siltstone, locally weakly calcareous and with some slumped beds, is exposed in the old quarries to the west of the farm.

On the eastern side of the Oaker Syncline, 48 ft of very hard, grey siltstone with a westerly dip of 43° are exposed in an old quarry [387825] 950 yd N. 10° E. of Oaker, and numerous exposures of similar beds are seen on the northern side of the small east–west valley 750 yd south of Hopesay church. B.A.H.

In an old quarry [39078367] 470 yd N. 19° E. of Hopesay church blue-hearted calcareous micaceous siltstones with sporadic limestone lenses are exposed. The beds are strongly sheared and cleaved and are affected by several faults, because of their proximity to the Church Stretton Fault (F1). In the main part of the quarry (Locality 311) the fauna, which includes *Conchidium knightii*, rare to the west of the Church Stretton Fault, is typical of the Aymestry Group of the Main Outcrop. The beds in the eastern part of the quarry (Locality 312), apparently separated from the main quarry by a fault, contain a fauna which includes *Monograptus dubius* and *M. tumescens*, species which are common in the Lower Ludlow Shales of the Main Outcrop. B.A.H., J.D.D.S.

Edgton and Ridgway. Grey calcareous flags and shale, exposed in the streams and tracks [387860] (Localities 287, 293, 300) about 400 yd north of Edgton church, include a bed with many brachiopods which has been seen at a similar horizon at several localities near Edgton. They are separated by the east-north-easterly Ridgway Hill Fault (p. 264) from other Ludlow rocks of this area.

At a spring [38408562] (Locality 288) north of the road, 290 yd W. 19° S. of the church, is a small exposure of greenish grey shale with occasional limestone nodules. The shale includes bands with abundant Chonetids and Rhynchonellids, like that seen in the stream exposure just described. These bands, in both localities, are about

100 ft above the top of the Edgton Limestone. Grey calcareous shales, like the Wenlock Shales of Plowden, crop out widely in the roadside banks at Edgton Cross [38528556] (Locality 317), 230 yd S. 40° W. of the church, and for 400 yd to the south-west (includes Locality 289). Fragments of Monograptids are common here and reddened polyzoa occur in places. Near Edgton Cross the beds dip at 27° to S. 30° E.

Grey flags and flaggy shales, weathering to a characteristic brown colour, form a prominent ridge south-west of Basford. They are well exposed on the footpath on the ridge [382849] between 1020 and 820 yd S. 28° W. of Edgton church, and in a low wooded cliff [387854] 400 yd S. 15° E. of the church. These calcareous beds locally show contorted lamination and contain small Chonetids and other brachiopods in some beds. They dip at 30° to 35°, to S.E. in the south-western part of the ridge, and to S. 15° E. in the north-eastern. In an old quarry [385850] on the eastern side of the ridge, 800 yd S. 11° W. of the church, current-bedded and slumped flags, exposed to a thickness of 15 ft, dip at about 40° to S. 40° E.

The grey calcareous shales exposed in a few places north-west of the road from Edgton to Ridgway [397864] give way upwards to increasingly flaggy beds, which form the ridge extending from Dunslow Coppice [389855], 200 yd north of Basford, to the head of Day Batch [398860], 500 yd S.S.E. of the Ridgway road junction. These flaggy beds appear to be stratigraphically equivalent to those described in the preceding paragraph. Grey flaggy shales, dipping at 30° to 40° to the south-east, are exposed at the roadside (Locality 316) up to about 70 yd north-east of Edgton church and beside the lanes 300 yd N. 23° E. [38768598] (Locality 315), and 310 yd E. 40° N. [38888591] (Locality 291), of the church. In the last of these exposures Monograptids and orthoconic nautiloids are common. Monograptids are abundant in the grey calcareous flags and shale extensively exposed beside the lane [39008578] (Locality 292) 380 yd E. 10° N. of the church.

Near Ridgway [397864] (Localities 294–299) the Edgton Limestone is succeeded by grey calcareous flaggy shales with a few bands and isolated nodules of limestone. Monograptids are locally common. Disturbed shales in the bank immediately west of the farmhouse, and in a small tributary 150 yd E. 24° S. of the road junction, are indicative of faulting on lines between E. and S.E., which may be associated with the disturbances observed in Ridgway quarry (p. 174). There are a few small exposures of grey and brown flags in the lane between Ridgway and Laplow Bank. Crushed shales in a stream [401861] some 550 yd S.E. of the Ridgway road junction have yielded a fauna with strong Wenlock affinities. This may indicate that the Wenlock–Ludlow boundary has been taken at too low a horizon, or that there is a faulted zone of Wenlock Shales close to the Church Stretton Fault on its western side. It is alternatively possible that certain elements of the Wenlock fauna have persisted into Ludlow time.

Dayia navicula, generally with Rhynchonellids and small lamellibranchs, is common in grey calcareous flaggy siltstones exposed at many points at the roadside for about a quarter of a mile east of Basford [389853], in the lane near Grist House, and in the old quarry [39308480] (Locality 285) 230 yd S. 35° W. of the house. Elsewhere nearby the rock is less conspicuously fossiliferous, as, for example, among the buildings [39058525] (Locality 319) 670 yd S. 41° E. of Edgton church, where rocks at about the same horizon are well exposed. Here and at the roadside to the north the dip is locally very steep to the south-east but it is mainly about 35°. In the Grist House quarry the beds are disturbed, cut by steep joints trending E. 20° N., and have a low variable dip, usually towards W. 30° S. Grey calcareous flags are seen in an old quarry [38848442] (Locality 318) 1430 yd S. 10° E. of Edgton church, and in the adjacent stream upwards of about 100 yd to the north-west. In the quarry they are much slipped and appear to dip at about 45° to the north-west, but in the stream they dip to south-south-east at angles which increase downstream from 30° to beyond verticality.

In the old quarries [395852] 200–270 yd N. 17° E. of Grist House the dip appears to be highly variable. The rocks here lie in the axial region of a sharp, somewhat faulted, syncline which plunges gently south-westwards towards the outcrops south of Basford just described. The straight steep-sided valley north-east of the latter outcrops is eroded along the line of this structure.

Dayia navicula occurs, along with *Monograptus leintwardinensis* and Rhynchonellids, in calcareous flags exposed in a footpath [382843] (Localities 301–310) about 1600 yd S. 16° W. of Edgton church and for about 150 yd down the hill. These beds, which dip at about 32° to S. 35° E. are on about the same horizon as the flags of Basford and Grist House. Small exposures of grey Ludlow shale (e.g. Locality 284) help to define closely the course of the Church Stretton Fault between Dunslow Hollow [398856] and Tadymoor [394843].

LOWER LUDLOW SHALES

Cwm Head. A small area of Lower Ludlow Shales with only a few very small exposures has been proved by the late Professor Whittard (personal communication) at Cwm Head. Diagnostic fossils were obtained by him from a stream exposure [422883] 390 yd S. 21° W. of the church. The nature of the boundary between these shales and the Wenlock Shales to the north-east (p. 174) is not clear. D.C.G.

MAIN OUTCROP (EAST OF CHURCH STRETTON FAULT COMPLEX)

LOWER LUDLOW SHALES

Clungunford to Stokesay. Between Clungunford and Rowton the lower part of the Lower Ludlow Shales is drift covered, and exposures are confined to the slopes of the Aymestry Group scarp. An intermittent section in weathered, olive-green siltstone with many thin, grey, calcareous bands is seen in the track running eastwards from The Crossways [409791]. These beds have a south-easterly dip of 8° to 10°. Two small quarries 300 yd E. 30° N. (Locality 54) and 450 yd E. 40° N. (Locality 55) of The Crossways show sections in similar beds, with abundant graptolites.

On the escarpment west of Brandhill and Gorst Barn are many exposures of the higher beds of the Lower Ludlow, mainly greenish brown and grey siltstone and silty flags, with thin bands of nodular limestone. The fauna includes Monograptids, ostracods, trilobites, and such brachiopods as '*Chonetes lepisma*', *Chonetoidea grayi*, *Leptaena rhomboidalis* and *Strophonella euglypha*. The dip is fairly constant at about 6° to the east-south-east, and at one locality steep joints trending E. 24° S. are very conspicuous. The principal exposures are in an old quarry [41517953] (Locality 57) 1020 yd W. 7° S. of Gorst Barn, where 17 ft of beds are seen, and 660 yd W. 11° N. of Gorst Barn, where 25 ft of beds are exposed [41867975] (Locality 56) just below a large old quarry in the Aymestry Group. Olive-green, silty mudstones with thin, grey, calcareous bands are exposed in the roadside [413796] on Goat Hill, 600 to 800 yd S. of Weo Farm.

Similar fine-grained flags with bands of nodular limestone are exposed in an old quarry [42028032] (Locality 59) 880 yd N. 33° W. of Gorst Barn, and in another [42298078] (Locality 60) 700 yd N. 42° W. of Newfoundland. The fauna is dominated by brachiopods and also contains *Dalmanites* cf. *myops*.

An intermittent section in buff, flaggy siltstone with bands of grey, nodular, finely crystalline limestone, near the base of the Lower Ludlow Shales, is seen in the stream [423816] 450 yd N. 25° E. of Clapping Wicket. Some 15 ft of grey, calcareous siltstone with many bands of nodular limestone, towards the top of the Shales, are exposed in a track [429809] in Stoke Wood, 1150 yd S. 40° W. of Stokesay church. D.C.G., B.A.H.

Onibury. Flaggy siltstone with thin limestone bands is exposed in two quarries, separated by a fault with northerly downthrow, at Stokewood Cottage [43608085] (Locality 63), 1120 yd E. 33° N. of Newfoundland, and in a wood [43518053] (Locality 64) 880 yd E. 18° N. of Newfoundland. In the former the beds dip at 7° to E. 7° N.

and contain an abundant fauna, including *Brachyprion arenaceus*, '*Chonetes lepisma*'. In the second, the beds, somewhat higher in the succession, are less fossiliferous and dip at 7° to E. 35° S. Similar beds in the lane [43488072] (Locality 62) some 200 yd north of the second quarry dip at about 15° to E. 15° S.

Grey micaceous flags with some calcareous bands, dipping at 9° to E. 40° S., form a long outcrop on the left bank of the Onny [44088087] (Locality 65) 950 yd north-west of Park Farm. The fauna includes *Skenidioides lewisii* and *Dalmanites myops*. Younger beds, with many bands of limestone nodules, are exposed (Localities 67–78) to about 60 ft in a 200-yd outcrop on the left bank a little north of west of the farm, and are also seen at the roadside above the bank. They dip to about E. 20° S. at 8° and are traversed in places by steep joints trending S. 36° E. and S. 22° E. The fauna, with large Strophomenids the most common fossils, and the lithology are characteristic of the uppermost of the Lower Ludlow Shales.

In the Onny the transition to the Aymestry Group takes place 140 yd S. 25° W. of Park Farm [44648010] (Locality 66). The change is very gradual and the criterion used for the top of the Lower Ludlow Shales is that proposed and used by Frances Alexander (1936, p. 109) viz. "the point where the shales between the limestone beds lose their even bedding and become irregular". However the boundary adopted by Alexander (1936, p. 109, pl. viii) is some 300 yd upstream, or 120 ft lower in the succession, than that now proposed. The faunal evidence is not convincing, species characteristic of both the Lower Ludlow Shales and the Aymestry Group occurring above and below the proposed boundary.

The calcareous siltstone with limestone nodules locally exposed (e.g. Locality 79) on the ridge of Whettleton Bank [444807] dips generally southwards at about 12°, but at the southern end of the ridge they dip at 8° to the south-east. They are here regarded as the highest beds of the Lower Ludlow Shales. A rather massive grey siltstone with limestone nodules is seen in an old quarry [444805] at the southern end of the ridge, dipping steeply to the east at the eastern end of the face. About 60 yd to the south, at the bend in the lane, 9 ft of siltstone with limestone nodules are exposed. All these beds, and those at the roadside and in the river bank 150 yd farther south, were included in the Aymestry Limestone by Elles and Slater (1906, p. 214, fig. 8), a classification which seems to be lithologically undesirable.

Calcareous siltstone with limestone nodules is seen in several small outcrops in the valley [445808] just east of Whettleton Bank. Much of it is probably at a rather lower horizon than the highest beds of the Bank. D.C.G.

Whettleton Bank to Westhope. Some 24 ft of olive-green, weakly calcareous, flaggy, micaceous siltstone with occasional bands of grey limestone nodules are exposed in an old working [447823] 600 yd E. of Whettleton, and 10 ft of similar beds, with an easterly dip of 10°, are seen in an old quarry [44898252] (Locality 80) 275 yd to the north-east, just outside Nortoncamp Wood. Northwards to the Bache Fault the beds are unexposed apart from an obscure section in the roadside [444836] 320 yd S. 19° W. of Ireland, where 15 ft of buff-grey, flaggy siltstone are seen.

Within the Bache Fault belt, Lower Ludlow Shales occupy a triangular area around Upper Dinchope [458834], bounded to the north and south-west by faults and to the east by the outcrop of the Aymestry Group. In the roadside [45608350] (Locality 82) 860 yd N. 27° W. of Greenway Cross 26 ft of pale greyish green, flaggy siltstone with rare thin calcareous bands are seen. These beds have an easterly dip of 11°. At 350 yd to the east, on the north side of the Dinchope Fault, a track section (Locality 81) in the upper part of the Shales, in buff, flaggy, micaceous siltstone with grey, fine-grained, limestone bands, has yielded an abundant fauna. There are several exposures in the Shales in the lane south of Lower Dinchope, and an old quarry [45418422] (Locality 83) 770 yd W. 7° S. of Siefton Forest shows 24 ft of beds. The basal beds of the Shales are exposed in the stream [45198476] (Locality 84) 1000 yd W. 19° S. of the Tower, Callow Hill. This section shows about 25 ft of olive-green, flaggy siltstone with occasional bands of limestone and limestone nodules. There is a steady increase in the proportion of limestone towards the base of the section (south-west end), and the basal beds may mark a passage from the Wenlock Limestone.

There are sections through a large part of the Shales between Westhope [470863] and the northern end of Siefton Batch [474852]. The basal beds are exposed in a track section [466862] 450 yd W. of Westhope church, where olive-green siltstones with bands of grey, fossiliferous limestone are seen. About 10 ft of olive-green, flaggy siltstones with occasional limestone bands and nodules, slightly higher in the sequence, are exposed 300 yd W. 8° S. of the church. Some 24 ft of similar beds are seen 260 yd W. 41° S. of the church. About 150 yd to the south a stream section [46718589] (Localities 85–96) shows beds about 150 ft above the base of the Shales. A thickness of 27 ft of olive-green siltstones is exposed, with some nodules of fine-grained limestone and five bands of white clay up to 4 in thick. These clay bands resemble those seen in the Wenlock Shales near Horderley (p. 173) and elsewhere. The basal beds of the section have a speckled appearance due to the presence of fragments of small brachiopods and other fossils. This lithology, together with a fauna which includes *Chonetoidea grayi*, *Dalmanites myops* and occasional *Monograptus colonus*, indicates a horizon close to the junction between the Lower and Middle Elton Beds of Holland, Lawson and Walmsley (1959). At 40 yd beyond the eastern end of this section the stream is joined by a tributary 350 yd S. 22° W. of Westhope church. An intermittent section (including Locality 97), covering about 150 ft of beds, extends downstream for about 300 yd from this point. The beds are olive-green, flaggy siltstones, gradually passing up into flaggy, micaceous mudstones with rare limestone nodules. The dip is south-easterly, from 12° to 22°. *Chonetoidea grayi*, *Dalmanites myops* and graptolites are common. For a farther 400 yd downstream there are few exposures, but 330 yd north-east of Hillend the stream [477856] swings sharply towards the south-west for 150 yd along a minor strike fault, and there are several exposures of strongly folded olive-green siltstones with limestone bands. The top 250 ft of the Lower Ludlow Shales are poorly exposed in the brook.

The top beds of the Shales are seen in several sections on the Aymestry Group scarp. A track-side exposure [468853] 330 yd W. 18° S. of Hillend shows 26 ft of olive-green, flaggy, calcareous siltstone with many bands of grey limestone nodules. Compound corals occur 4 ft from the base, and simple corals and *Strophonella euglypha* are common. There is a 6-ft gap above this section to an old quarry which shows 23 ft of similar beds. The limestone bands in this upper section become thicker, more common and less nodular towards the top, showing a transition into the Aymestry Group. *Conchidium knightii* was found 2 ft from the top of the section. The topmost beds of the Shales are seen in an old quarry [479863] 225 yd N. 12° E. of Titterel which exposes 44 ft of calcareous siltstone, locally shaly, with many thin bands of grey, crystalline limestone, commonly crinoidal. A track section [48578694] (Locality 98) 1170 yd N. 40° E. of Titterel shows 20 ft of calcareous, silty mudstone with bands and nodules of argillaceous limestone. These beds, which here form the crest of the scarp, are probably also very near the top of the Shales.

Several small exposures of the Shales are seen in roadsides and streams between Westhope and Burwood [484872], and in an old quarry 480 yd N. of Burwood. B.A.H.

Burwood to Blackwood. The basal beds of the Shales are exposed in several sections to the south of Middlehope [497884], particularly in the stream 420 yd S. 19° W. of the Motte and Bailey where grey calcareous mudstone with occasional small limestone nodules is seen to 6 ft. North of the hamlet the transitional nature of the Wenlock Limestone–Lower Ludlow Shales junction in this area is displayed in sections in the stream (p. 181).

The highest beds of the Shales crop out in an old quarry [49808678] (Locality 99) 1480 yd W. of Munslow Aston chapel, where 6 ft of bedded calcareous nodular siltstone with nodular bands of limestone are seen. The transition from the Shales to the Aymestry Group is seen in track sections and an old quarry east of Middlehope.

Here, a track section 1750 yd S. 18° E. of Eaton church shows about 12 ft of calcareous silty mudstone with occasional limestone nodules and bands of nodules. Rather more calcareous beds, with bands of shelly limestone, are exposed 40 to 90 yd to the south-south-west [50478844] (Locality 100), and have been taken as the highest beds of the Shales. An old quarry (Localities 200 and 201) 20 yd to the south shows about 25 ft of calcareous siltstone and argillaceous limestone, with shelly limestone bands (basal Aymestry Group). There is an unexposed gap of about 25 ft between the quarry and track sections. This transition from Lower Ludlow Shales to Aymestry Group by a sharp increase in the proportion of nodular argillaceous limestone bands is clearly seen in sections about 1000 yd north-west of Beambridge [532882]. In this group of sections the lowest beds, near the top of the Lower Ludlow Shales, are exposed in the stream 1370 yd N. 20° E. of Munslow church. They comprise brown and grey shaly mudstones with occasional limestone nodules. Slightly higher beds, the top beds of the Shales, comprising silty mudstones with occasional nodular bands and nodules of limestone, crop out in the gully 170 yd to the north-east. The basal beds of the Aymestry Group, bedded calcareous silty mudstones with nodular limestone bands, are seen in a track section [52618886] (Locality 204) 1360 yd N. 24° E. of the church. Higher beds, bedded calcareous siltstones with many bands of nodular limestone, are exposed in an old quarry [52718912] (Locality 203) 300 yd farther to the north-north-east (Shirley 1952, p. 205).

An excellent and very fossiliferous section (including Localities 101 and 102) in the Shales occurs in the banks of the stream at Upper Millichope for 600 yd upstream from a place [521893] 1800 yd N. of Munslow church (Shirley 1952, p. 205). The beds exposed comprise pale grey bedded mudstone, slightly calcareous when fresh, weathering to olive or buff. Doggers of fine-grained limestone up to 2 ft across and 9 in thick occur in the upper part of the sequence, usually in bands. Thin calcareous beds and beds of small limestone nodules also occur, particularly towards the base of the sequence, and there are some beds of soft, yellow, fine-grained silica-sand, usually about ½ in thick. There are many outcrops of grey and olive mudstone and silty mudstone, with some bands of grey nodular limestone, in the stream between the north-western end of the last section, and a place [526907] 470 yd S. 11° W. of Stanway Manor. R.H.H., B.A.H.

Stanway Manor. There are several good sections in the Shales in the Stanway district. Beds near the base of the sequence are seen in the track [533915] 750 yd E. 25° N. of Stanway Manor, where about 18 ft of grey, flaggy siltstone with many bands of nodular, greyish blue, fine-textured limestone are exposed. Some 60 ft of beds are intermittently exposed in a section, nearly parallel to the strike, which extends downstream for about 300 yd from a locality [531912] 500 yd E. 5° N. of Stanway Manor. They are olive-green, flaggy siltstones with sporadic thin calcareous bands and a fauna typical of the lower part of the Shales which includes *Chonetoidea grayi*, *Leptaena rhomboidalis*, *Skenidioides lewisii* and *Dalmanites sp.* About 120 ft of similar beds, dipping at 10° to E. 40° S., are exposed in a small tributary which joins the main stream at the north-east end of the previous section.

About 24 ft of olive-green, flaggy siltstones with many bands of grey limestone nodules and bands of tabular grey limestone, near the top of the Shales, are seen in an old quarry [534908] 900 yd E. 20° S. of Stanway Manor. The limestone bands are more common towards the top of the section, and contain abundant small brachiopods with crinoids and other shell debris. B.A.H.

Stanway to Presthope. There are many small exposures of grey mudstone with limestone nodules in the stream between Wilderhope and a locality [544909] 600 yd west-north-west of New House, but north of Wilderhope the Shales are poorly exposed. Buff, shaly weathering, grey mudstone and silty mudstone with occasional limestone nodules, near the top of the Shales, are seen in a track [553928] 830 yd E. 7° S. of Wilderhope. Grey silty mudstone is exposed beneath 8 ft of boulder clay in the stream [567947] 450 yd S. 20° E. of Easthope church. These beds are disturbed and have an east-south-easterly dip of 25° to 45° due to the proximity of the N.–S. fault which runs through Easthope and southwards to Shipton. A track section which extends for 200 yd to the south-east from a locality [571950] 480 yd E. 15° S. of Easthope church shows greyish buff mudstone with occasional thin argillaceous limestone bands. The highest beds of the Shales (included by Robertson 1927, p. 93 in the Aymestry Group) are exposed below the weir [57369426] (Locality 103) 330 yd E. 20° S. of Larden Cottage, where 12 ft of flaggy, buff and grey calcareous siltstone with occasional thin platy limestone and nodular limestone bands are seen. The limestones are generally argillaceous and smooth textured. This section has yielded a large fauna typical of the Lower Bringewood Beds of Holland, Lawson and Walmsley (1959). There are no significant exposures of the Shales north of Easthope, apart from a section in a small faulted trough in Lilleshall Quarry, Presthope (p. 180).

B.A.H., R.H.H.

AYMESTRY GROUP

Brandhill to View Edge. In the old quarry [418790] some 250 yd W. 36° N. of Brandhill Farm rather more than 30 ft of beds are exposed, consisting of nodular crystalline limestone with undulating bands of siltstone, the more persistent beds of the latter dividing the rock into posts from 3 to 10 ft thick. These posts maintain their thicknesses over the length of the quarry face, over 100 yd, and can easily be identified in the quarry a farther 100 yd to the south. *Dayia navicula* is abundant and other forms locally common include *Atrypa reticularis* and '*Camarotoechia*' *sp.* Corals and crinoids occur in places in the lower beds. It is clear that the beds in these quarries belong to the *Dayia* Shales of Elles and Slater (1906, pp. 197–8) and the Lower Leintwardine Beds of Holland, Lawson, and Walmsley (1959, p. 1038). In the northern quarry the strata dip at 5° to E. 35° S. and are cut by prominent steep joints trending N. and E. 20° S., the latter generally calcite-filled.

Lower beds protrude through the spoil of the quarry about 300 yd W. 24° N. of Brandhill Farm. The top of a band containing *Conchidium knightii* is exposed. A similar band, seen to 1 ft, occurs just west of the southern quarry, at a horizon estimated to be about 8 ft below the lowest limestone in the quarry itself. In an old working [416789] 350 yd W. of the farm a band of *C. knightii* 3 ft thick is exposed at the base of 30 ft of grey nodular limestone. The top of the Aymestry Group is taken at the transition from limestones to flaggy siltstones with calcareous bands, seen in a lane [419790] about 120 yd N. 13° W. of Brandhill Farm.

About 44 ft of grey crystalline nodular limestone are exposed in the main face of an old quarry [41967985] (Localities 107–109) 600 yd W. 24° N. of Gorst Barn. Three sub-equal posts of limestone are separated by bands of siltstone, with limestone nodules, each about 1 ft thick. Prominent steep joints trend N. and E. 20° S. as at Brandhill. *C. knightii* forms a 3-ft band at the base and occurs also, with *Leptostrophia filosa*, in the overlying 11 ft of limestone. Compound corals, crinoids and brachiopods are the main elements in the fauna. *Strophonella euglypha* and *Dalmanites sp.* occur in a lower outcrop of limestone within the quarry. The dip varies between 9° to E. 30° N. and 3° to S. 15° E., the latter attitude persisting northwards for some 250 yd.

These limestones of the lower part of the Aymestry Group cap the escarpment from south of the quarry, just described, to within about 100 yd of the road across View Edge. They are well exposed between 650 yd W. 35° N. [420800], and 700 yd N. 13° W. [423803], of Gorst Barn, where a band up to 12 ft thick, packed with *C. knightii* occurs near the top of the cliff. About 11 ft of sparsely fossiliferous limestone are seen below the *Conchidium* band, and above are some 6 ft of limestone with abundant crinoid ossicles and some corals and Strophomenids. Near the northern end prominent steep joints trend E. 15° N. and S. 20° E.

A. Band of *Conchidium knightii* in the Aymestry Group (A9535)

B. Slump bedding in the Upper Ludlow Shales (A9531)

A. CURRENT BEDDING IN DOWNTON CASTLE SANDSTONE (A9528)

B. SANDSTONE WEDGE IN CORNSTONE-CONGLOMERATE OF THE DITTON SERIES (A9525)

These outcrops are separated from those of the disused View Edge quarries by a fault, with an easterly downthrow of about 45 ft, which brings Lower Ludlow Shales to the west against beds above the band of *C. knightii* to the east. In the most southerly of the View Edge quarries [42608070] (Localities 110–112) about 27 ft of blue-hearted crystalline limestone are exposed, divided by undulatory partings of khaki siltstone. The lowest beds are relatively unfossiliferous but contain a varied fauna including *Favosites sp.*, *Ptilodictya lanceolata* (Goldfuss), *Sphaerirhynchia wilsoni*, *Strophonella euglypha* and gastropods. Between 9 and 16 ft from the base lies a band packed throughout with the valves of *C. knightii*, which has also yielded a compound coral and specimens of *Atrypa reticularis*, *Leptaena rhomboidalis*, and *S. euglypha*. Current bedding is locally developed and occasional lenticles of chert stand out on the weathered faces. Siltstone partings are better picked out in the highest beds, in which *A. reticularis* is especially common. *C. knightii* occurs near the base of these beds; compound corals and many crinoid columnals are seen near the top. The abundant fauna also includes *Gypidula galeata* (Dalman) and *L. rhomboidalis*. Prominent steep joints filled with calcite trend N. 2° W. and E. 18° S. In the adjacent quarries to the north [42628078] (Locality 113) an additional 2 ft of limestone are exposed below the *Conchidium* band, and the band is itself reduced to about 5 ft in the most northerly of the quarries. The beds dip generally to the south-east at about 3°, but in the western part of the most southerly quarry they dip at 5° to the south-west probably because of the proximity of the 45-ft fault referred to above.

Aldon. The Aymestry Group is well exposed in Springhead Gutter, the valley north and east of Gorst Barn. The band containing *Conchidium knightii*, about 6 ft thick, forms an inlier in the upper part of the valley. In an old quarry [424799] 260 yd N. 16° W. of Gorst Barn the base of the band is ill defined and *C. knightii* occurs in reduced numbers for about 2 ft below the main development. In a cliff section [42497978] (Localities 105 and 106) 170 yd N. 10° E. of Gorst Barn the band is separated by about 15 ft of strata, largely obscured, from 20 ft of limestones with *Dayia navicula* in abundance. In the floor of the Gutter [429798] 460 yd E. 28° N. of Gorst Barn the *Conchidium* band is seen below some 18 ft of limestone, about 40 yd west of a north-north-easterly fault, estimated to have a westward downthrow of about 40 ft. These beds all dip at about 8° to E. 25° S.

Some 50–150 yd E.S.E. of the fault, limestone, dipping at 10° to S. 40° E. is exposed to a total of 40 ft in the cliffs [42927979] (Locality 116) which delimit the valley floor. The *Conchidium* band, at the top, is at least 4 ft thick, and *Conchidium* also occurs sparsely in a thin band about 16 ft below. Prominent steep joints, filled with calcite, trend E. 25° S., approximately parallel to this part of the valley. Along the higher slopes are small outcrops of limestone and siltstone, generally packed with *D. navicula*.

The section given below is compiled from different parts of an old quarry [43107968] (Localities 114 and 115) 680 yd E. 5° N. of Gorst Barn:

		ft	in
Interbedded calcareous siltstone and limestone; *Dayia navicula* abundant .. .. grading to	c.	22	0
Grey crystalline limestone; *D. navicula* near top; crinoid columnals; some *Conchidium knightii* near base	c.	11	0
Limestone packed with *C. knightii*	c.	2	6
Blue-hearted crystalline limestone		10	0

Stoke Wood. The higher beds of the Aymestry Group occupy the ridge and the upper part of the dip slope between Brandhill and View Edge. Silty beds full of *D. navicula*, very near the top of the Group, are well exposed at Newfoundland [427803], and flaggy limestones, similarly fossiliferous, are seen beside the valley and the road to the north-east [430804], and again at the roadside [431803] 440 yd E. of Newfoundland, on the upthrow side of a north-north-easterly fault (p. 273). This fault explains the occurrence of the *Conchidium knightii* band, 5 ft thick, within 22 ft of limestone, in old quarries [432806] (Localities 120 and 121) some 630 yd E. 30° N. of Newfoundland. In an unfaulted sequence beds much higher in the succession would be expected to occur here. In these outcrops the dip is south-easterly at an average of about 8°.

Crystalline limestone rich in *D. navicula* is well exposed in Church Way [435804] and about 140 yd to the south-east (Locality 117), just within the southern boundary of Stoke Wood. Lower beds, including the *Conchidium* band at least 30 in thick, are exposed to 30 ft at an old working [43658035] (Locality 118) a farther 100 yd to the east. West of the Onny the *Conchidium* band is last seen at the top of a low cliff [439803] 900 yd W. 3° N. of Park Farm. Thin chert bands occur in the limestone just below it. There are small outcrops of limestone in and near the wood up to 270 yd to the east-south-east. D.C.G.

On the north side of the east–west fault in Stoke Wood a 200-yd cliff section at the top of the scarp, immediately north of the View Edge quarries, shows beds near the base of the Aymestry Group. Some 15 ft of grey nodular fine-grained limestone with many thin siltstone partings are exposed, the limestone becoming very nodular towards the base of the section. A few nodular chert bands are present. Beyond the eastern end of the cliff there is a slight embayment in the scarp, on the far side of which the cliff section is resumed after an interval of 250 yd. This second section (Locality 119) runs east-south-eastwards for 270 yd to a place [432808] 780 yd N. 40° E. of Newfoundland. It shows 4 ft of crystalline limestone above the *C. knightii* band, here 5 ft thick, with below a further 38 ft of similar beds with nodular chert beds. Thin siltstone partings become noticeably more prominent in the lower beds. The dip of the beds is uniformly about 7° towards S. 30° E., but locally towards the western end of the section it becomes markedly higher. At Church Way this outcrop of the Aymestry Group runs into the east-south-easterly fault, but it is seen again at an old quarry [439806] near some old limekilns, 1000 yd W. 20° N. of Park Farm. In the quarry *C. knightii* is abundant near the top of the 5 ft of limestone exposed. Here the beds dip at 20° to S. 35° E., but about 30 yd to the south, limestone with many specimens of *C. knightii* dips steeply towards N. 10° E., probably because of its proximity to the fault. Vertical limestone on the perimeter of a circular enclosure just to the south is probably not in place. D.C.G., B.A.H.

Park Farm. Two good sections (including Localities 122 and 123) in grey nodular crystalline limestone occur in the River Onny between 140 yd S. 24° W. [446801], and 240 yd S. [447800], of Park Farm. In the old quarry [44688005] (Locality 124) beside the cottage on the left bank, 15 ft of nodular limestone are exposed, divided into posts between 1 and 6 ft thick by detrital limestone, which in one case is full of crinoid columnals and shell fragments. *Conchidium knightii* is not very abundant in the river section and also present are *Isorthis* cf. *orbicularis*, *Sphaerirhynchia wilsoni* and *Dalmanites myops*.

The old quarry [44738006] (Locality 125) about 180 yd S. 18° E. of Park Farm shows about 18 ft of the *Dayia* Shales of Elles and Slater (1906, pp. 214–5), dipping at 7° to E. 30° S. The basal 12 ft are grey nodular limestone in two main posts and the remainder is highly fossiliferous calcareous siltstone with some limestone nodules. Species collected include '*Camarotoechia*' *nucula*, *Dayia navicula*, *Isorthis orbicularis*, *Leptostrophia filosa*, and *Sphaerirhynchia wilsoni*. Faunal classification places all the rock in the *Dayia* Shales or Lower Leintwardine Beds (Holland, Lawson, and Walmsley 1959) but on a lithological basis only the lowest 12 ft are included in the Aymestry Group.

Limestone has been extensively quarried [44708030] (Locality 126) 50 to 150 yd north of Park Farm, and in the wood northwards (Locality 127) from a place [44598060] 420 yd N. 14° W. of the farm. A total of some 65 or 70 ft of limestone is exposed. A band of *Atrypa reticularis*, near the top of the 25 ft of limestone in the southern quarry, can apparently be equated with a similar band, 30 ft from the top of the 44 ft exposed at the southern end of the northern quarries. Other common fossils are *I. orbicularis* and *S. wilsoni*. Beds, which are estimated by eye to extend to about 20 ft below the base of the 44 ft section, are exposed in the northern quarry at a point 440 yd N. 25° W. of the farm, and include, about 16 ft from the base, an 18-in band with *Conchidium knightii* in abundance. *Favosites sp.* is also common at about this horizon. The limestones are blue and grey, nodular, crystalline, and are divided by partings of buff siltstone into posts from 1½ to 6 ft thick. They are locally current-bedded and commonly coarsely crystalline.

The most southerly of the northern quarries extends to the south side of a footpath [446808], the lowest horizon at the northern end being about 12 ft below the base of the 44-ft section. Current-bedded wedges of coarsely crystalline limestone, with much crinoid debris, occur here and in the quarries north of the footpath, in zones up to about 1 ft thick. They weather out more readily than the thicker nodular limestones.

Altogether some 45 ft of limestones, of the same character and at the same general horizon as those just described, are exposed in the quarries north of the footpath. The occurrence of a specimen of *Favosites sp.* some 15 ft below a band with many *Atrypa reticularis* suggests comparison with the band of *C. knightii* in the lowest exposure south of the footpath.

In all these quarries the limestone dips at 7° or 8° to about E. 30° S. Prominent steep calcite-filled joints have often been used in quarrying, the main trends being about E. 20° N. and E. 35° S. Calcareous tufa locally covers the quarry face. The beds lack the persistence observed, for example, at Brandhill, and detailed bed-by-bed correlation along the outcrop is not self-evident. D.C.G.

Whettleton Bank to Titterel. Extensive quarrying of the Aymestry Group has taken place in the wood half a mile north of Park Farm. On the southern side of the valley through the wood a series of old quarries, largely inaccessible, links with those just described. At the north-eastern end [447810] of this line of quarries, 830 yd S. 24° W. of Campbarn, 32 ft of grey, crystalline limestone, locally crinoidal, with siltstone bands and partings, are seen, dipping at 8° to E. 30° S. At 100 yd to the west-north-west, on the opposite side of the valley, a large old quarry shows about 40 ft of similar beds, with well-developed steep joints trending E. 12° N. and E. 24° S. Another quarry 100 yd W. 18° S. of this section exposes beds lower in the Aymestry Group. About 25 ft 6 in of grey, nodular, crystalline limestone with occasional siltstone bands and partings are seen. *Conchidium knightii* occurs in two thin bands at 8 ft and 16 ft 6 in from the base of the section, abundant specimens of *Atrypa reticularis* occur 23 ft 6 in from the base, and a 3-in band of compound corals was noted at 11 ft 6 in from the base.

Limestones of the Aymestry Group are exposed in an almost continuous section, partly natural and partly due to quarrying, along the top of the scarp in Nortoncamp Wood. The dip of the beds varies from 8° to 12°, and the direction gradually swings from E. 30° S. at the southern end to due E. north of Norton Camp. At the southern end of the section, where it terminates in an old quarry [444811] with an east–west face, 900 yd S.W. of Campbarn, 19 ft of grey, crystalline, nodular limestone with buff siltstone bands and partings are exposed, with marked partings 6 ft 6 in and 11 ft 6 in from the top of the section. *A. reticularis* and *Sphaerirhynchia wilsoni* are common. The same beds are exposed for 250 yd along the scarp to the north (Localities 128–132), with local variations in the height of the cliff face causing an extra 6 ft of beds at the top and 12 ft at the base of the section to be seen. At the northern end of this section, the beds are disturbed by faulting for 150 yd, and individual beds cannot be traced across the zone. The continuous cliff section is resumed 100 yd farther to the north, 670 yd W. 7° S. of Campbarn, and continues northward for 320 yd. It is possible to trace individual beds and partings along the whole of this section, and in particular, two fossil bands near the base, the higher with *Strophonella euglypha* and compound corals, and the lower with *Conchidium knightii* and other brachiopods.

The most complete sequence is seen at the northern end of the section [44518186] (Localities 134–142), where the following beds are exposed:

		ft	in
Massive, grey, crystalline limestone, evenly bedded but locally nodular. *Atrypa reticularis*, '*Camarotoechia*' *nucula*, *Isorthis orbicularis* and *Sphaerirhynchia wilsoni* common .. .. seen		31	6
Alternate-bedded siltstone and platy limestone ($\frac{1}{2}$-inch bands)	Beds incut into face of cliff		6
Grey, crystalline limestone			6
Buff siltstone with thin bands of nodular and of platy limestone		1	0
Grey, nodular, finely crystalline limestone with irregular siltstone bands and partings, often wrapping around the nodules. Locally flaggy bedded with evenly bedded siltstone bands. '*C.*' *nucula* and other brachiopods. *Dayia navicula* abundant 3 ft from base..		9	6
Marked siltstone parting (incut) .. .. c.			3
Limestone as above, with *A. reticularis* and '*C.*' *nucula*		8	0
Band with many brachiopods and occasional compound corals .. to			6
Limestone as above, very irregularly bedded		6	6
Band with '*C.*' *nucula*, *Conchidium knightii*, *Gypidula galeata* and *Leptaena rhomboidalis*			3
Limestone as above, with *Leptostrophia filosa* and fragments of *C. knightii* 3 ft from base of section.. .. .. .. seen		6	6

North of this place the cliff section becomes intermittent and degraded. The band with *C. knightii* can be traced for a farther 250 yd to the north, and the 2-ft incut parting for 350 yd. Several small sections in the Aymestry Group are exposed on the northern slope of the hill [450923], about 700 yd north of Campbarn, but it is not possible to correlate individual beds with the sequence described above.

An old quarry [44578136] (Locality 133) 600 yd W. 33° S. of Campbarn shows 5 ft of greyish blue, crystalline limestone with siltstone bands and partings and abundant *D. navicula*. These beds, and also the upper part of the cliff section above, fall within the *Dayia* Shales of Elles and Slater (1906, p. 198) and the Lower Leintwardine Beds of Holland, Lawson and Walmsley (1959, p. 1038). Non-calcareous beds exposed within Norton Camp (p. 198) also fall within the *Dayia* Shales, so that in this area as at Park Farm (p. 198), the Aymestry Group includes only the lower part of the Shales.

East of Craven Arms a strong east-west valley cuts through the Aymestry Group scarp. However, the dip and strike of the beds on either side of the valley show that the outcrop is continuous from Norton Camp across the valley to a place [447837] 300 yd S.E. of Ireland. The east-west fault postulated by Alexander (1936, pl. viii) does not extend so far westwards as she indicated, and does not affect this part of the Aymestry Group outcrop.

On the north side of the Castle Road, old quarries extend along the scarp for 500 yd to the north-west from the disused lime works [454829] 600 yd W. of Greenway Cross. The most complete section (Locality 143), now becoming obscured by a rubbish tip, is 100 to 150 yd from the road, where 37 ft of grey, nodular limestone with siltstone bands and partings are exposed. The limestone is divided by incut partings into posts up to 6 ft in thickness. Some of these partings can be traced the full length of the quarries, but no detailed correlation is possible beyond this. North-westwards the limestone is unexposed for 350 yd, beyond which old quarries extend for a farther 350 yd.

The Aymestry Group outcrop is repeated by a fault with a south-easterly trend which branches from the Bache Fault south-east of Ireland (p. 185). The repeated outcrop extends northwards from the Castle Road, 300 yd west of Greenway Cross, to the Bache Fault. On the eastern side [45428319] (Locality 155) of a small hollow 710 yd W. 37° N. of Greenway Cross about 14 ft of grey, crystalline crinoidal limestone with occasional siltstone bands are exposed, dipping at 11° to the east. At 150 yd to the south, on the western side of the hollow, similar beds are exposed, dipping at 20° to S. 17° W., and the dip increases to 63° to W. 40° S. 50 yd farther to the south-south-west. This anticlinal folding within the Aymestry Group is probably due to drag on the south-easterly fault. At the southern end of some old quarries [45188345] (Localities 144–146), 1100 yd W. 38° N. of Greenway Cross, three sections are visible, exposing a total of 25 ft of beds comprising grey, crystalline, crinoidal limestone, commonly nodular, with many bands and partings of buff siltstone and calcareous siltstone. Between 100 to 200 yd to the north-west, beds lower in the Aymestry Group, with many *Conchidium knightii*, are exposed. At the northern end of these sections the strata are sheared and broken, with minor faults and anomalous dips.

North of the Bache Fault the Aymestry Group crops out over a small area south of Greenway Cross. The outcrop is isolated on three sides by faults, and overlain to the south-east by the Upper Ludlow Shales. The only exposure is in an old quarry [46068260] at the road junction 250 yd S. 30° E. of Greenway Cross. In the main part of the quarry (Localities 150 and 151) 6 ft of weathered, buff, calcareous siltstone overlie 14 ft of grey, crystalline, nodular limestone with many siltstone bands and partings. Brachiopods are abundant, particularly *Atrypa reticularis*, '*Camarotoechia*' *nucula* and *Sphaerirhynchia wilsoni*. On the north-eastern side of the quarry (Locality 242) these beds of the Aymestry Group are faulted against flaggy siltstone with a fauna including *A. reticularis*, *Shaleria ornatella* and *Sphaerirhynchia wilsoni*. Although *Dayia navicula* is absent these siltstones are probably near the top of the Lower Leintwardine Beds, and are here included in the Upper Ludlow Shales on lithological grounds.

The main outcrop of the Group is resumed 250 yd S. of Shawbank, north of the west-south-westerly cross fault (p. 274). At the quarry [46368319] (Locality 148) at Shawbank 25 ft of grey, crystalline, nodular limestone with siltstone bands and partings, and occasional bands of coarsely crystalline limestone, are exposed. Rotted chert occurs in a band 5 ft from the base of the section and locally elsewhere. The beds have an easterly dip of 11°, and joints trending E. 30° S., S. 40° E., and S. 17° W. are prominent.

North of Shawbank, the Aymestry Group outcrop is displaced dextrally for about 300 yd by the Dinchope Fault. There are several small outcrops (including Locality 149) in the track and old workings [464835] south-south-east of Buck House. Beds near the base of the Group crop out [45908379] (Locality 147) 1100 yd N. 4° W. of Greenway Cross, where 5 ft of grey, crystalline limestone with siltstone bands and partings has yielded a large fauna including *Conchidium knightii*. Similar beds, though without *C. knightii*, crop out (Locality 154) 90 yd to the north-east. A cliff section (Locality 152) at the top of the scarp to the east, 500 yd N.E. of Upper Dinchope, shows 35 ft of grey, flaggy, crystalline limestone, commonly crinoidal, with a large fauna typical of the Lower Leintwardine Beds.

The basal beds of the Group are exposed in a number of sections near Hillend. A track-side section [464852] (Locality 164) 740 yd W. 15° S. of Hillend shows about 4 ft of flaggy-bedded, olive-green siltstone and calcareous siltstone with bands of grey, finely crystalline limestone. The lithology and fauna, which includes *C. knightii*, *Sphaerirhynchia wilsoni*, *Strophonella euglypha* and *Dalmanites myops*, both indicate beds transitional from the Lower Ludlow Shales to the Aymestry Group. 30 yd east of this section 20 ft of alternating bands of grey limestone and buff, calcareous siltstone are exposed (Locality 163), slightly higher in the Group, and these beds are seen again 40 yd farther to the east. The dip in these sections is south-eastwards, 7° to 10°, and this direction is maintained along the outcrop towards the north-east.

Beds higher in the sequence are seen in an old quarry [465848] 920 yd W. 42° S. of Hillend, where 13 ft of nodular, grey limestone with thin irregular bands of buff siltstone are exposed. A degraded quarry [46548447] (Locality 157) 1150 yd S. 31° W. of Hillend shows 18 ft of flaggy-bedded, nodular, grey limestone with occasional thin siltstone bands. The fauna includes *Dayia navicula* and *Shaleria ornatella* and is typical of the highest part of the Group. The top of the Group is poorly defined on the dip slope. Exposures are few, and because of the transitional nature of the junction with the Upper Ludlow Shales no good feature is formed.

The Aymestry Group is well displayed in a number of sections in Siefton Batch. Beds near the base of the Group are seen in an old working [474850] 490 yd S. 42° E. of Hillend. 4 ft of greyish green, bedded, nodular limestone with bands of buff siltstone are exposed. Shirley (1952, p. 205) has recorded *Conchidium knightii* from this locality. A good section in the higher part of the Group is seen in an old quarry [47708477] (Localities 165 and 166) 940 yd E. 43° S. of Hillend, where the following beds are exposed:

	ft
Evenly bedded, greyish green, calcareous siltstone with abundant bands of grey limestone, locally nodular. Flaggy weathering towards the top	32
Parting	
Grey, nodular limestone with thin siltstone bands wrapping around the nodules	6
Evenly bedded, greyish green, calcareous siltstone and grey, crystalline limestone. Abundant brachiopods	7

Shirley (1952, p. 205) recorded rare *D. navicula* from this locality. A stream section 70 yd to the east and an old quarry 70 yd to the north-east show similar beds. A track section [47808468] (Localities 169–179) 100 yd to the south-east, 1040 yd S.E. of Hillend, exposes 12 ft of beds at the top of the Aymestry Group. The strata are buff-weathering, greyish green, flaggy, coarse-grained, micaceous siltstones with many bands of grey, crystalline limestone, especiallay at the base. *D. navicula* occurs throughout the section and *Sphaerirhynchia wilsoni* is common in the lower beds. *Shaleria ornatella* and *Monograptus* cf. *leintwardinensis* occur 40 ft from the road. The fauna indicates a horizon towards the top of the Lower Leintwardine Beds. The basal part of the Upper Ludlow Shales is exposed farther along the track (Localities 243–256) to the south-west (p. 199).

An old quarry [47548530] (Localities 158–162) 460 yd E. 8° S. of Hillend shows about 30 ft of bedded, grey, calcareous siltstone and grey, crystalline limestone. Above the basal 7 ft the limestones become thicker (up to 6 in) and less common. These beds are towards the top of the Aymestry Group, and similar strata are seen in numerous small quarries and degraded cliff sections (including Locality 180) along the top of the scarp between this section and a large old quarry [479857], 960 yd E. 27° N. of Hillend, which shows 52 ft of similar beds. Some 15 ft of bedded, nodular limestone and calcareous siltstone are seen in an old quarry [481861] 300 yd E. of Titterel, and 13 ft of similar beds are exposed 200 yd to the north. B.A.H.

Diddlebury Common to Henmoor Hill. The upper part of the Aymestry Group, and the transition to the Upper Ludlow Shales, is well displayed in a number of quarries near Halehead [489863]. Strata near the top of the Group, calcareous silty flags with evenly-bedded and nodular limestone bands, with many *Shaleria ornatella* and other brachiopods, are seen in an old quarry [48878611] (Locality 182) 2270 yd W. 19° N. of Diddlebury church. A second quarry, immediately above, shows 1 ft 6 in of silty and nodular limestone (Aymestry Group) overlain by 5 ft of buff silty flags with occasional thin limestone beds (Upper Ludlow Shales) (Locality 265). Another old quarry [48998597] (Localities 266 to 271), 200 yd S. 35° E. of the last, shows 25 to 30 ft of massive siltstone and calcareous siltstone with occasional limestone bands, the basal beds of the Shales. The highest beds of the Group are exposed in two more quarries 850 yd W. 9° N. and 950 yd W. 8° N. of Bache Mill chapel. About

12 ft of silty flags with bands of nodules and of shelly limestone, slightly lower in the sequence, are seen in an old quarry [49078656] (Locality 181) 2300 yd W. 33° N. of Diddlebury church.

The Aymestry Group is well exposed in several sections on the eastern side of the valley which cuts the scarp south of Middlehope. The highest beds of the Lower Ludlow Shales are seen in an old quarry [49808678] (Locality 99) 1480 yd W. of Munslow Aston chapel (p. 186), and the basal beds of the Aymestry Group, buff and grey silty limestone with some silty bands are exposed 25 yd to the south. An old quarry [49878693] (Localities 183–189) by the roadside, 1420 yd W. 7° N. of Munslow Aston chapel shows beds higher in the Group comprising 35 to 40 ft of bedded grey and buff limestone, nodular limestone and calcareous siltstone. The limestone shows cellular weathering on exposed joint surfaces. Another quarry [49928688] (Localities 196 and 197) 100 yd to the south-east, in still higher beds, shows 3 ft of massive limestone and silty limestone overlain by 15 to 20 ft of flaggy buff calcareous siltstone with bands and nodular bands of grey limestone. At the top of the eastern end of the quarry there are 3 ft of poorly exposed higher beds which appear to be more calcareous than those in the main body of the quarry, and may be the highest limestone development within the Aymestry Group. Shirley (1952, p. 204) recorded the faunas of these two quarries as being typical of the higher part of the Aymestry Group, and stated that the fauna of the upper beds of the second quarry compares closely with that of the *Chonetoidea grayi* Beds of Builth (Upper Leintwardine Beds). Upper Ludlow Shales are exposed in the quarry (Locality 272) 270 yd S. 30° E. of the last section (p. 199). A further old quarry [49938711] (Localities 190–193) in the Aymestry Group, 1390 yd W. 15° N. of Munslow Aston chapel, exposes 30 to 35 ft of nodular limestone, with calcareous siltstone beds towards the top of the section.

Beds near the top of the Group are seen in an old quarry [50618757] (Locality 194) 1080 yd N. 34° W. of Munslow Aston chapel, where buff silty flags with decalcified limestone bands containing *Salopina lunata*, *Shaleria ornatella* and other brachiopods are exposed. Slightly lower and more calcareous beds, with a fauna including *Isorthis orbicularis*, *Salopina lunata* and *Shaleria ornatella* occur (Locality 195) 50 yd higher up the lane. The basal beds of the Group are exposed on the scarp face east of Middlehope (p. 186). Strata low in the Group are also exposed in an old quarry [51308882] (Locality 202) 1490 yd N. 37° W. of Munslow church, where 6 ft of thin bedded calcareous silty mudstone with bands of limestone nodules and thin shelly limestone bands are seen. These beds appear to be at about the same horizon as those at Locality 204 (p. 187). Higher beds, bedded nodular limestone and calcareous siltstone, are seen in the old quarry [51728841] (Localities 198 and 199) 870 yd N. 30° W. of Munslow church, and in several small sections in the northern part of Millichope Park.

Between Millichope Park and Shipton there are numerous sections in the Aymestry Group. The transitional beds at the base of the Group near Upper Millichope have already been described (p. 187). The lowest beds of the Group are seen in the stream [533892] 1150 yd N. 8° E. of the road junction at Beambridge, and again in an old working [527901] immediately north of Topley, where 6 ft of silty calcareous mudstones with bands of limestone nodules and occasional thin shelly limestone bands are exposed. Higher beds are seen in the old quarry [53278921] (Locality 208) 2060 yd N. 36° E. of Munslow church, where 30 ft of bedded grey and buff nodular silty and rubbly limestone with occasional siltstone bands are exposed, and similar beds are seen in the old quarry 180 yd E. 20° N. of the last section. Strata within the Aymestry Group, dipping at 9° or 10° to the south-east, are exposed in two old quarries [530896, 531899] in Big Wood, and 25 ft of bedded argillaceous nodular limestone with some shelly limestone and calcareous siltstone bands are seen in the old quarry [533901] 1300 yd W. 12° N. of Broadstone church. 25 ft of similar beds are exposed in an old quarry [546912] 1850 yd W. 23° S. of Shipton church. An old quarry [555926] 1110 yd N.E. of Shipton church shows about 20 ft of bedded rubbly argillaceous nodular limestone and calcareous siltstone with some shelly limestone bands, within the main part of the Group. The highest beds of the Group and their relation to the Upper Ludlow Shales are seen in several sections between Upper Millichope and Shipton. A road cutting [52898899] (Locality 205) 1620 yd N. 30° E. of Munslow church shows buff bedded siltstone with several bands of shelly limestone and occasional bands of nodular and silty limestone. These beds contain a fauna including *Salopina lunata*, *Shaleria ornatella* and *Fuchsella amygdalina* and appear to be transitional to the Upper Ludlow Shales. Some 20 ft of beds of similar lithology and with a comparable fauna are seen in an old quarry [53608943] (Localities 206 and 207) 2460 yd N. 39° E. of Munslow church. Another old quarry [54629055] (Localities 209–213), 890 yd N. 11° E. of Broadstone church, shows 10 ft of bedded silty limestone and limestone with some siltstone bands, overlain by 20 ft of bedded calcareous flags with limestone bands and nodules. Beds similar to, and possibly at the same horizon as, those in the upper part of the quarry are also seen in a second section (Locality 214) 60 yd to the west-south-west. The faunas suggest that the upper beds of the first quarry are transitional from the Aymestry Group to the Upper Ludlow Shales. The highest beds of the Group are seen again in a small section [55269166] (Locality 215) at Upper Barn, 1000 yd W. 11° S. of Shipton church.

North of Shipton the Aymestry Group is well displayed in sections in the valleys which run south-eastwards from Moggforest [561939] and Larden Cottage [570944]. A section [563938] 260 yd E. 24° S. of Moggforest, near the base of the Group, shows 8 ft of bedded, buff, calcareous, nodular siltstone and nodular, argillaceous, smooth-textured limestone. About 300 yd S. 25° E. of this exposure an old quarry [56449353] (Locality 218) displays 18 to 20 ft of bedded calcareous siltstone and nodular siltstone with shelly and nodular limestone bands. Similar beds, slightly more calcareous, are exposed in the old quarry [56519342] (Localities 216 and 217) 150 yd farther down the valley, 1460 yd W. 15° S. of Brockton crossroads. The fauna from these two quarries resembles that from the *Dayia* Shales of Elles and Slater (1906, p. 198) of the Ludlow area. In the valley south-east of Larden Cottage, beds with a similar fauna are exposed in an old quarry [57309421] (Localities 219 and 220) 730 yd W. 40° N. of Brockton crossroads. The highest beds of the Lower Ludlow Shales are seen at the pool below this section (p. 188). Farther down the valley, in the stream [57579400] (Locality 221) 370 yd W. 43° N. of Brockton crossroads, buff calcareous silty flags with nodular limestone bands are exposed. These beds, at the top of the Aymestry Group, contain plentiful *Shaleria ornatella*. Small exposures of similar beds at the same horizon occur on the hillside 100 to 200 yd to the west.

North of Brockton much of the Aymestry Group is drift-covered, and exposures are few. An old quarry [594964] 350 yd W. 16° N. of Bourton church displays 20 to 25 ft of bedded nodular argillaceous limestone and nodular calcareous silty mudstone, with silty mudstone bands and shelly limestone bands. A poor section [592970], 880 yd N. 34° W. of the church, near the base of the Group, comprises 6 ft of bedded calcareous silty mudstone and mudstone with bands of limestone nodules. Traces of mudstone with limestone bands containing *S. ornatella* were seen in a road cutting [608975] 1680 yd N.E. of the church. B.A.H., R.H.H.

UPPER LUDLOW SHALES

Brandhill to the River Onny. Green flaggy siltstone with some sandstone, locally current-bedded, is exposed in and beside the lane [421789] east of Brandhill Farm. The fauna includes '*Camarotoechia*' *nucula*, *Chonetoidea grayi*, *Dayia navicula*, small Chonetids and *Beyrichia sp.* Beds with an occasional band of limestone and an abundance of *D. navicula*, occurring in the valley [424788] 940 yd S. 5° W. of Gorst Barn, must be very near the base of the Upper Ludlow Shales as here defined. The younger rocks in an old working just to the south have a much more diversified fauna. The calcareous siltstones exposed in the stream (including Locality 223) down to about 1200 yd E. 6° S. of Brandhill [431788] are all very low in the Upper Ludlow Shales. Buff flaggy siltstone from a higher horizon is seen in the old quarry [428785] 230 yd E. 24° S. of Coppy House, the fauna including *Campylites* [*Serpulites*] *longissimus*, '*C.*' *nucula* and *Orbiculoidea sp.*

Siltstones with calcareous bands, low in the Upper Ludlow Shales, are seen at Gorst Barn [42467962] (Locality 224). Higher siltstones, with many specimens of *Beyrichia sp.*, are exposed at the roadside between a place [427795] 280 yd E. 40° S. of Gorst Barn and the corner [432794] a farther 580 yd to the east, the road following the dip of the beds. In the 22 ft of beds at the corner, the fauna includes *Chonetoidea grayi* as well as ostracods. A thin bed of flaggy siltstone contains many rounded pellets of fine-grained phosphatic material up to about 1 inch in diameter. A similar bed occurs at about the same horizon at Shelderton (on Ludlow Sheet 181), about 1¼ miles to the south-west, and near Park Farm, Onibury (p. 190).

Olive-green siltstone, flaggy in part and cut by many curved joints, is exposed in the track [432789] (including Localities 227 and 228) east of the building 1070 yd S.E. of Gorst Barn. The fauna is dominated by small brachiopods, and includes also Orthoceratids and *Beyrichia sp.* These and *Fuchsella amygdalina* are common in similar beds in the lowest 300 yd of the Brandhill stream [436788] and near a well [434786] 860 yd E. of Coppy House. The beds at one place in the stream dip at about 35° to N. 30° W., representing a sharp flexure within the general east-south-easterly dip of about 7° or 8°.

Buff flaggy siltstones with bands of fine-grained sandstone are exposed in an old quarry [43757896] (Locality 226) 200 yd W. 43° S. of Aldon Court. The rock, high in the Upper Ludlow, is locally calcareous and shows spheroidal weathering above certain well-defined bedding planes. The fauna, abundant and varied, is normally concentrated in bands and includes '*Camarotoechia*' *nucula*, *Protochonetes ludloviensis*, *Salopina lunata*, horny brachiopods, gastropods and lamellibranchs. The beds at Aldon Court [439791] and in the lane to the south are similar, and in an old working [441786] 540 yd S. 21° E. of Aldon Court, *Chonetes sp.* and *S. lunata* are abundant. Siltstone and flaggy sandstone, estimated to lie 15 or 20 ft below the Ludlow Bone Bed, are seen in an old quarry [44297887] (Locality 229) 470 yd E. 28° S. of Aldon Court. All these beds dip at about 7° to points a few degrees south of east.

Flaggy siltstones with an abundance of '*C*'. *nucula* are widely exposed at Aldon, with an average dip of about 9° to E. 15° S. In the descending succession on the road [432796] to Brandhill the beds become less flaggy and more calcareous, containing *Dayia navicula* in increasing abundance, and local concentrations of lamellibranchs. Flaggy beds are exposed for about 150 yd north-west of the road junction [436796] north of Aldon. The basal flaggy siltstones of the Upper Ludlow are twice repeated on the road to View Edge by faults which downthrow to the north-west, and are exposed or dug up above and below outcrops of the Aymestry Group about 440 yd E. of Newfoundland (p. 189).

Flaggy siltstones and sandstones are exposed in ascending sequence for about 700 yd east-south-east of the road junction at Aldon. Fish fragments were observed at the entrance [441794] to Aldon Court but in general these highest beds are poorly fossiliferous in comparison with those nearer Aldon. They are exposed again beside a pond [438793] 150 yd N. 30° W. of Aldon Court. In an old quarry at Stokewood Farm [445797] micaceous siltstones with thin flags contain many small brachiopods. These beds dip at 5° to 7° in directions between east and south-east.

Siltstones near the base of the Upper Ludlow Shales are exposed in a re-entrant [441801] 580 yd W. 41° N. of Stokewood Farm and nearby (including Localities 230 and 231). The fauna includes *D. navicula*, lamellibranchs, and *Beyrichia sp.* Successively younger beds are seen in small outcrops between that point and an exposure in the wood [447797] 220 yd E. 11° N. of the farm where some 25 ft of calcareous rock contain many specimens of '*Camarotoechia*' *sp.*

Onibury to Upper Dinchope. The basal 6 ft of the Upper Ludlow Shales, calcareous siltstone with some limestone nodules, form the topmost beds in the old quarry [44738006] 180 yd S. 18° E. of Park Farm (p. 190). The abundant fauna, including *Isorthis orbicularis* and *D. navicula*, is that of the *Dayia* Shales of Elles and Slater (1906, pp. 197–8). Somewhat pelletty flaggy siltstones, like those on the road from Brandhill to Aldon (p. 197) are seen at a pond [44938008] (Locality 232) 300 yd E. 32° S. of the farm. The fauna includes *Chonetoidea grayi*, *D. navicula* and *Beyrichia sp.*, and the beds probably form part of the Upper Leintwardine Beds of Holland, Lawson and Walmsley (1959, p. 1039). Calcareous flags are exposed to about 25 ft in an old quarry [45117967] (Locality 234) 740 yd N. 43° W. of Onibury church. '*Camarotoechia*' *nucula* and *Protochonetes ludloviensis* are the most common fossils and the beds are assigned by Elles and Slater (1906, p. 215) to the upper part of the Lower Whitcliffe or '*Rhynchonella*' Flags. The Concretion Band of these authors (1906, pp. 198, 215) forms a distinct bed, about 2 ft 6 in thick, some 10 ft above the base of greenish brown flags exposed in a large old quarry [45257941] (Locality 233) 420 yd W. 36° N. of the church. The form of this bed is now considered to be due to slumping of the partly consolidated siltstone (Whittard and others 1953, p. 232). About 50 ft of flags, with a fauna like that in the last quarry, are exposed above the slumped bed.

Occasional small exposures of fossiliferous flags and shales are seen on the hillside north of these quarries. D.C.G.

There is much debris (e.g. Locality 236) of buff siltstone on the broad dip slope between Norton Camp [448820] and Norton [460816]. About 6 ft of olive-green, calcareous siltstone are exposed at a small pond (Locality 235) within Norton Camp, 350 yd W. 35° N. of Campbarn. These beds yield abundant *Dayia navicula*, and other fossils include *Atrypa reticularis*, '*Camarotoechia*' *nucula*, *Chonetoidea grayi* and *Leptaena rhomboidalis*. The fauna is typical of the *Dayia* Shales of Elles and Slater (1906, pp. 197–8) and the Lower Leintwardine Beds of Holland, Lawson and Walmsley (1959, p. 1038).

The top beds of the Upper Ludlow Shales are well exposed in track-side sections immediately west of Norton, and in sections in the roadside for 350 yd to the north of the hamlet. The beds, which have an easterly dip of 5° to 10°, are olive-green, flaggy, micaceous siltstones, and are locally calcareous. The sections are now less extensive than when described by Elles and Slater (1906, p. 218).

In the triangular outcrop of Upper Ludlow Shales between Greenway Cross and Upper Dinchope, an old quarry [45788318] (Locality 241) 460 yd N. 24° W. of Greenway Cross shows 12 ft of rather massive, olive-green, weakly calcareous siltstone. The extensive fauna is typical of the Lower Leintwardine Beds.

Bache to Siefton Batch. North of the Bache Fault the Upper Ludlow Shales are exposed in several sections at Bache. An old quarry [470819] 720 yd S. 20° W. of Medley Park shows 24 ft of massive-bedded, olive-green, micaceous siltstone, with many bands of grey, calcareous siltstone with strong vertical jointing. A 2-ft band showing well-developed slump structures (Pl. 10B) occurs 10 ft from the base of the section. This may be equivalent to the Concretion Band of Elles and Slater (1906, p. 198). The beds are affected by three marked shear planes trending E. 30° S., parallel with the dip and the Bache Fault. The amount of movement along these planes appears to be small, although there is slight brecciation at the south-eastern end of the quarry. The slumped bed is seen again in another section 100 yd E. 35° S. of the first quarry.

About 17 ft of olive-green, flaggy, micaceous siltstone with much false-bedding are exposed in an old quarry [472826] 50 yd N.W. of Medley Park. The central part of the quarry is affected by minor faults. Some 50 yd to the west, in the sides of a small pond, purplish brown-stained, micaceous siltstone is exposed. The staining appears to be associated with a minor fault, parallel with the Bache Fault, which runs immediately to the south of Medley Park. Beds near the base of the Shales (Locality 242) are faulted against Aymestry Group beds in a quarry [46068260] 250 yd S. 30° E. of Greenway Cross (p. 193).

North of the Dinchope Fault, buff flaggy siltstones and fine-grained sandstones are seen in an old quarry [471840] 700 yd W. 35° N. of Culmington Manor, and in the track for 300 yd to the north-west. About 12 ft of greyish green, calcareous siltstone, higher in the sequence, are exposed in an old quarry on the west side of Culmington Manor, and 20 ft of similar beds with occasional lenticular bands of brown, rotted, fossiliferous limestone are seen behind the lodge, 250 yd to the south.

The greater part of the Upper Ludlow Shales succession is seen in two parallel sections in Siefton Batch. The first is a track section [482837–475847] which runs at approximately N. 35° W. for about two-thirds of a mile from New House, and the second and more complete section is seen in the brook and adjacent exposures [483837–478847] in the Batch. The second section will be considered in detail. The basal beds of the Shales are exposed in a track section [47768466] (Localities 243–256) 1020 yd S. 44° E. of Hillend. About 32 ft of coarse-grained, buff, micaceous, flaggy siltstone with occasional bands of grey limestone are exposed. The limestone bands are more common towards the top of the section, but become more nodular. The lower part of the section contains abundant *Dayia navicula* and *Shaleria ornatella*, and the higher and more calcareous beds contain *Atrypa reticularis*. *Leptaena rhomboidalis* occurs throughout. This section shows clearly that the boundary of the calcareous Aymestry Group is not well defined, since calcareous beds tend to occur towards the top of the exposure, within the Upper Ludlow Shales. Beds within the Aymestry Group (Localities 169–179) are exposed farther down the track towards the road (p. 194). A section in the stream, 100 yd to the east-south-east also shows the Aymestry Group–Upper Ludlow Shales transition (Aymestry Group Localities 167 and 168, Upper Ludlow Shales Locality 262). Greyish green and buff, micaceous, flaggy siltstones with sporadic bands of grey, nodular limestone and an Upper Ludlow fauna are exposed intermittently downstream (including Locality 263) for about 500 yd. From a place [481841] 1720 yd S. 38° E. of Hillend, exposure is almost continuous to the main road at New House. In this section (which includes Localities 258–261) 118 ft of beds are exposed, comprising flaggy-weathering, greyish green, micaceous, calcareous siltstones with occasional thin, platy, limestone bands, usually with abundant brachiopods. Similar beds are exposed intermittently for about 200 yd below the main road, but the junction with the Downton Castle Sandstone is not seen. New House Quarry [481838], 150 yd N.W. of New House, shows about 24 ft of massive, pale greenish grey, coarse-grained, micaceous siltstone with occasional elongate, lenticular, grey, calcareous bands up to 3 in thick. These beds are equivalent to part of the stream section. B.A.H.

Corfton Bache to Callaughton. The Upper Ludlow Shales are well exposed in several sections around Corfton and Bache Mill. About 20 ft of brown and grey calcareous siltstone with occasional bands of silty limestone and limestone, some nodular, are exposed in an old quarry [491857] 1300 yd W. 18° S. of Bache Mill chapel. Beds of similar lithology, though higher in the succession, are seen in old quarries 1470 yd W. 7° S. [495852] and 2140 yd W. 10° S. [489850] of Diddlebury church, and in the cellar of the Sun Inn at Corfton. A lane section [49698522] (Locality 264), 1230 yd W. 8° S. of Diddlebury church, shows buff-yellow shaly calcareous flags, the highest beds of the Shales, overlain by the Ludlow Bone Bed (p. 219) and Downton Castle Sandstone. The transition from the Aymestry Group to the Shales, displayed in sections around Halehead [489861], has already been described (p. 194). The basal beds of the Shales are also seen in an old quarry [50048666] (Locality 272) 1230 yd W. 4° S. of Munslow Aston chapel, where about 18 ft of flaggy calcareous siltstone with occasional thin limestone bands are exposed. Higher beds, buff silty flags with sporadic limestone bands, are extensively exposed in old quarries on both sides of the valley between the south-eastern corner [500862] of Hall Coppice and the Much Wenlock–Craven Arms road.

The Shales are well exposed in sunken lanes in the village of Munslow Aston [511867]. At a locality 100 yd W. 32° S. of Munslow Aston chapel, buff flags with a number of delicately laminated bands, one of which shows slump structures, are seen. About 35 ft of buff massive-bedded silty flags with occasional bands of silty limestone and limestone, with a south-easterly dip of 10°, are seen in an old quarry [511868] 160 yd N. 25° W. of the chapel, and 20 ft of buff calcareous siltstones are exposed in another old quarry [505869] 720 yd W. 12° N. of the chapel. 5 ft of buff silty flags overlain by a thin impersistent Ludlow Bone Bed and fossiliferous Downton Castle Sandstone (p. 220) are exposed adjacent to the Swan Inn [512866], 110 yd S. 44° E. of the chapel.

Buff and greyish buff flaggy siltstone with occasional limestone bands, in the middle part of the Shales, are seen in two old quarries 230 yd W. 19° N. [519878] and 430 yd N. 26° E. [523881] of Munslow church. Higher beds, buff flaggy siltstones, are exposed in several sections in the village of Munslow. An excellent section occurs in a deep cutting [52838833] (Localities 273–275) 1010 yd E. 42° N. of Munslow church, where the richly fossiliferous beds consist of massive, greyish buff, calcareous and flaggy siltstones with occasional limestone and shelly limestone bands. Cliffs in similar beds occur by the pool 40 yd to the south-west. A section in the track [530880] 1040 yd E. 20° N. of Munslow church shows buff and yellow flaggy siltstone, of which the top foot is sandy and delicately bedded. Beds at this horizon, immediately below the Ludlow Bone Bed, are also exposed at the roadside [531882] 230 yd to the north-north-east.

Some 12 ft of greyish brown calcareous siltstone with strong open joints are exposed in an old quarry [535885] 1770 yd E. 32° N. of Munslow church, and 1 ft of yellowish brown flaggy siltstone, overlain by the Ludlow Bone Bed, is exposed in the brook [542890] 850 yd S. 16° W. of Broadstone church. Immediately north of this section the outcrop of the Bone Bed is displaced 750 yd to the west-south-west by a west-south-westerly trending fault. The fault dies out within the Shales, and their lower boundary is not displaced.

At Broadstone the lowest beds of the Shales and the top beds of the Aymestry Group are exposed in a stream section [542899] 330 yd W. 20° N. of the church, where 2 ft of bedded nodular limestone and calcareous siltstone with occasional shelly limestone bands are overlain by 3 ft of buff silty flags. The fauna of the lower beds includes *Atrypa reticularis*, *Dayia navicula* and *Leptaena rhomboidalis*. About 20 ft of higher beds, buff flaggy siltstone with limestone bands and occasional nodules, are seen in the old quarry 100 yd to the east-south-east, and 20 ft of greyish buff flaggy siltstone and calcareous siltstone, near the top of the Shales, occur in a track section [547900] 350 yd north-east of the church. The old quarry behind Rock Houses [54769067] (Locality 276), 1020 yd N. 17° E. of Broadstone church, shows about 18 ft of greyish buff flaggy calcareous siltstone with occasional shelly and nodular limestone bands.

Good sections in the Shales occur in the vicinity of Shipton. The basal beds, transitional to the Aymestry Group are seen in a stream section [558921] 400 to 500 yd W. 28° N. of Shipton church. Higher beds, buff flags and calcareous flags with occasional limestone bands, with a south-easterly dip of 10°, are exposed to 30 ft in an old quarry [56269199] (Locality 277) 165 yd N. 26° E. of the church and to 18 ft a farther 120 yd to the north-north-east. A section in the lane [56299174] (Locality 278) 170 yd E. 35° S. of Shipton church, described by Robertson (1927, p. 94), shows the top beds of the Shales, the Ludlow Bone Bed (p. 219) and the basal part of the Downton Castle Sandstone (p. 221). The Shales comprise an uncertain thickness of grey and buff calcareous siltstone with thin fossiliferous limestone beds and a typical Upper Ludlow fauna, overlain by about 1 ft 6 in of thinly bedded buff silty shales.

20 ft of buff flaggy siltstone with limestone bands, in the upper part of the Shales, are exposed in an old quarry [570929] 450 yd south-east of Larden Hall.

Sections through the greater part of the Shales occur around Brockton [578938]. The basal beds, buff and grey calcareous flaggy siltstones, are exposed in the stream [576939] 330 yd W. 43° N. of Brockton crossroads. The highest beds of the Aymestry Group (p. 196) crop out 40 yd farther upstream. Higher beds, flaggy calcareous siltstones, are seen in the old quarry behind the school, 200 yd N. 30° W. of the crossroads. The topmost beds of the Shales, together with the Ludlow Bone Bed and basal Downton Castle Sandstone, are exposed in the road cutting [579938] 80 yd N.E. of the crossroads (Robertson 1927, p. 93). Here the Shales comprise 2 ft of blocky calcareous siltstone overlain by 6 in of buff silty shales. The highest beds are seen in an old quarry [585948] 1350 yd N. 36° E. of Brockton crossroads where about 15 to 20 ft of flaggy calcareous siltstone are overlain by 5 ft of olive shaly mudstone. The Ludlow Bone Bed, only ¼ in thick, occurs 1 ft from the base of the mudstone, so that the topmost 4 ft are included within the Downton Castle Sandstone. About 25 ft of buff flaggy calcareous siltstone and limestone, with some nodular beds, dipping south-east at 8°, are seen in the roadside [595963] immediately to the east of the crossroads at Bourton, and 6 ft of flaggy siltstone, dipping at 10° to S. 15° E. are exposed [602968] 200 yd north-east of Bourton Grange. Flaggy siltstones are also seen in a road cutting [617976] west of Callaughton. B.A.H., R.H.H

All Stretton Area

The outcrop of Silurian rocks in the All Stretton area extends for about three miles in a north-north-easterly direction along the Church Stretton valley from near Church Stretton. Except towards the south-east its exact limits are uncertain due to a thick cover of drift in the valley bottom. The strata range from the *Pentamerus* Beds to the Upper Ludlow Shales.

The outcrop was mentioned by Murchison (1839, pp. 230–31) who referred to rocks of Lower Ludlow and Aymestry age. A more detailed account of its geology was given by Cobbold (1892) who suggested that rocks of Llandovery age might be present below the drift of the Church Stretton valley, a supposition which has been confirmed by the occurrence of the Hughley Shales and *Pentamerus* Beds in the Geological Survey Church Stretton No. 3 Borehole [45719447].

The south-eastern boundary of the outcrop is the F1 branch of the Church Stretton Fault Complex (p. 267), which throws the Silurian strata against Uriconian and Longmyndian rocks. To the north the outcrop terminates against the Coed-yr-Allt Beds of the Leebotwood Coalfield. The boundary is probably faulted, although the evidence provided by the Church Stretton No. 4 Borehole [47359653] (p. 257) on this point is not conclusive. The southerly extent of the outcrop is also uncertain but it is possible that a narrow outcrop may extend farther to the south, to link with the Silurian beds at Little Stretton (p. 168). To the west the Church Stretton No. 3 Borehole proved that the *Pentamerus* Beds rest unconformably on the Longmyndian, but again the exact limits of the outcrop are conjectural owing to the drift cover.

Only the higher Silurian beds, from the Wenlock Limestone upwards, are exposed at the surface. These beds, which crop out on the lower slopes of Caer Caradoc Hill, have a general easterly or east-south-easterly dip in the southern part of the area, and the occurrence of Llandovery rocks in the Church Stretton No. 3 Borehole suggests that this easterly dip is maintained in the lower beds. The outcrop of the Wenlock Limestone and higher beds is broken by several small cross faults, and to the north of the most northerly of these faults, two-thirds of a mile east of All Stretton church, the dip swings to a southerly direction. It is not known if this change in the direction of the dip affects the lower beds.

The Llandovery beds of the All Stretton area closely resemble those of both the Main and Long Mynd outcrops. The higher Silurian rocks are of the 'shelf' facies, despite their position on the western side of the Church Stretton Fault (p. 143).

The *Pentamerus* Beds, seen only in the Church Stretton No. 3 Borehole where they have an estimated true thickness of 140 ft, are entirely of the 'shales and limestones' facies. The Beds comprise grey siltstones with bands of sandstone and shelly limestone. Scattered fragments of grey shale, in limestone and mudstone, are present in the basal 6 inches, but no basal arenaceous beds are developed. About 120 ft (estimated true thickness) of Hughley Shales were proved in the No. 3 Borehole, but their full thickness is unknown. The Shales are maroon mudstones with green patches and occasional thin siltstone bands.

The Wenlock Shales, as seen in the Church Stretton No. 4 Borehole, are grey and greenish grey siltstones with occasional bands of white clay. Purple and purplish brown staining, similar to that seen in the ? Lower Ludlow Shales at Botvyle (p. 203), affects the upper part of the beds in the borehole. The thickness of the Wenlock Shales cannot be determined with any accuracy, but is probably less than Cobbold's estimate (1892, pl. vi) of 1350 ft. The outcrop of the Wenlock Limestone, as shown on the map, probably includes equivalents of both the Wenlock Limestone and Tickwood Beds of Wenlock Edge. The Limestone, which is about 250 ft thick, appears to be largely of Tickwood Beds facies (Cobbold 1900, p. 43) in its lower part, consisting of interbedded limestone and calcareous shale. The upper part of the Limestone is unexposed, but in the southern part of the area there is evidence that the interbedded shales and limestones pass upward into a nodular limestone (p. 203).

The Lower Ludlow Shales are poorly exposed, but where seen they are of normal 'shelf' facies, comprising grey shales with calcareous nodules and a shelly or shelly and graptolitic fauna. Purple and green staining affects the ? Lower Ludlow Shales of Botvyle (p. 203). Cobbold (1892, pl. vi) estimated that the Shales were about 750 ft thick, but the present survey has shown that this estimate is too great and that the true thickness is about 500 ft. The thickness of the Aymestry Group is also uncertain, but must be at least 200 ft. These beds are nodular limestones with bands of calcareous mudstone and an extensive shelly fauna. The Upper Ludlow Shales are not now exposed, though Cobbold (1900, p. 44) recorded sandy calcareous beds from above his Aymestry Limestone outcrop.

B.A.H.

DETAILS

Wenlock Shales

The Wenlock Shales outcrop probably covers a broad strip of the low ground to the west of the Roman Road between Church Stretton and Heath Cottage. There are no exposures, but the Shales were seen in 1904 in an excavation [45699409] (Locality 17) 660 yd N.E. of Church Stretton church, and in 1929 in several excavations

(Localities 18 and 19) in Essex Road, 470 yd E. 25° N. and 450 yd E. 27° N. of Church Stretton church. More recently (1963) green shaly mudstone, probably Wenlock Shales, was seen by the late Mr. T. H. Whitehead below 6½ ft of drift in a sewer trench [45659416] 680 yd N. 42° E. of Church Stretton church, and similar rock was thrown out of the trench a farther 270 yd to the S.S.E.

A Geological Survey borehole (Church Stretton No. 4) [47359653], 290 yd S. 18½° W. of Heath Cottage, showed 97 ft 3 in of drift and Coed-yr-Allt Beds, overlying Wenlock Shales dipping at about 60°. The nature of the Carboniferous–Silurian junction was obscure, but appeared to be an unconformity with associated faulting.

G.H.M., B.A.H.

Wenlock Limestone

At its southern end the outcrop of the Wenlock Limestone abuts against the Church Stretton Fault (F1). The most southerly exposure [46089408] (Locality 38) of the Limestone is in a small coppice on the eastern side of the Church Stretton by-pass, 700 yd W. 25° S. of New House Farm, where there is a small quarry in calcareous shaly mudstones with calcareous nodules. These beds pass eastwards into an impure nodular limestone. The strata are vertical, with a north-north-easterly strike. The outcrop of the Limestone, displaced by several minor faults, can be traced northwards, as a physiographical feature, to some indifferent exposures of interbedded limestone and calcareous shale, dipping at between 25° and 34° to the east, in the coppice half a mile west of Caer Caradoc Hill. Fossiliferous beds were found [46969564] (Locality 39) 850 yd E. 14° S. of All Stretton church. The Limestone outcrop terminates against a west-north-westerly trending fault 100 yd north of the latter exposure.

G.H.M., J.E.W.

Lower Ludlow Shales

Calcareous shale with graptolites, dipping at 27° to the east, is exposed in a stream [46699502] (Locality 52) 730 yd N. 2° E. of New House Farm. Some 150 yd north of this exposure an east-west fault displaces the outcrop dextrally by 250 yd, and north of the fault the shales occupy the slopes above the coppice half a mile west of Caer Caradoc Hill. There are no exposures.

The Shales appear to occupy the low ground around Botvyle, but the northern limit of the outcrop is conjectural due to the drift cover. Grey calcareous shales with limestone nodules are exposed 670 yd N. 24° W. of the summit of Caer Caradoc Hill [476960], and have yielded a Lower Ludlow fauna. An exposure (now obscured) in a pond [47679620] (Locality 53) near Botvyle, 880 yd N. 6° W. of the summit of Caer Caradoc Hill, showed green and grey mudstones with limestone nodules up to 5 inches in diameter. The mudstones have patchy purple staining, and the limestone nodules are green on the outside, with purple centres. The fauna from this locality suggests a Lower Ludlow age for the beds, rather than Llandovery or Wenlock, but is not absolutely conclusive.

Aymestry Group

Fossiliferous yellowish grey limestones of the Aymestry Group are exposed, close to the Church Stretton Fault (F1), in an old quarry [47219541] (Locality 104) 580 yd W. of the summit of Caer Caradoc Hill. The beds have an easterly dip of 25° to 30°. Some 300 yd to the north, the outcrop is shifted 500 yd to the west by a west-north-westerly trending cross fault. An excellent section in the Aymestry Group is provided by the old quarries [473958] 630 yd north-west of the summit of Caer Caradoc Hill.

The beds are as follows:

Beds above main quarry

		ft	in
Limestone, grey, thinly bedded and locally flaggy	c.	20	0
Gap	c.	15	0
Limestone, grey, cream and pink, shelly	c.	15	0
Main quarry			
Limestone, grey, nodular and fossiliferous; beds up to 3 in thick		6	6
Limestone, grey to cream, crystalline, shelly at base; beds up to 6 in thick; calcareous mudstone bands		8	8
Mudstone, calcareous—open joint			1
Limestone, grey to yellow, crystalline; beds up to 4 in thick; fossil band 10 in from base of unit		7	6
Mudstone, brown, calcareous			1
Limestone, grey, crystalline, somewhat nodular; beds up to 4 in thick, with calcareous mudstone partings		7	6
Mudstone, brown, laminated—prominent bedding surface			1
Limestone, grey, crystalline; slightly nodular, with calcareous mudstone partings, but weathering massive; some brown staining; prominent bedding surface 4 ft from top of unit		8	6
Mudstone, grey, calcareous, with crystalline limestone—prominent bedding plane			2
Limestone, grey, crystalline; beds up to 1 ft 4 in thick, with calcareous mudstone partings		5	8
Limestone, grey with some pink staining, crystalline, flaggy; in beds up to 4 in thick; calcareous shale partings up to 1 in thick		4	4
Mudstone, calcareous			1
Limestone, grey, crystalline, nodular and thin-bedded		1	5
Limestone, grey to brown, crystalline, fine-grained; beds up to 6 in thick		2	9
Mudstone, yellow, shaly, calcareous			1
Limestone, grey, crystalline, thin-bedded			4
Mudstone, calcareous, flaggy			1
Limestone, grey, crystalline			6
Mudstone			2
Limestone, grey to cream, coarsely crystalline, shelly			7
Mudstone parting			
Limestone, crystalline, well bedded; beds up to 9 in thick		4	3
Limestone, grey, nodular, crystalline			6
Mudstone, brown			1
Limestone, grey, nodular; beds generally less than 4 in thick; thin calcareous mudstone partings		6	8
Limestone, grey, crystalline; beds up to 6 in thick		1	8
Limestone, yellow, fine-grained, earthy; beds up to 1 in thick seen			6

The dip is from 62° to 80° in a southerly or south-south-easterly direction. The outcrop of the Group terminates to the east against F1.

Upper Ludlow Shales are probably present in the ground immediately to the south of the old quarries. G.H.M.

REFERENCES

ALEXANDER, FRANCES E. S. 1936. The Aymestry Limestone of the Main Outcrop. *Quart. J. Geol. Soc.*, **92,** 103–15.

BUTLER, A. J. 1939. The Stratigraphy of the Wenlock Limestone of Dudley. *Quart. J. Geol. Soc.*, **95,** 37–74.

COBBOLD, E. S. 1892. The Silurian Outlier West of Caer Caradoc. *Midland Naturalist*, **15,** 217–21.

—— 1900. in *Church Stretton*, 1. ed. C. W. Campbell-Hyslop. Shrewsbury.

CROSFIELD, MARGARET C. and JOHNSTON, MARY S. 1914. A study of ballstone and associated beds in the Wenlock Limestone of Shropshire. *Proc. Geol. Assoc.*, **25,** 193–224.

DAS GUPTA, T. 1932. The Salopian graptolite shales of the Long Mountain and similar rocks of Wenlock Edge. *Proc. Geol. Assoc.*, **43,** 325–63.

DAVIDSON, T. and MAW, G. 1881. Notes on the Physical Character and Thickness of the Upper Silurian Rocks of Shropshire with the Brachiopoda they Contain grouped in Geological Horizons. *Geol. Mag.* (2), **8,** 100–9.

DEAN, W. T. 1964. The geology of the Ordovician and adjacent strata in the southern Caradoc district of Shropshire. *Bull. Brit. Mus.* (*Nat. Hist.*) *Geol.*, **9,** 257–96.

DINELEY, D. L. 1960. Shropshire Geology: An outline of the tectonic history. *Field Studies*, **1,** 86–108.

DOWNIE, C. 1963. 'Hystrichospheres' (acritarchs) and spores of the Wenlock Shales (Silurian) of Wenlock, England. *Palaeontology*, **6,** 625–52, pls. 91, 92.

ELLES, GERTRUDE L. and SLATER, IDA L. 1906. The Highest Silurian Rocks of the Ludlow District. *Quart. J. Geol. Soc.*, **62,** 195–221.

EVANS, P. R. 1957. *The geology of the Longmynd–Clun area, South Shropshire.* Ph.D. Thesis, Bristol University.

HARPER, J. C. 1940. The Upper Valentian ostracod fauna of Shropshire. *Ann. Mag. Nat. Hist.* (11) **5,** 385–400.

HILL, DOROTHY, BUTLER, A. J., OAKLEY, K. P. and ARKELL, W. J. 1936. Report of "Coral Reef" Meeting at Wenlock Edge, the Dudley District and the Oxford District. *Proc. Geol. Assoc.*, **47,** 130–9.

HOLLAND, C. H. and LAWSON, J. D. 1963. Facies patterns in the Ludlovian of Wales and the Welsh Borderland. *Lpool. Manchr. Geol. J.*, **3,** 269–88.

HOLLAND, C. H., LAWSON, J. D. and WALMSLEY, V. G. 1959. A Revised Classification of the Ludlovian Succession at Ludlow. *Nature*, **184,** 1037–9.

—— 1963. The Silurian Rocks of the Ludlow District, Shropshire. *Bull. Brit. Mus.* (*Nat. Hist.*) *Geol.*, **8,** 93–171.

JOHNSON, HAZEL M. 1966. Silurian *Girvanella* from the Welsh Borderland. *Palaeontology*, **9,** 48–63.

LYELL, C. 1841. Some remarks on the Silurian Strata between Aymestry and Wenlock. *Proc. Geol. Soc.*, **3,** 463–5.

MURCHISON, R. I. 1833. On the sedimentary deposits which occupy the western parts of Shropshire and Herefordshire, and are prolonged from N.E. to S.W., through Radnor, Brecknock and Caermarthenshires, with descriptions of the accompanying rocks of intrusive or igneous characters. *Proc. Geol. Soc.*, **1,** 470–7.

—— 1834. On the Structure and Classification of the Transition Rocks of Shropshire, Herefordshire and part of Wales, and on the Lines of Disturbance which have affected that Series of Deposits, including the Valley of Elevation of Woolhope. *Proc. Geol. Soc.*, **2,** 13–8.

—— 1839. *The Silurian System*, Part I, London.

PITCHER, B. L. 1939. The Upper Valentian gastropod fauna of Shropshire. *Ann. Mag. Nat. Hist.*, **4,** 82–132.

POCOCK, R. W., WHITEHEAD, T. H., WEDD, C. B. and ROBERTSON, T. with contributions by WRAY, D. A., STUBBLEFIELD, C. J., CANTRILL, T. C. and DAVIES, W. M. 1938. Shrewsbury District. *Mem. Geol. Surv.*

ROBERTSON, T. 1927. The Highest Silurian Rocks of the Wenlock District. *Sum. Prog. Geol. Surv.* (for 1926), 80–97.

SALTER, J. W. and AVELINE, W. T. 1854. On the 'Caradoc Sandstone' of Shropshire. *Quart. J. Geol. Soc.*, **10,** 62–75.

SHIRLEY, J. 1952. The Ludlow Rocks north of Craven Arms. *Proc. Geol. Assoc.*, **63,** 201–6.

WHITAKER, J. H. MCD. 1962. The geology of the area around Leintwardine, Herefordshire. *Quart. J. Geol. Soc.*, **118,** 319–51.

WHITTARD, W. F. 1925. Notes on Valentian Rocks in Shropshire. *Proc. Geol. Assoc.*, **36,** 378–81.

—— 1928. The Stratigraphy of the Valentian Rocks of Shropshire. The Main Outcrop. *Quart. J. Geol. Soc.*, **83** (for 1927), 737–59.

—— 1932. The Stratigraphy of the Valentian Rocks of Shropshire. The Longmynd–Shelve and Breidden Outcrops. *Quart. J. Geol. Soc.*, **88,** 859–902.

—— 1938. The Upper Valentian trilobite fauna of Shropshire. *Ann. Mag. Nat. Hist.* (11), **1,** 85–140.

—— 1952. A Geology of South Shropshire. *Proc. Geol. Assoc.*, **63,** 143–97.

—— with contributions by BALL, H. W., BLYTH, F. G. H., DINELEY, D. L., JAMES, J. H., MITCHELL, G. H., POCOCK, R. W. and STUBBLEFIELD, C. J. 1953. Report of Summer Field Meeting in South Shropshire, 1953. *Proc. Geol. Assoc.*, **64,** 232–50.

—— and BARKER, GLENYS H. 1950. The Upper Valentian brachiopod fauna of Shropshire. Part I. Inarticulata: Articulata, Protremata, Orthoidea. *Ann. Mag. Nat. Hist.* (12), **3,** 553–90.

Chapter 7

OLD RED SANDSTONE

GENERAL ACCOUNT

OLD RED SANDSTONE rocks occupy an area of approximately 100 square miles in the southern and eastern parts of the Church Stretton district. They also form a small outlier at Oaker, near the south-western corner of the district. The classification adopted in this account is as follows:

	Approx. thickness in ft
UPPER OLD RED SANDSTONE	
Farlow Series	
Purple and green calcareous grits, often pebbly, and red marls; underlain by yellow fine-grained sandstone with bands of quartz pebbles and a massive conglomerate at the base	0 to 500
unconformity	
LOWER OLD RED SANDSTONE	
Clee Group	
Green and brown coarse-grained sandstones, often calcareous, often pebbly; a few thin bands of red marl and cornstone-conglomerate	0 to 1200
Ditton Series	
Red and green marls with many lenticular bands of purple and green, fine-grained, calcareous sandstones and cornstone-conglomerates, commonly current-bedded; nodular cornstones at many horizons, especially at the top and bottom of the Series, the Abdon and '*Psammosteus*'[1] limestones respectively ..	2000
Downton Series	
Ledbury Group. Red marls, sometimes with cornstone nodules, and a number of impersistent bands of purple micaceous sandstone	1500
Temeside Group. Green and purple marls and yellowish micaceous sandstones (Temeside Shales)	40 to 150
Yellow micaceous current-bedded sandstone (Downton Castle Sandstone)	20 to 50
Ludlow Bone Bed	

The history of the classification of these rocks is summarized by Ball, Dineley and White (1961, pp. 180–1) in a comprehensive account of the Old Red Sandstone of Brown Clee Hill. White (1950a, pp. 58–64) traced in detail the

[1]Taking its name from the fossil formerly called *Psammosteus anglicus* Traquair, now referred to *Traquairaspis symondsi* (Lankester). Tarlo (1961, p. 213) has lately stated that this fossil is not a Psammosteid but a close relative of the Pteraspids.

development of the controversy over the Siluro–Devonian junction, and strongly advocated the acceptance of the base of the Ludlow Bone Bed as the boundary between the two systems. Holland (1965, p. 213) considered the case for the use of the Ludlow Bone Bed as the Siluro–Devonian boundary and reviewed recent palaeontological evidence from several parts of Europe, relevant to the problem. The classification used here follows broadly that adopted by Whitehead and Pocock (1947, pp. 7–8), Ball, Dineley and White (1961, p. 180), and Mitchell and others (1962, p. 49). The most notable differences from Whitehead and Pocock are that the Downton Series is now included in the Old Red Sandstone and that it includes the Ludlow Bone Bed at its base. The Clee Group differs from the 'Clee Series' of Ball, Dineley and White in that it excludes the two Abdon limestones and the beds between them, and there are differences, of little practical importance, between their interpretation of the stratigraphy of the '*Psammosteus*' Limestones and that given in this account.

LOWER OLD RED SANDSTONE

DOWNTON SERIES

The Ludlow Bone Bed is a brown sandstone or sandy mudstone, packed with dermal plates, spines, and teeth of ostracoderms, and fragments of brachiopods, ostracods, and other fossils. It also contains small phosphatic concretions and is locally calcareous. In places it is seen to consist of several such bands, separated by up to 6 in of mudstone or sandstone, and the individual bands, between ¼ and 2 in thick, are often seen to be lenticular.

The Downton Castle Sandstone is a fine-grained, micaceous, yellowish buff sandstone, locally flaggy and current-bedded but often massive. The basal 3 to 6 ft are usually olive-green shales or mudstones, which rest on the Ludlow Bone Bed. The sandstone, which has been much quarried as a building stone south-west of Munslow, is sparsely fossiliferous, the characteristic species being *Lingula minima* J. de C. Sowerby. Fish fragments are also present, a 'fish bed' having been recorded at several localities by Elles and Slater (1906, pp. 209, 211, 216–8), and near the base is a band characterized by an abundance of *Platyschisma helicites* (J. de C. Sowerby), the 'Downton Bone Bed' (Elles and Slater 1906, pp. 209–10). This horizon is recognized from Ludlow to Much Wenlock (Robertson 1927, p. 97) but the characteristic gastropod is not recorded north-east of Shipton (op. cit., p. 94), and there, it is in the Ludlow Bone Bed.

The sandstone is nowhere fully exposed but it appears in general to be about 30 ft thick. In the Onibury area it increases to about 50 ft and near Shipton it may be as little as 20 ft.

The Temeside Shales are in some respects transitional between the Downton Castle Sandstone and the Ledbury Group. They consist of thin beds of green and purple blocky mudstone and yellowish green, micaceous, flaggy sandstone. The fauna is more abundant than that of the underlying sandstone but of the same type, the principal elements being *Lingula cornea* J. de C. Sowerby, fish fragments, ostracods, gastropods, and small lamellibranchs. A bone bed was recorded by Elles and Slater (1906, pp. 205, 209, 211, 217) at several localities between Ludlow and Onibury. In general the shales are poorly exposed and the variation in their thickness along the outcrop is probably due partly to the gradational nature of their upper boundary, and partly to differences in the level of the facies change which determine it. In the Onibury area the beds mapped as Temeside Shales are about 150 ft thick and they are about 100 ft thick near Munslow. Between these areas however, near Siefton, and farther north-east, near Brockton, their thickness is only about 40 ft.

The beds of the Temeside Group form a narrow outcrop along the foot of the dip slope of the Ludlow Series. In contrast the succeeding Ledbury Group occupies the broad valleys of the Corve and the Mor Brook, the low relief of which testifies to the predominantly argillaceous character of the constituent rocks. In many places, especially in the lower and middle parts of the group, the occurrence of thin lenses of micaceous flaggy sandstone leads to the development of minor scarps and dip slopes which diversify the topography, sometimes to a striking extent. Similar features, usually smaller, are sometimes caused by bands of calcareous nodules, 'race', or irregular vertical concretions, in the marl. In the higher part of the Group, sandstone and 'race' appear to be less common, but it may be that the features to which they might give rise are masked, as a result of the position of these beds on the lower slopes of the Ditton Series escarpment. In the Cleestanton area and at Balaam's Heath the rocks are largely concealed by solifluxion deposits from the Clee Hills to the south-east.

In general the rocks of the Ledbury Group are not well exposed. The sandstones have been quarried locally, for example near Onibury and at Holdgate, but many of the quarries are now overgrown and in them, as along the features, only debris of the rock is seen. The marls are predominantly unstratified, and red or purple in colour, bright red marls with green reduction spots, 'fish eyes', being particularly prominent at the very top of the group. The sandstones tend to be purple or dark greenish purple, and contrast with those of the Ditton Series in being more coarsely micaceous and less compact. In places they are current-bedded and sometimes contain marl pellets and are marked by 'fish eyes'. Cornstone-conglomerates (see below) occur but rarely in the Ledbury Group.

Fossils occur, as in the Ditton Series, in scattered pockets, and most often in arenaceous rocks. Usually fish fragments, they have been found mainly in pellety sandstones in the upper part of the group, often within about 100 ft of the '*Psammosteus*' Limestones.

DITTON SERIES

The extensive outcrop of the Ditton Series which surrounds Brown Clee Hill falls into two broad divisions. A lower group, including the *Cephalaspis* Sandstone of King (1921, p. 321), in which massive sandstones are common, forms the prominent escarpment on the south-eastern side of Corve Dale and south of the Mor Brook, and an upper group occupies an upland area of reduced relief round the steep upper slopes of the hill. Sandstones are again abundant in the highest beds. This distinction is not so clearly seen in the anticlinal outcrop round Cleestanton, nor in the Rea basin where, for example near Criddon, the lower beds form no well defined escarpment.,

The base of the series is taken at the base of the '*Psammosteus*' Limestones, a group of concretionary limestones, of varying number, which crops out on the Corve Dale escarpment and includes, and is overlain by, a succession of rocks in which the proportion of sandstone and cornstone-conglomerate is significantly higher than in the Downton Series below. The principal faunal change in the Lower Old Red Sandstone takes place at about this level, where the *Traquairaspis* fauna of the Downton Series is replaced by the *Pteraspis* fauna of the beds above (White 1950a, p. 56). White placed the '*Psammosteus*' Limestones largely within the zone of *Traquairaspis symondsi* (1950a, p. 55, fig. 1), but the mapping and collecting of Ball, Dineley and White (1961, p. 202) has shown that the species occurs most commonly below them, and that their base is accordingly the most suitable horizon, on both lithological and faunal grounds, at which to separate the Downton from the Ditton Series.

The base of the Ditton Series is considered by Allen and Tarlo (1963, pp. 146–8) to lie at the base of their '*Psammosteus*' Limestones Group. They point out that within this Group the individual limestones, where they are developed, occur at the top of the major units of sedimentation, and that logically the base of the Group should not be taken at a limestone but at the base of the earliest cycle of the fluviatile sediments characteristic of the whole Ditton Series. They further argue that there is no justification for the postulation of a distinct palaeontological 'Downtonian–Dittonian' classification, the division occurring within the Ditton Series.

Ball, Dineley and White (1961, p. 190) maintained that, although there are several '*Psammosteus*' Limestones, there is one main limestone which is in general more persistent, more compact, and thicker than the others, and it is at its base that they have drawn the base of the Ditton Series. A similar conclusion was reached by Taylor (*in* Mitchell and others 1962, p. 50) in the ground to the south-east of the Church Stretton district. The opinion now advanced is that in the Clee Hills area the main limestone does not maintain a constant horizon, that between some localities it may be seen to die out or become considerably attenuated and be replaced, as the main limestone, by a higher or lower band which is elsewhere of minor importance. This phenomenon is well displayed at Bouldon and Tugford (pp. 228, 229), where the change of level in the 'main' limestone is explained by Ball, Dineley and White (1961, map) as being due to faults. The comparative sections of Fig. 17 appear to indicate that, in relation to the basal quartz-bearing conglomerate and by consideration of the cycles of sedimentation, the horizon of the main limestone does in fact change, and that there is no single, continuous, main limestone over the whole area.

Because of the impersistence of the limestones, the base of the Ditton Series is difficult to define lithologically. The base of the lowest limestone which persists for any significant distance has been taken as the boundary and where it is not developed an attempt has been made to infer its horizon. The principal defect of this boundary is that the basal limestone cannot always be confidently correlated between adjacent sections, especially where they are separated by considerable lengths of outcrop in which the limestone appears to be absent. From Tugford to Monkhopton a quartz-bearing cornstone-conglomerate, believed to be a unique band, marks the base. It carries an abundant fauna of Downtonian aspect (p. 229), but with no element which excludes the possibility of its being Dittonian.

The rocks of the Ditton Series are predominantly red or purple marls. In a number of measured sections through the lower, more sandy, beds marl appears to constitute about 75 per cent of the succession. W. W. King's estimate for

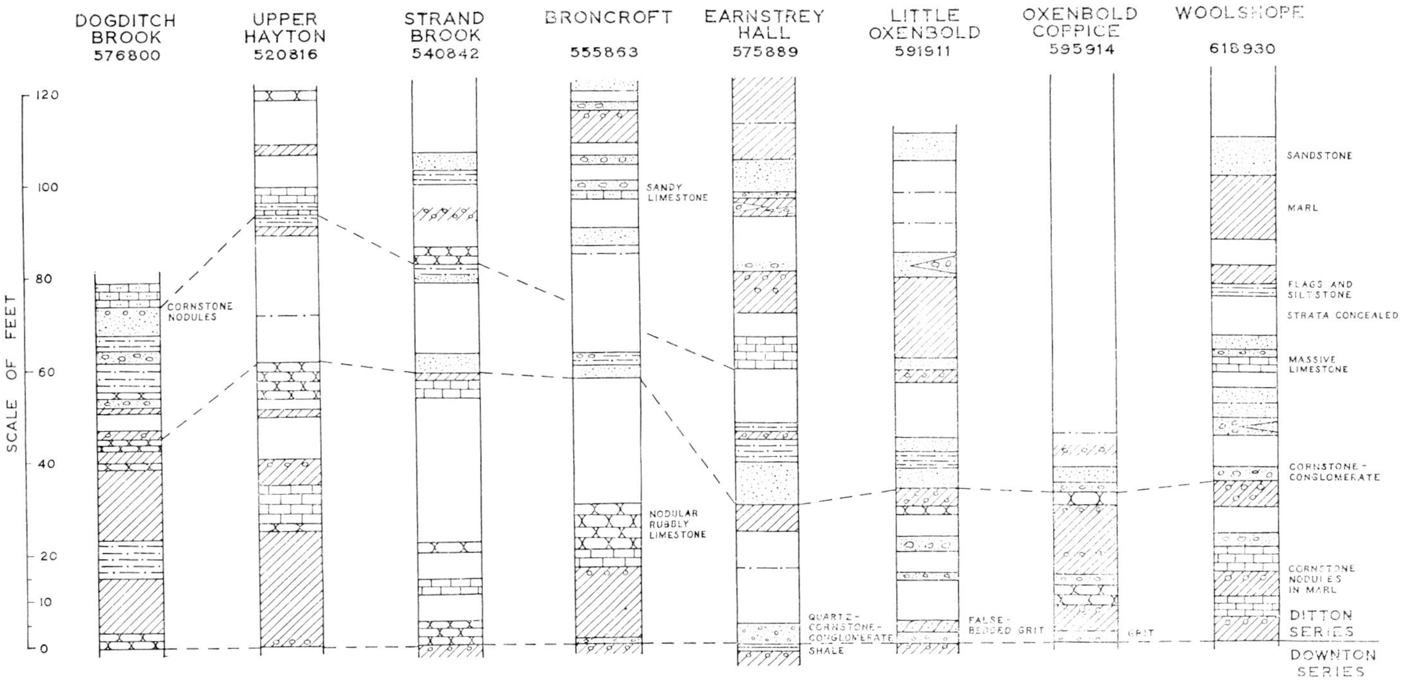

FIG. 17. Comparative sections in the 'Psammosteus' *Limestones. The National Grid References indicate the approximate location of the base of each section*

the marl content of the Dittonian, which excludes about 350 ft at the base of the Ditton Series, was 65 to 70 per cent (1934, p. 529). The marls are generally unstratified and commonly blocky, but locally grade into bedded rocks which might be described as shales. At outcrop they are usually not calcareous but where this is the case they are locally very strongly cemented. Small concretions of cornstone are not uncommon, especially towards the top of a marl, and these may become larger and more abundant to give a lateral or upward passage into a concretionary cornstone or limestone. The topmost inch or two of marl is commonly reduced to a greenish grey colour and bands and patches of this colour may occur elsewhere. Fine-grained sandstones, usually flaggy and calcareous, are second in abundance to the marls. They are usually reddish brown or purple, but greenish grey is a much more common colour than in the marls, both in isolated patches and beds and locally, as at Sutton Hill, in thick groups of beds. Current bedding is common and sandstone and cornstone-conglomerate are commonly inter-bedded. Sandstones with pellets of marl and fragments of siltstone and cornstone are transitional between the normal sandstones and the cornstone-conglomerates. Well-defined steep joints are common in the sandstones.

The type of rock most characteristic of the series is the cornstone-conglomerate. This is a conglomerate made up largely of sub-rounded pebbles, rarely reaching 2 in across, of calcareous siltstone, sandstone, and concretionary cornstone, in a calcareous sandy matrix which commonly contains much crystalline calcite. It occurs in lenticular beds up to about 20 ft thick but generally of less than 5 ft. Current bedding is common and the base of a bed cuts into the bed below. The rock fragments are not entirely of Ditton type. These conglomerates yield most of the fish fragments of the Ditton Series, although some, as well as most of the plants, are found in the sandstones. They are the 'cornstones' of local parlance and the term has been and is used in this sense by many geologists. The nomenclature, nature, and occurrence were discussed at length by Allen (1960), who concluded that 'cornstone' should be used only for concretionary rocks and that the fragmentary rocks should be called 'conglomeratic cornstones'. This is in essence the nomenclature adopted in this account.

The limestones or cornstones of the Ditton Series, of which the '*Psammosteus*' and Abdon limestones are the thickest and most persistent, are concretionary, chemical limestones of a high degree of purity. They exhibit all gradations from marl with many calcareous nodules, through rubbly limestone with discrete nodules, to massive compact limestone. They are most common near the base and towards the top of the series and are generally up to about 3 ft thick, the maximum thickness being about 15 ft.

The following sequence of sediments is frequently seen in the Ditton Series:

Cornstone-conglomerate
Cornstone
Marl, with cornstone nodules near the top
Sandstone, becoming flaggy and shaly upwards
Cornstone-conglomerate

The primary control on sedimentation is probably tectonic, but the effect of climate can be seen in the change from cornstone, a product of prolonged desiccation, to cornstone-conglomerate, deposited by fast-flowing water.

The cornstone is in fact usually absent, having presumably been eroded prior to the deposition of the conglomerate, or in many cases because the final stages of aggradation and desiccation, before the influx of conglomerate, were never attained.

CLEE GROUP

The main outcrop of the Clee Group is on the higher slopes of Brown Clee Hill where it is overlain unconformably by the Coal Measures. In the Farlow area the Group is overstepped by the Farlow Series, but it crops out in the Catherton area, and is seen beneath the Farlow Series a short distance to the east (Dudley Sheet 167) although not separately mapped in that district.

The Group consists predominantly of green and brown sandstones, commonly gritty and pebbly, and locally calcareous. The interbedded red marls are seldom exposed. It contrasts strongly with the Ditton Series in its higher proportion of sandstones and in their coarser texture, but some of the beds closely resemble the higher parts of the Farlow Series, and in faulted ground such as the Catherton–Ditton area their separate identification is liable to error. Cornstone-conglomerates are rare and concretionary cornstones are not recorded. No fossils have been found in the rocks of the Group. Allen (1961) divided the Group into the Clee Sandstone Formation and the Monkeys Fold Sandstone Formation, distinguished by the absence of marls in the latter, upper, division, and by the generally coarser grain of its beds.

The base of the Group has been taken at the top of the Upper Abdon Limestone since it is at that horizon that the change in the nature of the sandstones appears to take place. Also limestones of the Abdon type are not uncommon in the Ditton Series but are unknown in the beds above. The 'Clee Series' of Ball, Dineley and White (1961, pp. 195–8) begins with the Lower Abdon Limestone. This horizon was chosen as being possibly equivalent to the Lower Fynnon Limestone of the Black Mountains, which lies a short distance above the base of the Breconian Senni Beds (Croft 1953). The exact base of the Breconian in Shropshire was not determinable. Allen (1961) separated the 'Abdon Series', comprising the Abdon limestones and the beds between, from the underlying Ditton Series, because of its higher sandstone and carbonate content, but considered that the main lithological change, a change indistinguishable from that in South Wales, from the Dittonian to the Breconian, occurs at the top of the Upper Abdon Limestone. The Clee Group, Allen's 'Woodbank Series', was therefore correlated by him with the Senni Beds. The beds which have been mapped in the Droitwich area as Brownstones (Mitchell and others 1962, pp. 52, 66) are, in part at least, equivalent to the Clee Group.

UPPER OLD RED SANDSTONE

FARLOW SERIES

The outcrops of the Farlow Series lie at the north-eastern end of the Carboniferous outcrop of Titterstone Clee Hill. At outcrop the beds, with a basal conglomerate, rest unconformably on the Ditton Series, and are themselves overstepped in most of the area by the Carboniferous. The succession, seen in full in the Prescott area, is:

		ft
Green and purple coarse-grained sandstones, commonly pebbly and calcareous; at least one cornstone band; red marls, poorly exposed; some fish fragments	c.	400
Yellow medium-grained current-bedded sandstone, with bands of small quartz pebbles; easily weathered	c.	70
Conglomerate of quartz pebbles in a matrix of yellow sandstone	at least	10

Conditions of Deposition

The conditions of deposition of the Old Red Sandstone rocks have been discussed by Ball, Dineley and White (1961, pp. 214–9), and it is on their account that the following paragraphs are based.

The rocks of the Temeside Group represent the gradual change from the shallow-water marine conditions of Ludlovian times to the predominantly deltaic conditions of the Devonian. The bone beds at the base reflect alternately marine and brackish environments and are succeeded by the deltaic and lagoonal phases, with a few marine incursions, of the Downton Castle Sandstone and Temeside Shales. Later Devonian rocks were deposited in the delta of a river, or rivers, flowing southwards from the rising Caledonian mountains, the variations in grain size arising in part from the variations in the rate of uplift and in part from seasonal or more frequent variations in rainfall. Sudden floods deposited material from the uplands together with debris from the delta surface to form pellet beds and cornstone-conglomerates. The cornstones followed long periods of very limited uplift, when the supply of sediment was ultimately reduced to a minimum and the confined waters of sea level lagoons became over-saturated with calcium carbonate. The absence of stratification in much of the marl seems to indicate that they were formed by slow deposition from still, muddy water.

The red coloration of most of the rocks is probably due to the prevalence of a tropical humid climate in the area from which the sediment was derived. More arid conditions prevailed in Temeside times and during the deposition of the yellow sandstones of the Farlow Series.

From a detailed study of the sediments of the Downton and Ditton series of the Welsh Borderland, Allen and Tarlo (1963) have re-interpreted the conditions under which these rocks were deposited. The Downton Castle Sandstone is seen as a group of marine silts succeeded by beach and shallow-water sands, to be followed by inter-tidal and sub-tidal silts of the Temeside Shales and the Ledbury Group. Within this last formation the locally developed Holdgate Sandstones represent the channelling and infilling of the temporarily emergent tidal flats. The sediments of the Ditton Series are considered to be predominantly fluviatile. The chemical limestones and occasional mollusc bands in the lower beds, the '*Psammosteus*' Limestones Group, indicate respectively the occurrence of temporary lakes and of brief marine incursions.

A detailed study of the petrography and mineralogy of the beds of the Clee Series has been made by Allen (1962). He showed that the main sources of the rock-forming materials are the Lower Palaeozoic greywackes and volcanic rocks of North Wales, and that important contributions came also from the Wenlock Limestone of the Welsh Borderland, and from pre-existing rocks of the Old Red Sandstone within the Devonian basin of deposition. A directional study of sedimentation structures supports the theory of derivation from the north-west. Allen recognized a series of cyclothems within the formation. The nature and succession of the rocks led him to the conclusion that they were deposited on a slowly subsiding flood plain within a few miles to the south-east of a rising massif of Lower Palaeozoic and Dittonian rocks, and a detailed comparison was drawn with present conditions in the Colorado Delta.

Palaeontology

The vertebrate faunas and the palaeontological classification of the Old Red Sandstone have been described in detail by Ball, Dineley and White (1961) and the reader is referred to their paper for an up-to-date account and a comprehensive bibliography. Allen and Tarlo (1963, pp. 130–1, 142–3, 147–8) suggested that the main palaeontological break in the Lower Old Red Sandstone is between the predominantly brackish-water and the fresh-water faunas, and on this basis included the *Traquairaspis* fauna in the higher palaeontological division.

Collections of vertebrate fossils were made from many localities during the recent survey, mainly in the uppermost beds of the Downton Series and in the lower half of the Ditton Series. The Temeside Group also yielded vertebrates, principally from the Ludlow Bone Bed, and most of the invertebrates, of which the species of *Lingula* from the Downton Castle Sandstone and the Temeside Shales were the most common. The identification of these fossils, which are named in the succeeding pages (see Details, p. 218 et seq.) confirms the stratigraphical distribution ascribed by Ball, Dineley, and White (1961) to the several species.

Unidentifiable plant stems were collected at several localities from greenish grey flaggy sandstones in the lower part of the Ditton Series. D.C.G.

Petrography

The Old Red Sandstone of the Church Stretton district includes a number of distinct petrographical types, some of which occur throughout the sequence, and some of which are apparently more common at specific horizons. The main groups are: quartzitic sandstones; subgreywackes; calcareous sandstones; conglomeratic limestones and calcareous sandstones; and a group of limestones which vary from calcite-mudstones to calcarenite.

Quartzitic sandstones are found in the upper part of the Ditton Series below the Abdon limestones, and in the Farlow Series; subgreywackes are practically confined to the Downton Series (e.g. Holdgate Sandstone); calcareous sandstones are common in the lower part of the Ditton Series, the Abdon limestones, the Clee Group and locally in the Farlow Series. The other lithological types occur throughout the Old Red Sandstone.

Quartzitic sandstones. Quartzitic sandstones (protoquartzites of Pettijohn's classification, 1957, pp. 291–2) are the typical sandstone of the uppermost Ditton Series, and occur for instance 1140 yd W. 29° N. of the Gore [590829][1]

[1]Such numbers are National Grid References within the 100-km square SO (32).

(E 25728)[1] and 920 yd S. 12° E. of Cold Weston church [553823] (E 25726). They are medium-grained, buff to red in colour and consist dominantly of subangular to subrounded grains of quartz with minor amounts of feldspar (dominantly sodic plagioclase) and chlorite, together with rarer lithic fragments mainly of recrystallized rhyolite (E 25726), but also including basic volcanic fragments (E 25728). Metamorphic quartzites, chert, siltstone and mudstone are less common and rare organic phosphate fragments occur. The detrital heavy minerals include tourmaline, zircon and garnet. The cement is dominantly siliceous, although a small amount of interstitial calcite may be locally present (E 25728).

Similar quartzitic sandstones occur near the base of the Farlow Series 150 yd E. 39° N. of Farlow church [641807] (E 27103), together with abundant pebbles of vein quartz in the basal conglomerate of the Farlow Series 110 yd N. 4° E. of Farlow church [640807] (E 27104).

The Downton Castle Sandstone, at the base of the Downton Series, from the small quarry [619975] at the south side of Callaughton village (E 26326), is a similar quartzitic sandstone.

Subgreywacke. The term subgreywacke is used according to the definition of Pettijohn (1957, pp. 316–18) for a rock composed of quartz, more than 25 per cent of combined lithic fragments with minor feldspar, and less than 10 per cent of clay mineral matrix.

Subgreywackes are relatively common in the middle part of the Downton Series but are not represented in the collections from elsewhere in the Old Red Sandstone. A typical example from the Holdgate Sandstone 280 yd S. 22° W. of Holdgate church [560892] (E 26312) is a medium-grained limonite- and hematite-stained rock consisting of relatively well-sorted subangular grains of quartz, feldspar (mainly perthitic), lithic fragments and detrital heavy minerals including garnet, tourmaline, zircon and magnetite. The lithic fragments include chert, rhyolite and spherulitic rhyolite and fragments of low grade metamorphic rocks, comprising highly foliated chlorite-schists, muscovite-schist and muscovite-chlorite-schist. Metamorphic quartzites, consisting of an interlocking mosaic of elongated strain-shadowed quartz grains with sutured and crushed grain boundaries, are common, and finely foliated (nematoblastic) quartz-schists occur. Fragments of quartz with inclusions of pale green vermicular chlorite and rare fragments of granophyre showing micrographic texture are present. Similar, but rather coarser-grained subgreywackes of the Downton Series crop out in Beaconhill Brook 940 yd N. 24° E. [630942] (E 26319, 26321) and 1430 yd north-east [635944] (E 26328) of Monkhopton church. One of these specimens (E 26319) also contains relatively abundant fragments of organic phosphate.

Calcareous sandstones. These types are common in the lower part of the Ditton Series (e.g. E 25729, 25732, 26324–5) and also occur in the Abdon limestones (E 25727, 26304), the Clee Group (E 26303, 26476–9) and the upper part of the Farlow Series (E 27106). In hand specimen the calcareous sandstones may exhibit a lustre-mottling; in thin section they are seen to consist of a variety of subangular to subrounded lithic and crystal fragments in a matrix of large interlocking plates of poikilitic calcite (up to 1 cm across). The grain size is widely variable from fine (E 26325, 25732) to coarse (E 26476). Quartz

[1]Such numbers refer to rock slices in the Survey collection.

grains and fragments of metamorphic schistose quartzites constitute the largest group of fragments, but other low grade metamorphic and igneous fragments similar to those described from the subgreywackes (see above) are common (e.g. E 26303, 26324, 26476–7, 26479, 27105). Rhyolite, felsite and granophyre fragments and rare altered intermediate to basic volcanic rocks are particularly well developed in the calcareous sandstones of the Clee Group, for example at 1320 yd E. 18° S. of Heathamgate [583848] (E 26303); 830 yd north [654792] (E 26479), and 2220 yd N. 34° E. of the north-east corner of Catherton [665802] (E 26477); and between the Lower and Upper Abdon limestones 1660 yd east of Upper Norncott, Abdon [584859] (E 26305). Distinctive pale green vermicular chlorite inclusions in quartz grains occur as detrital lithic fragments in the Clee Group 2220 yd N. 34° E. of Catherton [665802] (E 26477) and in the Farlow Series 95 yd E. 33° N. of the north corner of Farlow church [641806] (E 27105). Organic phosphate fragments are relatively scarce but have been found in the lower part of the Ditton Series (e.g. E 26330, 25732, 25729). Detrital heavy minerals include sieve-textured garnets, dark green tourmaline, epidote and magnetite.

The flaggy *Cephalaspis* Sandstone (p. 231) in the lower part of the Ditton Series from 1050 yd W. 30° S. of Lodge Farm [514786] (E 25729) consists of calcareous sandstones containing organic phosphate fragments similar to those described above. Conglomeratic limestones and calcareous sandstones occur throughout the Old Red Sandstone of the area and constitute two of the commonest rock types.

The conglomeratic limestones are particularly well developed in the '*Psammosteus*' Limestones, for instance at 1575 yd E. 18° N. of the Moor [579804] (E 25720); 1510 yd west-north-west of Abdon church [562870] (E 26307); 1700 yd E. 32° N. of Stanton Lacy church [509797] (E 25737) and near the base of the Ditton Series 1000 yd W. 38° S. of Aston Eyre church [646935] (E 26317) and in the quarry [648932] 1120 yd N. 26° E. of the north corner of New House (E 26322).

These rocks are essentially calcite-siltstones or calcarenites in which sub-angular to rounded fragments of calcite-siltstone occur. The matrix of the rocks is composed of an interlocking mosaic of recrystallized calcite plates (0·06 mm-0·5 mm), which in some specimens show a radiating arrangement around the limestone fragments (e.g. E 26322). Locally a poikilitic texture is developed in the recrystallized calcite matrix (e.g. E 25720). Silt-grade fragments of quartz, lithic fragments and feldspar together with rare organic phosphate detritus is associated in some specimens with the calcite matrix (e.g. E 25720, 26317, 26322).

Similar conglomeratic limestones are interbanded with calcareous sandstone with included fragments of calcite-siltstone at the base of the Ditton Series 1000 yd W. 38° S. of Aston Eyre church [646935] (E 26317).

The limestone (calcite-siltstone) fragments in the conglomeratic calcareous sandstones are similar to those in the above group; but between these fragments there is a matrix of calcareous sandstone, which in some specimens is cemented by poikilitic calcite grains (e.g. E 26317, 26320). The suite of lithic fragments in the calcareous sandstones is similar to that described above from other lithological types in the area and includes granophyre and rhyolite fragments (e.g. E 26481, 29097–9) and grains of quartz with vermicular chlorite inclusions (e.g. E 26333, 26481). One particularly interesting fragment from the conglomeratic calcareous sandstone, at the waterfall 500 yd N. 43° E. of Bockleton Court [582836] (E 25715), is a devitrified welded tuff, similar to fragments found in the Longmyndian which have been derived from the Uriconian.

Limestones. The limestones are dominantly fine-grained (calcite-siltstones and calcite-mudstones) (e.g. E 25721, 25735–6, 26309–10, 26329) and may contain small silt-grade angular fragments of quartz (e.g. E 25735, 26300, 26309, 27100). A few specimens of limestone occur, crowded with partly recrystallized organic (?algal) fragments (e.g. E 25733, 26301, 26302, 26323).

The main limestone horizons are variable in lithology. The '*Psammosteus*' Limestones include conglomeratic limestones (E 25720, 25734, 25737, 26307), conglomeratic calcareous sandstones (e.g. E 26313, 26320), calcite-siltstones (E 25736, 26331) and organic (?algal) limestones (E 25733). The Abdon limestones consist of calcareous sandstones (E 25718, 25727, 26304–5), conglomeratic calcareous sandstones (E 25717, 27602), calcite-siltstones (E 25719, 26300) and organic (?algal) limestones (E 26301–2).

The suite of lithic fragments found in the Old Red Sandstone, including low grade metamorphic rocks, distinctive pale green vermicular chlorite inclusions in quartz grains, granophyre and rhyolite fragments, is clearly similar to that which occurs throughout the Longmyndian sediments.

Greenly (1919, pp. 250–2) and Woodland (1938) have described low grade schists of Mona Complex type in the Cambrian sediments of the Harlech dome, and Cummins (1957) found similar metamorphic fragments, including vermicular chlorite inclusions in quartz grains together with soda-granites, granophyres and rhyolites in the Denbigh Grits (Wenlock) of the Welsh Borders. Lithic fragments of the same suite are seen in the Ordovician iron ores of North Wales at Llandegai near Bangor (E 12732), which contain detrital fragments of chlorite-muscovite-quartz-schist and in the oolitic ironstones of Fford Ddu, near Dolgelly (E 16767) which contain fragments of quartz with pale green vermicular chlorite inclusions. The Ordovician Hoar Edge Grit, near Hamperley, (E 29066) also contains fragments of low grade metamorphic schists, rhyolite and granophyre.

Walder (1941) on the basis of a detailed study of the heavy minerals suggested that the Downtonian rocks of the Tenbury area were very probably derived from the Mona Complex. Wallis (1928) described an exactly similar lithic fragment assemblage from the Old Red Sandstone of the Bristol area and suggested derivation of that series from rocks of Mona Complex type.

The continued occurrence of the same suite of lithic fragments from Longmyndian to Old Red Sandstone times must indicate either the continued erosion of the original (or a similar) source, or erosion and redeposition of earlier sediments to provide the re-worked fragments for later deposits. R.D.

DETAILS

LOWER OLD RED SANDSTONE

DOWNTON SERIES

UNDIVIDED

Oaker. An outlier of Downtonian rocks occurs in the core of a syncline at Oaker [384816], half a mile west of Aston on Clun. The syncline plunges to the south, and the outlier is limited to the south by a fault along the valley of the River Clun. The rocks of the outlier are poorly exposed, and it is not possible to determine a succession within them. They are estimated to be about 600 ft thick.

Traces of greyish green, very micaceous siltstones and sandstones containing carbonaceous fragments, with purple silty mudstones and mottled purple and green silty mudstones were exposed under a fence [385822] 700 yd N. 6° E. of Oaker. The purple silty mudstones contained *Lingula cornea*, *L.* cf. *minima*, *Ctenodonta ?*, *Cyrtodonta ?*, Modiolopsid, *Myalina sp.*, *Nuculites ?*, *Palaeoneilo ?*, *Paracyclas ?*, *Kloedenia wilckensiana* (T. R. Jones) and *Leperditia sp.* Purple micaceous siltstones and coarse-grained, grey, micaceous siltstones were seen in an obscure track section 250 yd to the west-south-west. These beds yielded *Lingula cornea.* A small exposure of greyish green and purple, mottled, micaceous siltstone with *?Ledopsis barroisi* Reed, *Nuculites* aff. *oblongatus* Conrad, *N. ovalis* (J. de C. Sowerby), *Leperditia phaseola* (McCoy), was seen on the hillside 600 yd N. 26° W. of Oaker, with similar beds containing *Platyschisma helicites*, *Mytilarca sp.*, *Nuculites* cf. *antiquus* (J. de C. Sowerby), *N. ovalis*, *Vigorniella regis* Reed, and *Kloedenia wilckensiana*, 180 yd farther south. B.A.H.

The fauna of these four localities indicates a Downtonian age. Of particular significance is the occurrence of *Lingula cornea*, *Platyschisma helicites*, *Kloedenia wilckensiana*, *Leperditia sp.*, and a diverse lamellibranch fauna including *Vigorniella regis*. This latter genus and species were described by Reed (1934, pp. 578–80) from Field Brook, 1⅙ miles north of Abberley church, Worcestershire, from W. W. King's stage I.9; its range is unknown. J.D.D.S.

Temeside Group

Ludlow Bone Bed

Stokesay Court to Siefton. The Ludlow Bone Bed is rarely exposed in this area, but its outcrop can be traced topographically along the slight hollow or flattening in slope between the dip slope of the Upper Ludlow Shales and the feature formed by the Downton Castle Sandstone.

It was formerly seen at Norton [461816] (Elles and Slater 1906, p. 218), but this section is now obscured. A quarry [475815] on the west side of the road, 1100 yd S. 10° E. of Medley Park, shows the Bone Bed, up to ½ in thick and impersistent, overlying 12 ft of pale olive-green siltstone with a sparse Upper Ludlow fauna (Localities 239 and 240, Appendix 2). The Bone Bed is overlain by 5 ft of olive-green siltstone and mudstone, the lowest beds of the Downton Castle Sandstone. Traces of pale yellow sandstone occur at the top of the quarry.

Fragments of the Bone Bed were seen in a ditch [473823] 350 yd south-south-west of Medley Park, and it was found to be ½ in thick in a ditch [474832] on the north side of the road 860 yd E. 16° N. of Marsh Barn. B.A.H.

Siefton to Broadstone. The Bone Bed, ½ in thick, occurs near the base of a lane-side section [489847] 770 yd W. 3° N. of Hill House, and is up to 2 in thick beside the lane [497852] 1230 yd W. 9° S. of Diddlebury church. At Munslow Aston, 110 yd S. 44° E. of the chapel [512866], it occurs about 6 ft below the top of the section, only ¼ in thick and impersistent. It is ¼ in thick at Munslow, 4 in above the base of a roadside section [524876] 380 yd E. 35° N. of the inn, and 560 yd farther along the road to Beambridge it is seen to be 1 in thick. In the Trow Brook [542890] 840 yd S. 17° W. of Broadstone church the Bone Bed is impersistent, up to 1 in thick, and contains *Salopina lunata* (J. de C. Sowerby).

Broadstone to Callaughton. At Shipton, where the Bone Bed is next seen, it occurs in two impersistent bands, ½ in thick and 4 in apart in the lane [563918] 160 yd E. 38° S. of the church. It is exposed at Brockton, at the roadside [579939] 270 yd N. 32° W. of the Castle Mound, where Robertson (1927, pp. 93–4) observed a lower impersistent bed ¼ in thick, separated by 6 in of sandy mudstone from the main bone bed which consists of two ¼-in layers about 1 in apart. In an old quarry [586946] 1300 yd N. 27° E. of the Castle Mound the Bone Bed, ¼ in thick, is seen about 4 ft from the top of the 25-ft section. It was formerly visible (Robertson 1927, p. 86) in a stream at Callaughton [620975], 2200 yd E. 24° N. of Bourton Grange.

DOWNTON CASTLE SANDSTONE

Onibury. South-west of the Onny the Downton Castle Sandstone is best exposed at the roadside 1350 to 920 yd west of Onibury church, where *Lingula sp.* and ostracods were observed in the lowest beds visible. The Sandstone is poorly exposed in a large old quarry [445790] just south of the road, where debris contained *Lingula minima* and *Onchus sp.*, and its outcrop is easily followed by soil debris and feature from Stokesay Court to the river. There it is displaced by the Stokesay Fault about 500 yd to the south-east. About 25 ft of yellow micaceous flags and sandstone are exposed in an old quarry [455794] 250 yd N. 28° W. of Onibury church. In places they are strong current-bedded (Pl. 11A), the general dip of the strata being very low towards S. 40° E. The Sandstone can be followed by topographic feature, soil debris, and small outcrops to Upper Onibury, where 13 ft of yellow micaceous sandstone, dipping at 4° to E. 30° S., are exposed in an old quarry [456805] 150 yd west of the farm. D.C.G.

Culmington. The feature of the Sandstone continues from Upper Onibury to the road south of Culmington Manor and the rock has been quarried at several points. At Norton [461816] 8 ft of yellow sandstone, flaggy and false-bedded, are exposed, and buff micaceous sandstone with small irregular cavities is seen in another quarry [462821] 550 yd to the north. The outcrop is displaced by about 1600 yd to the south-east by a fault through Bache, and the Sandstone is well exposed in old quarries at Burley [476815]. An old quarry [475815] 1100 yd S. 10° E. of Medley Park (p. 219) yielded *Lingula cornea* 3 inches above the Bone Bed, with *Loxonema gregarium* (J. de C. Sowerby, *Platyschisma helicites*, *Modiolopsis complanata* (J. de C. Sowerby), *Nuculites sp.*, *Kloedenia wilckensiana*, and Thelodont denticles 1 to 2 ft above. Further quarries occur some 400 yd to the north-west [474818]. A second parallel fault just north of the latter exposure displaces the Sandstone by 300 yd to the north-west. Yellow sandstone with *Kloedenia wilckensiana* is seen on the northern side of the road [474832] 865 to 920 yd E. 16° N. of Marsh Barn. Just north of these exposures the outcrop is displaced by about 900 yd to the east-south-east and pale yellow, micaceous, fissile sandstone is exposed in the stream [483836] about 80 yd east of New House. B.A.H.

Diddlebury to Broadstone. Immediately east of Pedlar's Rest [487843] the top 4 ft of the Sandstone are exposed at the foot of a quarry face. The basal beds of the Sandstone, to a thickness of up to 25 ft, are seen above the Ludlow Bone Bed at three localities between the above locality and Munslow Aston, and again north-east of Munslow (p. 219). The lowest 3 ft are usually siltstone or silty mudstone. An exposure very near the base of the Sandstone, in the lane [501857] 840 yd W.N.W.of Diddlebury church, shows 3 ft of grey and olive-green silty shales beneath yellow sandstone, with 1 ft of grey silty limestone below them. A roadside cutting [496850], 60 yd south-east of the Inn at Corfton, has yielded *Lingula cornea*, *L. minima*, *M. complanata*, ostracods indet. and fish spines. The lowest beds of the Sandstone, comprising 3 ft of yellow flaggy sandstone overlying 3½ ft of siltstone and silty mudstone with some fine sandstone bands, are seen adjacent to the Swan Inn [511866], 110 yd S. 44° E. of the chapel, Munslow Aston, where they overlie the Bone Bed. The sandstone has yielded an extensive fauna. One-inch above the Bone Bed *Orbiculoidea rugata* (J. de C. Sowerby), ostracods indet., and Thelodont denticles were found, and 6 inches higher the beds yielded *L. cornea, Loxonema obsoletum* (J. de C. Sowerby), *Modiolopsis sp.; Kloedenia wilckensiana* and a eurypterid fragment. *Loxonema obsoletum*, *Platyschisma sp.*, and *Modiolopsis complanata* were found 2 ft above the Bone Bed, and *Lingula minima* occurred 2 ft 6 in higher in the section. In an old quarry [535884] 1650 yd E. 27° N. of Munslow church 4 ft of yellow flaggy sandstone are exposed. Yellow flaggy sandstone is seen above the Bone Bed (p. 219) in Trow Brook [542890], but the next exposure is 270 yd S. 26° W. of Broadstone church [543895], the outcrop having been displaced by some 700 yd to the west-south-west along a fault passing close to the Trow Brook exposure.

Broadstone to Brockton. The Downton Castle Sandstone is poorly exposed between Broadstone and Shipton where [563918] the basal 8 ft are seen overlying the Ludlow Bone Bed (p. 219). Buff silty shales, 3 ft thick, are succeeded by yellow flaggy sandstone. North of Shipton a strike fault with a north-westerly downthrow of about 200 ft crops out about 350 yd north-west of the road to Brockton. The Sandstone forms a narrow outcrop immediately north-west of the fault and is exposed to a thickness of 2 ft in a stream [563922] 430 yd N. 13° E. of Shipton church. The main outcrop is followed, primarily on the evidence of soil debris, to Brockton, where 6 ft of olive-coloured sandy mudstone are exposed at the base of the formation (Robertson 1927, pp. 93–4). D.C.G.

Brockton to Callaughton. There is a section in buff and brown, flaggy, micaceous sandstone with carbonaceous impressions, [586947] 700 yd S. 3° E. of Patton Grange. At Callaughton there is a small exposure of grey sandstone on the western side of the fault trough [612973] 1900 yd east-north-east of Bourton Grange. The best exposure here is in an old quarry [619975] 2100 yd east-north-east of Bourton Grange, where about 15 ft of greyish yellow sandstone are seen, partly flaggy but thickly bedded towards the top, and with carbonaceous layers in the lower beds. R.W.E.

TEMESIDE SHALES

Onibury. The Temeside Shales are poorly exposed south-west of the River Onny, being seen at several points in the stream about ¼-mile east-north-east of Stokesay Court. The best exposure in the district is the well known section in the lane [458798] about 800 yd N. 15° E. of Onibury church, embracing some 63 ft of rock of which about 35 ft were exposed at the time of the survey. The rock varies between micaceous blocky mudstone and coarsely micaceous sandstone, mainly pale yellowish green, but often with purplish brown and black staining. *Lingula cornea* is common in differing lithologies at several horizons. It is accompanied by fish fragments in a fine-grained sandstone 17 ft from the top of the section. Very abundant ostracods occur in a weathered layer within sandy beds 34 ft from the top of the section. The section was described in detail by Elles and Slater (1906, p. 217–8, fig. 10) but time and the metalling of the road have altered its appearance, and it is difficult to effect an exact correlation between the present section and their description. A 3-in band of hard, coarsely micaceous sandstone, about 32 ft from the top of the section, is thought to be part of the 15-in bed of 'hard grey micaceous sandstone' (Fb) of that description. The highest beds exposed at the time of the survey, on the western side of the road just north of the sharp bend and some 22 to 32 ft above this sandstone, are greenish marl and sandstone, of Temeside lithology but apparently assigned by Elles and Slater to the base of the 'Old Red Sandstone' (Ledbury Group). Similar green sandstones forming the feature 200 yd to the east are the highest beds included in the Temeside Group. Small exposures of greenish and yellowish sandstones and purple mudstones occur between the lane section and Upper Onibury [458805] and between the farm and the stream 600 yd to the east. D.C.G.

Culmington. The shales are seldom exposed in the vicinity of Culmington, and their upper boundary is ill defined. Sandstone bands give rise to small topographic features, often with soil debris of yellowish green sandstone. Six inches of purplish green sandstone are exposed in a stream [477826] 500 yd E. 20° N. of Medley Park, and traces of purplish brown marl are seen a farther 80 yd upstream. In a quarry [487843] 50 yd south-east of Pedlar's Rest, 4 to 5 ft of thinly-bedded greenish sandy shales were formerly seen (Robertson 1927, p. 95) and these beds yielded *Beyrichia wilckensiana* and *Leperditia elongata* Weller. B.A.H.

Diddlebury to Broadstone. In the Diddlebury and Munslow areas no notable outcrops of the Shales are known, the upper boundary being taken at the top of a lenticular sandstone which locally forms a feature 100–200 yd south-east of the outcrop of the Downton Castle Sandstone. This topmost bed has a wide outcrop in the Trow Brook [546891] west of Thonglands, where it is seen to be a grey and reddish brown micaceous sandstone. The outcrop beyond the fault (p. 221), some 300 yd to the north-west, is marked by debris of yellow medium-grained sandstone. Two feet of yellow sandstone are exposed in the bridle road 215 yd E. 23° S. of Broadstone church.

Broadstone to Brockton. North-east of Broadstone the upper boundary of the Shales is largely conjectural, the rocks being very poorly exposed. At Brockton, in the stream [580937] up to 90 yd west-north-west of the Castle Mound, grey micaceous shales with thin sandstone bands are seen. D.C.G.

Brockton to Callaughton. Weathered grey mudstone of the Temeside Shales is seen in a small pit [592955] 630 yd E. 10° N. of Patton Grange, and again at a locality [615976] some 1730 yd east-north-east of Bourton Grange, on the eastern side of the fault trough. In this area the top of the shales is marked in many places by an impersistent sandstone which is exposed by the side of the lane [620973], 2280 yd E. 20° N. of Bourton Grange. Two feet of green micaceous sandstone are exposed here, and the outcrop can be followed elsewhere by topographic feature and soil debris. R.W.E.

LEDBURY GROUP

Cleestanton. In the Cleestanton area the exposures of rocks of the Ledbury Group occur mainly in the banks of the Ledwyche Brook, and to the north-west, the ground on the south-eastern side being largely covered by head.

Fine-grained grey and purple flags, of Ditton Series aspect but probably high in the Downton Series, crop out in the small stream [537786] some 600 yd W. 16° S. of East Lodge. Purple micaceous flags, on the left bank of the Ledwyche Brook [545784] about 570 yd S.E. of East Lodge and 150 yd upstream near the mouth of the Hopton Brook, may belong to one band, displaced by the fault trending E. 25° S., exposed on the left bank [546785] 600 yd E. 32° S. of East Lodge. These beds dip gently to the north and north-west and are interbedded with purple marls, locally with cornstone nodules. A higher band of coarse-grained sandstone crops out in the banks of the brook [550788] west of Roundthorn and forms the steep left bank. Marl is seen in places between Roundthorn and Upper Ledwyche, where a still higher sandstone is exposed. Ten feet of soft, coarse-grained, brown sandstone are visible in the lane [554793] just north of the buildings, and current-bedded sandstone and thin cornstone-conglomerate crop out on the right bank of the brook about 150 yd to the east-north-east. Pellety sandstone at the latter locality is somewhat fossiliferous.

Red marl is seen in places between Gibbridge and Bank Farm, generally about 2 ft below the surface. The prominent hill [566789] some 500 yd north-east of Kitesnest may mark the outcrop of sandy or nodular beds, but no such beds were seen. Coarse-grained purple flaggy sandstone has been quarried to build a cottage [566781] 790 yd S. 42° E. of Kitesnest (Ludlow Sheet 181). The rock, approximately horizontal, forms an outlier of which the northern limit is some 600 yd E. 22° S. of Kitesnest [567785]. No solid rock of the Downton Series is exposed between Cleestanton and Langley.

Onibury to Munslow. Between Onibury and Walton [467795] the lower beds of the Ledbury Group form several distinct features trending somewhat east of north. Occasional ditch and quarry exposures indicate that these are mainly due to the presence of lenticular sandstones in the predominantly marly sequence.

At Padmore [464794] about 6 ft of purple and grey micaceous sandstone, resting on purple marl, are exposed at the base of a wall. Sandstones below and above are seen in a ditch, 270 yd to the west-north-west, and in the gateway 100 yd to the east-south-east respectively. Extensive old workings in the lower beds occur about 400 yd east-north-east of Onibury church [460792]. The upper sandstone forms a strong feature extending northwards for about 1500 yd from the railway south of Padmore. In a ditch [466794] 290 yd E. 8° N. of Padmore the basal 2 ft of a higher sandstone are exposed. A low dip to the east-south-east is observed in all these outcrops. In many instances the features continue to Vernolds Common but the presence of the underlying sandstone is shown almost entirely by soil debris and the existence of old workings.

A series of sandstone groups, higher in the succession, forms several prominent west-facing escarpments between Walton and the River Corve. Exposures are few but debris of purple micaceous sandstone is often abundant. The rock is exposed in old workings on scarp crests 530 yd S. 43° W. [471794], and 1350 yd S. 3° E. [475785], of Ayntree chapel. Cornstone-conglomerate, a rock type uncommon in the Downton Series, occurs in the debris on the prominent feature which lies some 500 yd east of High Walton. Exposures of sandstone are common near Holt and Stocking, especially in the lane up to 300 yd to the north [479804], dipping at up to 8° towards the east. The sudden failure of several features suggests the presence of a fault, following a sinuous east-north-easterly course from west of Ayntree chapel to east of Holt. About 10 ft of coarse-grained micaceous purple sandstone are seen in the bank of a pond [486802] 540 yd E. 16° S. of Ruckley. D.C.G.

At Vernolds Common the sandstone features trend northwards, but they swing to north-east in the faulted area around Burley, and return to northward between Burley and Siefton. Purple and grey marl is seen in an old clay pit [467811] 900 yd S.E. of Norton. Two feet of purple, micaceous, flaggy sandstone are present on the south side of the pool [471817] 250 yd E. 27° S. of Bache, and purple sandstones and red marls, dipping at 11° to S. 30° E., are exposed in the stream 300 yd below the pool. Twelve feet of purplish brown, micaceous sandstones and marls, dipping at 8° to E. 16° S., are exposed in a stream [481831] 750 yd E. 43° S. of Culmington Manor, and 12 ft of similar beds are seen 50 yd upstream. B.A.H.

Between Siefton and Elsich sandstones low in the Ledbury Group form a number of prominent features. Sandstone debris is locally abundant and in a stream [492840] immediately south of Elsich several feet of closely jointed flaggy sandstone are exposed, dipping at 9° to the east-south-east. The sandstone features continue north-eastwards to Diddlebury but exposures are few and insignificant. Four feet of purple, micaceous, fine-grained sandstone are exposed [506853] 250 yd W. 1° N. of Diddlebury church. Beyond the village the features are fewer and exposures are infrequent. In the right bank of the Corve [527871] 1650 yd E. 14° N. of Munslow Aston Chapel, 2 ft of red and greenish grey marl overlie 5 ft of purple, flaggy, micaceous sandstone with some soft shaly bands. Five feet of sandstone, overlying marl, are exposed in the right bank 980 yd E. 10° S. of Munslow church. At both these localities the beds dip at 10° to the south-east.

Stanton Lacy to Peaton. Between the flood plain of the Corve and the road from Stanton Lacy to Lower Hayton the solid rock is very seldom seen, either at outcrop or as soil debris. Thin terrace deposits, with few stones, may extend for some distance east of the flood plain, and there has been some westward solifluxion of material from the escarpment of the Ditton Series. A topographic feature, extending for some 1300 yd to the north-north-east from a point [499794] 750 yd N. 26° E. of Stanton Lacy church, may be formed by cornstone nodules. Marl with many nodules is exposed to 3 ft in a ditch on the crest of the feature [503805] and nodules are common in the soil along its length.

East of the road, features are better developed and outcrops more common. Coarse-grained, micaceous, purple flags are seen 1170 yd E. 18° S. of the church [506785], on the south side of the road to The Hope. Rocks of this horizon, in the stream 360 yd to the south-east (Ludlow Sheet 181), are highly fossiliferous. Only purple and red marl, locally with small calcareous nodules and 'fish eyes', is exposed in this stream for 500 yd, comprising some 150 ft of strata, below the '*Psammosteus*'

Limestones at the road bridge [511786]. Intermittent exposures of the highest rocks in the Downton Series occur in the streams descending the escarpment south-west of Woodlands and east of the inn at Lower Hayton. The topmost 70 ft, consisting mainly of blocky and shaly marls and broken by minor faults, are almost continuously exposed east of the inn [508802]. At the base of this section are 2 ft of well-bedded purple cornstone-conglomerate. Small calcareous nodules occur in some of the marls and pronounced lustre-mottling is seen locally. Flags and marls from the top of the Group are exposed in the old lane [510807] west of Titterhill and there are a few small outcrops in streams and farmyards at Lower Hayton.

Purple micaceous flags with 'fish-eyes' are exposed to 5 ft at Upper Hayton [516812], 380 yd N. 43° W. of the chapel, and form a feature for about 350 yd to the north. From here to Strand Brook the highest beds of the Downton Series are seldom exposed. Marl is seen locally, and cornstone nodules and sandstone debris occur in places in the soil.

A prominent feature extends north-eastwards from about 1570 yd E. 13° N. of Culmington church to about ¼ mile beyond Great Sutton. On a bank [516831] south-east of the house at Great Sutton, purple micaceous flags and massive sandstone with 'fish eyes', are exposed to 8 ft, dipping at a low angle to the east-south-east. Higher beds of marl and flags are exposed 50 yd to the north-east and, still higher, around the buildings east of the road [518831]. The prominent feature about ¼ mile south of Great Sutton is formed by the highest of these beds. Cornstone nodules in the marl seem to be the cause of a feature, locally very prominent, which trends north-eastwards for about 1900 yd from a point [518826] 570 yd E. 10° N. of Little Sutton.

Sandstone bands form a number of features trending north-eastwards between the River Corve and the Pye Brook. The Camp Ring [497821] stands on one, apparently separated by a dip fault from those to the north-east. Exposures are few, but over 13 ft of purple micaceous sandstone, with marly and pellety bands, are seen in the left bank of the Corve [500829], 420 yd E. 2° S. of Sparchford, and flags from a higher horizon crop out near the track [504829] 745 yd E. 6° S. of that farm. The higher beds form a prominent feature just to the north and again west of Lawton. An intervening feature is formed by sandstone overlying marl with abundant cornstone nodules. The marl is exposed in two ponds on the crest some 600 yd south-east of Sparchford. The buildings at Lawton stand on the dip slope of a group of flags and sandstone which is well exposed in their vicinity. All these beds between the brooks dip at low angles to the south-east.

The sandstone features die out north-west of Peaton. Debris of coarse-grained sandstone is seen about 70 yd south of Rosamond's Well [518843] and on a prominent feature [525853] 300 yd north of Corfham Castle.

Peaton to Brockton. Sandstone bands which form two features near Broncroft Parks appear to die out in the head-covered ground of Balaam's Heath, but the four bands mapped north-west of Holdgate are at about the same general horizon. The rock does not crop out but its debris is locally abundant. Red and grey marl at the base of the Ledbury Group is exposed in the Trow Brook [548891] north-west of Thonglands.

The features of higher and more persistent sandstones arise in the ground south-west of Broncroft. The lowest bed is exposed to 10 ft beside the brook [540866] 550 yd W. 19° S. of Broncroft Castle and has been extensively quarried at the roadside some 600 yd to the north-east. Sandstones and marls from higher horizons are exposed in the brook immediately north-west of the Castle, and again 300 yd upstream [547866] where 53 ft of beds are visible. They dip at about 10° to the south-east. The two lower sandstones can be followed with some difficulty to beyond Stanton Long. At Holdgate the lowest forms a prominent feature on which the buildings stand, and the rock is exposed in one or two places to a thickness of up to 8 ft. This is the Holdgate Sandstone of King (1934, p. 528), distinguished by him from the other Downton Series sandstones by its coarseness, a distinction of doubtful validity.

West of Holdgate, between the Corve and the Trow Brook, thin sandstones and marls near the base of the Group are exposed in places. These beds lie on the north-western, downthrow, side of the fault which had been seen (p. 221) to displace the Downton Castle Sandstone.

Between Shipton and The Leath [5889][1] the topography and occurrence of rock debris indicate that sandstone bands occur repeatedly, and fairly regularly, from the base of the Ledbury Group to within some 300-400 ft of the top. Many of these features continue north-eastwards to the vicinity of Weston [5992], displaced at intervals by minor faults. East of the Corve at Shipton three small faults with westerly downthrow follow approximately north–south courses. The remaining faults cross the strike of the beds more or less at right angles. Five feet of reddish brown and grey, fine-grained, micaceous sandstone are seen in the left bank of the Corve [587933] 785 yd N. 14° W. of Skimblescott, and cornstone nodules and sandstone debris from a much higher horizon are seen beside the pond at Little Oxenbold [586915], but otherwise the evidence for the sandstones depends mainly on topography and debris in the soil. A well drilled at Weston [599929] proved 38 ft of sandstone in the topmost 40 ft of strata, resting on 174 ft of mainly marly beds. The sandstone is exposed amongst the buildings [598930] north-east of the road junction.

The topmost beds of the Group are seldom exposed and where they are, as for example, in the Strand Brook [5384], in the brook at Tugford [5687], and at The Leath [584899], they consist generally of red marl, locally with bands of small cornstone nodules. D.C.G.

Brockton to Shirlett and Aston Eyre. Near Bradeley and Corve Barn [5993] numerous sandstone bands can be followed by feature, some, especially in the higher beds, for a considerable distance. The beds are thin and are not exposed.

The lowest part of the Ledbury Group, with generally thin, impersistent sandstones, is separated from the highest part, in which sandstones are more prominent and persistent, by a thick zone of marl which crops out along the River Corve, in the plantation south-west of Woodhousefield, and in Spoonhill Wood [6195]. The marl is appreciably thicker in this wood, and there are occasional exposures of red, and red and green, marl with calcareous concretions. Its outcrop continues to the north-east, forming the low ground north of Acton Round and along the Mor Brook. Purple sandstones of the lowest part of the Ledbury Group are exposed in the dingle [612965] some 1100 yd W. 20° N. of Spoonhill Hall. Three beds, up to 4 ft thick, are seen, interbedded with red marl.

The sandstones above the marl are well exposed, and form prominent features which can be traced for distances of up to 3 miles. The lowest of these crops out in the bank of the stream [621955] 1510 yd W. 9° S. of Acton Round church, where about 12 ft of sandstone are seen, predominantly red but with bands of brown calcareous grit and thin bands of marl. It is exposed again in a stream [627956] 790 yd W. 9° S. of the church. Mottled reddish blue marl with 'race' is seen at the base of this section, overlain by 6 ft of red marl with an irregular upper surface, on which rest 3 ft of fairly massive, green, calcareous sandstone containing fish fragments and bands rich in lenticles of green marl.

The next sandstone above is not well exposed between Corve Barn and Acton Round where, in an old quarry [639956] 450 yd E. 5° S. of the church, about 9 ft of beds are visible, mainly finely bedded flags with ribs of greenish massive sandstone, with pellets and thin intercalations of marl. It is also exposed in Beaconhill Brook [641952], 850 yd E. 38° S. of the church, and in the stream flowing northwards from Aston Eyre. A higher band of sandstone, exposed to 9 ft in Beaconhill Brook, is well exposed in an old quarry [640950] 980 yd S. 44° E. of Acton Round church. At the base are 6 ft of massive purple sandstone, locally pellety and locally finer-grained, separated by 6 in of highly micaceous red siltstone from micaceous, green and purple, flaggy

[1]Such numbers are four figure National Grid References within the 100-km square SO (32).

sandstone, seen to 2 ft at the top. The sandstone feature can be followed to the north and east, and an extensive dip slope is developed east of Ouseley, where the sandstone is thrown down to the east by a fault.

The 2 ft of micaceous flaggy sandstone exposed at the small pond [616946] 1660 yd N. 41° W. of Monkhopton church lie at a higher horizon, and can be traced north-eastwards and eastwards to a small pit [637949] 840 yd S. 17° E. of Acton Round church, where 5 ft of purple micaceous sandstone are exposed. The most easterly exposure of this bed is in Beaconhill Brook. A higher bed forms a prominent feature north of Monk Hall. The rock is well exposed in an old quarry [621947] 200 yd north-east of Monk Hall Grange, where about 15 ft of purple and green sandstone are seen, mainly flaggy and micaceous but with a massive central band, 3 ft 4 in thick, showing contorted structures.

Over 10 ft of purple and green flaggy sandstone, with two thin bands of marl, are exposed on the left bank of Beaconhill Brook [630942] 945 yd N. 24° E. of Monkhopton church. The sandstone is commonly pellety and contains abundant fish fragments in two bands about 9 ft from the top of the section. The junction between marl and overlying sandstone is very irregular. The section has yielded *Ischnacanthus sp.*, *Poraspis sp.*, *Pteraspis sp.*, *Tesseraspis sp.* and *Traquairaspis sp.* This sandstone also crops out at several adjacent localities, both upstream and downstream. Some 650 yd upstream, marl from a higher horizon is exposed, with large nodules of cornstone.

The highest rocks of the Downton Series exposed in the scarp streams at Monkhopton are nearly all marl, but some bands of sandstone and cornstone-conglomerate occur near the top of the Series. Three bands of red, calcareous, hard, silty mudstone with fish fragments occur below the '*Psammosteus*' Limestones in the coppice north-west of Woolshope. R.W.E.

Two sandstones in the lower part of the Ledbury Group form features around the spur [640966] ½ mile north-north-east of Acton Round. The same sandstones also give rise to features north-north-east of Beggarhill Brook Farm and around The Hawthorn [652969]. Grey flaggy sandstone with a southerly dip of 15° is exposed at a spring [665966] 1300 yd E. 10½° S. of The Hawthorn. Sandstones, probably at a similar horizon, give rise to features ¼ to ¾ mile north of Poplar Cottage.

In the higher part of the Ledbury Group, sandstones form the ridge which runs east-north-eastwards from Aston Eyre towards the Mor Brook. Five feet of lilac, micaceous, flaggy sandstone dipping at 3° to S. 20° E. are exposed at 800 yd east of Aston Eyre church [660940]. B.A.H.

DITTON SERIES

'PSAMMOSTEUS' LIMESTONES

Cleestanton to Strand Brook. Limestone, apparently from two bands, has been worked between Stocking and Langley. Old quarries in the lower band occur between 180 yd W. 21° S. [579783], and 550 yd N. 7° E. [582789], of Stocking. Three feet of fine-grained, grey, nodular limestone, overlain by purple and grey marl with many cornstone nodules, are exposed at the latter locality, but elsewhere only debris is seen. The higher band has been extensively quarried and burnt between 320 and 500 yd north of Stocking [581787]. At the southern end of the workings 2 ft of grey fine-grained massive limestone are exposed, dipping at a low angle to north of east. A short distance to the north-east the features formed by the two bands run together. There is an exposure of 1 ft 6 in of massive grey limestone in an old quarry [588795] 510 yd S.S.W. of Langley. A few small exposures, old workings, and patches of rock debris mark the outcrop of the apparently single limestone to about 100 yd south of Langley. The beds in this area being nearly horizontal, this outcrop must be separated by a fault from those in the stream west of Gibbridge, some 1500 yd to the west and at an altitude some 200 ft lower.

Three groups of cornstones crop out in the stream between 250 and 1030 yd east-north-east of the road bridge south of Lackstone. All are purple and grey nodular rocks, the highest being somewhat sandy. The middle one is in two bands, 3 ft apart. About 35 ft of beds, mainly purple flags, separate the lower groups, and there are about 28 ft of beds between the upper groups. The top group is also exposed in an old quarry [579804] 630 yd E. 15° S. of Lackstone where, as in the stream, grey nodular cornstone, 5 ft thick, is seen to grade downwards into less calcareous sandstone and flags. The outcrops of the highest and lowest cornstones are extrapolated from the fault at Lackstone to The Moor, where they appear to converge, but there is little significant evidence in this area. A 2-ft band of grey, fine-grained, massive limestone, 470 yd W. 5° S. of The Moor [561799], is the only rock in the Ledwyche Brook which can be referred to the '*Psammosteus*' Limestones. The presence of a lime kiln suggests the proximity of old workings, but none are now visible. Nodules of fine grey limestone in the fields to the north-east probably come from a higher horizon. South-westwards the position of the limestone is based largely on topography and the occurrence at intervals of definite Downton Series or Ditton Series rocks. Fine-grained nodular limestone is poorly exposed in the brook [548791] 720 yd E. 31° N. of East Lodge and is 6 in thick in the brook [536787] 620 yd W. 10° S. of East Lodge.

The outcrop of the Limestones continues south-westwards for a short distance into the adjacent district (Ludlow Sheet 181) before swinging north-westwards, due to the north-easterly plunge of the Brown Clee Syncline (p. 272), towards The Hope. In the woods about ¾ mile south of Downton Hall, in the Ludlow district, the limestones were quarried and mined in the mid-19th century by the landowner, and substantial kilns were built.

At The Hope two cornstones are exposed in the stream immediately below the road bridge [511786]. Both are fine-grained, nodular, and mainly grey, the upper 4 ft thick and the lower 3 ft, separated by 5 ft of purple and greenish grey marl with cornstone nodules. Some large quartz grains occur in the lower band. A similar cornstone, exposed to 3 ft in an old quarry [509785] 1540 yd E. 11° S. of Stanton Lacy church, has been analysed (Harvey 1957, p. 56–7) and is estimated to contain some 88–89 per cent of calcium carbonate. Outcrop-working of this cornstone is continuous for about 250 yd to the north-north-east of this locality but the rock is not exposed. A lower band of grey fine-grained cornstone is seen in an old working [508785] 1400 yd E. 13° S. of the church and may persist northwards for at least 600 yd. Ultimately, both northwards and southwards, it is considered to coalesce with the higher band, but it may pass laterally to a marl. Abundant cornstone nodules in places suggest that a third, intermediate, band may occur.

Two cornstone horizons assigned to the '*Psammosteus*' Limestones are recognized in the stream south-west of Woodlands. The lower, tentatively correlated with the highest at The Hope, is a very calcareous band of purple fine-grained sandstone, some 15 in thick, occurring 1625 yd E. 32° N. of Stanton Lacy church [509796]. Some 80 yd upstream and about 100 ft higher in the succession, purple and grey cornstone is exposed to 18 in. In the stream [509801] east of the inn at Lower Hayton the Limestones include three bands of cornstone and several strongly nodular marls. The sequence is broken by faults and the intervals between the cornstones cannot be accurately measured. The two lower bands are purple and grey mottled rocks, the lowest a calcareous nodular siltstone, the higher apparently a more typical cornstone. The highest bed is a fine-grained grey limestone. None of these beds is exposed to a thickness of more than 8 in but large fragments of grey limestone are seen in an old quarry immediately north of the stream. The intervening beds are mainly marls but include an 18-in band of well bedded cornstone-conglomerate.

At Titterhill 5 ft of marl, packed with nodules of grey cornstone, are exposed at the roadside [510807] 850 yd N. 36° E. of Lower Hayton inn. Some 70 yd to the south-south-east, and at a considerably higher level, nodular cornstone was dug out in the construction of a small reservoir. Two cornstone bands, which have been correlated with the two at Titterhill, crop out some 400 yd W. 15° N. of Upper Hayton chapel [514810]. The lower band is very sparsely exposed between here and Strand Brook and its continuity is therefore somewhat suspect. In a ditch [528830] 1350 yd N. 38° E. of Lydehole it is perhaps represented by 3 ft 6 in of nodular cornstone lying 15 ft below what is thought to be the upper band of Titterhill. Outcrops and debris of this band occur in old workings [517813] 480 yd N. 12° W. of Upper Hayton chapel and about 250 yd to the north-east, and show that there are at least 3 ft of massive, fine-grained, purple and grey cornstone. Cornstone nodules in the soil indicate the occurrence of a slightly higher band in this area, and this is confirmed in the stream nearby.

The succession through the '*Psammosteus*' Limestones is well displayed in this stream, some 600 yd north-north-east of Upper Hayton chapel, and is illustrated in Fig. 17. The band quarried just to the west is the lowest in the stream, a fine-grained grey limestone 10 ft thick. The band above, inferred from soil debris just to the west, is about 8 ft thick, a mottled, purple and grey, marly, nodular limestone. The highest important band is similar, but only about 4 ft thick. These beds are thicker, individually and in total, than any seen farther south within the area surveyed. Marls predominate between the limestones, but calcareous flags and a thinner limestone also occur.

Between the stream and a locality [524822] 420 yd N.E. of Lydehole the 10-ft limestone has been extensively quarried and is exposed to 8 ft in two places. At the top of a high waterfall [522819] 180 yd E. 42° S. of Lydehole the rock consists of 5 ft of fine-grained, massive, grey cornstone, resting on 1 ft 9 in of nodular cornstone. No higher limestones were seen in this stream.

Beyond and on the upthrow side of a small dip fault, in a brook [525823] some 1340 yd east of Little Sutton, purple and grey nodular cornstone is developed at four horizons in a group of marls some 28 ft thick. The thickest band, the highest, is only 2 ft 9 in thick, and it is impossible to correlate any one band with the cornstone so widely quarried south of the fault, although one of them probably is its equivalent. Massive grey limestone up to 5 ft thick has been quarried and is exposed at intervals for about 350 yd to the north, and 3 ft of mottled nodular cornstone, thought to be at the same horizon, are exposed in a ditch [528830] 1350 yd N. 38° E. of Lydehole (see above). It is seen again, 3 ft thick, in each of two streams, a few yards above their confluence [534833] 1550 yd E. 16° S. of Patch House, and 310 yd north of the confluence [534836] where a field track passes through a hedge. Three feet of grey limestone, from the same horizon, are exposed in the wood [537841] 710 yd S. 4° W. of Peatonstrand chapel, and a higher band of purple limestone is seen in the quarry a few yards to the south-west.

The succession well displayed in the Strand Brook [540842] is shown in Fig. 17. One or both of the two basal limestones is perhaps equivalent to the lower band at Titterhill (above). The lower is somewhat nodular and mottled, the upper a massive grey limestone. The massive grey limestone, 55 ft up, correlated with the bed which has been followed from Titterhill, is poorly exposed in the brook. The grey limestone about 30 ft higher is in a sequence disturbed by faulting and is somewhat fragmentary in places, like a cornstone-conglomerate. This is the higher bed seen in the quarry south of Peatonstrand chapel.

Strand Brook to Monkhopton. Each of the three limestones exposed in the Strand Brook has been quarried at intervals between the brook and Bouldon, but only the middle band, 3 ft of massive grey limestone, is exposed in this area. The two higher bands are the main limestones in the Clee Brook at Bouldon [548850], cropping out some 310 and 340 yd south-east of the church. The lowest band is probably represented there by a 2-ft sandstone, calcareous and pellety. Each of these limestones dies out within 300 yd to the north of the brook. The purple and grey limestone, exposed to 2 ft, 270 yd N. 14° E. of the church, at the southern end of a line of old quarries [547854] which extends intermittently to the valley south of Tugford, occurs in the brook at Bouldon as nodular marl some 25 ft above the 2-ft sandstone. South of the brook it is not recognized at all. It appears therefore that the main limestone south of Bouldon is at a higher horizon than that to the north, both being identifiable in the Clee Brook where they are about 20 ft apart.

In the valley (Kidnall Gutter) [555863] 870 yd S. 15° W. of Tugford church the limestone quarried to the south-west is seen to be about 14 ft thick. Nodular marl below, some 2 ft thick, passes upwards into grey and reddish brown, massive limestone, 5 or 6 ft, on which lie some 9 ft of mottled marly limestone, becoming more nodular upwards. The position of the lowest limestone is recognized some 15 ft below, where 9 in of grey gritty cornstone-conglomerate rests on red marl with cornstone nodules. None of the three highly calcareous bands in the 90 ft of beds above the thick limestone can be confidently correlated with any of the higher limestones of Bouldon or farther south. Only the thick limestone can be followed, by feature, soil debris, and occasional old quarries, from here to Tugford. In the brook [563870] at Tugford, where the rocks are slightly disturbed by a small dip fault, three limestone bands, between 480 and 750 yd east of the church, are assigned to the '*Psammosteus*' Limestones. The lowest, 4 ft thick, is equivalent to the 14-ft band of the last section, and is separated by only 2 ft of marl from a 2-ft band of marly cornstone-conglomerate with occasional rounded quartz pebbles, taken to be the basal bed of the Ditton Series. Quartz pebbles are uncommon in the Downton and Ditton Series conglomerates of this district and are used to identify this band at other localities. The thickness of the higher limestones is not more than 1 ft, and that of the group as a whole, with intervening beds of marl, cornstone-conglomerate, and sandstone, is about 120 ft.

These three limestones can be seen to be present as far as the vicinity of Lower Earnstrey Park [566879] but north-eastwards from there only the middle band is seen. At Lower Earnstrey Park it is exposed as 3 ft of massive limestone, and a short distance below is a locally developed sandy limestone, 2 ft 6 in thick, correlated with the thickest limestone in the important stream section north of Earnstrey Hall ('Earnstrey Hall' Fig. 17). Thus once again the position, in the succession, of the main limestone has changed. In a stream [575889] 600 yd N. 12° E. of the Hall a 4 ft 6 in band of grey cornstone-conglomerate, of which the top 6 or 9 inches is a coarse calcareous grit, forms a waterfall. Quartz pebbles and fish fragments occur throughout. Beneath lies 1 ft 6 in of grey silty shale, with black streaks composed of fish fragments, and three lenticles of soft gritty conglomerate, up to 1½ inches thick, locally passing laterally into cornstone-conglomerate. The shale also contains cornstone nodules. Purple marl, with cornstone nodules at the top, is at least 20 ft thick below. The conglomerate was described by King (1934, p. 528) as "very calcareous coarse sandstone", 5½ ft thick, and by Wills (1935, p. 427) as "very rough calcareous pellet rock and conglomerate", 4 ft thick. King considered it to be the basal bed of his subdivision I. 8, the 'Psammosteid Limestones', and Wills, describing the fauna, showed that it contained the genera *Corvaspis* and *Phialaspis* (now *Traquairaspis*), whereas the shale below yielded *Anglaspis* and the types of *Tesseraspis tessellata* Wills. However White (1946, p. 210) found that the faunas were so much alike that the shale should be taken as the basal bed of I. 8. Later work by White (1950a) and Ball, Dineley and White (1961) showed that the fauna is of Downtonian type, probably from the Zone of *Traquairaspis symondsi* which White equated broadly with the diachronous '*Psammosteus*' Limestones (1950a, fig. 1), but which underlies the 'Main "*Psammosteus*" Limestone' of Ball, Dineley and White (1961, p. 202). The base of the conglomerate is here taken as the base of the Ditton Series, on lithological grounds and because the bed is a recognizable marker. The 'main' limestone of Ball, Dineley and White is 60 ft higher (Fig. 17); it consists of 5 ft of grey limestone, rubbly in the top foot and massive below. This bed is poorly developed in the stream 200 yd to the north-east, and can only be followed for about 400 yd along the strike. The marl lying about 10 ft above, with cornstone nodules up to 4 inches across, appears to be correlative with the middle limestone of Tugford. It can be followed for another 100 yd to the north-east and is exposed at a point [579891] 380 yd S. 38° E. of New Buildings as a red and grey rubbly limestone 3 ft thick, apparently the only limestone in the succession at this point. The basal cornstone-conglomerate of the Ditton Series is exposed 305 yd S. 35° E. [578891], where *Anglaspis macculloughi* (A. S. Woodward) and *Traquairaspis symondsi* (Lankester) were collected, and 330 yd E. 13° S. [580893], of New Buildings.

Quartz grains occur in a 1-ft band of grey cornstone-conglomerate exposed 820 yd E. 27° N. of New Buildings [584897], considered to be the basal bed of the Ditton Series. In the stream [585899] at The Leath, 1000 yd E. 35° N. of New Buildings, fossiliferous cornstone-conglomerates occur at this horizon just below an outcrop of sandstone and nodular marls. King's description of this locality, the type locality of *Pteraspis (Simopteraspis) leathensis* White, is quoted by White (1950b, p. 70), but clearly places these outcrops slightly higher than the basal 'hard band' of his 'Psammosteid Limestones'.

In the stream [591911] 630 yd E. 41° S. of Little Oxenbold the basal bed of the Ditton Series consists of 3 ft of grey and brown, flaggy, current-bedded, fossiliferous grit resting on 2 ft of grey quartz-conglomerate. No good limestones occur in the overlying beds but some of the marls carry many cornstone nodules ('Little Oxenbold', Fig. 17). A band about 1 ft 6 in thick, some 30 ft above the base of the Series, contains many nodules up to 4 inches across. The basal conglomerate, with limestone and quartz pebbles, is 2 ft 6 in thick in the stream [595914] 630 yd S. 25° E. of Great Oxenbold. Two bands of nodular limestone, 4 ft and 3 ft thick, occur in the overlying 30 ft of marls, sandstones, and cornstone-conglomerates ('Oxenbold Coppice', Fig. 17), and debris higher in the stream probably indicates a third unexposed band. The lower band is exposed in the stream 150 yd to the north-east, and both appear to have been worked in places in Oxenbold Coppice and again [604925] some 550 yd W. 26° N. of Upper Netchwood chapel. Evidence of at least three limestone bands is seen in exposures [606926] some 480 yd N. 28° W. of the chapel. All of them are nodular, the highest being possibly 6 or 8 ft thick, and they are interbedded with cornstone-conglomerates and unusually coarse sandstones. They occur near the lower horizon of Oxenbold Coppice and the lowest is taken as the basal bed of the Ditton Series. At the same horizon, in the stream [618930] 1320 yd E. 41° N. of the chapel, lie two bands of grey rubbly limestone each 5 ft thick and 5 ft apart, overlain by about 8 ft of cornstone-conglomerate and resting on red marl at least 8 ft thick. The higher limestone of Oxenbold Coppice has been worked [612928] 850 yd W. 9° S. of Woolshope, but only separate exposures of 6 ft of red, silty marl and of 2 ft of lilac, flaggy sandstone with some fragments of reddish brown cornstone-conglomerate are now visible. At the same horizon a small exposure, [614930] 600 yd W. 6° N. of Woolshope, shows 1 ft of soft, flaggy sandstone overlain by 4 in to 1 ft of grey, slightly gritty cornstone-conglomerate and 1 ft of grey, fine-grained flaggy sandstone. Traces of grey limestone are seen 80 yd to the north-east. D.C.G.

Monkhopton to Aston Eyre. North and east of Woolshope [619929] occasional exposures of limestone occur, showing a thickness of about 6 ft. This limestone, near the lower horizon of Oxenbold Coppice (above), also crops out in the stream [627928] 760 yd S. 11° E. of Monkhopton church where, about 40 yd upstream, a second limestone, about 8 ft thick, is also exposed. In another stream (Hudwick Dingle) [631927], 500 yd to the east, the only limestone is about 2 ft thick and locally dies out, being replaced by marl with rubbly limestone concretions.

East of Monkhopton in a section [636937] 1970 yd W. 14° S. of Aston Eyre church the basal limestone is 10 ft thick, but at the south-eastern edge of the coppice [641934] 1545 yd W. 32° S. of the church and in the stream east of the coppice it thins to 1 ft. In an old quarry 100 yd west of this stream a second, higher, limestone is exposed. On the scarp [650935] 745 yd S. 28° W. of Aston Eyre church the basal limestone is 2 ft thick, and a small exposure of rubbly grey limestone occurs about 10 ft higher in the sequence. R.W.E.

Six feet of limestone, in marl, are seen in the stream [660935] 1090 yd S. 40° W. of Ash Bridge. The most easterly outcrop within the Church Stretton district is in the next stream [664935], 875 yd S. 21° W. of the Bridge.

Pocock (*in* Whitehead and Pocock 1947, p. 22) described an outcrop of the topmost bed of the '*Psammosteus*' Limestones between the bridge, 900 yd east of Criddon (Dudley Sheet 167), and 700 yd upstream. It seems likely that the outcrop extends a farther 600 yd, to [663921] 1000 yd N. of Criddon, where over 10 ft of massive sandstone and cornstone are recorded. The beds exposed in the stream below the cornstone have been assigned by Ball, Dineley and White (1961, p. 190 and map) to the Downton Series. These authors also divided the cornstone outcrop into two parts, separated by a small fault, and ended it about 900 yd up from the bridge. The attitude of the beds in these exposures is very varied and the interpretation of the structure is debatable.

Main Part of Ditton Series

Downton Hall, Stoke St. Milborough, and Clee St. Margaret. At intervals along the prominent escarpment between The Hope and the Strand Brook the beds which overlie the '*Psammosteus*' Limestones are exposed in stream sections and old quarries. In an old quarry [514786] 360 yd W. 21° S. of the chapel at The Hope, one of several nearby, the following section occurs, typical of the lower part of the Ditton Series:

	ft
Marl, shale, and soft flags; plant horizon 5 ft down	16
Grey and purple massive sandstone; in 4-ft bands	8
Interbedded sandstone and flags	2
Massive sandstone	4
Cornstone-conglomerate, current-bedded; lenticles of sandstone ..	6

In the stream there is almost continuous exposure of the marls and sandstones, with cornstone-conglomerates becoming more common upwards, from the '*Psammosteus*' Limestones up to [521798] 1000 yd east of Woodlands. The beds generally dip eastwards at about 10° but are locally disturbed by faults.

Grey and buff calcareous sandstone has been quarried in places at the top of the steep wooded escarpment between The Hope and Woodlands, and higher beds, mainly cornstone-conglomerate, on the slopes between the stream and Woodlands. The beds above the '*Psammosteus*' Limestones are well exposed for about 140 ft in the streams south-west of Woodlands and east of the inn at Lower Hayton. Blocky and shaly marls predominate, in beds up to at least 10 ft thick, flags, massive sandstones and cornstone-conglomerates, seldom more than 3 ft thick, making up about 25 per cent of the succession. Most of the arenaceous rocks are greenish grey. A highly fossiliferous conglomerate, 8 ft thick, occurs 13 ft above the highest cornstone in the latter stream [510801]. Exposures on the scarp slope are otherwise uncommon, but just east of the crest near Titterhill are a number of old workings in which up to 20 ft of grey flags and sandstone are seen.

Many old sandstone quarries lie on the spur west of Upper Hayton chapel. Massive and flaggy, grey, calcareous sandstone is exposed to 13 ft, with grey cornstone-conglomerate below, in the largest [517808], 200 yd S. 33° W. of the chapel. The beds hereabouts dip at some 8° to between south-east and north-east. Debris of grey calcareous sandstone and cornstone-conglomerate is found near what appears to be an old shaft [518811], 180 yd N. 17° W. of the chapel. The rock bears a film of malachite in places and was worked intermittently for copper until 1919. According to local information adits were also opened at the roadside and the mine is said to date back to mediaeval or even Roman times. Murchison (1839, p. 188) made brief reference to it. The horizon of the rock must be very near the '*Psammosteus*' Limestones.

Old quarries in purple and grey sandstone occur near the top of the escarpment north and east of Upper Hayton chapel. The 250 ft of beds overlying the '*Psammosteus*' Limestones, mainly purple marls and shales, with subsidiary flags and sandstones and very few cornstone-conglomerates, are almost continuously exposed in the stream north-east of Upper Hayton. They dip generally at about 7° to the south-east. There is good exposure also, in the stream south-east of Lydehole, of some 300 ft of beds above the Limestones. The exposed rock has approximately 45 per cent each of marl and sandstone, in beds on average 4 ft thick. Thin beds of nodular marl and cornstone-conglomerate make up the remainder. Purple is again the usual colour but the higher sandstones are often grey. On the scarp north of this stream grey sandstone is seen in one or two old quarries.

In the scarp stream east of Little Sutton about 140 ft of beds above the '*Psammosteus*' Limestones are almost continuously exposed. Purple marls predominate in the lowest 80 ft, but the higher beds consist mainly of grey flags and calcareous sandstone, which appear to constitute the bulk of the succeeding 400 ft of strata cropping out eastwards to beyond the road. The south-easterly dip of 5 to 8 degrees is maintained in this area. Grey flags and cornstone-conglomerate are seen in several large old quarries on the scarp as far as 350 yd north of the stream.

Although outcrops on the scarp slope are infrequent, many of the harder rocks can be followed by feature along the slope between Woodlands and the Strand Brook. Such features are especially prominent between Upper Hayton and Lydehole, and east of Great Sutton. In the stream west of Hollybush grey flags and sandstones, with subsidiary cornstone-conglomerates and purple and grey marls, are extensively exposed from the '*Psammosteus*' Limestones up to [538830] 480 yd W. 26° S. of Hollybush. In the Strand Brook the rocks are extensively exposed for a thickness of the order of 700 ft above the Limestones, up to the road north-east of Redfurlong [547830]. Below a locality [544833] some 200 yd E. 23° N. of Hollybush grey sandstones and cornstone-conglomerates predominate in the lower 400 ft of beds, in contrast to the mainly purple beds above. As elsewhere the marls are poorly exposed but probably form the bulk of the succession. The individual beds are generally of the order of 5 ft thick, but one band of grey sandstone with interbedded cornstone-conglomerate is exposed to a thickness of 18 ft. Sandstones and cornstone-conglomerates are also seen in a number of small old quarries on both sides of the valley. In the highest of these [547830], 530 yd E.S.E. of Hollybush, 6 ft of coarsely pebbly conglomerate and thin flags are exposed. A calcareous bed at about this horizon has been worked and burnt at many points between Lesser Poston [538820] and New House [557837].

Between the Strand Brook and Bouldon one or two old quarries are situated on the scarp slope. In a quarry [548849] 350 yd S. 38° E. of Bouldon church about 18 ft of beds are exposed, including 5 ft of grey cornstone-conglomerate and 8 ft of massive sandstone. The sandstones and conglomerates which form the escarpment are well exposed in the Clee Brook and its tributaries where they are seen to be predominantly grey, in beds up to about 12 ft thick, and to dip at about 10° to between east and south-east.

The beds above the '*Psammosteus*' Limestones form a steep slope between France [5378] and Langley [5879], overlooking the marl valleys of the Ledwyche and Dog-ditch brooks, but the crest of the escarpment is not so clearly defined as it is in Corve Dale. South of Langley the crest is occupied by the intrusive dolerite of Titterstone Clee Hill (Ludlow Sheet 181) and the Lower Ditton beds make no separate feature.

Predominantly purple marls, sandstones and cornstone-conglomerates are well exposed in the streams between France and Meesons, dipping at about 20° to the north-north-west, and in several old quarries. In a quarry [533784] 210 yd S. 39° E. of France 24 ft of beds are exposed, nearly all grey and purple sandstone and flags. A plant layer occurs at the base of a 3-ft sandstone, 4 ft from the top, and some 5 ft lower a ½ in band of highly carbonaceous material lies below 2 in of buff and yellow coarsely pellety sandstone. Numerous small outcrops between France and Downton Hall show a gradual decrease in dip from over 25° to less than 10° and a change in direction from north-westerly to north-easterly. Purple and grey calcareous sandstones and cornstone-conglomerates are well exposed in old quarries near the Hall. Marls predominate in the streams south-west of Cuckoopen and are exposed to 25 ft beneath a footbridge [536799] 180 yd W. 11° S. of that cottage. Several bands of sandstone and cornstone-conglomerate at slightly higher horizons give rise to small distinct features, and have been quarried, in the vicinity of Hoptongate. In a quarry [530804] some 500 yd E. 19° S. of the cross roads 18 ft of purple and grey current-bedded conglomerate are exposed. Rounded pebbles of calcareous siltstone and cornstone are up to about 1 inch in diameter, and there are a few wedges and lenses of sandstone up to about 1 ft thick. The rock appears to die out rapidly towards the south-east.

Apart from the stream sections already mentioned exposures are few on the high ground between Hayton's Bent and Witchcot. Occasional old quarries, patches of rock debris, and small topographic features facilitate the understanding of the structure and succession of the rocks. In the old quarries [530820] about 350 yd north-west of Witchcot about 2 ft of grey calcareous flags overlie about 4 ft 6 in of grey cornstone-conglomerates, higher and lower beds being unexposed. The conglomerate yielded fragments of fish. A number of small lime kilns occur nearby. A grey cornstone-conglomerate at about the same horizon has been worked beside a barn [533827] 980 yd N. of Witchcot but no other good exposures are seen on Sutton Hill.

The streams which unite north of Hopton Hall [542812] expose successively younger beds as they are followed downstream, but the attitude of the beds is often such that one group of them may have a very extended outcrop in the banks. The beds are mainly purple marls and flags, cornstone-conglomerates being uncommon although they are often exposed in old quarries between the streams. Grey calcareous flags are widely exposed on the extensive dip slope at Witchcot and in the dingle to the north-east. Cornstone-conglomerates are exposed near the junctions of the streams and on the higher ground to the west and north. In an old quarry [537813] 600 yd N. 40° W. of Hopton Hall and in adjacent workings wedges of current-bedded sandstone and conglomerate are closely interbedded. The purple cornstone-conglomerate and flags which crop out at Lesser Poston and on the ridge to the south-east are at the same general horizon. In the old quarry [540821] about 130 yd north-east of the buildings 8 ft of current-bedded conglomerate are exposed, with several wedges of calcareous sandstone up to about 1 ft thick (Plate 11B).

Predominantly purple calcareous flags and cornstone-conglomerates are commonly exposed in Hopton Brook below Hopton Cangeford. The lower beds dip gently to the north-north-west, the upper, near the village, at even lower angles to the north-north-east. Five bands of nodular cornstone, up to 5 ft thick, crop out in the Brook or its tributaries between 315 yd S. 25° W. of the church [547801] and 420 yd E. 5° S. of Hopton Hall [545808]. A higher marly band is exposed at the roadside [549804] 60 yd east of the church and the highest in this area forms an outlier on the hill north of Meesons. It is best exposed in an old quarry [542798] 450 yd N. 9° E. of Meesons where 3 ft 6 in of grey and purple nodular cornstone underlue cornstone-conglomerate and flags, the beds being approximately horizontal. This bed is at about the horizon of the cornstone-conglomerates of Lesser Poston, the other cornstones being somewhat below it. Higher beds of mainly purple flags and conglomerates are seen at many points south-west of Greater Poston, near the farms south-west and south of Poston Coppice, and near Greater Poston.

Purple flags and cornstone-conglomerates, dipping at 15° to 30° to the north-north-west, are exposed at many points along the features which they form between the Hopton Brook, south of Hopton Cangeford, Wain Hill, and Bank House [5881]. The lower beds, mainly marls and sandstones, are well exposed in and near the stream north-west of Upper Ledwyche, in the Ledwyche Brook, and between the Brook and The Moor. From Stoke Lodge to south of Bank House the Newton Dingle follows approximately the outcrop of a mainly marly group of beds. Six inches of cornstone are seen at the foot of a small outcrop [578815] 1250 yd above the Stoke Lodge road bridge and again, apparently at a lower horizon, 1770 yd above the bridge [583815].

The beds between, and those in the Ledwyche Brook north of Stoke Lodge, are predominantly arenaceous with several bands of cornstone-conglomerate. The higher beds, seen in the Ledwyche Brook, are extensively quarried all along their outcrop between Hopton Brook and Bank House. Interbedded purple and grey sandstones and cornstone-conglomerates are particularly well exposed in old quarries 550 yd N. 19° E. [570812] and 740 yd N. 26° E. [571814] of Stoke Lodge, and again up to 250 yd south and west of Bank House.

Purple sandstones and conglomerates low in the Series are well exposed south and south-east of Cleedownton. Part of a head shield of a fish was obtained from flags at the roadside opposite the milestone [580808]. The stream from Bromdon exposes successively older rocks as it flows westwards. Purple marls appear to constitute about 75 per cent of the succession, purple and grey sandstone 20 per cent, and cornstone-conglomerate 5 per cent. The beds dip at between 6° and 14° to the east and south-south-east, lying on the south-eastern limb of the Ledwyche Anticline (Fig. 2), and are sparsely exposed on the ridge east of Gibbridge. South of Langley exposures are generally small and isolated, especially on the open slopes of Titterstone Clee Hill. Calcareous flags and cornstone-conglomerate, dipping at 11° to S. 40° E., are seen at a locality [591784] 1030 yd S. 26° W. of Callowgate, partly enclosing what appears to be a small swallow hole.

The rocks are very well exposed in the Newton Dingle and its tributaries above the confluence south of Bank House. In general terms the younger rocks occur upstream, the dips being low towards the north-west. In the more south-easterly outcrops, north-eastward and south-eastward dips occur in places, under the influence of the Ledwyche Anticline. A vertical fault striking N. 25° E. cuts the rocks in the stream [585814] 890 yd W. 20° N. of Starvecrow, separating beds dipping at 20° to the east-north-east from others dipping at up to 40° towards W. 20° N. Intersecting faults, with strikes of E. 10° N. and E. 20° S., are seen 70 yd upstream. Grey calcareous flags and cornstone-conglomerate, with a very low easterly dip, and seen in the stream to be at least 17 ft thick, have been worked in the covert south-east of Starvecrow.

An instance of the lateral variability of the Ditton Series is seen in the stream [584817] 690 yd S. 44° W. of Newton where a wedge of marl varies from zero to 3 ft in the width of the stream. Faulting on an east-north-easterly line may explain an abrupt change of dip 300 yd E. 16° S. of Newton [592820], from high angles to the north-north-west to a low angle to the south-east. Fragments of fish were obtained from a 2-ft band of cornstone-conglomerate here and a band of cornstone, 2 ft thick, caps a waterfall about 20 yd upstream. Flags and cornstone-conglomerate, 20 ft thick, are seen in the stream [599822] 250 yd N. 22° E. of Coldgreen and a group of purple and grey calcareous flags, at least 16 ft thick, is well exposed in the lane [599826] and in the streams just above it some 200 yd north of The Gore. Two feet of purple cornstone-conglomerate occur near the base. This type of rock is absent from the succeeding 150 ft of flags and marls which crop out upstream to the road. A good exposure of flags and marl, about 40 ft thick, is seen in a sunken lane [591823] about 300 yd north-east of Newton. North of the road at Lower Bush the beds below the Lower Abdon Limestone are seen to be mainly sandstones, commonly coarser in grain than the normal Ditton sandstone.

Nodular cornstones occur at several horizons in the higher Ditton rocks of Stoke St. Milborough and Weston Hill, at a level rather higher than those at Hopton Cangeford. The lowest is exposed as 3 ft of marl, packed with nodules, in the stream [566821] 180 yd S. 29° W. of Stoke St. Milborough church, possibly at the roadside [568823] 170 yd E. of the church as 9 inches of fine-grained grey limestone, and again in the brook [574826] west of New House. If correlations are correct the beds between this horizon and the Lower Abdon Limestone are 50–100 ft thick here, in contrast to about 300 ft on the north side of The Thrift [5683]. Purple and grey nodular cornstone occurs in an old quarry [562816] 60 yd N.E. of Furlong, and a higher band in the lane [564820] 410 yd S. 37° W. of the church. Two higher beds of grey massive limestone appear to have been quarried on the western face of Weston Hill, and the highest of all, possibly an outlier of the Lower Abdon Limestone (Ball, Dineley and White, 1961, p. 196), has been quarried some 75 yd N. and 200 yd E. of the triangulation point.

The calcareous sandstones and marls which make up the bulk of the higher Ditton succession can be followed by feature between the outcrops in Bockleton Brook and the streams at Stoke St. Milborough and Poston Coppice. Purple cornstone-conglomerate is well exposed in these beds at Greater Poston but only at lower horizons at Cold Weston and below The Thrift. They occur however on or near the crest of The Thrift, down to about 200 ft below the Lower Abdon Limestone.

In the Clee Brook and its tributaries grey calcareous sandstones and cornstone-conglomerates are well exposed up to 720 yd below Clee St. Margaret church [558843]. Fossiliferous sandstones occur in the tributary 620 yd N. 40° W. [561848], and 720 yd N. 20° W. [562850], of the church. The former has yielded *Pteraspis sp.* The higher beds include many bands of cornstone-conglomerate, usually purple or brown, but are well exposed only near the village. The general dip in this area is south-eastwards at about 10°.

Heath, Abdon, Ditton Priors, and Neenton. The beds which succeed the '*Psammosteus*' Limestones continue to form a prominent escarpment north-east of Bouldon. In the stream ½ mile south of Tugford church, Kidnall Gutter, the basal 290 ft of the Ditton Series are well exposed. Reddish brown massive and flaggy sandstones are fairly common in the mainly marly sequence, and there are a number of grey beds and beds of cornstone-conglomerate. Fish fragments occur at several horizons, notably in a conglomerate about 60 ft from the top of the section (probably locality No. 48 of Ball, Dineley and White 1961) and plants and *Pachytheca* were observed in a grey marl about 90 ft lower. Fragments of *Pteraspis sp.* were obtained from a 5-ft band of grey cornstone-conglomerate, from a slightly higher horizon, in an old quarry [555859] 380 yd N. 42° W. of Heath church (Ball, Dineley and White 1961, locality 34, probably the same bed).

The brook at Tugford, like the Clee Brook at Bouldon, makes a major breach in the escarpment. Approximately the basal 240 ft of the series are intermittently exposed up to the bridge [568867] 1170 yd E. 17° S. of the church, the beds seen being largely red or grey sandstone, with two bands of cornstone-conglomerate. Fish fragments were obtained from 2 ft 6 in of grey calcareous current-bedded sandstone 100 yd north of the bridge. An Acanthodian symphysial tooth, *?Pteraspis rostrata* (J. R. L. Agassiz) [cf. Tarngrove form] and *Pteraspis sp.* were obtained from a 15-in band of grey pellety conglomerate in the stream [565871] 870 yd S. 6° W. of Lower Earnstrey Park, and '*Onchus*', from the base of a 7-ft sandstone, possibly the same horizon, 220 yd upstream. At this locality *Modiolopsis complanata* and *Modiolopsis sp.* were found about 2 ft below. This is the highest known horizon in the Old Red Sandstone of the Clee Hills from which lamellibranchs have been found. (Ball, Dineley and White 1961, p. 204; Eyles 1953, p. 26). A slightly lower band of cornstone-conglomerate, 150 yd upstream from the first locality, yielded fish fragments. One of these horizons may be the correlative of a conglomerate in the main stream [567870] 270 yd below the bridge. Upstream to Lower Earnstrey Park the stream follows approximately the strike of the beds, which are mainly red marls and sandstones but include a variable band of cornstone-conglomerate up to 4 ft 6 in thick. Younger beds crop out in the higher part of the stream and in the two eastern tributaries. They are similar to those below, containing in addition one or two bands of nodular cornstone. Fragments of *Pteraspis sp.* were obtained from calcareous sandstone 1110 yd W. 37° S. of New Earnstrey Park [570872], and other fossiliferous beds, with fish fragments, are located 240 yd S. 16° E. [579876], 580 yd W. 33° N. [574881], and 670 yd N. 38° E. [583883] of that farm.

The scarp-forming beds are not well exposed between Earnstrey Hall and The Leath, but appear to be more often grey than in the Tugford area. Purple sandstone and cornstone-conglomerate have been quarried [585895], just west of Ashfield, and apparently marl, from a similar horizon, some 350 yd north of this farm. Current-bedded, flaggy, brown sandstones are exposed in places in the steep road above The Leath, and have been quarried at intervals on the scarp to the north. Good sections in the beds above the '*Psammosteus*' Limestones are seen in the streams some 750 yd south-east, and 1000 yd east-south-east, of Little Oxenbold (Fig. 17). The beds are mainly purple but the highest seen are grey. In an old quarry [598917] 1200 yd E. 10° N. of Little Oxenbold the steep escarpment is seen to be capped by a 10-ft band of fine-grained brown sandstone. At least 6 ft of a rather higher sandstone have been quarried [608925] 270 yd N. 10° W. of Upper Netchwood chapel.

Higher beds of the Ditton Series, flaggy sandstones and some cornstone-conglomerate, are exposed in several places at Heath, but in general they are poorly exposed between the Clee Brook and the brook west of Abdon. In the brook between Abdon and the bridge ½ mile west of the church, beds of purple sandstone and cornstone-conglomerate, up to 5 ft thick, are often seen in a mainly marly succession, but outcrops are rare in the stream flowing north-west from Cockshutford. Fragments of fish were obtained from a 1-ft band of purple conglomerate 910 yd W. 13° S. of Abdon church [567864]. Roadside exposures are also common in the vicinity of Abdon. Some 650 yd S. 30° E. of the church [578861] the following section was observed:

	ft	in
Grey and reddish brown flaggy sandstone; some hard calcareous bands, especially near base .. .. c.	20	0
Grey-brown, cornstone-conglomerate, gritty	2	0
Buff marly sandstone with carbonaceous spines and flakes near base and some marl pellets	2	6
Grey marl		6

Between the escarpment, west of Ashfield, and Ditton Priors the ground is generally featureless, and the rocks are seldom exposed except in the upper waters of the Rea Brook. From Powkesmoor Coppice to the bend below Ruthall the Brook follows approximately the strike of the beds and cuts down gradually into lower beds in that direction. Between the Coppice and the bridge 300 yd below [591892] fish fragments were obtained from three outcrops of calcareous sandstone and cornstone-conglomerate. A band of brown current-bedded sandstone, 7 to 10 ft thick, forms a waterfall [597898] 450 yd above the confluence of streams at Ruthall. Near the top is a lenticular pellety band crowded with fish remains, which also occur less abundantly in pellety bands elsewhere in the sandstone The fauna includes *Cephalaspis sp.*, *Pteraspis* (*Belgicaspis*) *crouchi* Lankester and *Weigeltaspis sp.* indicative of the Zone of *P.* (*B.*) *crouchi*. The sandstone is exposed to 12 ft in an old quarry about 150 yd to the north-east. The lower beds, down to the confluence, include two bands of nodular cornstone about 1 ft thick. Flaggy sandstone in the stream 350 yd west of the confluence dips at a high angle. The general dip of the beds in this area is towards the south-east or south-south-east at between 7° and 10°. Downstream to Derrington and in the tributary from the north-west the beds dip more towards the south, and, north-west of Derrington, at very low angles.

Fragments of fish were obtained from a band of cornstone-conglomerate in the Rea Brook [605901] 120 yd N. 4° E. of Lower Ruthall. This bed is up to 2 ft 6 in thick but is locally attenuated by the down-cutting of the overlying current-bedded flags. Grey sandstone with bands of cornstone-conglomerate is widely exposed at Derrington. In the streams to the west and north purple marl and grey and purple flags are intermittently exposed. Just above the junction of this stream with the Rea Brook a band of nodular cornstone is exposed, up to 4 ft thick. In this part of the stream the dip reached 30°, and a minor east–west anticline was observed. A 9-in lenticle of cornstone-conglomerate, in the brook [611905] 60 yd below the bridge at Derrington, yielded fragments of *Pteraspis* (*Pteraspis*) *dairydinglensis* White. *Pteraspis sp.* was collected from a cornstone-conglomerate, probably the same bed, 20 yd farther downstream. A 1-ft band of cornstone lies below the conglomerate at the first mentioned locality, and an 18-in band, possibly at the same horizon, is seen a farther 50 yd downstream. Cornstone is exposed at several other localities in the stream nearby.

Southerly dips prevail in the well exposed beds, often grey, down to Rea Bridge, but from there to below Birches Farm, and in the stream from Middleton Priors, the direction of dip is very variable. In particular the more south-westerly outcrops in the brook often show north-eastward dips, opposed to the more usual direction, but southward and south-eastward dips are also common away from the main valley. Highly localized minor folds occur at a few places. Nine hundred yards upstream from the bridge at Neenton the general direction of dip changes from south-west to north-west as the axial region of the Brown Clee Syncline is crossed. Between Rea Bridge and Neenton the rocks are well exposed, comprising marls and flags with several cornstone-conglomerates, usually reddish brown but locally grey, in beds seldom more than 3 or 4 ft thick. Similar rocks with a general southward dip of about 8° are less well exposed in the tributary from Ditton Priors. It is from this village that the Ditton Series takes its name, but the rocks are not well exposed in its immediate vicinity. A band of limestone, formerly quarried [609891] some 100–150 yd E.S.E. of the church, is exposed to about 4 ft. It is a grey and reddish brown mottled rock, with nodules from 3 to 12 in across, and locally has the appearance of a conglomerate with rounded pebbles of grey limestone in a grey sandy matrix. Debris of the same rock is present in an old working [618889] 1150 yd E. 17° S. of the church. A 3-ft band exposed at a pond [620875] 250 yd S. 31° E. of Great Leasowes may be the same bed.

The solid rock is seen at a number of small scattered outcrops between Ditton Priors and Neenton. About 7 ft of sandstone, dipping at 7° to the south-east, are exposed in an old quarry [629888] 580 yd S. 34° E. of Birches Farm, the lower half being massive and locally pellety. Two bands of cornstone-conglomerate, 6 ft and 2 ft 6 in thick, have been quarried 1150 yd N. 32° E. [629879], and some 1400 yd N. 40° E. [632880] of Cleobury North church. These beds dip at 15° and 10° to the north-west.

Marls, sandstones, and cornstone-conglomerates are well exposed in the stream which flows south-eastwards from Netchwood Common and there are several outcrops in the roads and tracks around Middleton Baggot. A nodular limestone, 18 in thick, is seen in the stream [620911] 800 yd above the Middleton Priors bridge and traces of a similar rock occur [627908] 480 yd N. 41° E. of the bridge. Over 5 ft of sandstone are seen in an old quarry [629898] 750 yd N. 28° E. of Birches Farm. The stream south-west of New Farm cuts only gradually across the succession of southward-dipping beds. Fragments of fish were obtained from the upper part of a 4-ft sandstone 580 yd E. 17° S. of Birches Farm [631891]. A band of cornstone-conglomerate has been extensively quarried 500 yd S. 39° W. [634894], and immediately north-east [638898], of New Farm. At the former locality the rock, 2 ft thick, yielded fish fragments. Ten feet of 'cornstone' were recorded at the top of the borehole 50 yd south-east of the quarries at the farm. In the stream [636892] 420 yd W. 31° N. of Park Farm is a 2-ft band of nodular cornstone. A similar rock 130 yd upstream may be at the same horizon, as the descending stream cuts very gradually into younger rocks. Fragments of cornstone occur also in an old quarry [640892] 250 yd N. 7° E. of Park Farm where 6 ft of purple cornstone-conglomerate with scattered fish fragments are exposed. Exposures are good in the upper part of the stream south of Park Farm. Fragments of fish were obtained from a 2-ft band of calcareous pellety sandstone, [641886] 420 yd S. 12° E. of the farm. Cornstone-conglomerate forms two prominent features just south-east of here, exposed to 6 ft at one locality. Conglomerate is also well exposed at many points around New House Farm [639881].

Monkhopton, Upton Cressett, and Chetton. Cornstone-conglomerate is exposed about 50 ft above the '*Psammosteus*' Limestones, 1150 yd S. 39° W. of Ash Bridge [660935]. This section is quoted by White (1950b, pp. 74–5) as a locality for *Pteraspis* (*Simopteraspis*) *leathensis* White. Sandstone is exposed in a few places on the escarpment between this point and the limit of the Church Stretton district. D.C.G.

In the stream [656928] 100 yd south-south-west of Upper House 3 ft of rubbly limestone are exposed, and grey-green limestone is also seen 50 yd upstream. R.W.E.

Cornstone, with an anomalous south-eastward dip of 12°, is exposed [666930] 1200 yd E. 31° N. of Upton Cressett church. A slightly higher band has been mapped from an old working [663931] 1000 yd E. 41° N. of the church to another [667922] 200 yd S. of Meadowley, and for a farther 1400 yd into the adjacent district. In the brooks below Upton Cressett and east of Upton Park outcrops are not very numerous, the beds dipping at low angles in all directions. They are rather more common in the brook north of Chetton, dipping generally to the south-east or east, but locally to the south-west or west. Three cornstones were recorded, at localities 490 yd N. 14° W., 450 yd N., and 650 yd E. 35° N. of the church. Fragments of fish were obtained from depressions in the base of the third cornstone. Exposures are common in the stream south of Chetton, the strike of the beds being again very variable and the dips ranging up to 35°. A few exposures of marl and sandstone are seen in the streams east of Faintree Hall, but in general the ground between Chetton and Overton is poorly exposed.

Reddish brown sandstones and cornstone-conglomerates, dipping variably, but generally to the south, are commonly exposed in the headwaters of the Winterburn Brook, west and south-west of Upper House Farm [6590]. Exposures are also abundant downstream to the bridge north of Lower Faintree. The individual beds seldom exceed 2 ft, and marl and marly flags are the most common lithologies with subsidiary hard sandstones and cornstone-conglomerates. The strike is variable, and the dip, usually about 10°, locally reaches 25°. Low northward dips prevail near the bridge. Sandstone and cornstone-conglomerate are exposed at The Hay and at the buildings [644896] 300 yd to the south-west.

Neenton, Aston Botterell, and Prescott. Reddish brown sandstone is exposed in some deeper road cuttings at Cleobury North and in places in the streams south-west and south-east of the village. In the Cleobury Brook and on its left bank, between Neenton and Charlcotte, there are a few outcrops of westward-dipping flags and cornstone-conglomerate. Such beds appear to be thicker, and are much better exposed, at Charlcotte and Wrickton and in the right bank tributary at Charlcotte, dipping generally to the north-west at about 10°. Plant remains occur in 10 ft of buff sandstone 130 yd N. 40° W. of Wrickton Hall [641857]. Extensive old workings [636856] some 650 yd W. of the Hall show 5 ft of reddish brown conglomerate with sandstone bands. In another working [650856] at The Thorns conglomerate dips at 10° to E. 25° S., on the south-eastern limb of the Ledwyche Anticline. The solid rock is very seldom seen between Wrickton and the Rea Brook, or on the rounded ridge between the latter and the Moor Brook.

Flaggy sandstone and marl are exposed at the roadside [617849] west of Burwarton church and there are a few outcrops in deeply cut streams nearby. Flags and cornstone-conglomerate occur in the stream [630856] and railway cutting [631859] north-east of Chatmore, dipping to the north-west at up to 35°. Exposures, nearly all of sandstone, are common in the Moor Brook [634846] west of Cockshall Cottage, but are scarce downstream to the Rea Brook or in the tributary south of Aston Botterell church. Dips in this area are low to the east-south-east.

Alluvium obscures the solid rock in the Rea Brook below Neenton except at a very few localities. Fossiliferous cornstone-conglomerate crops out 1080 yd E. 31° N. of Moorbrook [654842], and again 80 yd downstream. Exposures are abundant in the left bank tributary, 'Dairy Dingle' of Ball, Dineley and White (1961, map), between the road west of Walkhamwood Farm and 800 yd north of Walkerslow. The beds, which dip north-westwards at about 10°, are mainly marls and marly flags, but some hard sandstones, up to 5 ft thick, and cornstone-conglomerates are also developed. A fauna including Arthrodire plates and *Pteraspis sp.* was obtained from a 2-ft reddish brown conglomerate, [650876] (Ball and Dineley 1961, locality 20) 1420 yd E. 5° S. of Neenton church. Southward and south-westward dips are locally developed between 400 and 650 yd downstream but at Walkerslow, near the crest of the Ledwyche Anticline, the dips are westward.

Exposures are scarce between 'Dairy Dingle' and the Winterburn Brook but cornstone-conglomerate and sandstone are well exposed at Lower Faintree [657883] and The Heath [658860]. A specimen of a ?Cephalaspid was obtained from the conglomerate at Lower Faintree. In the Winterburn Brook outcrops are abundant from 600 yd above Winterburn Bridge to east of The Heath. The beds usually dip to east or south-east, but westward dips and local structural complications are common. Thus, just below the Bridge dips of 70° to the east and 8° to the west are recorded within a few yards, and in a tributary [663871] 470 yd E. 38° S. of the Bridge, where the dip is generally gently southwards, an inclination of 60° to the north-east was observed. The rocks are generally red marls and flaggy sandstone, and there are a few bands of nodular marl and cornstone. The most southerly outcrop in the Brook [661861] is of sandstone, vertical on a north-easterly strike. There are very few exposures between Faintree Hall and Overton, but in an old quarry [665869] (Ball, Dineley and White 1961, locality 108) some 820 yd E. 42° S. of Winterburn Bridge a rubbly cornstone is exposed.

Brown and greenish grey sandstones and cornstone-conglomerates are exposed in a few small old quarries, and in a stream, between Wallfurlong and Duddlewick. Cornstone 1 ft thick has been quarried [664842] 950 yd S. of Wallfurlong. Debris of similar rock possibly at the same horizon, can be followed for about ¾ mile from 600 yd E. of Duddlewick [662834] to 80 yd W. of Hardwick [665823]. Sandstone and marl are well exposed in the lane [656833] below Duddlewick. At the base of the section is a 1-ft band of purple cornstone-conglomerate with many small fish fragments. Exposures are numerous in the stream east of Duddlewick as far downstream as the railway. The beds dip generally at about 10° to between north-east and south-east. Greenish sandstones, dipping at 15° to 20° to the south-east, crop out in the stream [667817] some 600 yd south of Hardwick, and cornstone-conglomerate has been quarried beside the track [664814] 490 yd S. 22° E. of New House Farm.

Farlow, Silvington, Wheathill, and Loughton. Between Prescott and Farlow, south of the Farlow Brook, the rocks of the Ditton Series are seen in only a few small ditch exposures. Three feet of cornstone-conglomerate, overlain by purple flags, crop out at the entrance to Well Farm [642811], and 4 ft of conglomerate, with 3 ft of flags above, are exposed at the roadside some 200 yd to the north-west. These beds have a low variable dip and the relation between the two outcrops is uncertain. The marls and sandstones at the roadside [641808] some 300 yd N.E. of Farlow church, dipping at 21° to S. 15° W. are at a higher horizon. The Ditton rocks are poorly exposed south of the Brook between Farlow and Cleeton St. Mary but there are a few outcrops and old quarries near Silvington.

Purple sandstone and cornstone-conglomerate, with low variable dips, are exposed at only a few points in the Farlow Brook below the bridge north-west of Well Farm. The best exposures are N. of Upper House and just below the bridge. At 170 yd above the bridge [639812] 3 ft of calcite-veined cornstone-conglomerate dips steeply to the east-north-east, and vertical faces trending E. 30° S. carry horizontal slickensides. A fault striking N. 20° E. crops out 460 yd above the bridge [637810], immediately west of an 8-ft bed of sandstone dipping at about 70° to W. 15° N. Between 140 and 750 yd upstream a succession of steep dips to the south-east, up to nearly 90°, and west-north-westward dips of up to 37°, marks the south-south-westward course of a sharp, possibly faulted, anticlinal wrinkle in the south-eastward dipping beds. The rocks involved are grey and purple flags and marl, locally with many cornstone nodules.

In the Brook [631805] above Lowe Farm purple and grey flags and cornstone-conglomerate are very well exposed for about 350 yd, with little change of horizon. The dip here is low to the south-east and this attitude is in general maintained upstream. Similar beds, at lower horizons, are seen in the Brook near Silvington, but exposures become less common as it is followed towards Cleeton St. Mary. Above a locality [611792] about 600 yd north of Cleeton St. Mary church the stream is deeply incised, and purple and grey calcareous flags and marls are widely exposed up to 400 yd W. 29° S. of the church [607784]. Except near the road bridge [609787], where steep east-north-eastward dips probably indicate the proximity of a fault trending north-north-west, the beds dip gently to the south-south-east. Only a few old workings and small exposures occur on the slopes between Cleeton St. Mary, Cleeton Court, and Callowgate.

Two sharp anticlines are present near Goldthorn. About 80 yd S. 11° W. of the farm [603797] the dip of the sandstones in the stream changes from 53° to N. 25° W. to 65° to S. 35° E., within a cross-strike distance of 7 ft. The anticline is thought to be fractured here and all along its crest, which can be followed towards E. 15° N. to a locality [607797] some 450 yd below Goldthorn. A similar structure, trending N. 25° E., occurs about 300 yd downstream and in a nearby ditch [610800] 1250 yd W. 8° N. of Silvington church, the dips varying in 10 yd between 44° to the north-west and very steep to the south-east. These structures, which may be one, appear to be similar to the faulted anticline above the bridge north of Farlow. In the streams above Goldthorn, and downstream towards Silvington, the rocks are only locally exposed, mainly sandstones and marls. The dip is variable, but in general younger beds occur downstream.

The ground between the Farlow and Batch brooks consists essentially of a scarp and dip slope, the several beds rising to the west on the north-facing scarp, and crossing the crest in slowly descending succession as it is followed from Farlow to Bromdon. Much of the exposed rock, which has often been quarried, is cornstone-conglomerate, dipping generally to the south-east at about 10°. The dip slope is well marked between The Lowe and the road north of Silvington and is littered with debris of flags and conglomerate. The rock is exposed to 11 ft in an old quarry [627807] 570 yd S. 21° W. of Ingardine. Between Silvington and Bromdon the detailed stratigraphy is obscure, but in general the younger beds succeed towards the south and east. This trend is visible in a ditch exposure of buff flags and marl [611802] some 1200 yd west-north-west of Silvington church, and in the way in which the features run south-westwards across the road on the ridge. In one of the many old quarries [615802] 750 yd W. 33° N. of the church, 13 ft 6 in of current-bedded purple cornstone-conglomerate and thin flags are exposed. The beds are disturbed and are traversed by a zone of crushed rock, about 1 ft wide, with a strike of E. 30° N. and a steep hade to the north-west. Slickensides pitching about 8° to the north-east are seen on the fault plane. Exposures are rare on the high ground north-west of Goldthorn and on the col at Bromdon.

In the Batch Brook there is little change of horizon up to the plantation [621806] north of Silvington, a low dip to the south-east bringing in slightly older beds upstream. Specimens of *Kujdanowiaspis* cf. *anglica* (Traquair), *Pteraspis sp.* and *Protaspis* (*Europrotaspis*) *crenulata* White were obtained from a 2-ft band of cornstone-conglomerate 600 yd S. 37° W. of Ingardine [626808]. Massive and flaggy sandstones and cornstone-conglomerates are well exposed in the brook, seldom more than 5 ft thick. Between the plantation and the confluence [608807] south of Coveridge Fields the attitude on the beds is very variable and there is evidence at several localities of faulting of lines between east and south-south-east. The general dip is probably at a low angle to the south-east, but north-north-easterly dips of 10° to 17° prevail for about 350 yd, south of Bradley's Farm [614807]. Above the confluence the sandstones and conglomerates are usually grey and dip generally southwards at low angles. A group of flags and sandstone about 30 ft thick crops out in the stream [606811] west of Coveridge Fields.

There are few exposures between the Batch and Ingardine brooks but topographic features and some outcrops in minor streams indicate that the younger rocks lie towards the south. North and west of Bradley's Farm the rocks dip almost imperceptibly to the south-east and the younger rocks occupy the higher ground, towards the west, where grey cornstone-conglomerate has been extensively quarried for lime some 450 yd N. 36° W. of Coveridge Fields [606814].

Exposures are few in the lowest part of the Ingardine Brook and the southward-dipping beds are affected at one point by faults, as in the Farlow Brook some 200 yd to the south. Where the Brook is joined by the Wheathill Brook the beds for a few hundred yards are thrown into a syncline, whose axis plunges at perhaps 15° to the west-south-west. These well exposed rocks are mainly flags and cornstone-conglomerate, in beds up to 14 ft thick, but 220 yd above the confluence [633814] is a grey nodular limestone, apparently about 2 ft thick. Grey and purple, marly, nodular limestone, up to 4 ft thick, occurs in the brook [626815] 400 yd W. 35° N. of Ingardine nad elsewhere nearby in the brook and, as soil debris, in the fields just to the north. A 2-ft band at about the same horizon was seen in a stream [612816] 1185 yd W. 30° S. of Wheathill church. This is a grey nodular cornstone, lustre-mottled, and apparently more siliceous, with wavy bedding, in the basal few inches. From Ingardine to about 750 yd upstream thin beds of marl, flags, and cornstone-conglomerate are well exposed. From a 3-ft band of conglomerate, 550 yd W. 25° N. of Ingardine [624815] (near locality 53 of Ball and Dineley 1961), were obtained fragments doubtfully referred to the Pteraspid genera *Althaspis* and *Cymripteraspis*. A farther 550 yd upstream fish fragments were obtained from a similar bed at about the same horizon. These beds in general dip at about 7° to the south-east but a highly localized dip of 40° to W. 10° S. occurs at the ford [618814] 900 yd S. 28° W. of Wheathill church. Some large outcrops of sandstone and marl are present in the Brook north and north-west of Bradley's Farm and in a northern tributary [611815] south and south-west of The Green where cornstone-conglomerates are also common.

Cornstone-conglomerate, exposed to 6 ft in an old working [618818] 200 yd W. of Doctor's Cottages, forms a prominent scarp and dip slope across the lane. A number of old quarries lie on the slopes west and south-west of The Green. In one [608819], 1600 yd W. 9° S. of Wheathill church, cornstone-conglomerate and sandstone are exposed to about 20 ft. A rich fauna, indicative of the Zone of *Pteraspis* (*Cymripteraspis*) *leachi* White, was recorded from this locality by Ball, Dineley and White (1961, pp. 203, 220). The high ground of this area and the escarpment between Besom Farm and the Burwarton road are formed by grey sandstones and conglomerates at about this horizon, perhaps preserved as an outlier amongst older rocks. The conglomerates have been quarried at each end of the escarpment, [598814] and [606825], and locally elsewhere.

In the lower part of the Wheathill Brook the attitude of the beds is very variable but low dips to the south-south-west are prevalent except near the Ingardine Brook where the strike swings clockwise and the beds ultimately dip north-north-westwards at 43°. In the south bank tributary, Green Dingle, successively older rocks crop out upstream to Wheathill, but the change in horizon is gradual, the dip being low to the east-south-east. Purple and grey flags are the most common rocks but no thick beds are seen. Above Wheathill, where flags and cornstone-conglomerate are well exposed in the Dingle, the dip is south-westward for some 350 yd before reverting to the south-east. Fragments of fish were obtained from a 1-ft conglomerate 540 yd W. 15° S. of the church [617820]. A higher conglomerate, 2 to 4 ft thick, crops out in places between 60 and 530 yd upstream. Higher beds, which are well exposed to about 1700 yd west of the church [607822], dip generally at a low angle to the south-west.

One of the larger quarries in the Ditton Series rocks [620821] lies some 200 yd W.S.W. of Wheathill church. About 18 ft of beds are exposed, principally purple and greenish grey calcareous flags, with 2 ft 8 in of massive sandstone near the base and two bands of cornstone-conglomerate of varying thickness up to 3 ft. Other outcrops occur nearby and in the lane [619825] between 450 and 550 yd north-west of the church.

The general southward dip, with frequent variations to south-west and south-east, is maintained in the well exposed rocks in Wheathill Brook. Between the confluence with Green Dingle and the bridge east of Wheathill the rocks are mainly calcareous flags and cornstone-conglomerate, one band of the latter being unusually coarse, with large fragments of purple calcareous sandstone. Upstream the beds consist of purple and grey marls and sandstones, with several thin bands of cornstone-conglomerate up to, exceptionally, 8 ft thick. Marls predominate for about 500 yd, north of Wheathill. The oldest beds in the brook crop out in the northward bend south-east of Loughton, and are also exposed in the lane on the north bank, dipping at 5° to the south. Fragments of *Pteraspis* (*Belgicaspis*) *crouchi* were obtained from a 3-ft band of cornstone-conglomerate 440 yd S. 11° W. of Loughton church [615826]. Outcrops are numerous up to 800 yd W. 29° S. of the church [610826].

Rocks of the Ditton Series are well exposed in the ground north of these brooks, between the Rea Brook and the vicinity of Loughton. Minor streams, old quarries, and well developed topographic features, indicate the nature and sequence of the rocks. The structure is broadly simple, the rocks dipping gently to the south-east, but there are local complications, notably in the stream [647818] 1000 yd east of The Down where steep dips to the north-east and south-west occur in a narrow zone, and the interplay of topography and geological structure, as at The Down, is often difficult to interpret.

From the Farlow Brook, south-east of The Down, the rocks become successively younger towards Hinton. The sharply folded grey sandstone and cornstone-conglomerate in the stream 1000 yd east of The Down are probably cut by a north-westerly fault. Old conglomerate workings follow a strong feature parallel to the stream and about 100 yd north-east of it. Flags and conglomerates crop out in a few places and form some good features near Hinton. South-east of Hinton and near Hardwickforge the beds dip to the east and north-east, rather than to the south-east, and from here to Farlow the dip and strike are variable. Beds rather lower than those at Hinton crop out near The Down and Gobbit. They have been quarried at several points, notably 300 to 400 yd south of The Down where up to 6 ft of cornstone-conglomerate are exposed. A lower group of conglomerate and flags can be followed from the Wheathill Brook some 900 yd east of Wheathill church to the Moor Brook about 300 yd south-east of Moorbrook farm. It is well exposed in an old quarry [635823] 375 yd W. 36° N. of Gobbit, where a fish fragment was obtained from 12 ft of current-bedded purple cornstone-conglomerate and flags, and again in the stream [640824] some 500 yd N. 37° E. of Gobbit where only 1 ft of conglomerate occurs near the base of a thick succession of purple calcareous flags. Yellow and grey cornstone-conglomerate from a slightly lower horizon is exposed to 4 ft in an old quarry [630822] 880 yd E. 4° N. of Wheathill church.

Only a few small exposures are present near Heywood, where marl was apparently worked in the past for bricks, and at Loughton, where sandstone has been quarried. The rock is seen at intervals in the stream between Loughton and the Moor Brook, principally near the latter, at Brook House, and near the road north-east of Loughton. North and west of Loughton are extensive spreads of drift but 8 ft of buff flags are exposed at the roadside [614833] 390 yd N. 35° W. of the church, and yellowish cornstone-conglomerate is exposed to 1 ft in an old working [609835] 950 yd W. 33° N. of the church.

From the limit of the drift some 600 yd north of Old Lodge the trend of the outcrops can be followed by features and occasional old quarries, in a direction just west of south towards the upper reaches of Newton Dingle.

Catherton. Rocks high in the Ditton Series occupy a small area in the south-eastern corner of the district. They are part of the extensive Ditton Series outcrop of the Cleobury Mortimer area and outside the present district are continuous with the main outcrop of Brown Clee Hill. In general the oldest rocks lie to the south-east but the structure is complicated by subsidiary folding, the axes of which are divergent towards the south and form arcs concave towards the east.

The rocks are well exposed in the brook east of Catherton, purple and green marls, sandstones, and cornstone-conglomerates, dipping to the north-east and, downstream, to the north-west at from 20° to 35°. About 2 ft of nodular cornstone are exposed 670 yd E. 3° S. of Catherton [660784] and again in an old working [658785] 450 yd E. 2° N. of the farm. This rock may be one of the Abdon limestones. It occurs at the top of a group of rather coarse-grained sandstones and flags, exposed to 20 ft in the lane [655785] about 100 yd north-east of Catherton and elsewhere near the farm. The beds here dip at up to 32° towards the north-north-west. Higher sandstones, locally coarse and conglomeratic, are present at the sharp bend in the Rea Brook [660787] north-east of Catherton and on the slopes just to the west. They dip to the north-north-east at angles of up to 53°. Lower sandstones and cornstone-conglomerates are well exposed in the Rea Brook down to the foot-bridge, dipping at some 20° to between north-west and west. In the lane and brook [667792] some 1600 to 1700 yd E. 30° N. of Catherton they dip at up to 40° to the north-north-west.

D.C.G.

Abdon Limestones and Associated Beds

The Lower Abdon Limestone crops out at the pool [576832] south-west of Bockleton Court, where a pale grey nodular limestone is exposed to 1 ft 6 in, and the top 3 in of it are seen at a spring [573836] 640 yd W. 33° N. of the farm. Springs occur at four places on the Limestone outcrop between these exposures. The slope east of the road, between here and the crest of The Thrift, is essentially the dip slope of the Limestone and its debris is locally abundant. A 6-in bed of marl with many cornstone nodules, exposed in a pit [567829] 350 yd E. 10° S. of Pel Beggar, may be the Lower Abdon Limestone, possibly forming an outlier separated from the main outcrop to the east.

North-eastwards from The Thrift the Lower Abdon Limestone crops out at Cockshutford, in the brook [581850] 215 yd W. 7° N. of the chapel, where it occurs as a 2-ft band of purple calcareous sandstone, with limestone nodules, and a 2-in lenticle of grey limestone at the base. Exposures are more numerous northwards and old quarries mark the outcrop east of Abdon. In one quarry [582858] 1180 yd S. 39° E. of the church, 5 ft of grey limestone are exposed, locally with a brecciated appearance. The quarries continue north of the fault, east of the church, to about 700 yd S. 15° W. of Powkesmoor Farm [591876]. Limestone debris is abundant but the only exposure occurs [589875] 940 yd S. 21° W. of the farm, where fine-grained cementstone is overlain by a hard fine-grained sandstone. Just south of the road at Hillside chapel, old lime works extend for some 600 yd, but no outcrop of limestone is visible. Limestone debris is present about 70 yd south-east of Hillside Farm [604877] but from there to Loughton Enclosure the course, and even the existence, of the Lower Abdon Limestone are conjectural.

In the stream [600835] 510 yd N. 28° W. of Old Lodge 3 ft of purple and yellow limestone, marly and siliceous, are exposed beneath marl and on top of 15 ft of flags. From its position about 220 ft below the Upper Abdon Limestone this band may well be the Lower. Calcareous nodules in grey flags, exposed in the valley [591830] 800 yd W. of Blackford chapel, are at about the same horizon, and a 10-ft marl band with many cornstone nodules, exposed within grey flags and cornstone-conglomerate 300 yd W. 33° N. of the chapel [596832], appears to be just a little higher. An old working north of the road [586827] at Red Hall is also at about the horizon of the limestone.

Grey and purple flags between the Abdon limestones are exposed at Bockleton Court, and in an old quarry [574839] 1080 yd E. 26° S. of Clee St. Margaret Church where they include a thin pebbly grit and a lenticle of cornstone-conglomerate. Brown and purple grit and sandstone are exposed at intervals on the steep slope between here and Cockshutford. Some 35 ft of grit, locally pebbly, are exposed in an old quarry [576848] 800 yd W. 12° S. of Cockshutford chapel. The uppermost beds here bear the problematical markings referred to by Ball, Dineley and White (1961, pp. 195, 198), known also from the Senni Beds of South Wales. A 3-ft band of limestone, locally sandy and grading down into calcareous grit, is exposed in the stream [583850] 70 yd N. 27° E. of the chapel but cannot be traced for any distance from the outcrop. At about the same horizon, [583859] 1180 yd E. 42° S. of Abdon church, is a 2-ft sandstone with large limestone nodules, overlain by 3 ft of limestone, sandy at the base. Flaggy gritty sandstone crops out at intervals all along these steep slopes but is less well exposed at the northern end of the hill and very poorly exposed on the eastern slopes.

Grey calcareous flags and sandstone are well exposed between the limestones in the stream north-west of Old Lodge, on the slopes of The Toot, and here and there between that cottage and Red Hall.

The Upper Abdon Limestone was seen at the southern end of a line of old quarries [578836], now filled in, 350 yd N. 7° W. of Bockleton Court, 8 ft of massive grey limestone, sandy at the base, being exposed. Six inches of purple limestone, at the foot of an old quarry face about 100 yd N.E. of the farm, are thought to be the top of the same bed. The outcrop is followed by a line of old quarries from near the cottage [576842] 600 yd south of Nordybank to about 300 yd east of the fort. The Limestone is best exposed in the last quarry [579848], some 400 yd W. 29° S. of Cockshutford chapel, where 6 ft of grey and reddish brown limestone form the lowest rock visible. It is variable in character, being massive above and apparently conglomeratic below; very sandy and rubbly varieties also occur. The overlying 3 ft 6 in of marl carry limestone nodules near the base, and are succeeded by 6 in of cornstone-conglomerate. Old workings are common on the valley sides at Cockshutford and round the spur to the north, but the rock is not well exposed in the streams. Small outcrops of limestone are present on the slope east of Abdon, but north of the fault at several points west of Abdon Burf, the rock is well exposed. At 1275 yd [587867] E. 6° N. of Abdon church, 8 ft of limestone, sandy at the base, are exposed, and 930 yd S. 14° W. of Powkesmoor Farm [590874] the Limestone is at least 10 ft thick. Small exposures and old quarries mark the outcrop of the Limestone as far as Park Gate [605873] where it is cut by faults.

Between Park Gate and Burwarton Park the position of the Upper Abdon Limestone is unknown. Grey limestone, with nodular weathering and highly calcareous sandstone is exposed 1005 yd W. 40° N. of Burwarton church [610855], and 2 ft of sandy limestone are present at the northern end of a 150-yd line of old workings

A. Aerial view of unconformity between Old Red Sandstone and Carboniferous, Brown Clee (*Cambridge University Aerial Photograph* No. 0025 EY)

B. Farlow Bank, an escarpment capped by the Carboniferous Limestone (A9509)

Geology of Church Stretton
(*Mem. Geol. Surv.*)

PLATE 13

A. Dry valley on west side of Burrow Hill (A9457)

B. Possible relic of pre-glacial valley bottom, Little Stretton (A9444–9445)

C. Cwmdale, All Stretton, a glacial drainage channel (A9431)

[605847] 1415 yd W. 9° S. of the church. Old workings are very common south-west of this point but the Limestone itself is rarely exposed. In an old quarry [587830] 1210 yd W. 3° S. of Blackford chapel, 5 ft of sandy limestone are visible. The line of quarries ends in a deep dry valley [580829] about 400 yd S. 33° E. of Bockleton Court, between which point and the farm the outcrop, believed to be fault-displaced, is conjectural.

CLEE GROUP

Brown Clee Hill. In the stream north-west of Old Lodge yellow sandstones are poorly exposed above the Upper Abdon Limestone, but green sandstones and pebbly grits are well exposed on the slopes to the west and in the deep valleys [592836] from 200 to 400 yd north-west of The Toot, where they dip to the north-north-west at nearly 20°. Exposures of similar rocks are fewer on the gentler slopes between Shortwood and Bockleton Court, the beds here dipping at 10° to 20° to the north-east. In the stream north-east of Bockleton Court yellow and green grits are well exposed, commonly with quartz pebbles or marl pellets and usually calcareous. A dip of 22° to E. 10° S. was recorded at the highest exposure [587841] 1330 yd E. 44° N. of the farm.

The individual sandstone bands form many prominent features between Bockleton Court and Woodbank. Greenish grey flaggy sandstones and grits, locally pebbly or pellety, are exposed in places, especially in the streams south-east of Cockshutford and at Woodbank. The general dip is south-eastwards at about 10°. The basal beds of the Clee Group are well exposed above the Upper Abdon Limestone in the old quarry [579848] 400 yd W. 29° S. of Cockshutford chapel. The 22 ft of beds visible are mainly flaggy sandstones, brown, yellow, or buff, but include two lenticles of cornstone-conglomerate, up to 1 ft thick, and, at the base, 3 ft 6 in of purple and grey silty marl with cornstone nodules in the lower part.

Large-scale current bedding, or possibly a more widespread unconformity within the group, is shown by the divergence of features at a point [589847] 700 yd E. 30° S. of Cockshutford chapel. The Five Springs [596853] occur on the outcrop of the highest of three marl bands which can be mapped in places on the higher slopes of Clee Liberty and Abdon Liberty. The features on Abdon Liberty display the angular unconformity between the Clee Group and the Coal Measures (Pl. 12A). Twenty five feet of yellow marly sandstone, with gritty bands, are exposed in an old quarry [587861] 925 yd W. 40° S. of the triangulation station, Abdon Burf. The highest beds, exposed in places to the south-east, are generally pebbly. Recorded dips range between 10° and 30° to the south-east. Only a few exposures of sandstone and marl have been recorded on the extensively wooded northern slopes of the hill.

On the western side of Brown Clee Hill the Clee Group is best exposed in the stream west of Burwarton Hall, but brown, flaggy, pebbly sandstone, with some marl pellets up to 4 in across, is exposed to 20 ft in an old quarry [607863] 1350 td N. 37° W. of the Hall. In the stream, up to 1620 yd W. 9° N. of the Hall [599855], brown and green sandstones are well exposed. They are commonly coarse-grained, locally pebbly and locally calcareous. Red marls occur in places, sometimes with calcareous nodules. The beds dip north-westwards at between 10° and 30°, the higher beds being more gently inclined. Bands of cornstone conglomerate up to 2 ft thick are interbedded with pale grey calcareous grit in an old quarry [606854] 870 yd W. 7° N. of the Hall. Yellow sandstone and red marl, just above the Upper Abdon Limestone (above), are seen at a few localities [605847] about 1200 yd south-west of the Hall, but in general the rocks of the group are poorly exposed between the stream west of the hall and that north-west of Old Lodge.

Catherton. In the Catherton area the Clee Group is represented by coarse-grained purple and green sandstones, with some marls, conformably succeeding the finer-grained rocks of the Ditton Series. The rocks are well exposed in the brook south-east of Catherton Marshes and on the ridge between it and Catherton. In the brook they dip steeply with a north-easterly strike and are folded into a sharp syncline between 1000 and 1250 yd north-north-east of Catherton [658794]. In a distance of 1150 yd west of the Rea Brook gritty and pebbly flags in the stream [662797] show a clockwise rotation of dip from northwards to east-north-eastwards, across the crest of an anticline.

In the Rea Brook [662788], 880 yd E. 23° N. of Catherton, coarse pebbly sandstones near the base of the group are present, and 8 ft of strongly current-bedded gritty and pebbly sandstone are exposed in an old quarry [664792] 1350 yd E. 39° N. of Catherton. Similar pebbly sandstone crops out at the roadside [664795] west of Detton Hall just east of the railway, and on the left bank of the brook [665797] for about 200 yd east of the railway bridge. Brown pebbly coarse-grained sandstone, calcareous in part, is exposed 810 yd N. of Detton Hall [667803], nearly vertical with a north-easterly strike. Although mapped as Upper Old Red Sandstone in the adjacent district (Dudley Sheet 167) this sandstone is now, on lithological grounds, assigned to the Clee Group. D.C.G.

UPPER OLD RED SANDSTONE

FARLOW SERIES

Two small outcrops of the Farlow Series occur at Cleeton St. Mary. In the stream [607784], between 420 and 480 yd south-west of the church, are outcrops of coarse- and fine-grained yellow sandstones and a massive yellow and green quartz-conglomerate. They are locally disturbed by the proximity of a north-north-easterly fault, which cuts the Carboniferous rocks upstream. The hamlet stands on the second outcrop, soft yellow sandstones with bands of small quartz pebbles being exposed in the road [610787] some 80 yd N.W. of the church. They dip at 22° to the south, but a short distance to the west the strike is thought to swing counter-clockwise, as it does in the Ditton Series, and the base of the Farlow Series to run southwards under the Carboniferous cover. No exposures were found to substantiate this hypothesis.

Rocks of the Series are not exposed elsewhere near Cleeton St. Mary and are believed to be overstepped by the Carboniferous within a few hundred yards to the north-east. They are next exposed at the foot of the steep slope [639807], 110 yd N. 5° E. of Farlow church, where 4 ft of quartz-conglomerate are overlain by 16 ft of yellow pebbly sandstone. Flags of the Ditton Series are exposed only a few feet below. Yellow quartz-conglomerate is exposed to 10 ft at the roadside [641807] 210 yd E. 28° N. of the church, overlain by 13 ft of yellow pebbly sandstone which becomes less pebbly upwards and is probably altogether about 70 ft thick. It dips at 29° to S. 20° E. Purple and greenish grits, sandstones, and marls, overlying the sandstone, are exposed at the roadside [640806] about 100 yd north-east of the church, and may be about 40 ft thick here.

At Upper House [652809] soft yellow sandstone, about 20 ft thick, is well exposed on the southern side of the road. The rock is current-bedded, fine-grained, and in places coarsely pebbly and, despite the absence of a conglomerate, must be very near the base of the series. These basal yellow sandstones, dipping at 24° to the south-south-east, form cliffs beside the Rea Brook [662810] south-west and west of Prescott and are well exposed at the roadside [665812] between the Brook and about 150 yd east of Prescott.

Green and brown sandstone and conglomerate, locally highly calcareous, are present in a railway cutting [659806] and in the Brook about 650 yd south-west of Prescott. About 200 yd upstream [661808] yellow gritty sandstone, fine-grained and very calcareous at the base, dips at 15° to S. 40° E. These higher beds crop out in the lane [663809] south of Prescott, and consist of calcareous quartz-grits with some pebbles of jasper and green sandstone. In the stream [666805] some 800 yd south-east of Prescott grey and yellow calcareous grits and sandstones crop out, and are thought to belong to this upper part of the Farlow Series. They dip southwards at about 20°. On the northern bank [667806] 640 yd S. 35° E. of Prescott debris of purple cornstone occurs near a small old working. D.C.G.

REFERENCES

ALLEN, J. R. L. 1960. Cornstone. *Geol. Mag.*, **97**, 43–8.

—— 1961. The Highest Lower Old Red Sandstone of Brown Clee Hill, Shropshire. *Proc. Geol. Assoc.*, **72**, 205–19.

—— 1962. Petrology, origin and deposition of the highest Lower Old Red Sandstone of Shropshire, England. *J. Sediment. Petrol.*, **32**, 657–97.

—— and TARLO, L. B. 1963. The Downtonian and Dittonian Facies of the Welsh Borderland. *Geol. Mag.*, **100**, 129–55.

BALL, H. W., DINELEY, D. L. and WHITE, E. I. 1961. The Old Red Sandstone of Brown Clee Hill and the Adjacent Area. *Bull. Brit. Mus. (Nat. Hist.) Geol.*, **5**, 175–310.

CROFT, W. N. 1953. Breconian: a Stage Name of the Old Red Sandstone. *Geol. Mag.*, **90**, 429–32.

CUMMINS, W. A. 1957. The Denbigh Grits; Wenlock Greywackes in Wales. *Geol. Mag.*, **94**, 433–51.

ELLES, GERTRUDE L. and SLATER, IDA L. 1906. The Highest Silurian rocks of the Ludlow District. *Quart. J. Geol. Soc.*, **62**, 195–221.

EYLES, V. A. 1953. District Report: Wales and West Midlands District. *Sum. Prog. Geol. Surv.* for 1952, 23–6.

GREENLY, E. 1919. The Geology of Anglesey. *Mem. Geol. Surv.*

HARVEY, C. O. 1957. In *Sum. Prog. Geol. Surv.* for 1956, 56–7.

HOLLAND, C. H. 1965. The Siluro-Devonian Boundary. *Geol. Mag.*, **102**, 213–21.

KING, W. W. 1921. The Geology of Trimpley. *Trans. Worcs. Nat. Cl.*, **7**, 319–22.

—— 1934. The Downtonian and Dittonian Strata of Great Britain and North-Western Europe. *Quart. J. Geol. Soc.*, **90**, 526–70.

MITCHELL, G. H., POCOCK, R. W. and TAYLOR, J. H. 1962. Geology of the Country around Droitwich, Abberley and Kidderminster. *Mem. Geol. Surv.*

PETTIJOHN, F. J. 1957. *Sedimentary Rocks*. 2nd Edition. New York.

REED, F. R. C. 1934. Downtonian Fossils from the Anglo-Welsh Area. *Quart. J. Geol. Soc.*, **90**, 571–84.

ROBERTSON, T. 1927. The highest Silurian rocks of the Wenlock District. *Sum. Prog. Geol. Surv.* for 1926, 80–97.

TARLO, L. B. 1961. Psammosteids from the Middle and Upper Devonian of Scotland. *Quart. J. Geol. Soc.*, **117**, 193–213.

WALDER, PHOEBE S. 1941. The petrography, origin and conditions of deposition of a Sandstone of Downtonian age. *Proc. Geol. Assoc.*, **52**, 245–56.

WALLIS, F. S. 1928. The Old Red Sandstone of the Bristol District. *Quart. J. Geol. Soc.*, **83** (for 1927), 760–89.

WHITE, E. I. 1946. The genus *Phialaspis* and the "*Psammosteus* Limestones". *Quart. J. Geol. Soc.*, **101**, 207–42.

—— 1950a. The vertebrate faunas of the Lower Old Red Sandstone of the Welsh Borders. *Bull. Brit. Must. (Nat. Hist.) Geol.*, **1**, 51–67.

—— 1950b. *Pteraspis leathensis* White a Dittonian zone-fossil. *Bull. Brit. Mus. (Nat. Hist.) Geol.*, **1**, 69–90.

WHITEHEAD, T. H. and POCOCK, R. W. 1947. Dudley and Bridgnorth. *Mem. Geol. Surv.*

WILLS, L. J. 1935. Rare and new ostracoderm fishes from the Downtonian of Shropshire. *Trans. Roy. Soc. Edin.*, **58**, 427–47.

WOODLAND, A. W. 1938. Petrological Studies in the Harlech Grit Series of Merionethshire. II. The Petrography and Petrology of some of the Grits. *Geol. Mag.*, **75**, 440–54.

Chapter 8

CARBONIFEROUS

GENERAL ACCOUNT

At the end of Lower Old Red Sandstone times changes in the shape of this Salopian part of the old Welsh cuvette can be related to the structural evolution of St. George's Land, which was a land area covering most of Central and North Wales in Carboniferous times. Farther east there appears also to have existed the Midland Barrier land mass (Wills 1948) and during part of Carboniferous time a narrow isthmus may have linked these two land areas. At all events a synclinal gulf from the southern sea apparently extended northward over this Clee–Farlow–Prescott area during Upper Old Red Sandstone and Lower Carboniferous times, for the sediments laid down during this interval have closer affinities with those of the South-West Province (S. Wales, Bristol and Forest of Dean area) than with those of the more northerly province over N. Wales and Derbyshire.

Erosion of the Lower Old Red Sandstone deposits preceded the deposition in this gulf of a thick series of fine-grained yellow sandstones with conglomeratic bands comprising the local Upper Old Red Sandstone, known as the Farlow Sandstone. The relationship of these sediments with the marine Devonian sediments farther south in south and south-west England is uncertain. A revival of movements in the area resulted in a further local influx of conglomeratic material being washed into the gulf, to form the Lower Carboniferous basal conglomerate. This was followed by a sequence of silty shales and silty limestones succeeded by shallow water organic limestone muds and finally calcareous shales with subordinate limestone and chinastone bands; these deposits, exposed in the Farlow–Oreton area, are the most northerly remnant of Tournaisian sediments found in the South-West Carboniferous province (Trueman 1947). The surveyors of the Old Series one-inch geological map 55 N.W. showed all the rocks between the Carboniferous Limestone and the Coal Measures as Millstone Grit. In 1911 Dixon, while using the same designation for these predominantly conglomerates and sandstones, maintained they were conformable to the underlying Carboniferous Limestone but he also considered they were older than the Millstone Grit proper and suggested a non-committal place-name be applied to the formation. By 1917 Dixon had applied the term 'Cornbrook Sandstone' for these rocks which he classified (following Vaughan 1905) as Lower Carboniferous (Upper Avonian). George (1955, pp. 308–9) claimed that the Cornbrook Sandstone overlapped the underlying Lower Carboniferous beds in the Farlow–Oreton area (also confirmed by the writer's six-inch survey in 1945) and 'belongs to the Millstone Grit', but recorded no palaeontological evidence. Jones and Owen (1961) recorded a Westphalian B macroflora from a thin mudstone band about

250 ft below the top of the Cornbrook Sandstone, in the type section in Cornbrook Dingle, on the south side of Titterstone Clee in the Ludlow (181) Sheet. These authors considered the thickness of the Cornbrook Sandstone amounted to no more than about 700 ft against Dixon's figure (1911) of 1000 ft. Sullivan (1964) maintained there was no palaeontological evidence for the age of the basal portions of the formation and it is possible that strata equivalent to the Edgehills Sandstone (Forest of Dean)—basal part of the Westphalian—may be present.

In the south-east corner of the Church Stretton district the Cornbrook Sandstone comprising thick conglomerates, coarse sandstones and shales rests with marked unconformity on Tournaisian beds; from exposures in the adjacent Ludlow (181) Sheet, it has recently yielded plant spores which have been identified by Dr. B. Owens. These indicate its age as being in the main, Millstone Grit Namurian A (E_1) with the highest sandstones being Coal Measures of Westphalian age. The succeeding Coal Measures rest unconformably on the Cornbrook Sandstone. They consist of a thick fine-grained quartzitic sandstone with an overlying sequence of sandstones, shales and coal seams.

Throughout this long interval of post-Lower Old Red Sandstone time the Brown Clee area farther north may have been part of the land barrier referred to above, for no indications of any Middle and Upper Old Red Sandstone deposits, Lower Carboniferous limestones or Cornbrook sandstones have been found overlying the Lower Old Red Sandstone in that area. However, the area was eventually submerged in Coal Measures times and a relatively small sequence of shales and coals was deposited locally, at the same time as the fuller sequence was being formed in the basin to the south.

Rocks of Carboniferous age occur in four areas of this sheet, viz. a faulted area north-east of All Stretton; two small patches in the Shirlett area in the extreme north-east corner of the map; two large areas on Brown Clee; and a considerable area in the south-east corner of the sheet lying to the south and south-west of Farlow, Oreton and Prescott. The latter area is part of a Caledonoid synclinal trough and includes rocks ranging from Lower Carboniferous Limestone up to and including the Coal Measures; in the other areas Coal Measures only are present.

The physiographical conditions of Upper Old Red Sandstone times closed with the deposition, in the early Carboniferous, of the dark reddish brown, coarse conglomeratic sandstone up to 35 ft or more in thickness which lies unconformably on the Farlow Sandstone and precedes the shallow, quiet local conditions of limestone mud deposition which characterize the immediately overlying Lower Carboniferous rocks. This basal conglomeratic sandstone is predominantly composed of subangular to rounded pebbles of white and variegated quartz and quartzite pebbles, occasional jaspers and coarse sand grains. It passes upwards into calcareous strata which crop out over a distance of about 2½ miles from near Silvington eastward through Farlow and Oreton until they are faulted out near Prescott. These strata comprise calcareous shales, impure buff limestones, hard grey crystalline limestones, oolites and crinoidal limestones and are thickest between Farlow and Oreton. Previous estimates of thickness have been given as 40 to 50 ft, though Morris and Roberts (1862) estimated around 80 ft; more recent investigations lead to a figure of 150 ft or more for these beds.

Murchison (1839) first pointed out that the limestones form "a narrow broken zone thinning out at each end"; this has again been emphasized more recently by George (1955). When Murchison visited this area the limestones were being actively worked in numerous quarries and he noted many small faults which affected the outcrop and produced a variety of southerly dips of between 20° and 70°, with a predominant value in the central part of around 34°. Most of the features he observed are now no longer visible on account of the large amount of stone which has been extracted from this district.

A blue shelly oolitic limestone, locally known as 'The Jumbles' or 'Clee Hill Marble' was formerly worked as an ornamental stone and fine examples of it are found in the mansions of Oakley Park, Downton Hall and Hopton House. The higher calcareous shales and limestones up to 80 ft thick near Springvale Cottages, Oreton, are overlain unconformably by the Cornbrook Sandstone. This attenuated sequence of coarse and pebbly sandstones with red and grey clays overlaps part of the underlying limestones towards the east, and completely overlaps these beds and the underlying Farlow Sandstone westward towards the northern slopes of Titterstone Clee Hill. These beds crop out on the hillside south of the village of Cleeton, near the southern edge of the present district, and as the sequence thickens towards the north-east so the outcrop widens through Silvington Common until it reaches its maximum spread over Oreton Common, before being faulted against Old Red Sandstone on Catherton Marshes. The rocks are not well exposed except for occasional small natural bluffs and stream sections and the general dip is to the south-east; much of the ground is sandy and pebbly and only supports rough heath-land vegetation.

The north-eastern part of the Titterstone Clee synclinal trough of the Coal Measures extends into the south-east corner of this district. At the base is a thick massive quartzitic sandstone with a conglomeratic base which forms a pronounced unconformable feature on the Cornbrook Sandstone. It crops out to the south of Cleeton and extends north-eastward across Silvington Common to the vicinity of Hill Houses where it turns southward towards the southern edge of the map. The coal-bearing measures above it form a smaller moorland area extending from the Hill Houses neighbourhood to the Crumpsbrook. The surface of this irregular area is covered with extensive old shallow coal excavations, now overgrown, and no sections are visible or old records extant. The seam which was formerly worked hereabouts is the Gutter or Bluestone Coal, $2\frac{1}{2}$ to $4\frac{1}{2}$ ft thick, which lies about 50 ft above the base of the shales. It is overlain by measures which, farther to the south-west in the main part of the trough, total over 400 ft. Farther east, a small faulted area of Coal Measures occurs half a mile south of Prescott House.

Within 6 miles north of the Farlow–Oreton ridge the Farlow Sandstone, Tournaisian limestones, and Cornbrook Sandstone are all cut out and, on Brown Clee, two outliers of Coal Measures lie unconformably on the Clee Group of the Lower Old Red Sandstone. V.W.

Brown Clee. The outcrop of the base of the Carboniferous rocks is marked by an abrupt change in vegetation. The Coal Measures sustain a thick carpet of ling with patches of bent and sedge in the wetter hollows, all growing on a thin peaty soil. The base of the measures is marked by a sandstone which has been referred to the Millstone Grit, probably because of its lithological similarity to that rock where it is coarse and pebbly. However, as there is no positive evidence of age the view is here adopted that this sandstone is the basal bed of the Coal Measures comparable with similar beds of that age in other Midland coalfields. The mapping of the unconformity shows that the Old Red Sandstone was folded into a shallow syncline whose axis trends north-north-eastwards through Clee Burf, before the Coal Measures were deposited. The resulting unconformity may be compared with that at the base of the Coal Measures on Titterstone Clee, about 6 miles farther south, which is general throughout the Midlands. Thus on Brown Clee there are no representatives of the Carboniferous Limestone and Millstone Grit. G.H.M.

DETAILS

LOWER CARBONIFEROUS

Farlow–Oreton area. Lower Carboniferous Basal Conglomerate. The first indications of the conglomeratic base of the Lower Carboniferous are seen about [63408009][1] 850 yd S.W. of Farlow church, where reddish blocks of coarse sandy conglomerate occur in the hedgerow below the feature of the limestone. Other blocks of this conglomerate are also seen at about the same level north-eastward along the hillside towards Farlow. In the east bank of the road immediately east of Farlow school [64068063] the conglomerate is 6 ft thick; eastward it expands to 12 ft and forms a well-defined feature, with numerous craggy outcrops, through The Rough to Oretonbank [65208075], beyond which it is again unrecognizable at the surface as far as the River Rea. East of the river it again thickens rapidly to over 30 ft and locally forms a considerable outcrop over the small ridge [66548085] 350 yd S.E. of Prescott Mill.

Lower Carboniferous Limestones. Debris of thin flaggy crinoidal limestone and old shallow pits on the fields [63488013] west of Lane's End, 800 yd S.W. of Farlow church, are the first field evidence of the presence of these attenuated beds west of Farlow. In the east bank of the road immediately east of Farlow school about 25 to 30 ft of the lowest beds overlie the basal conglomerate; they are made up of 12 to 15 ft of hard grey, massive limestones overlain by 12 to 15 ft of buff, thinly bedded gritty limestones with occasional shaly bands. A few feet higher than this section and along the W.S.W.-E.N.E. strike is a small section of 3 to 4 ft of thinly bedded, light grey oolitic limestones in the road and the small old quarry [63988060] opposite Farlow church. Another small 3-ft section at the north-east corner of the Vicarage garden wall [64118059], 180 yd E.N.E. of St. Giles's Church, shows buff flaggy limestones which must be stratigraphically only a few feet above the sequence in the roadside near the school. On the hillside [63828067] north-west of Farlow there is a considerable mass of light grey oolitic limestones which appears to have broken away from the main outcrop and slipped some yards down the slope.

In the fields adjacent to the corner in the road [63698035] 450 yd S.W. of St. Giles's Church, old excavations indicate that a thin group of limestones was formerly worked hereabouts; these old workings extend eastward along the north side of the footpath running to the old lane from Farlow southward to The Foxholes. Occasional small exposures in these old workings indicate that the sequence consisted of about 6 ft of buff, impure crystalline limestones with silty bands, overlying light grey oolites and crinoidal limestones, of which about 5 ft are seen in one place. At the top of the oolite band evidence of penecontemporaneous erosion was pronounced. From the increased size of the old workings eastward towards the lane it is evident that these beds rapidly thicken in that direction.

The old quarry [64098048] adjacent to the south side of the Vicarage garden, 210 yd S.E. of Farlow church, shows a face of about 15 ft of buff, wavy bedded flaggy limestones with occasional thicker grey limestones. The sequence has yielded the following

[1]Such numbers are eight figure National Grid References within the 100-km square SO(32).

fauna: *Conularia* cf. *quadrisulcata* J. Sowerby, *Fenestella sp.*, '*Camarotoechia*' *mitcheldeanensis* Vaughan, *Cleiothyridina ?*, *Pugilis* cf. *vaughani* (Muir-Wood), *Reticularia sp.*, *Schizophoria ?*, a smooth Spiriferid, *Syringothyris sp.* and fragments of gastropods. In the absence of more diagnostic fossils this fauna can only be regarded as being high in the *Cleistopora* (K) Zone or low in the *Zaphrentis* (Z) Zone.

East of the old lane running south-eastward from Farlow to The Foxholes, the steep hill slope extending through Oreton towards the River Rea has been largely quarried away in the past, leaving rough overgrown ground in which scattered relics of old quarry faces are frequently difficult to examine and often obscured by scree.

About 560 yd E. 23° S. of Farlow church the remains of an old face [64448040] shows the following details:

	ft	in
Hard grey crystalline limestone, weathering buff	3	0
Soft buff silty marl	–	5
Hard grey crystalline limestone	1	6
Buff silty marl with hard calcareous bands	1	3
Thickly bedded grey oolites with fine shell debris	20	0

Farther east and near the top of the hillside an old quarry face [64818057] 250 yd N.W. of New Inn, Oreton, shows 12 ft of buff, grey-centred crystalline crinoidal limestones containing *Michelinia* cf. *favosa* (Goldfuss), *Fenestella sp.*, *Fistulipora incrustans* (Phillips), *Rhabdomeson ?*, '*Camarotoechia*' *mitcheldeanensis*, *Cleiothyridina* cf. *royssii* (Davidson), *Reticularia ?*, *Schizophoria sp.*, *Spiriferellina sp.*, *Syringothyris cuspidata* J. Sowerby sp. (*cyrtorhyncha*/*exoleta* North group), and *Tetracamera ?* About 120 yd S.E. of this last face and some 100 to 200 yd W. of the New Inn a long irregular old quarry [64898052] shows 24 ft of massive crystalline grey limestones with much crinoid debris; the following list of fossils from this section includes many collected in 1930–31 by the late R. W. Pocock and named by Sir James Stubblefield: Stromatoporoid, *Michelinia* cf. *favosa*, *Syringopora* cf. θ Vaughan, *Fenestella sp.*, *Rhabdomeson ?*, Athyrid, '*Camarotoechia*' aff. *mitcheldeanensis*, *Cleiothyridina sp.*, *Dielasma sp.*, *Eumetria sp.*, *Lingula sp.*, *Orbiculoidea ?*, Orthotetid, Productids, *Schizophoria sp.*, *Syringothyris cuspidata* (*cyrtorhyncha*/*exoleta* group), *S.* cf. *principalis* North, *Tylothyris laminosa* (McCoy), *Tetracamera ?*, *Cladodus mirabilis* Agassiz, *Ctenacanthus major* Agassiz [=*C. salopiensis* J. W. Davis], *Orodus ramosus* Agassiz, *Psephodus salopiensis* A. S. Woodward, *Petalodus sp.*, and *Sandalodus morrisii* J. W. Davis.

There are two large old quarries, near Bank House, to the west and east respectively of the road from Oreton north-north-eastward over Oreton Bank; in the first one [65198068] 270 yd N.E. of the New Inn the details are: massive limestone, thin and flaggy in the top foot, 6 to 7 ft, on a 4-ft bed of soft layered calcareous silty shale split into three 2- to 3-in harder bands at the base, resting on 27 ft of massive grey finely oolitic limestones. The only fossils from this section are *Conularia sp.*, '*Camarotoechia*' *mitcheldeanensis*, *Syringothyris cuspidata* (*cyrtorhyncha*/*exoleta* group) and a smooth Spiriferid. The deep face on the east side of the road [65188062] shows 32 to 36 ft of very thickly bedded compact oolites which have yielded: *Michelinia* cf. *megastoma* Z_2 Vaughan, indeterminate Zaphrentids, *Fenestella sp.*, *Rhabdomeson ?*, '*Camarotoechia*' *mitcheldeanensis*, *Dielasma sp.*, *Reticularia sp.*, *Schellwienella* cf. *aspis* Smyth, *Syringothyris cuspidata* (*cyrtorhyncha*/*exoleta* group), *Psephodus* cf. *dubius* A. S. Woodward and fragments of fish spines. Commenting on this fauna Mr. Mitchell notes that "*S. aspis* has its maximum in Z_2, but it also occurs above and below this horizon. The Zaphrentid corals are indeterminate, but the presence of *M.* cf. *megastoma* Z_2 indicates that the beds in this quarry belong to the upper part of the Z Zone".

About 180 yd E.S.E. of the last section and 350 yd E.N.E. of the New Inn another small section [65358060] shows 6 ft of these limestones from which the fauna found in the last quarry has again been recorded along with *Eumetria sp.*, at present only known from the K Zone of the S.W. Province.

Limestones are again exposed in the narrow lane [65858048] 900 yd E. of the New Inn, Oreton, and 200 yd W. of the Factory Cottages near the River Rea, and from them '*Camarotoechia*' aff. *mitcheldeanensis*, *Cleiothyridina sp.*, *Eumetria sp.*, *Orbiculoidea sp.*, and *Schellwienella sp.* have been collected. Sir James Stubblefield who named this fauna considered that it was probably of K Zone age.

The limestones again crop out east of the River Rea where they form the southern flank of the small ridge 350 yd S.S.E. of Prescott Mill. In this ground they appear to be locally somewhat thicker than was apparent in the narrow lane referred to above. They consist of thick crinoidal limestones in the upper part, with brown earthy limestones below, and dip to the south at 42°; they have been formerly worked in numerous old quarries hereabouts.

East of the old lane running southward from Prescott and about 750 yd S.S.E. of Prescott Mill they are faulted against the Old Red Sandstone. A small old quarry [66578044] about 40 yd from the fault has yielded fragments of *Michelinia sp.*, *Fenestella sp.*, *Rhabdomeson sp.*, and *Syringothyris sp.*, from these beds. The highest shales and thin limestones of this limestone sequence appear to be best developed along the lower part of the hillside to the immediate south of the New Inn, Oreton. They extend westward through Springvale Cottages and appear to thin out in the vicinity of Lane's End about 450 yd S.S.W. of Farlow church. In an old pit [64668035] about 60 yd west of Springvale Cottages they are estimated at about 80 ft thick; excavated samples from these beds yielded a sparse microfloral assemblage identified by Dr. B. Owens as being of Tournaisian age. A few feet above the top of this quarry a line of strong springs marks the base of the overlying Cornbrook Sandstones. V.W.

Palaeontology of the Carboniferous Limestone of the Farlow–Oreton Area

The Carboniferous Limestone of Oreton is of Tournaisian age, but only parts of the K and Z Zones are represented, while the upper part of the Tournaisian and the whole of the Viséan is missing. The presence of K and Z Zone faunas at Oreton was first recognized by Vaughan, and recorded in his classic paper on the Carboniferous Limestone of the S.W. Province (1905, pp. 252–3).

Vaughan dismissed the *Cleistopora* (K) Zone, "as showing the normal character", but listed the *Zaphrentis* (Z) Zone fauna in detail and suggested that the beds were "low down in Z_2".

The Oreton outlier was briefly mentioned by Dixon (1917, p. 1064) and Whitehead and others (1928, p. 42), but no further detailed work on the area was published until T. N. George in his paper on 'The Namurian Usk Anticline' discussed the Oreton Limestones in relation to the age of the Cornbrook Sandstone (1955, pp. 307–8). George described the lower part of the succession as the Lower Limestone Shales of *Cleistopora* Zone age and listed the important elements of the fauna. The upper part of the Oreton limestone succession, including the beds exposed in the now disused Oreton Quarry were grouped as the Main Limestone. He did not comment on the age of the upper beds, but cited Vaughan's fossil list (1905, p. 253) together with a few additional forms not recognized by Vaughan.

Cleistopora Zone. Many of the characteristic K Zone fossils have not been recorded from Oreton, one of the more striking absences being that of the Chonetids. The correlation of the lower beds at Oreton with the K Zone largely rests on the identification of *Eumetria sp.* This is based on fragments too poor for specific determination, but having the distinctive costate and punctate ornament of this genus. This fossil, which occurs commonly in the S.W. Province, is at present only known from beds of K Zone age. No other fossil diagnostic of K has been recorded from Oreton. The rest of the fossils listed from the lower beds, including '*C.*' *mitcheldeanensis* Vaughan, *Pugilis vaughani* (Muir-Wood) and *S. aspis* Smyth, all range through from K into Z Zone.

Zaphrentis Zone. No Zaphrentid corals of zonal value have been recorded from Oreton, although Vaughan included *Zaphrentis* aff. *phillipsi* Milne Edwards and Haime in his fossil list (1905, p. 253). Carruthers subsequently redefined several of the Tournaisian Zaphrentids, (1908, pp. 25–31, 63–73) and included part of *Z.* aff. *phillipsi* of Vaughan 1905 in *Zaphrentites delanouei* (Milne Edwards and Haime) a species common in the lower part of the Z Zone, and part in *Fasciculophyllum omaliusi* (Milne Edwards and Haime) which is common in the upper part of Z and in γ C_1 Subzone.

The specimens of *Michelinia* and *Syringopora* are more helpful, however, with the forms *M.* cf. *favosa* (Goldfuss), *M.* cf. *megastoma* Z_2 Vaughan and *S.* cf. θ Vaughan all being characteristic of the upper part of the Z Zone in the S.W. Province.

The brachiopods, including '*C.*' *mitcheldeanensis*, *T. laminosa* (McCoy) and *S. cuspidata* (J. Sowerby) subspp. are all long-ranging forms, and although characteristic of the Tournaisian, do not give any accurate indication of zone. Most of the specimens of *Syringothyris* have a concave area and belong to the two forms *S. cuspidata cyrtorhyncha* and *S. cuspidata exoleta* of North. These two mutations are not easily distinguished without complete specimens and are here combined as one group which is known to range from the top of K, throughout Z and into the Upper *Caninia* (C_2S_1) Zone.

The main part of the Oreton limestone succession is of Z age as described over 50 years ago by Vaughan, and little can be added to his work on the Oreton faunas. The additional evidence from the specimens of *Michelinia* and *Syringopora* only helps to confirm his observations.

The fish fauna and its comparisons with the faunas at Bristol have been discussed by Woodward (1904, p. 431) and Vaughan (1905, pp. 252–3). M.M.

The highest beds of the Carboniferous Limestone consist of shales with thin nodular, concretionary limestones and chinastones. Two samples taken from the old clay pit [64668035] about 60 yd W. of Springvale Cottages (see Vaughan 1905, p. 253) have yielded relatively sparse microfloral assemblages. From the north side of the pit the following have been named: *Calamospora sp.*, *Cyclogranisporites sp.* cf. *Lophotriletes rugosus* Naumova, *Lycospora ?*, *Perotrilites ?* and *Retusotriletes spp.* A sample taken from a slightly higher level on the south side of the same pit yielded the following assemblage: *Acanthotriletes sp.*, *Calamospora sp.*, *Cyclogranisporites sp.* cf. *Lophotriletes rotundus* Naumova, *Dictyotriletes sp.*, *D.* cf. *trivialis* Naumova, *Granulatisporites sp.*, cf. *Hymenozonotriletes punctatus* Andreyeva var. *minor* Kedo, *Hymenozonotriletes sp.*, *H. sp.* cf. *H. submirabilis* (Luber) Kedo, *Leiotriletes sp.* cf. *L. inermis* Ishchenko, *Planisporites sp.*, *Punctatisporites spp.*, *P. irrasus* Hacquebard, *Retusotriletes spp.*, *R. incohatus* Sullivan, *R. sp.* cf. *R. septosus* Kedo or *R. brevidenticulatus* Chibrikova, *Trachytriletes medius* Naumova and *T. solidus* Naumova.

The commonest forms in both assemblages are representatives of the genus *Retusotriletes* Naumova. Some of the specimens may be assigned to described species, but the majority appear to be new, hitherto unrecorded forms. The genus as a whole is commonly recorded in the Devonian and the Tournaisian, but is normally a rare component of Viséan assemblages. Of the remaining spores which appear to have some stratigraphical significance, three, *Dictyotriletes* cf. *trivialis*, *Hymenozonotriletes sp.* cf. *H. submirabilis* and cf. *Hymenozonotriletes punctatus* var. *minor*, closely resemble forms described by Kedo (1963) from the Tournaisian deposits of the Pripyat Depression in the Byelorussian SSR, whilst *Punctatisporites irrasus* has been recorded from the Horton Group (Mississippian) of Nova Scotia, Canada, by

Hacquebard and from the Lower Limestone Shale (Tournaisian) of the Forest of Dean by Sullivan (1964).

Although a precise age-determination is not possible, due mainly to the sparse nature of the assemblages recovered and partly as a result of the lack of adequate well-documented assemblages from elsewhere in Western Europe, there is sufficient evidence to suggest tentatively a Tournaisian age for the samples. It would be necessary to examine additional material before any further refinement could be made. B.O.

UPPER CARBONIFEROUS

CORNBROOK SANDSTONE AND LOWER AND MIDDLE COAL MEASURES

Small friable and rounded outcrops of coarse sandstone occur on the heath [616790] between Cleeton and Silvington Common, and one such block contained a red quartzite cobble 4 in long. Small banks of red clay are also common. Low crags of coarse pebbly sandstone occur in and around the small plantation [63888005] west of the Workhouse some 600 yd S. by W. of Farlow church. Farther east, small crags of coarse brown sandstone form a bluff [65948051] about 140 yd N.W. of the Factory Cottages near the River Rea. On the east side of the river a few feet of coarse sandstones and pebbly beds, dipping southward off the underlying limestones, are exposed in the footpath [66218051] 180 yd E.N.E of the Factory Cottages. They appear to be cut off to the south by a fault which locally throws down a small area of Coal Measures.

At the site of the old Brick Kiln [64658002] 650 yd S.W. of New Inn, Oreton, 8 ft of bluish grey clays were proved; similar clays associated with coarse sandstones were formerly dug in the ground farther to the north-east at The Claypits [650802] and the old Brick Works [65238032], 300 yd S.E. of the New Inn. Red and grey clays and sandstones are occasionally indifferently exposed in the Cramer Gutter [65457932] along the south side of Oreton Common. Where the coarse sandstones are deeply weathered and friable they have been used for building purposes.

Titterstone Clee: Crumpsbrook area. The measures above the Cornbrook Sandstone are from 50 to 100 ft thick and, in the lower part, consist mainly of hard white sandstones, up to about 60 ft, with a conglomeratic base largely composed of white quartzite pebbles. They form an easily distinguishable feature. Shales predominate in the higher beds up to the Gutter Coal. The group is readily traceable from the vicinity of The Birches [616784], east of Cleeton, through Crumpsbrook [627784] to Catherton Common [638787]. The presence of the Gutter or Bluestone Coal, 3 to 4½ ft thick, above is revealed by the extensive old workings on or near its approximate outcrop. Except for an occasional few feet of shale or sandstone seen in Crumpsbrook these measures are rarely seen and there are no old records of former workings in this district.

Prescott area. Some 1300 to 1400 yd E. of Oreton and about ½ mile S. of Prescott House there is a small area [663805] immediately east of the River Rea where 2-3 ft of coal were formerly worked about 1860. The size and extent of this area are largely obscure as also are its structural relations to surrounding rocks. On the south-east it is cut off by the boundary fault while to the west it is flanked by the wide alluvial belt of the River Rea. North and north-east lies the ridge of Carboniferous Limestone on the southern flank of which up to 100 ft of coarse conglomeratic sandstones may represent part of the Cornbrook Sandstone, but the junction between these rocks and the Coal Measures is probably faulted though the relationship is obscure and cannot be accurately shown on the map. The coal seam may have been the Gutter Coal or it may even have been a local development of a coal seam within the Cornbrook Sandstone group. V.W.

Brown Clee. Though exposures are few the basal Coal Measures sandstone appears generally to be a fine-grained ganister-like sandstone, but in the south it becomes a coarse pebbly grit where it is seen [599841] west of Banbury and on the moor to the south-west. Its thickness varies from about 10 ft up to 50 ft and where well developed it forms an appreciable feature backed by a shelf up to several hundred yards wide. In other places as to the west of Clee Burf [590842] it dwindles to insignificance.

The position of the outcrop of the Bottom Coal shown on the map has been plotted by reference to the marked line of old workings ending abruptly at a slight slack in which the coal evidently crops. Sometimes it is thought to rest directly on the basal sandstone but more generally it is separated from it by up to about 18 ft of measures.

In some places, about 9 ft above the Bottom Coal there is a rider coal, named the Foot Coal where it was workable (Jones 1871), and within the 21 ft of measures above the Bottom Coal occurs the richest nodular ironstone which was also formerly worked. The nodules were reputed to vary from a few pounds in weight up to half a ton.

The higher coals were only indifferently seen as recently as 1945 in artificial cuts north and south of the basalt of Abdon Burf and immediately below that rock. The two coal smuts at the northern end are about 12 ft apart, separated by grey clay; the upper seam was seen to be immediately below the weathered basalt.

The shales, clays and ironstone debris which make up the spoil-heaps of the old pits yield nothing of more diagnostic value than a few poor plant fragments. G.H.M.

Shirlett area. In the S.E. corner of the Shrewsbury (152) Sheet there is a conspicuous faulted outlier of the lower beds of the Middle (Productive) Coal Measures covering the Shirlett Common–Shirlett High Park area. This occurrence has been adequately described by Robertson (Whitehead and others 1928) and requires no further mention here; suffice it to add that two small southerly extensions of this area enter the north-east corner of the present district ending in the vicinity of Hawthorn Bank. There are no workings or sections of the measures in these areas.

UPPER COAL MEASURES

COED-YR-ALLT BEDS

Southern extension of Leebotwood Coalfield. The only recent information on this small area of these beds has been obtained from the Geological Survey Church Stretton Borehole No. 4 which was sunk in 1961 at a locality [47359653] (O.D. 566·7) 1465 yd N. 58½° E. of the north end of St. Michael and All Angels Church, All Stretton. It was hoped that the full sequence of these beds would be proved but unfortunately the borehole encountered a fault, undetected beneath the Drift, at 96¾ to 98 ft, beyond which it entered Wenlock Shales. The abridged log of the borehole is as follows:

	Thickness		Depth	
	ft	in	ft	in
Pleistocene: Boulder Clay and Glacial Sand and Gravel	38	0	38	0
Carboniferous: Coal Measures: Coed-yr-allt Beds:				
Pale grey seatearth with rootlets and plant debris ..	7	0	45	0
Pale grey, micaceous sandstone	6	9	51	9
Soft grey, micaceous siltstone	–	4	52	1
Pale grey, laminated, micaceous siltstone	1	5	53	6
Pale grey, laminated, micaceous sandstone with carbonaceous layers	6	6	60	0
Grey micaceous mudstones and siltstones with carbonaceous layers and coal fragments	7	10	67	10
COAL	2	2	70	0
Silty, micaceous seatearth	5	0	75	0
Fine-grained, brown, laminated, micaceous sandstone	5	6	80	6
Grey silty mudstones with occasional sideritic nodules and slickensides	16	3	96	9
Faulted and unconformable junction with underlying Silurian siltstones and mudstones.				

The coal seam encountered at 67 ft 10 in was submitted to Dr. Mavis A. Butterworth at the Coal Survey Laboratory, Chester, for miospore examination, who reports that it contains "a *Verrucososporites obscurus* assemblage such as occurs in seams from the Newcastle, Coalport and Coed-yr-allt beds and also from the Etruria Marl of North Staffordshire. The assemblage is dominated by species of *Lycospora* and contains very few specimens of the genera *Laevigatosporites*, *Crassispora* or *Florinites; Endosporites* is relatively common. The following species, which are restricted to such horizons, are present: *Cardiospora magna* Kosanke, *Cirratriradites annuliformis* Kosanke, *Gravisporites sphaerus* (Butterworth and Williams), *Latosporites globosus* (Schemel), *Schopfites dimorphus* Kosanke, and *Triquitrites protensus* Kosanke". The nearest seams at a comparable horizon to the present coal would be those in the Coed-yr-allt Beds of the southern part of the Denbighshire Coalfield and in the Coalport Beds of Coalbrookdale. V.W.

REFERENCES

BALL, H. W. and DINELEY, D. L. 1961. The Old Red Sandstone of Brown Clee Hill and the adjacent area. I. Stratigraphy. *Bull. Brit. Mus. (Nat. Hist.) Geol.* **5**, 175–242.

CANTRILL, T. C. 1908. Chapt. I, Geology, *in* A History of Shropshire. *The Victoria History of Counties of England*, 26–34.

CARRUTHERS, R. G. 1908. A revision of some Carboniferous corals. *Geol. Mag.* (5), **5**, 20–31, 63–74, 158–71, pls. iv-vi.

DIXON, E. E. L. 1911. The Geology of the Titterstone Clee Hills. *Rep. Brit. Assoc.* for 1910, 611–13.

—— 1917. The Geology of the Titterstone Clee Hill Coal Field. *Trans. Roy. Soc. Edin.*, **51**, 1064–71.

GEORGE, T. N. 1956. The Namurian Usk Anticline. *Proc. Geol. Assoc.*, **66**, 297–316.

—— 1958. Lower Carboniferous Palaeogeography of the British Isles. *Proc. Yorks. Geol. Soc.*, **31**, 227–318.

JONES, D. 1871. On the co-relation of the Carboniferous deposits of Cornbrook, Brown Clee, Hurcott and Coalbrookdale. *Geol. Mag.*, **8**, 363–71.

—— 1873. On the Carboniferous deposit of the Brown Clee Hill and its relation to contiguous coalfields. *Geol. Mag.*, **10**, 348–51.

JONES, D. G. and OWEN, T. R. 1961. The Age and relationships of the Cornbrook Sandstone, *Geol. Mag.*, **98**, 285–94.

KEDO, C. I. 1963. Tournaisian spores from the Pripyat Depression and their stratigraphic significance. *Palaeont. Stratigr. BSSR*, **4**, 3–121.

LAPWORTH, C. 1891–2. *Trans. Fed. Inst. Min. Eng.*, **3**, 10.

MORRIS, J. and ROBERTS, G. E. 1862. On the Carboniferous Limestone of Oreton and Farlow, Clee Hills, Shropshire. *Quart. J. Geol. Soc.*, **18**, 94–102.

MURCHISON, R. I. 1839. *The Silurian System*. London. 122–4.

POCOCK, R. W., WHITEHEAD, T. H., WEDD, C. B., and ROBERTSON, T., with contributions by WRAY, D. A., STUBBLEFIELD, C. J., CANTRILL, T. C. and DAVIES, W. M. 1938. Shrewsbury District. *Mem. Geol. Surv.*

SULLIVAN, H. J. 1964. Miospores from the Drybrook Sandstone and associated measures in the Forest of Dean Basin, Gloucestershire. ***Palaeontology*, 7**, 351–92.

TRUEMAN, A. E. 1947. Stratigraphical Problems in the coalfields of Great Britain. *Quart. J. Geol. Soc.*, **103**, lxx-lxxii.

VAUGHAN, A. 1905. Palaeontological Sequence in the Carboniferous Limestone of the Bristol area. *Quart. J. Geol. Soc.*, **61**, 181–307, pls. xxii-xxix.

WHITEHEAD, T. H., ROBERTSON, T., POCOCK, R. W. and DIXON, E. E. L. 1928. The Country between Wolverhampton and Oakengates. *Mem. Geol. Surv.*

WILLS, L. J. 1935. The Palaeogeography of the Birmingham country. *Proc. Geol. Assoc.*, **46**, 221–4.

—— 1948. *The Palaeogeography of the Midlands*. London. 30–7.

WOODWARD, A. S. 1904. Notes on the Geology and fossils of the Ludlow district. *Proc. Geol. Assoc.*, **18**, 429–42.

WRIGHT, R. 1840. Notes on the Geology of the Brown Clee Hill in the county of Salop. *Trans. Geol. Soc. Lond.*, **5**, 125–6.

Chapter 9

STRUCTURE

THE TWO principal structural elements in the Church Stretton district can be traced into it from the north-east and have been described in the relevant Geological Survey Memoirs (Pocock and others 1938 and Whitehead and Pocock 1947). The broad structural pattern follows a north-easterly, Caledonoid, alignment. The western part of the district consists of the Long Mynd–Wrekin anticlinal uplift, to the east of which, covering about three-quarters of the district, lies the East Shropshire–West Staffordshire Syncline. Both structures are modified by a number of major internal features, which have a parallel alignment, and there are abundant minor structures bearing various relationships to them. Within the present district the Long Mynd–Wrekin zone of uplift comprises the Longmyndian Ridge, locally concealed by Palaeozoic rocks, and the Eastern Uriconian Axis. The East Shropshire Syncline comprises the broad asymmetrical folds which Ball and Dineley (1961, p. 209) call the Brown Clee and Titterstone Clee synclines and the intervening Ledwyche Anticline. The last is more spectacularly developed south-west of Ludlow (Sheet 181) where, as the Ludlow Anticline, it gives rise to the prominent horseshoe-shaped outcrop of the Aymestry Group.

THE LONGMYNDIAN RIDGE

The Longmyndian Ridge is separated from the Eastern Uriconian Axis by the southward extension of the Leebotwood Trough (Pocock and others 1938, p. 169) to the vicinity of Church Stretton, the Palaeozoic rocks filling the trough being underlain by a continuation of the Longmyndian rocks exposed to the west. The Church Stretton Fault, F1 of Cobbold's notation (1927, p. 565), is the essential boundary between the Longmyndian Ridge and the Eastern Uriconian Axis, although Longmyndian rocks lie east of it at Church Stretton and elsewhere. A horizontal section across the southern part of the Long Mynd, as shown on the one-inch geological map, indicates that, beneath the Silurian rocks of the Plowden–Little Stretton outcrop, Longmyndian rocks occupy all the ground west of F1. D.C.G.

DETAILS

FOLDING

The Longmyndian rocks are thought to be folded into a large isoclinal syncline, the core of which lies in the Bridges Group. This hypothesis is based largely on the evidence of stratigraphical sequence discussed above (p. 49). The eastern limb of the fold comprises the Stretton Series, the Bayston outcrop of the Bayston–Oakswood Group and the eastern part of the Bridges Group. The western limb is composed of the Oakswood outcrop of the Bayston–Oakswood Group and the western part of

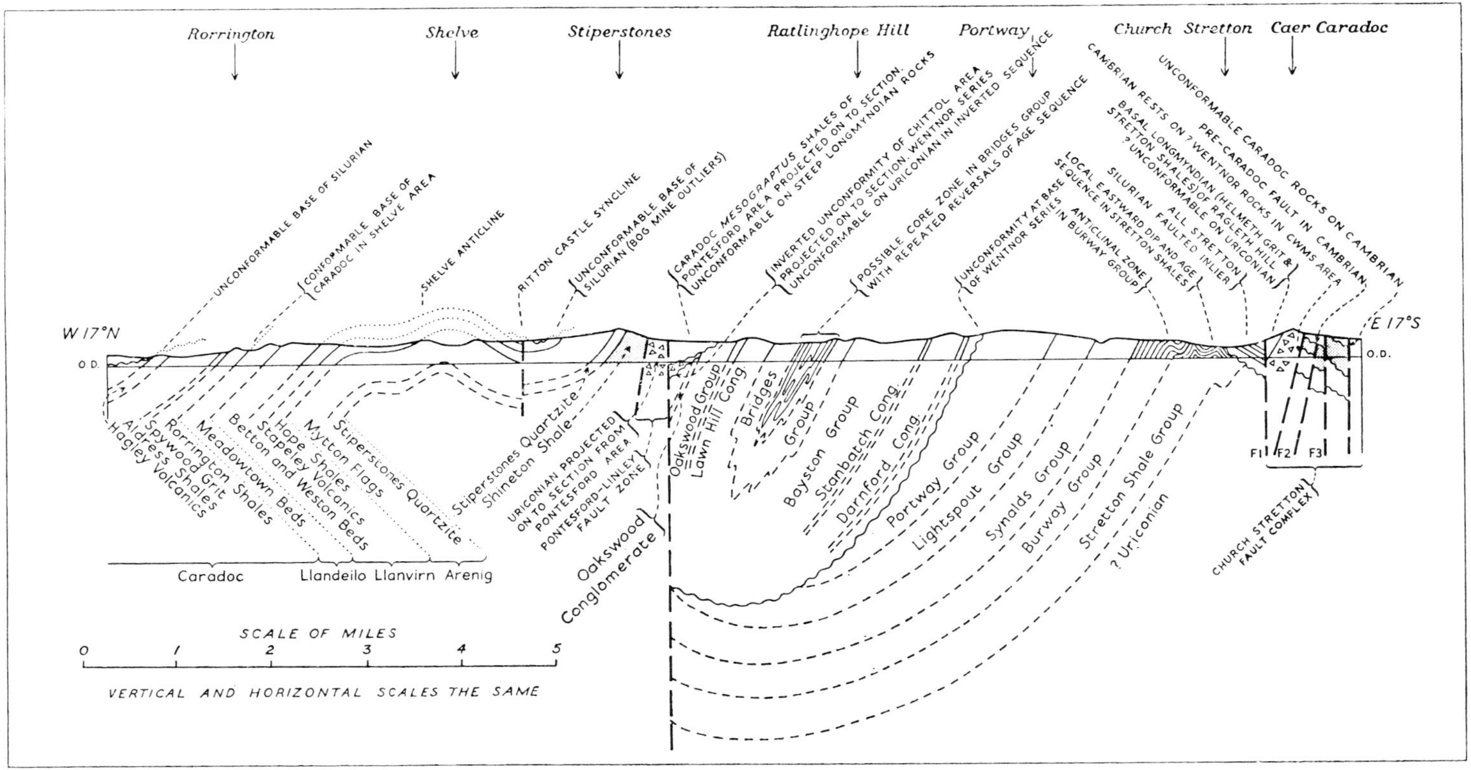

FIG. 18. Diagrammatic composite section from Rorrington to Caer Caradoc to compare the inferred folding of the Longmyndian with folding of the Ordovician of the Shelve area and to illustrate various unconformities. Structure of Longmyndian based on work of J. H. James (1956) with some modification. Structure of Ordovician of Shelve area based on work of C. Lapworth and W. W. Watts with some modification

the outcrop of the Bridges Group. Within the Church Stretton district the regional dip of the Longmyndian is steep to the west-north-west. The absence of the Stretton Series on the western limb of the fold is attributed to unconformity at the base of the Wentnor Series.

This idea of a major syncline has been suggested by Whittard (1952, p. 149) and James, the latter author publishing a diagrammatic cross-section to illustrate the structure (1956, fig. 4). That section, with minor modification, is incorporated in Fig. 18 which compares the structure of the Long Mynd with that of the Shelve area to the north-west. James (1956, p. 333) suggested that a downfold to a depth of about 3½ miles on the base of the Wentnor Series is required to account for the absence of the Stretton Series on the western limb of the fold.

The axial plane of the syncline extends along the outcrop of the Bridges Group, striking about N.N.E.-S.S.W. but, as noted by Whittard (1952, p. 149) and Whitehead (1955, p. 469), with a steep dip swinging slightly about the vertical from steep to the west-north-west in the Long Mynd to mainly steep to the east-south-east in the Shrewsbury area. James (1956, p. 333) suggested that the northward narrowing of the outcrop of the Bridges Group indicates a southerly plunge of the longitudinal fold axis. The expected northward closure of the fold, around the base of the Bridges Group however, is hidden by a cover of Upper Carboniferous and Triassic rocks in the Hanwood area, west of Shrewsbury. Within the Church Stretton district, minor linear features such as the intersection of cleavage and bedding, or the b-axes of minor folds, commonly strike parallel with the trend of the major structure but may plunge at moderate angles to the north-north-east or south-south-west. J.E.W.

The Silurian rocks west of the Long Mynd from Plowden to Criftin [3891][1], dip at angles of about 25° to the south and west, i.e. away from the adjacent outcrops of Longmyndian rocks. The synclinal effect thus produced is probably essentially superficial, in the sense that it represents in part an original seaward depositional dip, subsequently exaggerated by more intense compaction of the Silurian rocks where they were most thickly developed. Similar effects are observed south and south-east of the Long Mynd, but as the rocks are followed away from the Pre-Cambrian outcrop the regional north-easterly strike becomes dominant. North of Edgton the general dip of these Silurian rocks is south-eastwards, but dips to the south-west prevail in the Hopesay area. The trough of the syncline thus indicated can be followed from near Grist House [3984] to west of Coston [3880]. The southward plunge of this fold is most clearly demonstrated by the occurrence of Downton Series rocks near Oaker [3881]. The trough of the fold runs parallel to and about 600 yd west of, the main fault F1 of the Church Stretton Complex and it is probable that it owes its origin to movements along that line. Holland (1958, pp. 468–9) suggests such an origin for similarly located folds in the Knighton area, some 7 miles to the south-south-west. D.C.G.

FAULTING

The one-inch map shows that the dominant direction of faults in the Longmyndian is about W.N.W.-E.S.E., i.e. about normal to the regional strike of the bedding. Faults trending in this direction show lateral displacements of the steeply dipping beds, commonly of the order of tens of feet, but in some instances as much as 1200 ft. These displacements are both dextral and sinistral and are considered to result from movements in which the major component was horizontal as the movements, if vertical, would have to have been extremely large to account for the observed displacements of the steeply dipping or vertical Longmyndian. It should be noted that the trends of these W.N.W.-E.S.E. faults may really be more variable than is shown on the geological map. The fault planes very rarely give rise to surface features and

[1]Such numbers are four figure National Grid References within the 100-km square SO (32). They refer to general areas rather than specific localities.

the presence of a fault has usually been deduced from observed displacements of a horizon such as the Stanbatch Conglomerate or the Batch Volcanic beds. Because of the generally restricted nature of the outcrops there is commonly an area between known exposures on either side of the fault, in which the line of the fault can be drawn in various directions within certain limits. If the rocks were more continuously exposed, it might therefore be found that these faults exhibit a greater variation in trend than the map shows.

In addition to the dip faults described above there are several major faults in the Longmyndian which trend more nearly north-south. These are as follows:

The Pock Fault. This fault was named by James (1956, p. 332) and trends about N. 30° E.–S. 30° W., running from just north of Medlicott [400949][1] to Shaw Wood [406962] near Belmore Farm. Along the line of the fault there are several exposures of hard, grey, silicified sandstone with abundant quartz veins. Some of the quartz veins show slickensided surfaces, the striations plunging in various directions. J.E.W.

Long Mynd Scarp Fault. This fault was so named by James (1956, p. 331). It had been described as the West Longmynd Boundary Fault by Whittard (1932, p. 883), who remarked that there is no evidence of its persisting to the north of Stanbatch [402937]. Whittard (1932, p. 884) also drew attention to the sinuous course followed by the fault south of Myndtown. In that area additional evidence of post-Llandovery movement is afforded by the occurrence of small areas of conglomeratic *Pentamerus* Beds on the upthrown eastern side of the fault (p. 163). Between Asterton and Stanbatch there is little evidence for the scarp fault apart from the strong scarp feature. At 550 yd S. 20° E. of Stanbatch Cottage [404933] there are exposures of hard silicified sandstone and siltstone which appear to be brecciated and which may be connected with the fault. James (1956, p. 331) remarked that to the north of Medlicott [400946] the existence of the fault is inferred from topographical features combined with widely separated exposures of brecciated quartzite. From the present survey it is concluded that there is no satisfactory evidence for the extension of the fault northward from Stanbatch, particularly as the strong scarp slope of the Long Mynd dies out to the north of this locality. D.C.G., J.E.W.

An important fault traverses the rocks in the plantation [384875] west of the Rabbit Warren on a trend of N. 33° E. It throws the Portway and Lightspout groups together and cuts out the Huckster Conglomerate, and may be regarded as a tear fault with a horizontal displacement of at least some 300 yd. A north-westerly downthrow of at least 650 ft would give the same effect.

Black Knoll Fault. The Huckster Conglomerate is sinistrally displaced by about 150 yd by this fault which is here named the Black Knoll Fault. Associated strike-faulting increases the total cross-strike displacement in the vicinity of Black Knoll [3989] to about 300 yd. Along the eastern side of the Rabbit Warren [390875] the fault separates rocks of the Lightspout Group from those of the Synalds Group. The displacement does not appear to be very significant, but as the fault here runs approximately parallel to the strike of the beds there may be an important element of vertical movement. D.C.G.

Yewtree Bank Fault. This fault, here named the Yewtree Bank Fault, has been mapped from the top of the Long Mynd [405924], north-east of Asterton, southward across Yewtree Bank [408916], 1150 yd E.N.E. of Asterton, as far as Mount Gutter [402883] 900 yd N. 28° E. of Hillend Farm, on the south-east side of the Long Mynd. It is much less well defined than the Ashes Hollow Fault (see below), but a considerable sinistral lateral fault displacement is required to explain the proximity of the Huckster and Darnford horizons near the head of Minton Batch [406920]. In this area these two horizons are separated by only 1400 ft across the strike of the beds whereas it would be expected that they would be at least 2500 ft apart in a succession not disturbed

[1]Such numbers are National Grid References within the 100-km square SO (32).

by faulting. It is thought that on the western side of the Long Mynd [405924] the Yewtree Bank Fault may swing to the north-west, giving rise to a sinistral displacement of about 600 ft in the Darnford Conglomerate.

Ashes Hollow Fault. This fault extends from the upper part of Ashes Hollow, southward to the mouth of Minton Batch where it disappears beneath the cover of Upper Llandovery rocks. It was first named by James (1956, p. 332), who mapped it as far south as Round Hill [423928], while the present survey has confirmed James's observations and has extended its recognition to the south. It is one of the best-defined faults in the area and can be delineated from the mapping of contrasting groups of the Longmyndian on either side of the line of movement. There are however no exposures of the fault plane nor does it give rise to a well-marked topographic feature. There is a sinistral lateral displacement of about 4000 ft along the line of this fault. J.E.W.

Although it is founded on essentially the same evidence, the suggested fault pattern in the Silurian rocks between Minton and Edgton differs in many respects from that postulated by Whittard (1932, pp. 882–3). Evidence for each of the faults shown on the one-inch map is visible in the field, in the form of broken strata or anomalous disposition or width of outcrops. In many instances the topography suggests the more exact line of a fault. The most important of Whittard's faults which is not here recognized is his 'Plowden boundary fault', which he depicted as following an easterly course from the Long Mynd Scarp Fault immediately adjacent to the southern tip of the Long Mynd. No positive evidence for it has been observed and there appears to be no need to suppose that it exists. The *Pentamerus* Beds seem to have a less varied thickness if they are not faulted, and the outcrop anomalies can be adequately explained by a fault near the line of Whittard's 'Plowden Woods Fault' (1932, pl. lix) but continuing east-north-eastwards to meet the Church Stretton Fault, 1400 yd south-west of Cwm Head church [414878]. This fault, and some new evidence as to the boundary between the Wenlock and Hughley shales, obviates the need for Whittard's 'Churchmoor Dingle Fault' (1932, p. 882). His 'Minton boundary fault' is considered to swing to the south between Minton and Minton Batch, rather than westward into the Longmyndian, and explains the very narrow outcrop of the *Pentamerus* Beds in that area.

The throw of the south-south-easterly fault, which branches from the Long Mynd Scarp Fault just south of the Onny, appears to be about 600 ft. The limestones at the top of the Wenlock Series are thrown down to the west against shales about 400 ft above the base of the Series. The fault which follows the line of Ridgway Hill from Horderley [409871] to west of Edgton [380857] has a small throw south-west of the hill, but just north-east of it the downthrow towards the north-north-west may be as much as 350 ft. The south-south-easterly fault at Edgton is exposed below a building [386858] 70 yd N. 20° W. of the church. Limestones of Wenlock age west of the fault are thrown against Ludlow shales of a horizon some 650 ft higher. The continuation of this fault south of Basford [388853] is not established. D.C.G.

The outcrop of Downton Series rocks in the core of the Oaker Syncline is terminated to the south by a fault, with a northerly downthrow, which runs under the alluvium of the River Clun and joins the Church Stretton Fault north of Beambridge [389816]. B.A.H.

The Eastern Uriconian Axis

The Eastern Uriconian Axis forms a narrow zone of steeply dipping rocks, essentially of Uriconian and Longmyndian age but including also, on its eastern flank, the overlying rocks of the Cambrian and Ordovician. Faulting is the dominant structural feature of the axis, the more persistent faults being parallel to the axis itself, and Lower Palaeozoic rocks are preserved within the zone at several localities. Complication is intensified in three areas along the axis, the Uriconian culminations of Wart Hill and Caer Caradoc, and the intervening area of Brokenstones. A significant north-westerly, Charnoid, trend is developed in the Uriconian rocks of the Caradoc area. D.C.G.

DETAILS

FOLDING

The Hoar Edge Grit of the Coston district appears to be folded into a southward-plunging anticline with its axis parallel to the Church Stretton Fault, F1. Beds adjacent to F1 at Coston [386803] and The Hollies [387806] dip westwards at up to 60°, and on the eastern side of the outcrop southerly or south-easterly dips are seen to the north of Coston Hall [391801]. As in the case of the Oaker Syncline (p. 262) this folding may be associated with movement along F1.

North of the River Clun the Ordovician rocks dip steeply or are overturned on a north-north-easterly strike, and no fold pattern is seen. The structure of the Longmyndian rocks of Hopesay Hill [4084] is not known, due to lack of exposures. B.A.H.

Between Wart Hill and Marshbrook the rocks generally dip at very high angles on a north-north-easterly strike, parallel to the major faults. Variations are in places clearly related to proximity to other faults and seldom develop into a consistent pattern of folding. The Llandovery rocks which lie between F1 and F2 at New House, Marshbrook [4389] dip consistently towards the north-west, in the opposite direction from those of the same series immediately north-west of F1.

The line of F3 south of Marshbrook has been considered to separate the steeply dipping Caradoc rocks west of the fault from those to the east which are much more gently inclined. The easterly and south-easterly dip of the Caradoc rocks increases eastwards towards the fault, and in its vicinity at several points between the Onny and Woolston the beds are overturned towards the north-east. D.C.G.

In the faulted area between Marshbrook [442897] and Ragleth Hill [4591] there is no large-scale folding in the Palaeozoic strata. The beds dip to the south or south-east at 20° to 35°, apart from local disturbances near the faults. The only interruption to this steady dip is a syncline in the Acton Scott Beds on an east-north-easterly axis which extends for about $\frac{2}{3}$ mile to the east-north-east from the north–south fault 100 yd east of Marshbrook station. Dips on the southern limb range from 15° to 35°, and a minor anticline on this limb is exposed at the pool [446899], 400 yd E. 14° N. of Marshbrook station. B.A.H.

Little major folding has been recognized in the Uriconian rocks of the Stretton hills and such folding has not been observed on The Lawley [4997]. A north-westerly plunging syncline is an important feature of the structure of Caer Caradoc Hill [4795], the axis passing about 600 yd south-west of the summit. This structure is most readily followed by tracing the outcrops of the Ragleth Tuffs backwards and forwards across the watershed. On the northern limb the dip reaches about 50° to the south-west, but in the steeper southern limb the beds are locally vertical. The Ragleth Tuffs at the southern end of Caer Caradoc Hill may be folded sharply, but no instance of the turn-over of the beds was observed. Folding may also exist on Ragleth Hill [4591], but again this cannot be demonstrated on the ground where, apart from some local deviations, the strike is remarkably constant in a south-westerly direction.

The Ordovician rocks between the Lawley and Cwms–Hoar Edge faults are folded into a southward-plunging syncline to the north of Cwms Farm. In this area the strike of the Cambrian rocks is west of north (Cobbold 1927, p. 567), and a similar direction is taken by the axis of the fold on Hope Bowdler Hill, which is shown by the mapping of the conglomerates there (p. 19). A corresponding syncline lies to the north-east on Willstone Hill, again shown by the shape of the outcrop of the conglomerates and the disposition of the dips.

The only other major fold structure which can be recognized in the area is the north-westerly trending syncline which extends across Cardington Hill [4994] from near Woodgate Cottage. Its recognition depends on the identification of the two separate outcrops of the Woodgate Batch Dacites and Andesites with opposing dips (p. 21). Dips on the western limb are at about 30° to 45° to the north-east; those on the easterly limb range from 25° to 60° in a south-westerly direction.

In the Hazler Hill area the evidence suggests a north-westerly dip without folding.

Although major folding appears to be slight, there is a great deal of minor folding accompanied by shearing, crushing and faulting of the rocks. It is only possible to demonstrate this in ground where there are good and fairly continuous exposures. Thus the presence of minor folding on The Lawley or on much of the ground occupied by the Little Caradoc Basalts is only suspected. A striking feature of the many exposures of Caer Caradoc Rhyolites is the intense folding which the flow banding displays. Although the general dip of the rocks is shown, by the mapping and by the direction of the majority of readings taken on the flow banding, to be towards the south-west, individual bands, when traced for even a short distance, are commonly found to be greatly folded. Thus minor anticlines and synclines with all manner of directions of plunge are found. Moreover the junction of the Caer Caradoc Rhyolites and the overlying Caer Caradoc Andesites appears to be affected by the same sort of minor folding. This folding is apparently due to considerable earth movements and is closely associated with shear planes, generally dipping to the south-west and often polished or slickensided. Also present are crush breccias in which the rhyolite has been broken into angular fragments set in a crushed matrix, in places aligned along definite planes, but also occurring in patches without obvious relationship to shear planes. These features suggest earth movements directed chiefly from the south-west but sometimes from the south-east. The fact that the base of the Caer Caradoc Rhyolites maps as a plane surface dipping to the south-west, whereas the Rhyolites themselves are seriously disturbed, suggests the presence of a thrust plane or reverse fault between them and the underlying Little Caradoc Basalts. The smashed and folded rhyolites would thus lie within the imbricate zone above the thrust. At the southern end of its course the northward displacement of the line of the lowest outcrops of the Caer Caradoc Rhyolites suggests that the thrust is slightly folded. The alternative explanation of cross faulting does not seem to satisfy the evidence on the ground.

Though the Ragleth Tuffs are not so crumpled as the Caer Caradoc Rhyolites they also are traversed by shear planes and include crush breccias. In places the andesites and basalts show comparable features, so that clearly the whole mass of the Uriconian has yielded to the earth movements in a similar manner.

Good examples of slickensided fault planes, believed to be minor thrusts, occur 190 yd W. 11° N. of Caer Caradoc summit [476954], where the dip is 37° S. 10° W. and the slickensides aligned W. 10° S.; 150 yd S. 2½° E. of Caer Caradoc summit where the surface dips S. 60° E. at 36° and 370 yd S. 12½° W. of Caer Caradoc summit with dip W. 20° S. at 40°. There are many others, particularly in the Caer Caradoc Rhyolites and in the Ragleth Tuffs.

Particularly good exposures for the study of these features are to be found in the crags forming the summit of Caer Caradoc and those 420 yd south of the summit. The remarkable breccia, 250 yd S. 13° W. of the summit, at the north side of the camp entrance, composed of the normal vesicular rhyolite set in a rhyolite matrix, may also be a crush breccia, but in many respects it recalls a flow breccia. The fragments in the breccia are up to 6 in across, which is distinctly larger than usual in the crush breccias, though they are set in a brown matrix not unlike that of the crush breccias.

On the large crags south of the camp entrance, shear planes inclined to the east show horizontal slickensides and may well be tear planes.

An instructive example of minor thrusting occurs 840 yd S. 35½° W. of Caer Caradoc summit [473948] in the Ragleth Tuffs. Bedded tuffs dipping at 27° N. 50° W. have a thrust at their base with the same inclination. Below them other tuffs dipping at 32° N. 60° W. are closely broken by joints which dip 55° S. 40° W. and are themselves traversed by another thrust dipping at 20° to the south-west. Below this are other similar tuffs with a dip approximately the same as that of the highest beds in the section. G.H.M.

FAULTING

Cobbold (1927, pp. 565–6) recognized three major faults in the Church Stretton Complex, which he described, from west to east, as the Church Stretton Fault (F1), the Lawley Fault (F2), and the Cwms–Hoar Edge Fault (F3). F1 can be followed with little difficulty across the area of the map, but F2 and F3 are clearly identified only in the 'type-area' east of Church Stretton and in the ground between Wart Hill and Brokenstones.

Church Stretton Fault. From the south-western corner of the Church Stretton district to Wart Hill the Church Stretton Fault follows a fairly straight course, trending about N. 15° E., with broad, very slight sinuosities. These sinuosities do not give any indication of the hade of the fault. It throws down strata of Ludlow age ('basin' facies) to the west against the Wenlock Shales ('shelf' facies) and Hoar Edge Grit in the south and the Wentnor Series of Hopesay Hill farther north. No estimate of the throw of the fault is possible. The fault is not exposed, although sections in strongly sheared Ludlow and Longmyndian rocks north of Hopesay indicate its proximity to these outcrops. B.A.H.

Between Wart Hill and Marshbrook the fault continues to follow a generally straight course with only one significant bend, about ¼ mile west of Lower Carwood [399858], where it veers from about N. 17° E. to N. 35° E. No consistent evidence is afforded by the line of outcrop as to the direction of hade, but the fault plane is clearly very steeply inclined. At the one locality in the area where the fault is exposed [414877], 950 yd S. 16° E. of Churchmoor Hall, it is reversed, hading towards the south-east at an angle of about 15° from the vertical. As observed below, different conditions prevail in the vicinity of Caer Caradoc. D.C.G.

Between Marshbrook and Caer Caradoc the line of the fault is obscured by alluvium and glacial deposits, and nothing is known of its hade. From Marshbrook to Little Stretton it throws Silurian rocks on the western side against Caradoc Series and Pre-Cambrian strata to the east. Northwards from a locality [453930] about ½ mile south of Church Stretton church it forms the eastern boundary of the Silurian area of All Stretton. B.A.H.

Along the western flank of Caer Caradoc the fault forms the western limit of the Uriconian outcrop and farther north, on the west side of The Lawley, it is believed to consist of two closely-spaced parallel fractures. This fault has been traced to the north for many miles, and in the latitude of Wellington, Triassic rocks are thrown down westwards against Uriconian. In the Caradoc area the hade of the fault is to the west and there is no reason to suspect that it is other than a normal tensional fault. Cobbold (1927, p. 565) estimated the throw to be between 3000 and 6000 ft but, as Whittard (1952, p. 187) has pointed out, these figures are invalidated by stratigraphical breaks in the faulted rocks. Near New House, Marshbrook, where rocks of a conformable Llandovery sequence are thrown together, he has calculated the displacement to be about 600 ft. D.C.G., G.H.M.

The Lawley Fault. From Upper Carwood [4085] almost to Brokenstones the Lawley Fault, F2, separates the Stretton Shale Group from Wentnor grits. Its course is marked by only sporadic exposures of the contrasting rocks, but such evidence is strongly reinforced in the woods north of the Onny by a prominent broad bench or shallow groove at the foot of a distinct west-facing escarpment. This feature stands out extremely clearly in aerial photographs. The scarp is composed variously of Wentnor and Caradoc grits, the rocks to the west being usually Stretton Shales. North of the cross-fault at Brokenstones (below) the fault considered to be F2, and marked by a continuation of the topographic feature just described, separates Wentnor grits on its western side from Silurian shales. The outcrops of F1 and F2 converge ¼ mile north-east of Cwm Head church [426890]. At the southern end of Ragleth Hill the outcrops of F2 and F3 are coincident so that within the nexus of faults near Marshbrook F2 must run from F1 into F3. However, none of these faults, which affect only Palaeozoic rocks at the surface, has the south-easterly downthrow of F2, and it is impossible to identify that fault among them with any certainty. D.C.G.

The outcrop of F2 diverges from that of F3 at the north-eastern end of Hazler Hill. The fault separates rocks probably of Wentnor age from the Uriconian on its western side, but northwards Cambrian rocks take the place of the former group. Along the north-eastern flank of Caer Caradoc Hill it passes through Cambrian and Ordovician rocks. Northwards from the northern end of this hill the rocks to the west are either Uriconian or Lower Cambrian, and to the east Lower or Upper Cambrian or Ordovician. The outcrop of the fault swings westwards into the low ground between Caer Caradoc Hill and The Lawley, whence it follows an east-north-easterly and north-easterly course along the eastern side of this hill. Such sinuosities of outcrop led Cobbold (1927, p. 565) to postulate that the fault had a westward hade of roughly 45°, the Uriconian and Lower Cambrian rocks having been pushed over the younger rocks to the east. G.H.M.

Cwms–Hoar Edge Fault. Steep and reversed dips in the Chatwall Flags immediately west and north-west of Sibdon Carwood are the most southerly indications of the occurrence of the Cwms–Hoar Edge Fault, F3. In the valleys south of Hammonds-green the westerly downthrow is made manifest by the outcrop of Harnage Shales to the east of eastward-dipping Chatwall Flags. North-eastwards, towards Marsh-brook, the fault causes repetition of strata at various levels in the Caradoc Series with a consequent widening of the outcrops of the formations affected. The beds west of the fault dip at high angles towards it and are locally, as in the old railway cutting [413860] south of Glenburrell, overturned beyond the vertical. Immediately east of the fault the dips decrease to the order of 15°, a change which takes place within 5 yd in the old cutting, and is also well displayed at Woolston [424872]. The westerly downthrow is anomalous and unless the fault be interpreted as a reaction to the release of horizontal compression it is difficult to regard it as a normal fault. It may alternatively be a high-angle reverse fault, hading towards the east, the gently dipping beds on the hanging wall having been pushed upwards by westward pressure against the stable Pre-Cambrian horst (Fig. 19).

Between the Onny and Marshbrook the effect of the fault is not known. Its position is mapped partly on the evidence of change of dip and partly on topographic feature. In the brook west of Marshbrook the Cheney Longville Flags are thrown into a vertical attitude against it. D.C.G.

F3 is not clearly defined in the complex of faults between Marshbrook and Ragleth Hill. It can probably be identified with the fault which forms the western boundary of the narrow outcrop of Harnage Shale which ends 150 yd north-west of The Hough [450913]. This fault has a downthrow to the north-west, and the beds on this side of the fault are more steeply dipping than those on the south-eastern side. Crushed Harnage Shales adjacent to the fault are seen 475 yd W. 15° S. of The Hough. B.A.H.

In the north of the district F3 extends along the eastern side of Ragleth Hill, across Hazler Hill, and along the western edge of Hope Bowdler Hill. Thence to Hoar Edge it cuts Cambrian and Ordovician rocks, with an apparent downthrow towards the south-east. Cobbold (1927, pp. 565–6) regarded it as a 'septal' fault since along its course "the Caradocian rocks approach from the east at a gentle angle and are

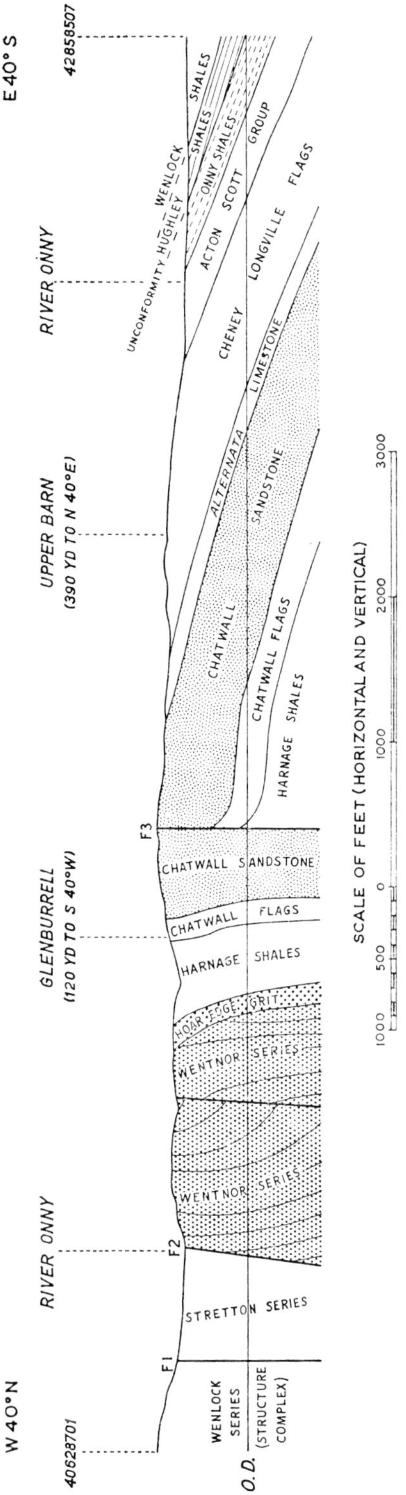

FIG. 19. Section along the Onny valley from Horderley to Wistanstow

turned up suddenly into a vertical or even reversed position just at the line of the fault". The explanation is however more complicated than that. In the Ordovician country to the east of Caer Caradoc the throw of the fault is to the east and the Hoar Edge Grit is cut out for some distance. Approaching the Sharpstones Thrust its throw must be small, for Harnage Shales appear on both sides of the fracture. South of this thrust the throw is to the west, for Ordovician, Cambrian, and probably Wentnor rocks appear on that side, with Uriconian to the east. This attitude continues to Hazler Hill, where a further change in the direction of throw occurs and persists along the south-eastern side of Ragleth Hill, causing Harnage Shales to be thrown down on the eastern side of the fault against Ragleth Tuffs. Such behaviour suggests that there is a considerable sinistral horizontal component in the movement of the fault, probably intimately connected with the Willstone Hill and Sharpstones thrusts. Indeed this diversity of throw as between Uriconian and later rocks would be wiped out if the rocks on the eastern side could be moved southwards so that the faulted ends of the base of the Harnage Shales on Ragleth Hill were brought together—a lateral shift of nearly a mile. A reasonable explanation would be that the Uriconian of Cardington Hill has moved northwards over Caradoc strata along the line of the Sharpstones Thrust, and the accompanying tear to the west is the Cwms–Hoar Edge Fault.

G.H.M.

Subsidiary Faulting. A minor fault with a north-easterly downthrow is exposed in a track section [387803] in the Hoar Edge Grit, 60 yd east of Coston. Several other exposures in the vicinity shows strong slickensiding and small-scale faulting, probably associated with the Church Stretton Fault.

North of Oldfield [406831] a north–south strike fault, parallel to F3, forms the eastern margin of the Longmyndian of Hopesay Hill. The Hoar Edge Grit is cut out by this fault, which throws down Harnage Shales against the Longmyndian.

B.A.H.

In the vicinity of Wart Hill [400847] the distribution of the Uriconian and Longmyndian rocks is ascribed to faulting, but few of the boundaries can be certainly shown to be faulted. In addition to the major north-north-easterly faults are several on north-north-westerly lines, which may be compared with the many cross-strike faults on the Long Mynd (p. 262).

West of Cheney Longville the Caradoc rocks are cut by a west-north-west fault which appears to cause an essentially horizontal displacement. Although the rocks are displaced sinistrally the zones of high and low dips are unaffected, and the folding and the movement along F3 are clearly later phases in the development of the structure. A smaller parallel fault cuts the rocks between Cheney Longville and the Onny but probably dies out east of F3.

Faults on north-north-westerly lines appear to delimit the small outcrop of Silurian rocks which lies west of the Onny just south of Horderley (Fig. 16). The more easterly fault can be precisely located on the river bank [408865] 820 yd N. 37° E. of Lower Carwood and its outcrop can be followed to its junction with F1, about 400 yd to the north-north-west. At the river bank exposure it can be seen to be offset dextrally by F2. The western fault follows a less well-defined course, which can be located to within about 20 yd in the tributary [407865] 810 yd N. 33° E. of Lower Carwood. It is uncertain whether the outcrop is limited towards the south-east by the convergence of the two faults now described or by the north-north-easterly fault which cuts the Longmyndian rocks 430 yd west of Glenburrell [409862].

The structural pattern in the Brokenstones area [4288] is exceedingly complex and, in view of the scattered nature of the outcrops, the interpretation presented on the map is in the nature of a hypothesis which attempts to reconcile the many anomalies of outcrop distribution in this small area. A north-westward valley divides the area into two parts which are in many ways geologically distinct. The most striking difference is the absence to the north of the Hoar Edge Grit which forms very prominent outcrops on the south side of the valley. In addition the outcrops of rocks of the Stretton and Wentnor series are differently disposed on the two sides of the valley. These effects are ascribed to a north-north-westerly fault, approximately along the line of the valley, cutting all the structures between F1 and the parallel fault about 700 yd to the south-east. About 150 yd south of Blakemoor a similar, but smaller, fault on an approximately easterly line cuts the Ordovician rocks lying west of F3. Their cross-cutting relationship to the other structures suggests that these two faults are the result of later movements, probably extending into post-Ludlow times. The other faults in the area run approximately parallel to the main Church Stretton faults and are considered to be more localized movements associated with the more persistent fractures. If the Wentnor rocks are indeed faulted against those of the Stretton Series, and not unconformable on them, as Whittard (1953, p. 247) believed, then the displacements in these steeply dipping beds must be essentially horizontal. On the other hand the faults which limit the Ordovician and Silurian outliers may have had a major vertical component.

An important fault on a north-westerly line cuts off the outcrop of the Chatwall Sandstone about 300 yd south-west of Whittingslow [430887]. Its north-easterly downthrow is demonstrated by the occurrence in that direction of rocks of the Cheney Longville Flags. The displacement in these steeply inclined beds, be it vertical or horizontal, amounts to at least 600 ft and may exceed 1200 ft. D.C.G.

The main fractures F1 and F3 are linked by three subsidiary faults between Marshbrook and Ragleth Hill. Of these the central fault is the most significant, with a northerly downthrow of some 300 ft.

South-east of F3 trough faulting has let down the Cheney Longville Flags and Acton Scott Group, cutting out the lower beds of the Caradoc Series between Ragdon [458915] and The Hough. This trough is bounded to the north-west partly by F3 and partly by a branch fault from F3, between The Hough and the railway 300 yd north of Marshbrook station, with a south-easterly downthrow of 150 to 350 ft. The southern side is bounded by a fault which branches from F3, 700 yd north-east of Whittingslow [437894], passes to the south-east of Ragdon, and rejoins F3 near Dryhill [459923]. The north-westerly downthrow of this fault is about 100 ft, north of Acton Scott, increasing to over 300 ft east of Ragdon. Associated minor folds occur in a stream section 1000 yd north of Acton Scott church. A small fault trending westwards through Marshbrook Station throws down beds at the top of the Cheney Longville Flags on the north side of the road against beds rather lower in the sequence to the south.

The outcrop of the Acton Scott Limestone west of Acton Scott is broken up by small-scale block faulting, the faults having north-east and south-east trends and throws of up to 50 ft. A fault under the alluvium of the valley between Henley and Marshbrook has an easterly downthrow of about 100 ft. B.A.H.

Thrust faults of Caer Caradoc. The evidence for the presence of a thrust fault beneath the Caer Caradoc Rhyolites has already been given (p. 266) in the description of the minor folding and shearing which accompanies it. The small outcrop of rhyolite 600 yd north of the summit of Caer Caradoc probably owes its position to a thrust between it and the underlying Little Caradoc Basalts. Shearing and crush-breccias are prominent features in the crag of rhyolite exposed there.

A similar cause is invoked to explain the presence of the 'cakes' of Cwms Rhyolites and Caer Caradoc Andesites which rest discordantly on the Ragleth Tuffs and the associated dolerite intrusion, west and north of Three Fingers Rock, the prominent crag [47089474] 1030 yd south-west of Caer Caradoc summit. The rhyolites, which are better exposed, are everywhere sheared and crushed near the junction, and the feature, made by weathering along the outcrop of the thrust plane on which they rest, is clearly seen when looking north from Three Fingers Rock or south-west from the slopes west of Caer Caradoc summit. The thrust plane is steeply inclined to the south-west and the shearing within the rhyolite suggests movement from that direction.

Willstone Hill Thrust. The dissimilarity in appearance of the rocks of Hope Bowdler Hill and Cardington Hill and the discordance of their structures are explained by the presence of an important fault, here called the Willstone Hill Thrust (Fig. 4). This fracture seems likely to belong to the same group as the Sharpstones Thrust and Cwms–Hoar Edge Fault, and to be a thrust with movement from the south.

Sharpstones Thrust. The Sharpstones Thrust, here so named, is a major displacement, for Caradoc strata on its north side are thrown against Uriconian on the south. If this were a normal fault the displacement would be at least 4000 ft near North Hill, a figure which casts some doubt on such an explanation of its nature, particularly as its strike is at right angles to that of the Church Stretton Fault Complex which is later in age and against which it apparently terminates. The possibility that it is a thrust finds support in the form of the trace of the fault plane on the ground, so far as it can be accurately mapped by feature lines in the absence of actual contacts. The age of the thrust is certainly pre-Silurian as the basal bed of that system passes unbroken across its course.

Hill End Thrust. The same reasoning applies to the Hill End Thrust, which is here so named, as to the Sharpstones Thrust. The Hill End Thrust bounds the Uriconian and Cambrian rocks of Hill End on the west and north and can be regarded as a branch of the Sharpstones Thrust.

Hazler Hill faults. Two faults are shown on the map, running in a north-westerly direction on Hazler Hill and in the valley to the east. They are inferred in badly exposed ground between areas with entirely distinct successions, which cannot be correlated.

G.H.M.

The East Shropshire Syncline

In contrast to the Long Mynd–Wrekin uplift the East Shropshire Syncline is, within the present district, an uncomplicated area of open folds and few important faults. A marked feature of the constituent folds is that the north-western limbs of the anticlines are invariably steeper than the south-eastern. The Caledonoid trend is preserved even into the Carboniferous rocks of the Clee Hills. The Titterstone Clee Fault is the most important fault in the area of the Syncline and can be followed for several miles to north-east and to south-west. Significant dip faults affect the Silurian rocks near Craven Arms, and, owing to the lack of 'marker' horizons, the Old Red Sandstone is probably considerably more faulted than it appears to be.

DETAILS

Folding

In this account each of the constituent folds of the East Shropshire Syncline is described from south-west to north-east, the more westerly being described first.

Brown Clee Syncline. In the area between Clungunford [3978] and Craven Arms [4382] the Silurian rocks dip eastwards and south-eastwards at an average angle of 8°. Similar dips prevail in the Old Red Sandstone to the east until the axial region of the Brown Clee Syncline is approached (Fig. 2). In the central part of the area the Silurian and Old Red Sandstone dip south-eastwards at about 10° on this limb of the fold, and the Ordovician at about 15°. In the Wistanstow area these figures are increased by 5° or 10° and in the extreme north, between Willstone [4995] and Chatwall Hall, dips of between 30° and 75° to the south-east are recorded in the Cheney

Longville Flags and older rocks. Near Downton Hall [5279] the axis of the fold plunges north-eastwards at rather less than 10°, but from Cuckoopen [5480] to Bockleton Court [5783] it is essentially horizontal. North-east of Brown Clee Hill the plunge is to the south-west at a low variable angle. In the northern part of the outcrop of the Ditton Series the dips are much more variable and the fold axes are ill defined. Southerly dips of up to 30° are indications of a local irregularity near Derrington [6090]. Northwards from Middleton Priors a subsidiary syncline is developed for about 1 mile, indicated primarily by persistent westerly dips near Netchwood [6291]. East of Monkhopton it is clear that the general dip is southwards although individual observations show much variation.

Ledwyche Anticline (Ball, Dineley and White 1961, p. 210). Localized disturbances of bedding, sometimes associated with faulting, are common between Stoke Lodge [5680] and Coldgreen [5981], near the crest of the Ledwyche Anticline. This structural feature is marked by a zone of very low dips in the Ditton Series, extending north-eastwards from Gibbridge [5780]. In the south-west it appears to be essentially horizontal, but between Besom Farm [6082] and Loughton a south-westward plunge of up to 5° may be observed. A low plunge to the north-east is indicated in the ground between Aston Botterell and Neenton, beyond which the crest of the fold is not clearly recognized. The eastward flexure of the crest in this area (Fig. 2) suggests that the fold has developed in at least two stages. The south-eastern limb is gently inclined but it is more complicated by minor folding and faulting than the ground between it and the Church Stretton disturbance. In addition to a number of broad open flexures there are several narrow zones of sharp folding, usually accompanied by faulting. Of these the most prominent are the faulted anticline which is followed by the stream south of Goldthorn [6079] and the broader westward-plunging syncline at the junction of the Wheathill and Ingardine brooks [6381].

Titterstone Clee Syncline (Ball, Dineley and White 1961, p. 209). The Titterstone Clee Syncline affects the Lower Old Red Sandstone and the unconformably overlying Farlow Series and Carboniferous strata. Paucity of outcrops makes the trough of the fold difficult to locate precisely. It appears to plunge north-eastwards and, as Ball and Dineley (1961, p. 210) have pointed out, to run into the Titterstone Clee Fault in the vicinity of the Rea valley. South-east of the fault the Old Red Sandstone rocks dip more steeply than on the north-western limb of the syncline, abnormally high dips being developed close to the fault. The occurrence of a subsidiary anticline on a curving northerly and north-north-easterly axis is indicated on the geological map by the shape of the outcrop of the base of the Clee Group. This fold plunges northwards at upwards of 20°. D.C.G.

FAULTING

The distribution of outcrops on View Edge [4280] demonstrates the occurrence of a number of minor faults. A north–south fault passes beneath the farm south of Stoke Wood and accounts for the absence of the Aymestry Group at the top of the escarpment in the western part of the wood. The easterly downthrow at the farm appears to be about 45 ft. In the eastern part of Stoke Wood the outcrop of the Aymestry Group is duplicated by an east-south-east fault with a northerly downthrow of up to about 70 ft. In Church Way [435806] the more northerly outcrop of the Group is completely cut out by this fault, which also causes minor dislocation of the beds in the northern part of the old View Edge quarries. Between the wood and the railway line anomalous steep dips occur in the two outcrops of the Aymestry Group. A north-north-east fault cuts off the western end of the southern outcrop about 300 yd west of Church Way. In the valley [429798] 500 yd east-north-east of Gorst Barn the band of *Conchidium knightii* is thrown down to the west by about 40 ft. About 75 yd to the east a north-north-west fault occurs with a westerly downthrow of 12 ft. On the northern side of the valley a sinuous fault branches off to the north-east from the 40-ft fault. It has a north-westerly downthrow of up to about 100 ft.

The River Onny has breached the limestone barrier of the Aymestry Group along the line of a N.W.–S.E. fault which is here called the Stokesay Fault. The limestone is clearly displaced to the south-east on the north-eastern side and there is a comparable displacement in the outcrop of the Downton Castle Sandstone but there is no direct evidence as to the direction of movement. The effective vertical downthrow of the top of the Aymestry Group is about 270 ft to the south-west. D.C.G.

Between Broome and Craven Arms the outcrop of the Tickwood Beds is affected by three strike faults. The most westerly is ½ mile east of Broome [409810] where the basal beds of the Tickwood Beds are thrown down about 25 ft to the west. A second fault follows the line of Park Lane, trending N. 15° E. and with a westerly downthrow of about 65 ft. The third fault diverges from the second 450 yd west of Clapping Wicket [418812] and ends against the Stokesay Fault 450 yd north of Paddock [427824]. This fault has a westerly downthrow of about 60 ft, north of Clapping Wicket, increasing to about 100 ft near its junction with the Stokesay Fault. The Stokesay Fault, at Paddock, has a westerly downthrow of about 250 ft.

In the area between Grove [436845] and Culmington [4982] the Wenlock, Ludlow and Downton series are affected by a number of dip faults. The total effect of these faults is a south-westerly downthrow of about 700 ft, measured on the displacement of the Aymestry Group outcrop. It is possible that there may be a horizontal component in the throw of some of the faults, but there is no direct evidence of this.

The most important of these is the fault here named the Bache Fault. This has a south-westerly downthrow of 475 ft, measured on the Downton Castle Sandstone at Bache [4782]. The outcrop of the Ludlow Bone Bed is shifted dextrally by 1650 yd. The fault gives a very strong feature at Bache, where the Downton Castle Sandstone and the hard siltstones of the Upper Ludlow Shales are thrown against the soft Old Red Sandstone marls. Near Greenway Cross [460828] the Bache Fault is shifted dextrally 475 yd by a cross fault. North of this cross fault it throws Upper Ludlow Shales against Lower Ludlow Shales, and there is no feature. About 200 yd north-east of the Bache Fault there is a sub-parallel subsidiary fault with a north-easterly downthrow of about 100 ft. This fault ends against the cross fault at Greenway Cross.

Another important fault runs from Siefton to join the Bache Fault between Upper Dinchope [458834] and Ireland [445839]. This fault, here named the Dinchope Fault, has a southerly downthrow, displacing the Downton Castle Sandstone by about 200 ft and the Aymestry Group by 100 ft.

A third fault runs from the cross fault 300 yd west-south-west of Greenway Cross to join the Bache Fault 275 yd east-south-east of Ireland. This fault, which is oblique to the strike, also has a south-westerly downthrow, and causes a repetition of the Aymestry Group outcrop. Five hundred yards south-west of Upper Dinchope beds of the Aymestry Group are anticlinally folded, probably as a result of drag on this fault.

North-west of Ireland the Bache Fault splits into two parts and causes a double repetition of the outcrop of the Wenlock Limestone. The total downthrow on the Limestone is about 600 ft to the south-west. The fault system cannot be traced for any distance in the Wenlock Shales to the north-west, nor in the Old Red Sandstone beyond the Culmington area, owing to lack of marker horizons.

A dip fault with a downthrow of about 50 ft to the north-east causes a breach in the Wenlock Limestone scarp near Lower Dinchope [446846]. A less pronounced break in the scarp is caused by a fault ½ mile S. of Harton [480878] with a south-easterly downthrow of 75 ft, which runs obliquely across the strike. Two hundred yards north-east of Hillend, near Westhope [473855], a strike fault of small throw gives rise to minor folds and vertical bedding in the Lower Ludlow Shales exposed in the brook. B.A.H.

In addition to the major faults which extend from the Silurian into rocks of the Downton Series near Culmington, are a number of minor cross-strike faults which are invoked to explain the abrupt termination of some of the sandstones of the latter formation. Faults of this type cut the '*Psammosteus*' Limestones in places and occur throughout the length of Corve Dale. The direction of displacement is in most cases indeterminable.

Several minor faults occur in the stream east of the inn at Lower Hayton [5080], the overall effect being a south-westward downthrow of one limestone of the '*Psammosteus*' group by some 20 ft. About ¼ mile north-east of Lydehole [5282] a similar displacement is indicated by the disposition of the limestone outcrops. The trend of the fault is suggested by a small dislocation of a cornstone in the Downton Series about 700 yd to the north-west. A small fault trending probably to the east-south-east is exposed in the brook [562870] 500 yd east of Tugford church, the lowest of the '*Psammosteus*' Limestones, on the south side, being thrown down a few feet against Downton marls. Minor displacements of these Limestones, as shown on the map farther to the north-east, are caused by faults which are not exposed.

The absence of the '*Psammosteus*' Limestones in the stream south of Measons [5479] is attributed to a fault, with north-easterly downthrow, which cuts Downton Series rocks in the stream [546784] just north-west of Crowleasows. The throw is estimated to be of the order of 200 ft. An apparent displacement of the Limestones across the road south of Lackstone [5780] is thought to be due to a fault along a south-south-easterly line which passes just west of the farm. Between Langley [5879] and Gibbridge [5780] the cornstone marking the base of the Ditton Series is thrown down to the north by nearly 200 ft, the fault being considered to follow an easterly course along the valley south of these farms.

Within the outcrop of the Ditton Series the paucity of marker horizons makes the recognition of faults difficult, and where they are visible in outcrops it is seldom possible to evaluate the extent, or even the direction, of the displacement. Such exposures of fractured rocks are not uncommon but they do not reveal any major structures or broad tectonic pattern. The faults are predominantly at high angles, with trends in the directions of east-north-east and west-north-west. Calcite veining and near-horizontal slickensiding are common. An unusual concentration of such faults occurs in the Batch Brook, north-west of Farlow.

Three faults are recognized in the vicinity of Bockleton Court [5783] by the dislocations observed in the Abdon limestones. Their north-easterly trends are approximately parallel to the axis of the Brown Clee Syncline (p. 272). The most southerly fault is marked by the termination, on the south side of a deep dry valley, of a line of workings in the Upper Abdon Limestone. It probably has a south-eastward downthrow of about 50 ft. The south-westward extension of the fault line has been taken along the foot of the steep slope, facing south-east, which runs from Bockleton Court to Furlong [5681]. The middle fault is indicated primarily by anomalous northerly dips immediately south of the farm buildings, supplemented by a probable small displacement of the Lower Abdon Limestone. The evidence is weak, but a downthrow of some 25 ft to the north-north-west is suggested. East of the farm a narrow dry valley has been taken to define the outcrop of the fault. The fault north of the farm throws down the Upper Abdon Limestone by about 80 ft to the south-east. Of the two east-north-easterly faults shown on the map at Stoke Lodge [5680] the more northerly is exposed at three localities in the streams. The more southerly is extrapolated along a topographic feature from a stream exposure of disturbed sandstone just west of Stoke Lodge. The direction of throw is unknown in both cases.

The north-easterly fault on Clee Burf has a north-westerly downthrow in the Coal Measures, and apparently also in the highest beds of the Clee Series, of some 25 ft. This fault appears to be the same as that which effects a south-easterly downthrow in the Upper Abdon Limestone north of Bockleton Court. On the eastern side of Abdon Burf a north-south fault has an easterly downthrow of the order of 60 ft in the Coal Measures. It clearly displaces the Upper Abdon Limestone near Park Gate

[6087] but the size and direction of the throw are uncertain. The excellent development of topographic features on the western slopes of Brown Clee Hill, and the frequent quarrying of the Abdon limestones, facilitate the recognition of minor faults in that area. The faults themselves are not exposed.

Faulting is more common in the ground between Cleeton St. Mary and Farlow. Farlow Series and Carboniferous rocks are thrown down to the west on a north-north-easterly fault along the line of the brook west of Cleeton St. Mary, and indication of a north-north-westerly fault is given by the almost vertical attitude of rocks of the Ditton Series just above the bridge [609787]. The faulted anticline at Goldthorn [6079] (p. 240) can be followed for about 700 yd along an east-north-easterly line. Minor faults on north-north-easterly and north-north-westerly lines are exposed in the Farlow Brook west of Well Farm [6481].

Fossils collected by Ball and Dineley (1961, pp. 219–39) show that rocks in the brooks ¾ mile west of Silvington and in Batch Brook, south of Wheathill church, belong to a lower horizon than those forming the high ground near Besom Farm and in the Ingardine Brook ½ mile south-south-east of Wheathill. The mapping of individual beds suggests a contrary relationship and the anomaly is probably to be explained by faulting along an east-north-easterly line, passing near the farm [622811] west of Ingardine, perhaps a continuation of the east–west fault at Langley (p. 275). No clear evidence of this faulting has been observed in the field.

The Carboniferous rocks of Titterstone Clee Hill are bounded to the south-east by the Titterstone Clee Fault (Ball, Dineley and White 1961, p. 211), by which they are thrown down against various formations of the Lower Old Red Sandstone. Because of unconformities in the sequence of rocks affected it is not possible to calculate the throw of the fault, but west of Catherton [6578] it must be at least 400 ft. Less than ½ mile beyond the southern border of the district, rocks of the Downton Series lie against the fault, indicating a north-westerly downthrow of over 1500 ft. North of Detton Hall [6679] the fault splits, and the ground between the fault components to the north-east (Dudley Sheet 167) is considerably broken. The north-north-easterly branch continues into the adjacent district as the Deuxhill Fault (Whitehead and Pocock 1947, p. 131), but within the present district it has but a small throw in the Farlow Series. Ball and Dineley (1961, p. 211, map) considered that the continuation of the fault in the Dudley district runs about 400 yd farther east. It seems equally possible that along that line the Farlow Series rests unconformably on the Clee Group and is not faulted against it. The north-easterly branch has a greater downthrow, towards the north-west, which varies considerably along its length. North-eastwards (Dudley Sheet 167) it splits into several parallel faults but the major displacement of the Titterstone Clee Fault is not developed in any of them. An important east-west fault lets down a small outlier of Coal Measures sandstone in the angle between the two main branches of that fault. At the southern end of Catherton Marshes [6478] a south-south-westerly fault separates beds of the Clee Group to the west from those of the Ditton Series. It is exposed in the stream about 200 yd south of the limits of the one-inch sheet.

Analysis of Earth Movements

Cobbold (1925) and Whittard (1952, p. 185) have drawn attention to the large number of unconformities which break the sequence of rocks in the Church Stretton area, and to the fact that this indicates a succession of orogenic disturbances, usually of comparatively minor intensity. Recurrent movements have taken place along some of the major structures, a notable example being the Eastern Uriconian Axis.

Pre-Cambrian movements. Two periods of pre-Cambrian folding are recognized in the Longmyndian rocks, and it is possible that the Uriconian rocks were themselves folded prior to the deposition of the Longmyndian. The evidence for this earliest movement is derived from the possible unconformity between the basal Longmyndian Helmeth Grits and the underlying Uriconian rocks at the southern end of Caer Caradoc (p. 37), and from the Charnoid folding which affects the Uriconian rocks alone on Caer Caradoc and Cardington Hill (p. 265). The unconformity and the Cardington Hill syncline are not conclusively proved and it has been suggested by Dr. Gilbert Wilson (personal communication) that the folding on Caer Caradoc may be a volcanic structure of a local and adventitious nature. In addition the Uriconian rocks of Ragleth and Hazler hills exhibit a Caledonoid trend.

The Wentnor Series oversteps the Stretton Series westwards to rest directly (Montgomery Sheet 165) on the Western Uriconian and is much more widespread east and south of the Long Mynd than the Stretton Series (Whittard 1952, p. 149). The angular discordance between the two series is not very great in the area of the Long Mynd, but if the junction between them at Brokenstones (p. 77) is not faulted it is probably highly discordant. The low angular discordance between the Caradoc and Wentnor rocks at Glenburrell (p. 269) shows that there at least the latter were not steeply inclined by pre-Cambrian movements. D.C.G.

Outside the present district in the Habberley Brook area, Caradoc shales rest with strong unconformity on steeply inclined rocks of the Wentnor Series (Pocock and others 1938, p. 91, figs. 14 and 15) and it is thus inferred that the intense folding of the Longmyndian is at least of pre-Caradoc age. To the north-west of the present district, in the adjacent Shelve area, there is a continuous stratigraphical sequence from the Arenig into the Caradoc, with no evidence of folding in this area during this period of deposition. Comparatively small breaks have been detected in the palaeontological record at the base of the Stiperstones Quartzite (Arenig) in the Shelve area (Whittard 1931, pp. 324 and 344), and also between the Upper and Middle, and Middle and Lower Cambrian in the Comley area (Cobbold 1927). The mapping in the ground immediately east of Caer Caradoc indicates a minor phase of folding and faulting of the Cambrian rocks of the Comley area prior to the deposition of unconformable Caradoc sediments. Apart from the minor movements noted above, there is, in the Cambrian and lower Ordovician rocks of Shropshire, no indication of folding of such magnitude as is proposed for the Longmyndian syncline. It is therefore suggested that the overfolded syncline in the Longmyndian was formed in pre-Cambrian times.

Stille (1958, p. 31) described the Pre-Cambrian of Shropshire as being of Algonkian age. He suggested that the Uriconian is the product of Algonkian acid vulcanism of 'subsequent' type, possibly referable to the Algoman orogeny, which might be represented by the Malvern gneiss. The folding of the Longmyndian sediments was considered by Stille to be of late pre-Cambrian (Assyntic) age. J.E.W.

Pre-Cambrian faulting on a major scale is not clearly recognized but it is probable that the Church Stretton disturbance was active before and during the deposition of the Longmyndian. Some fault movement may have accompanied the intense folding of the Longmyndian, but the faults mapped in these rocks are probably to be assigned to post-Ordovician orogeny (p. 278). The thrust-faults of Caer Caradoc are probably of pre-Cambrian, even of pre-Longmyndian, origin. D.C.G., J.E.W.

Intra-Cambrian movements. Cobbold (1927, pp. 569–71) recognized no fewer than seven breaks in the sequences of faunas and sediments of the Cambrian system east of Caer Caradoc. Of these all but one appear to result from epeirogenic movements. The exception is that which separates the Lower from the Middle Cambrian. Cobbold showed that the Lower Cambrian rocks were folded, uplifted, and differentially eroded before the deposition, on an irregular surface, of the Middle Cambrian, the later members of which overlap the earlier (1927, fig. 9). No additional information has been obtained about the age of the post-Middle Cambrian, pre-Caradoc faults described by Cobbold (1927, pp. 566, 571).

Post-Cambrian movements. The progressive subsidence of Upper Cambrian times was followed by a period of uplift and erosion, accompanied by minor folding and faulting, which kept the sea away from the present district until the Caradocian epoch, when the Hoar Edge Grit, or in places the Harnage Shales, was deposited on a floor of possible Uriconian, Wentnor Series or Cambrian rocks usually without marked disconformity. In the Shelve area, west of the Pontesford–Linley Disturbance (Welshpool (151) and Montgomery (165) Sheets) older Ordovician rocks, from the Arenig upwards, rest with slight unconformity on the Tremadoc shales, but the Caradoc Series alone is developed immediately east of the disturbance near Pontesford (Shrewsbury Sheet 152). Thus during pre-Caradoc time the Long Mynd–Wrekin Axis appears to have formed the western limit of a stable area adjacent to the Welsh Geosyncline, of which the Shelve area formed the eastern margin. During the Caradoc it continued to preserve a relative stability to which the extremely shallow-water deposits to the east of it bear witness.

Post-Ordovician movements. In the interval between the deposition of the Caradoc Series and that of the Upper Llandovery important orogenic movements affected the rocks of the Church Stretton district. It was probably at this period that differential uplift along the Eastern Uriconian Axis tilted the Cambrian and Ordovician rocks east of it into the steeply dipping attitude which they occupy to-day. The Upper Llandovery rocks overstep the earlier formations, their younger beds overlapping the older towards the south and west. Despite the clear evidence of a pre-Upper Llandovery unconformity the present attitude of the more easterly Ordovician rocks patently owes much to post-Llandovery movements, and it is possible that these caused much of the uplift farther west. D.C.G.

The faulting in the Longmyndian has been discussed by Whitehead (*in* Pocock and others 1938, p. 176) and Whittard (1952, p. 186). Whitehead pointed out that the faulting is most probably later than the folding because of the effects of fault displacements on the steeply dipping Longmyndian. He also noted that near Longden Manor [SJ/422060][1], a transverse fault in the Wentnor Series probably, although not certainly, affects Caradoc shales. He concluded from this and other evidence that the transverse faults are probably of late Ordovician (post-Caradoc) or early Silurian age.

Whittard described powerful northerly tear faults in the Ordovician of the Shelve inlier accompanied by a complementary set of shear faults. These fault movements took place before the deposition of the Upper Llandovery rocks and Whittard noted that the pattern of the Shelve fault system is similar to

[1]This National Grid Reference lies within the 100-km square SJ (33). References without initial letters lie within the 100-km square SO (32).

that in the Longmyndian. He referred the greater part of the faulting in the Longmyndian to an orogenic phase in late Ordovician times. This was considered to be an early phase of the Caledonian orogeny and is tentatively correlated with the coeval Taconian movements of North America.

The survey of the Church Stretton district has found nothing to conflict with the conclusions set out above. It appears that the north–south faults are the major features while the faults trending west-north-west to east-south-east seem to mark smaller adjustments consequent upon the greater north–south movements. J.E.W.

By analogy much of the transverse faulting within the Eastern Uriconian Axis, and considerable movements of the north-easterly faults of the Church Stretton disturbance, took place during this orogenic phase. It is clear however that there was considerably activity along many of these lines at later periods.

Intra-Devonian movements. Williams (1953, p. 201) and Jones (1956, pp. 332–3) have demonstrated that in Wales, where the Lower Old Red Sandstone is unconformable upon the Silurian, considerable folding of the Lower Palaeozoic rocks took place after the deposition of at least most of the Lower Old Red Sandstone. Throughout Britain there is a marked unconformity between the Upper and Lower Old Red Sandstone and it is to this interval that, in the present district as in Wales, the principal Caledonoid folding of the Palaeozoic rocks is assigned. East of the Church Stretton Fault, in the south-east foreland of the Caledonian geosyncline (Bailey 1938, p. 47), this has resulted in the broad simple folds of the Clee Hills area.

The Church Stretton Fault Complex was active during Wenlock and Ludlow time (p. 143) as a boundary between the shelf-sea to the south-east and the more rapidly subsiding basin to the north-west. Post-Silurian movement is clearly responsible for the distribution of Silurian rocks immediately west of, and within, the Fault Complex, and for the folding and faulting exhibited by these western outcrops. Much of it may have taken place in the intra-Devonian interval, but the post-Carboniferous and post-Triassic fault movements, of which there is clear evidence in the Wellington area (Shrewsbury Sheet 152), may have been significant in the present district as well.

Armorican movements. The interpretation of the orogenic phases which have affected the Carboniferous rocks of the Clee Hills depends on the age, Namurian A–Westphalian B (see p. 250; Jones and T. R. Owen 1961, p. 292) assigned to the Cornbrook Sandstone. If this age be correct then folding of pre-Namurian (Sudetic) and of pre-Westphalian C age is implied. The effect of mid-Dinantian folding, especially evident in South Wales, has been erased by the later Sudetic movements, although Jones and Owen (1961, p. 293) interpret it as the main pre-Westphalian movement. George (1956, p. 310, fig. 1) attached great importance to the Sudetic movements in the Usk Anticline and interpreted the pre-Cornbrook Sandstone structure in the area of Titterstone Clee as a southerly plunging syncline, continuous or homologous with the Forest of Dean Syncline.

In the absence of rocks younger than Middle Coal Measures (Jones and Owen 1961, p. 293) the relative importance in the Clee Hills area of pre-Upper Coal Measures and late Coal Measures movements cannot be gauged. The effect of the earlier phase is seen in the Coalbrookdale Coalfield (Dudley Sheet 167) where the Upper Westphalian Coalport Group rests unconformably on the 'Middle' Coal Measures (Whitehead and Pocock 1947, p. 121). The importance of the latter phase in the West Midlands generally has been pointed out by Wills (1948, pp. 56–8), and a particular instance in the Abberley Hills (Droitwich Sheet 182) has been described by Mykura (1951). Lower Westphalian rocks are cut by the Titterstone Clee Fault, and to the north-east (Dudley Sheet 167) faults of the same group displace Upper Westphalian and Triassic rocks. The fault may have exercised some control on the deposition of the Clee Group and the Farlow Series. Movement along this line has thus probably been recurrent from at least Devonian until post-Triassic times.

The Armorican folding in the Church Stretton district followed the Caledonoid trend already established in the older rocks and it is probable that the earlier structures were accentuated by these later movements, at least in those areas where the rocks had previously suffered little compression. The principal folding of the Silurian and younger rocks south-east of Church Stretton has been assigned by some authorities to this period (Pocock and Whitehead 1948, pp. 7–8), but general stratigraphical considerations indicate that it probably took place in Devonian times (p. 279). D.C.G.

References

BAILEY, E. B., *in* BAILEY, E. B. and HOLTEDAHL, O. 1938. North-western Europe Caledonides. *Regionale Geologie der Erde*, Leipzig, **2**, Abt. 2, 1–76.

BALL, H. W., DINELEY, D. L., and WHITE, E. I. 1961. The Old Red Sandstone of Brown Clee Hill and the Adjacent Area. *Bull. Brit. Mus. (Nat. Hist.) Geol.*, **5**, 175–310.

COBBOLD, E. S. 1925. Unconformities in South Shropshire. *Proc. Geol. Assoc.*, **36**, 364–7.

—— 1927. The Stratigraphy and Geological Structure of the Cambrian Area of Comley (Shropshire). *Quart. J. Geol. Soc.*, **83**, 551–73.

DEAN, W. T. 1964. The geology of the Ordovician and adjacent strata in the southern Caradoc District of Shropshire. *Bull. Brit. Mus. (Nat. Hist.) Geol.*, **9**, 257–96.

GEORGE, T. N. 1956. The Namurian Usk Anticline. *Proc. Geol. Assoc.*, **66** for 1955 297–316.

HOLLAND, C. H. 1958. The Ludlovian and Downtonian Rocks of the Knighton District, Radnorshire. *Quart. J. Geol. Soc.*, **114**, 449–82.

JAMES, J. H. 1956. The Structure and Stratigraphy of part of the Pre-Cambrian Outcrop between Church Stretton and Linley, Shropshire. *Quart. J. Geol. Soc.*, **112**, 315–37.

JONES, D. G. and OWEN, T. R. 1961. The Age and Relationships of the Cornbrook Sandstone. *Geol. Mag.*, **98**, 285–94.

JONES, O. T. 1956. The Geological Evolution of Wales and the Adjacent Regions. *Quart. J. Geol. Soc.*, **111** for 1955, 323–51.

MYKURA, W. 1951. The Age of the Malvern Folding. *Geol. Mag.*, **88**, 386–92.

POCOCK, R. W., WHITEHEAD, T. H., WEDD, C. B., and ROBERTSON, T. with contributions by WRAY, D. A., STUBBLEFIELD, C. J., CANTRILL, T. C., and DAVIES, W. M. 1938. Shrewsbury District. *Mem. Geol. Surv.*

—— and WHITEHEAD, T. H. 1948. The Welsh Borderland, 2nd Edit. *British Regional Geology, Geol. Surv.*

STILLE, H. 1958. Die assyntische Tektonik in geologischen Erdbild. *Beitr. z. Geol. Jahrb.*, **22.**

WHITEHEAD, T. H. 1955. The Western Longmyndian Rocks of the Shrewsbury District. *Geol. Mag.*, **92,** 465–70.

—— and POCOCK, R. W. 1947. Dudley and Bridgnorth. *Mem. Geol. Surv.*

WHITTARD, W. F. 1931. The Geology of the Ordovician and Valentian rocks of the Shelve country, Shropshire. *Proc. Geol. Assoc.*, **62,** 322–39.

—— 1931. Easter Field Meeting (Extension) to Minsterley. *Proc. Geol. Assoc.*, **62,** 339–44.

—— 1932. The Stratigraphy of the Valentian Rocks of Shropshire. The Longmynd-Shelve and Breidden Outcrops. *Quart. J. Geol. Soc.*, **88,** 859–902.

—— 1952. A Geology of South Shropshire. *Proc. Geol. Assoc.*, **63,** 143–97.

WILLIAMS, A. 1953. The Geology of the Llandeilo District, Carmarthenshire. *Quart. J. Geol. Soc.*, **108** for 1952, 177–207.

WILLS, L. J. 1948. *The Palaeogeography of the Midlands.* Liverpool.

Chapter 10

GLACIAL, POST-GLACIAL AND RECENT DEPOSITS

GLACIAL DEPOSITS

Introduction

The glacial deposits of the Church Stretton district occur chiefly in the north-western part of the district and consist of boulder clay locally overlain by sand and gravel. The sequence of the deposits indicates only one period of glaciation which is tentatively correlated with a maximum of Irish Sea and Welsh ice in Shropshire. There is, however, no local evidence that the maxima of the Irish Sea and Welsh ice were coincident in the Church Stretton district. The deposits of sand and gravel occur at various topographic levels and are considered to mark retreat stages of the valley glaciers. There are many striking river channels and dry valleys which were probably cut by melt water from the retreating glaciers.

Boulder Clay

The most extensive areas of boulder clay occur in the valley of the River East Onny, in the Church Stretton–Horderley valley, in Ape Dale, and in its southern continuation towards Clungunford [400787].[1]

The boulder clay of the East Onny valley is continuous with that in the low ground west of Asterton [398912] and extends northwards from Wentnor [384927] to near New Leasowes [404998] (Shrewsbury Sheet 152) where it reaches a height of about 1050 ft O.D. South and east of Wentnor, boulder clay covers the large hollow of Asterton and Wentnor Prolley Moors below the 1000-ft contour.

In the Church Stretton–Horderley valley boulder clay covers most of the ground below the 800-ft contour, but on the north-east shoulder of the Long Mynd ice from the Shropshire Plain apparently over-rode the higher ground up to about 1100 ft O.D. and left thin deposits of boulder clay around Plush Hill and Jinlye [452963] (Fig. 20).

In the valley between Clungunford and Craven Arms boulder clay is confined mainly to the ground below 700 ft but there are some small deposits between 800 ft and 900 ft O.D. near Barlow [383840]. In the southern part of Ape Dale and in the Acton Scott–Whittingslow area boulder clay reaches a height of about 770 ft O.D., and in one small area to the south of Alcaston [457871] it occurs on top of Wenlock Edge at about the same level. To the north of Rushbury [514918] there are extensive spreads of boulder clay, which reach a

[1]Such numbers are National Grid References with the 100-km square SO (32).

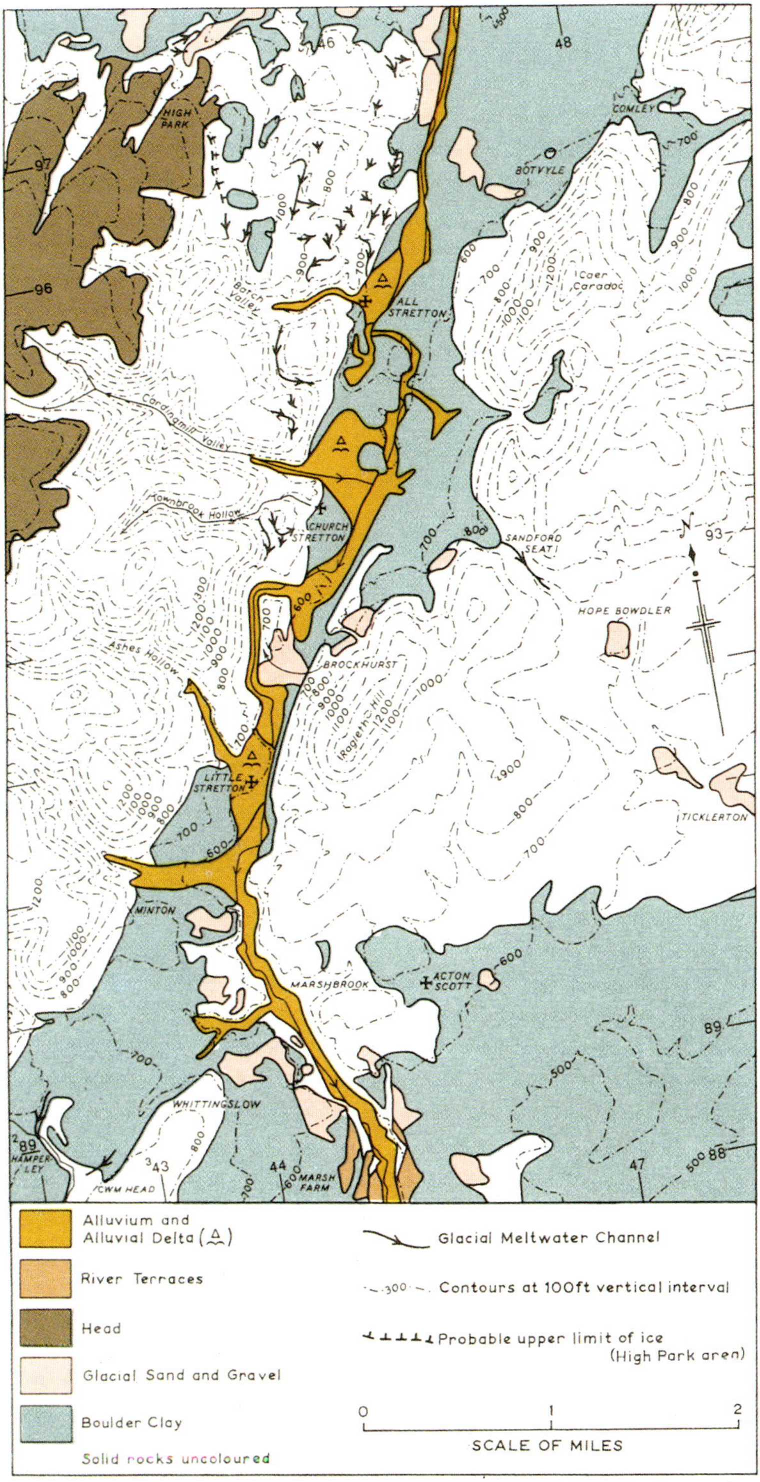

FIG. 20. – Map of drift deposits of the Church Stretton–Marshbrook area.

height of about 970 ft O.D. on the east side of Yell Bank [506970], and which lie at about 900 ft O.D. on the top of the Aymestry Group escarpment south-west of Bourton [595963]. In this area it appears that northern ice easily surmounted the Silurian limestone ridges, and a small patch of boulder clay south of Brockton [578938], in the upper part of Corve Dale, indicates that this ice also penetrated down the valley from Easthope to Brockton.

There are no deposits of boulder clay in the greater part of Corve Dale but to the south of Culmington [493820] there are small patches of till that are probably relics of a sheet deposited from western ice, which came through the gaps to the north and south of Norton Camp [448820], but which did not penetrate far up Corve Dale.

The boulder clay is variable in texture and colour, with erratic stones of local origin, of far-travelled rock types such as granites of Scottish and Lake District origin, and of dolerite similar to that occurring in the Breidden Hills. In the area between Shrewsbury and Ludlow, the distribution of outcrops of the Lower Palaeozoic and Pre-Cambrian rocks is such that erratics of local origin are not diagnostic in deciding whether the boulder clay of the Church Stretton district was derived from northern or western glaciers. Erratic blocks of northern origin, for example the Bourton boulders (Mackintosh 1879, p. 442) have been found in the northern part of the Church Stretton district, close to the Shropshire plain which was occupied by Irish Sea ice. Excavations for sewer trenches, made in 1963 revealed boulders of granite in the boulder clay a short distance south of Little Stretton. The occurrence of pebbles of northern granites and Breidden Dolerite in gravels of the Ludlow area (Dwerryhouse and Miller, 1930, p. 121) indicates only that these rocks were transported across the Church Stretton watershed by ice, while their further movement southward was probably accomplished by river transport.

Glacial Sand and Gravel

Deposits of sand and gravel are mainly small and scattered and probably originated as outwash spreads and morainic ridges in the retreat stage of the glaciation.

Gravels occurring at about 800 ft O.D. on the north-west side of Ragleth Hill were probably deposited at an early stage in the retreat in a marginal glacier lake, which also spilled eastwards across the col at Sandford Seat [468933] and deposited gravels at Hope Bowdler, Soudley and Ticklerton (Fig. 20). An irregular area of clayey gravel around Cardington village [505952] is probably the remnant of a moraine, terminal to northern ice which was obstructed in its southward movement by the high ground of the Cardington hills. A small deposit of gravel, lying north-west of Rushbury [514918], may mark approximately the junction of northern and south-western ice in Ape Dale and was possibly laid down at an early stage in the retreat of the ice.

The more extensive gravel deposits of the Marshbrook area lie on the pre-glacial erosion surface above the lip of the Marshbrook channel (Fig. 20). They were probably deposited at an early stage in the northward retreat of the Church Stretton glacier prior to the cutting of the Marshbrook channel. The gravel ridges of Brockhurst [446926] and of the Botvyle area [476962] are moraines marking later halts in the withdrawal of the Church Stretton glacier (Figs. 20 and 21). The gravel ridges of Myndmill Bridge [381886] and Oaker

[382816] mark stages in the westward retreat of the Plowden and Clun glaciers respectively (Fig. 21).

HEAD DEPOSITS

Head deposits are discussed more extensively below (p. 298) in connection with post-glacial and recent deposits. It is thought, however, that some of the accumulations of local rock debris on high ground, particularly on the Long Mynd, were formed during the glacial period when the valleys were occupied by ice.

INTERPRETATION OF GLACIAL HISTORY

With the development of glacial conditions, glaciers from the ice sheet of the Shropshire Plain extended southwards along Ape Dale and the Church Stretton valley, while ice from the north-west moved into the head of the East Onny valley north of Wentnor (Fig. 21). Possibly at the same time Welsh ice from the Camlad valley advanced around the southern end of the Long Mynd, across the low ground of the Wistanstow area, and invaded the southern part of Ape Dale. This ice was probably confluent with a glacier issuing eastwards from the Clun valley, and the combined stream apparently advanced eastwards through the valleys to the north and south of Norton Camp [446820] into the lower part of Corve Dale in the Culmington–Onibury area. It is not certain whether the maxima of the Welsh and Shropshire Plain ice occurred at the same time.

At the peak of the glaciation the major valleys were probably occupied by ice although the intervening ridges were apparently not overwhelmed except in one or two places. In the Easthope area northern ice surmounted the Wenlock and Aymestry limestone ridges and penetrated down the Easthope–Brockton valley at least as far as the Corve. At the southern end of Wenlock Edge ice evidently over-rode the escarpment to the south-east of Alcaston [457871]. At the north end of the Church Stretton valley the north-eastern shoulder of the Long Mynd was apparently covered with ice up to about 1100 ft O.D. in the area north and east of Plush Hill [451964]. The late T. H. Whitehead (personal communication) suggested that a north–south topographic feature, about 200 yd west of Plush Hill, may mark the upper limit of ice in this area (Figs. 20 and 22). There seems to be no definite evidence of ice having covered the higher parts of the Long Mynd. Cobbold (1900, p. 21) described boulders of quartzite (?Stiperstones Quartzite), occurring near Pole Bank [415945] on the summit of the Long Mynd, as being of erratic origin, but there is a possibility that these boulders originally formed a barrow or similar man-made mound. Nevertheless the top of the Long Mynd may have been covered by ice and in this case the topographical feature near Plush Hill could be interpreted as marking a halt stage in the waning of the Shropshire Plain ice sheet. If the margin of the ice remained in the Plush Hill area for a relatively long period, intensive frost-action on the exposed upper parts of the Long Mynd might have produced the thick cover of rock-debris which occurs above the Plush Hill feature (Fig. 20).

Relics of the erosion surface which existed prior to the glacial episode under discussion are preserved around the southern end of the Long Mynd from Little Stretton (Plate 13B) [443992] to Myndtown [391896] and in the East

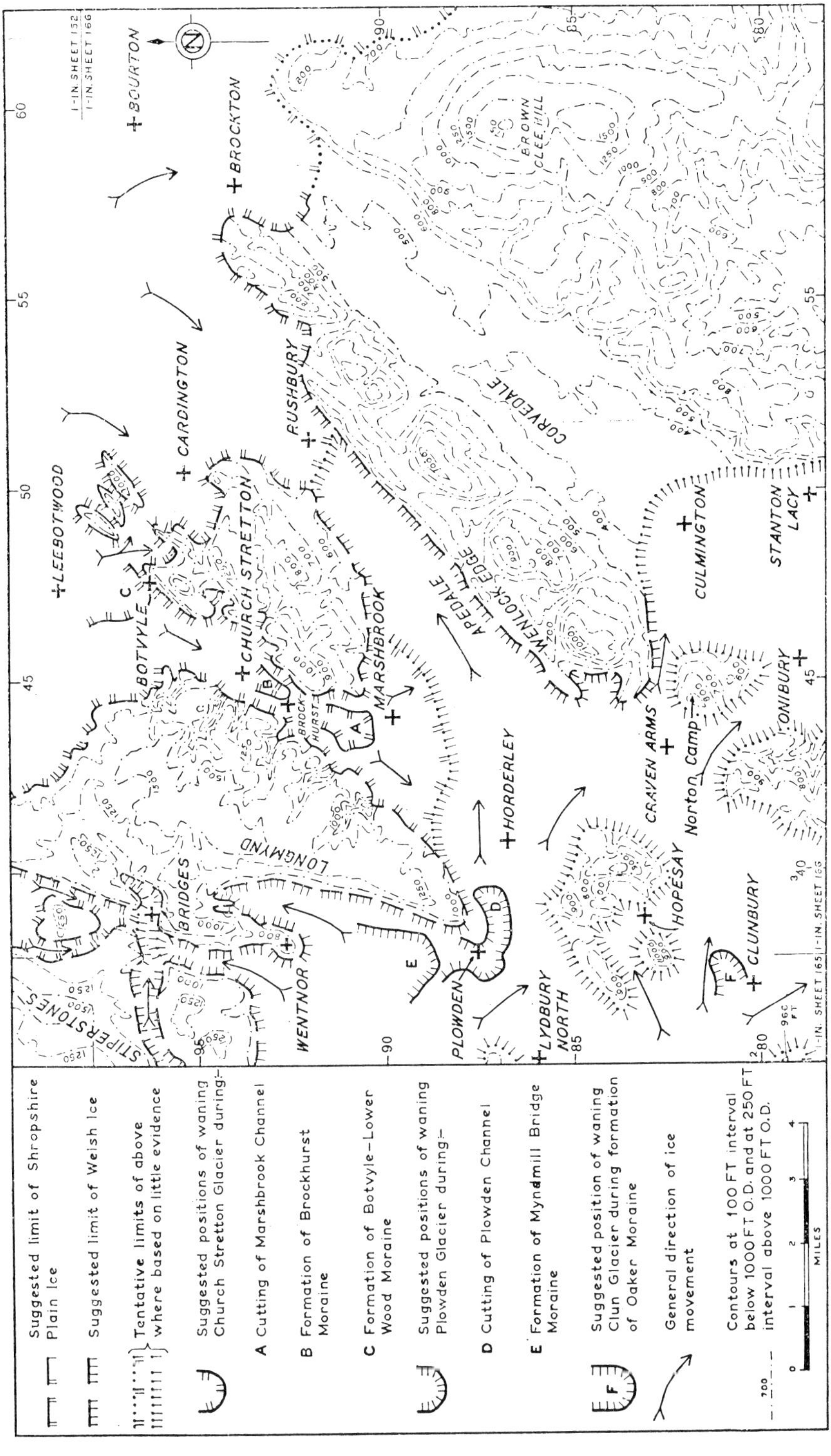

FIG. 21. Suggested limits of glaciers in the Church Stretton district

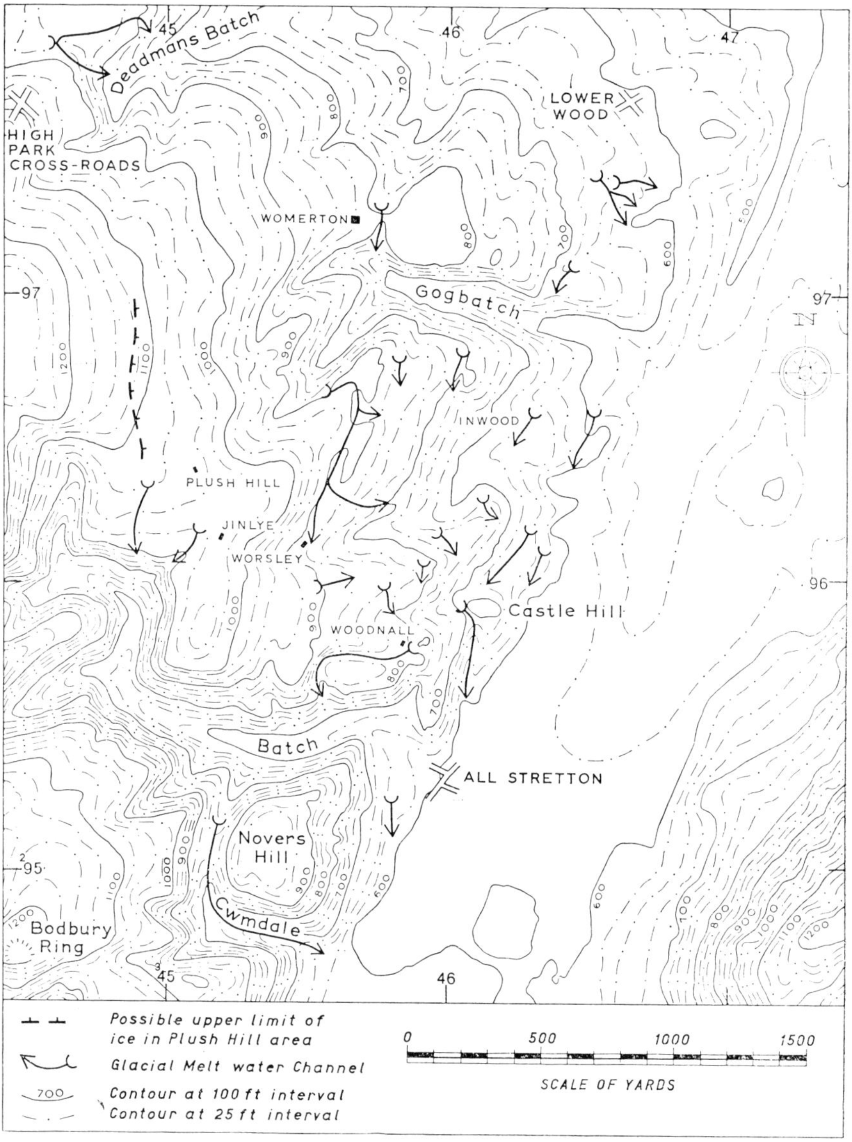

FIG. 22. Glacial melt water channels in the All Stretton area

Onny valley north of Wentnor. There was probably a pre-glacial water parting at the south end of the Long Mynd between Plowden and Horderley. The dry valley between Plowden [380877] and Lydbury North [352860] is thought to be part of the course of the pre-glacial drainage. In the Plowden area, the Geological Survey Church Stretton No. 1 Borehole near Eaton Farm [37238978] found 25 ft of drift, mainly boulder clay, at 523 ft O.D. This is possibly 50 ft or more below the pre-glacial level of the area. It is considered that ice, advancing eastwards from the Camlad valley, eroded the pre-glacial surface west of the Long Mynd and later deposited boulder clay on the deepened

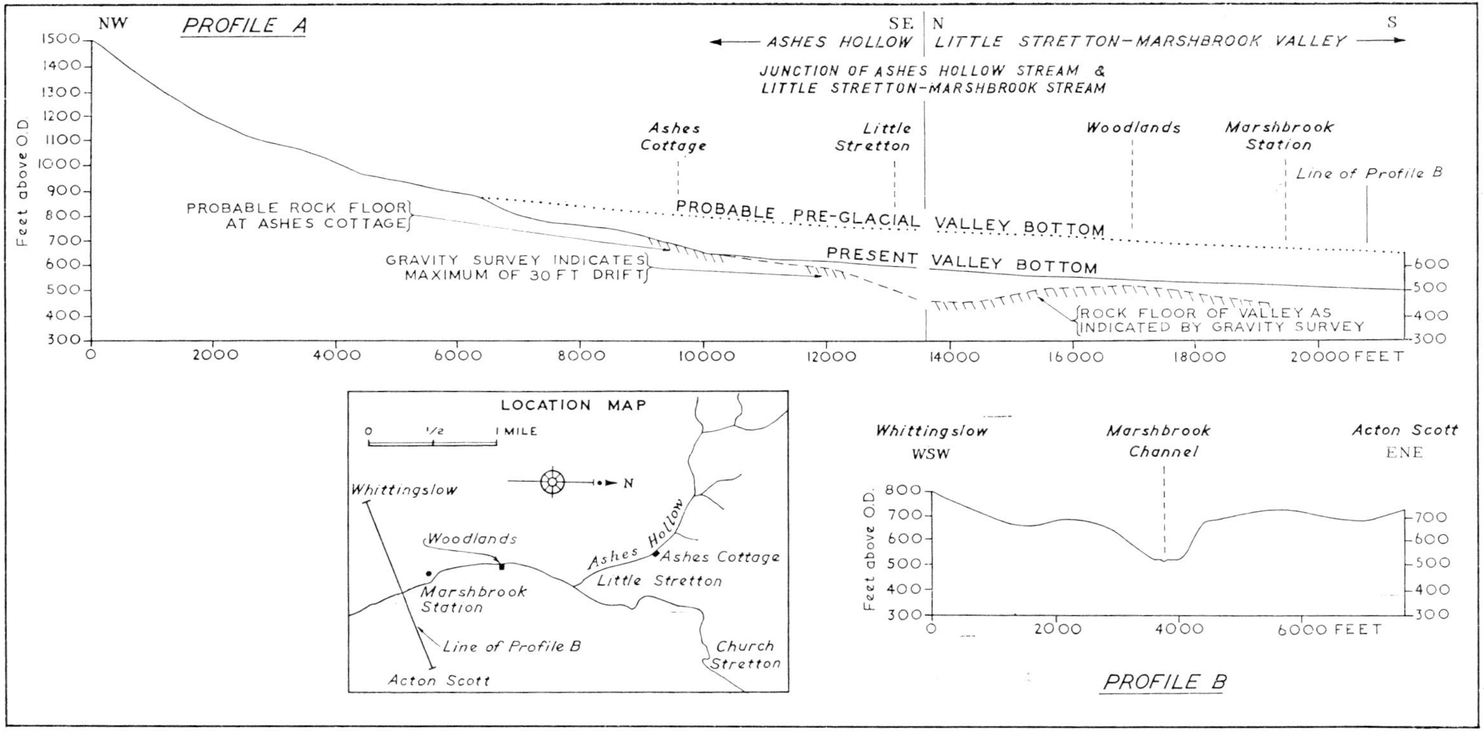

FIG. 23. A. Longitudinal profile along Ashes Hollow and the Little Stretton–Marshbrook valley, comparing the present profile with the probable pre-glacial profile and showing the solid rock surface as indicated by gravity survey (see Chapter 11)
B. Transverse profile across the Marshbrook valley from Whittingslow to Acton Scott

valley floor. It seems that the ice was less effective in downward erosion of the pre-glacial surface east of Plowden.

In the Little Stretton area the pre-glacial surface is preserved north-west of the village as a platform at about 720 ft O.D. (Plate 13B). The shape of the present profile across the Marshbrook valley [444895] (Fig. 23) indicates that the pre-glacial valley bottom here lay at about 650 ft O.D. The Geophysical Department of the Geological Survey has carried out a detailed gravity survey of the Church Stretton valley from Marshbrook to Botvyle [477961], the results being given in Chapter 11. Figure 27 indicates a maximum thickness of 155 ft of drift near Little Stretton (traverse 15) and only a small thickness of drift at Woodlands [441907], 1000 yd south of Little Stretton (traverse 19). There appears to be a culmination of the rock floor (at c.520 ft O.D.) at Woodlands whence the surface descends sharply to c.450 ft O.D. at Little Stretton and thence more gently towards the north. Figure 23, which shows the probable position of the rock floor of the valley south of Little Stretton, indicates that there has been about 270 ft of downward erosion of the pre-glacial valley floor at Little Stretton. It is suggested that the downcutting at Little Stretton and in the valley to the north was accomplished by the Church Stretton valley glacier in its advance southwards. The glacier later deposited boulder clay in the eroded hollow. It seems that little downward glacier erosion took place south of Woodlands [441907] and the cutting of the present narrow Marshbrook channel is attributed to melt water from the retreating valley-glacier (see p. 289 under Glacial Retreat Phenomena).

The late T. H. Whitehead (personal communication) suggested another explanation of the occurrence of boulder clay near Little Stretton and Plowden at levels below the pre-glacial valley-floors. He thought it unlikely that the local glaciers, being close to the limit of their maximum extension, were capable of such downward erosion of solid rock as has been postulated. He suggested that the pre-glacial surface, represented by the bench at Little Stretton, was eroded by river action prior to the glaciation and that the boulder clay was deposited in the down-cut river valleys.

A feature concerning the arguments set out above is the rejuvenation of a number of the valleys in the Long Mynd, notably Ashes Hollow and the Cardingmill Valley. This feature and its significance were brought to the notice of the authors by Mr. Whitehead. The upper stretch of Ashes Hollow is of relatively mature aspect with narrow alluvial flats. The middle part of the valley is being eroded at the present day and has an incised cross section; the lower stretch is mature and possesses narrow alluvial flats consisting mainly of angular rock debris, which probably accumulated largely under solifluxion conditions. The upper part of Ashes Hollow appears to be graded to the pre-glacial valley level at Little Stretton and this gradient also coincides with the possible level of the pre-glacial valley bottom at Marshbrook (Fig. 23). Hence it is suggested that the pre-glacial drainage from Ashes Hollow flowed southwards along the line of the Marshbrook valley. This contradicts the idea of Cobbold, expressed by Dwerryhouse and Miller (1930, p. 121), that there was a pre-glacial watershed at Marshbrook.

If, as suggested by Mr. Whitehead, the overdeepening of the main valley were due to river erosion it might be expected that there would be a continuous southward descent of the sub-drift surface in the lower part of Ashes Hollow and the Little Stretton–Marshbrook valley. The results of the gravity survey illustrated in Figures 26 and 27 indicate that this is not the case, and the idea of overdeepening by glacier erosion is thought to be the more likely. Glacier excavation of the main valley would leave Ashes Hollow as a hanging tributary. The results of a gravity traverse across the mouth of Ashes Hollow, at a point [441920], 250 yd W. 20° N. of the inn at Little Stretton, indicate a maximum thickness of 30 ft of drift (Chapter 11 p. 315). This is shown on Figure 23 and should be compared with the indicated maximum of 155 ft of drift in the main valley, only 500 yd to the south-east. It is thus apparent that the rock floor of Ashes Hollow hangs above that of the main valley even though the land surface of these drift-filled hollows is maturely graded at the present day.

No detailed evidence about the sub-drift surface is available from the Plowden area and in this instance it is admitted that the pre-glacial level of the area could have been lowered by river erosion. This idea would imply pre-glacial excavation of the Plowden channel (see p. 290 under Glacial Retreat Phenomena) almost to its present sub-drift level, and the abandonment of the Plowden–Lydbury North valley as a result of pre-glacial capture from the east, of the River East Onny.

Glacial Retreat Phenomena

Church Stretton to Horderley valley. It is thought that the Church Stretton glacier advanced at least as far south as Hamperley [420893] and moved southwards from Marshbrook towards Henley [451883]. It is likely that the hill on which Whittingslow [432890] stands emerged from the ice at an early stage of the retreat and then, with the decay of ice in the lower ground of the Wistanstow area, melt streams from a glacier snout standing on the col north of Hamperley cut the two dry valleys at Hamperley and Cwm Head (Fig. 20), and probably began the remarkable incision of Scaldbank Gutter and its southward continuation to Horderley. Whittard (1952, p. 184 and 1953, p. 244) suggested that the Cwm Head channel was excavated by spill from a lake, ponded up by ice standing to the north in the Church Stretton valley. Such a lake (spilling at about 760 ft O.D. at Cwm Head) would require a confining barrier across the Marshbrook valley. From the present profile of this valley it appears that the pre-glacial valley floor lay at about 650 ft O.D. (Fig. 23) so that a 100-ft ice barrier would be necessary here to dam up a lake to a height of about 760 ft O.D. The ice forming this supposed barrier would most probably have come from the Plowden gap via Horderley (Fig. 21), thus blocking the southward exit of any water flowing from the Hamperley–Cwm Head col. It seems more likely therefore that the Hamperley and Cwm Head channels were cut by melt water issuing directly from a northern glacier standing on the col. A modern example of this phenomenon in East Greenland has been admirably illustrated by Cowie and Adams (1957, fig. 26, p. 66).

The northward retreat of ice from the Hamperley col led to the abandonment of the Hamperley and Cwm Head channels, while the withdrawal of ice northwards from the Marshbrook valley was apparently accompanied first by the deposition of spreads of sand and gravel on the higher ground flanking the Marshbrook channel, and later, with the concentration of melt water into one stream, by the downcutting of the channel which carries the present southward drainage of the Marsh Brook.

Deposits of bedded sand and gravel which rise to about 830 ft O.D., near Plock's Coppice [463932] and in Coles Wood [455927] on the north-west side of Ragleth Hill, indicate the existence of a small lake, marginal to the Church Stretton glacier, at an early stage in the retreat. This lake probably spilled eastwards through the Sandford Seat col at 840 ft O.D. (Fig. 20), the spill water depositing small spreads of sand and gravel at Hope Bowdler, Soudley and Ticklerton.

The cutting of the numerous melt water channels on the Long Mynd to the north of All Stretton (see p. 298 for details) probably took place at an early stage in the retreat, before the ice withdrew from its high position around High Park and Womerton [456972].

The gravel deposit at Brockhurst Castle [446926] is probably a terminal moraine marking a halt in the northward retreat of the glacier. The late T. H. Whitehead (personal communication) suggested that, when the glacier front stood at this position, marginal melt water on the north-west side of the glacier breached the spur connecting Brockhurst Hill and the Long Mynd and formed the channel which carries the present southward drainage. The north–south part of this valley may have originated as a tributary to the main valley but its present form is probably due to erosion by glacial melt water.

A discontinuous gravel ridge running from Lower Wood [468976] to Botvyle [476962] is interpreted as a moraine, marking a halt stage prior to the northward withdrawal of the ice into the more open ground of the Leebotwood area.

Plowden area. Dwerryhouse and Miller (1930, p. 101) suggested that the blocking of the Kemp valley at Clunbury [370805] by the Clun glacier resulted in an overflow of water across the watershed east of Plowden, cutting the Plowden channel. This implies that the Kemp valley and the Plowden–Lydbury North valley were inundated by a lake to a height of about 600 ft O.D. The area south-west of Plowden has not yet been surveyed in detail but a cursory examination of the northern end of the Plowden–Lydbury North valley has not revealed any strand line at this height. However, blue clay of possible lacustrine origin was found by Mr. Whitehead in the Kemp valley, southwest of the Red House [364860] at about 500 ft O.D. The Plowden channel may have been cut by melt water issuing directly from the snout of a retreating glacier as has been suggested in the case of the Marshbrook channel. As the Plowden glacier retreated westwards from Horderley the melt stream at the ice front could have excavated the present channel, the head of which would have been cut back as the snout withdrew westwards (Fig. 21). After this channel was cut back beyond and below the lip of the Plowden–Lydbury North valley, drainage from the ground to the north and west of Plowden would continue to excavate along the line of the present Onny valley. It has been suggested by Mr. Whitehead that the present Onny channel at Plowden follows the line of a pre-glacial stream.

Between Myndtown [390896] and Plowden there are several well-marked dry valleys which run parallel to the western scarp slope of the Long Mynd. These were most probably cut by melt water streams marginal to the ice which had receded a small distance westward from the steep slope of the Long Mynd. The formation of these channels probably took place after the initiation of the Plowden channel.

A small ridge of gravel near Myndmill Bridge [381886] probably marks a minor halt in the retreat of the ice westward from Plowden. Dwerryhouse and Miller (1930, p. 101) described this deposit as a moraine, lateral to a glacier which flowed southwards down the East Onny valley to Plowden. From the results of the present survey it appears more likely that the low ground west of Myndtown was occupied by ice which moved eastwards from the Severn valley via Lydham and which probably moved northwards up the East Onny valley towards Bridges (Fig. 21).

Mr. Whitehead has reported the occurrence of erratics of Stiperstones Quartzite and boulders of dolerite, similar to that of Squilver (Disgwylfa) Hill [324933], on Oakeley Mynd [350875] to the west of Plowden. This indicates that at some stage of the glaciation there was a south-eastward movement of ice across the Lydham valley.

Clun Valley. At an early stage in the glaciation it appears that ice, possibly from the Clun valley, was banked up against the west side of Burrow Hill and passed over the col at Barlow [382835] leaving boulder clay in the Hopesay valley. During the retreat phase it is probable that ice remained against the west side of Burrow Hill while melt water flowing southwards along the front of the ice, cut the deep valley along the west side of Burrow Hill.

A low mound of unsorted gravel, 300 yd west of Oaker [382816], is the remnant of a moraine which probably marks a halt stage in the westward retreat of the Clun valley glacier (Fig. 21). J.E.W.

DETAILS

Boulder Clay

East Onny Valley, Plowden to Bridges. Around Plowden and Myndtown there are few exposures of boulder clay but at 1460 yd W. 42° N. of Myndtown Church [380904] there is a 3-ft section in brown stony clay with grey patches. Towards the Criftin Brook there are exposures of brown clay with many bands of poorly sorted and unstratified gravelly clay. Erratics in this area consist largely of Lower Palaeozoic and Longmyndian greywackes but limestone with *Pentamerus oblongus* is also present. D.C.G.

The Geological Survey Church Stretton No. 1 Borehole sited 400 yd N. 40° W. of Eaton Farm [37238978], found 20 ft of boulder clay below 5 ft of alluvium. Erratics in the boulder clay include Longmyndian grits and greywackes together with silt-stones, mudstones and impure limestones, of Silurian type. Occasional pebbles of much altered igneous rock, possibly Uriconian, and erratics of fine-grained por-phyritic igneous rocks and of quartzite, may indicate derivation from the Shelve area.

The large hollow of Asterton and Wentnor Prolley Moors is entirely covered with brown-weathering boulder clay. The Geological Survey Church Stretton Bore-hole No. 2 situated 450 yd W. 25° S. of Robury Ring [39409302] found 14 ft of brown sandy boulder clay with erratics of Longmyndian greywackes and grits, probable Silurian mudstones and siltstones, possible Arenig quartzites, and much-altered igneous rocks of Uriconian aspect. At about 400 yd west of Medlicott [400946] a ploughed field showed erratics of crinoidal limestone, sandstone with *Pentamerus*, and fine-grained porphyritic igneous rocks. The varieties of erratic pebble in this area suggest northward movement of ice up the Prolley Moor hollow as far as the high ground near Medlicott.

In the East Onny valley to the north of Wentnor, boulder clay occurs close to the stream as an irregular bench which probably represents the pre-glacial valley floor and which has been incised by the present stream to a depth of about 30 ft. There are few good exposures, but a ditch section [382942] 550 yd S. 30° W. of Ashgrove farm exposed 10 ft of brown silty boulder clay containing erratics of sandstone with *Pentamerus*. A ditch [385960] 1000 yd S. 16° W. from Stedment farm revealed about 4 ft of reddish brown boulder clay while a similar section [395974], 1000 yd N. 5° E. from Bridges comprised about 3 ft of mottled grey and purple boulder clay with erratics of dolerite, quartzite, sandstone with *Pentamerus*, and fine-grained feldspar-porphyritic igneous rocks. According to the distribution of outcrops in the Shelve–Longmynd area, these erratics could have been transported by ice coming from the north or the south.

According to Cobbold and to Whitehead (*in* Pocock and others, 1938, p. 197) the relatively flat ground between Bridges and Stedment was probably the site of an ice-dammed lake. There are no exposures of undoubted lacustrine clay in this area but auger holes show that the clay, although relatively stone-free, does contain a few pebbles. The last exposure of boulder clay noted above lies within the area which might have been occupied by this lake. The drift deposits around Bridges have accordingly been mapped as boulder clay but it is admitted that there may be some relics of glacial lacustrine deposits in this area. J.E.W.

Plowden to Minton, Church Stretton valley. Brown gravelly boulder clay, poorly exposed, occupies much of the ground between the Long Mynd and the Onny up to about 1000 yd east-south-east of Plowden. To the north-east of Horderley [408868] there is a general cover of boulder clay on the relatively gentle slopes between the Long Mynd and the Church Stretton Fault. These slopes, which are deeply incised by several streams, are a relic of the pre-glacial valley floor. The boulder clay contains much local material, yellowish, purple and buff clay derived from the *Pentamerus* Beds, Hughley Shales and Wenlock Shales respectively, and many fragments of Longmyndian rocks. Worn fragments of buff quartzite are locally conspicuous. Up to 20 ft of boulder clay are exposed in a stream section [401878] 370 yd E. 29° N. of Hillend Farm but in many places it is very thin, particularly on the Wenlock outcrop between the Onny and Hamperley. Over this stretch the soil is a stiff grey clay with many small erratics, but debris of Wenlock Shales is abundant locally. In a ditch section [417884], 770 yd W. 21° S. of Cwm Head church, there are 10 ft of brown stony clay. Drainage trenches in the fields north of Churchmoor Hall [412886] showed about 3 ft of brown, yellow and grey stiff clay with many fragments of Longmyndian rocks and some blocks of buff quartzite. In this area the surface deposits are probably contaminated to some extent by solifluxion material from the adjacent steep slope of the Long Mynd.

Boulder clay is thin but continuous over the Hamperley watershed [426888] and is apparently thicker to the north-east where it is at least 8 ft thick and locally very stony on the lower ground. Similar stony clay covers much of the ground between New House [437899] and Minton. D.C.G.

Boulder clay is continuous between Minton and Little Stretton although locally it is thin. However some ditch sections show up to 6 ft of reddish brown boulder clay. Between Little Stretton and Church Stretton boulder clay is confined to the valley bottom and is largely covered by alluvial deposits. To the east of Church Stretton reddish brown boulder clay covers the slopes leading up to the Sandford Seat col (Fig. 22) and is exposed to a depth of 12 ft in the deep sunken lane [464940] between Church Stretton and New House Farm.

The Geological Survey Church Stretton Borehole No. 3 near Springbank Farm, Church Stretton [45719447] found the following downward succession in the drift: angular gravel 6 ft; sand and gravel 3 ft; boulder clay 10 ft; sand and gravel 15 ft; boulder clay 14 ft; sand and gravel 9 ft; boulder clay 12 ft; sand and gravel 11 ft; solid rocks (Hughley Shales). The sequence of three layers of boulder clay interbedded with sand and gravel presumably indicates minor phases of advance and retreat of the Church Stretton valley glacier. Erratics from the Springbank Farm Borehole include greywackes, subgreywackes, grits and siltstones of Longmyndian type, mudstones and siltstones of Silurian type, quartzite of possible Cambrian origin, a few specimens similar to the Hoar Edge Grit, some dolerites including one similar to the Breidden dolerite, and several altered igneous rocks of possible Uriconian origin. In the upper 30 ft of the drift, Pre-Cambrian rocks are most abundant but below about 35 ft most of the erratics are of Palaeozoic rocks.

Boulder clay covers the undulating valley floor to the north of Church Stretton and is overlain by alluvium in the hollows close to the streams.

In the Batch valley, All Stretton [457955] the stream section shows 3 ft of coarse alluvial gravel on a reddish brown clay with small pebbles, probably boulder clay. The gravel contains rounded erratic boulders and in particular one of coarse pegmatoid granite, probably of Scottish or Lake District origin. In Gogbatch [461969] the stream section shows 3 ft of bluish grey clay with subrounded stones, resting on a stiff stony gravel. These occurrences indicate that the Church Stretton valley glacier penetrated up the tributary valleys for short distances, although in the case of Gogbatch it is possible that ice moved down the valley from Womerton.

To the north-west of All Stretton there are patches of boulder clay at about 1000 ft O.D. around Jinlye, Plush Hill [451964] and Bullocks' Moor [452970]. There are practically no exposures of these deposits but they give rise to clayey soils in which there are scattered erratic pebbles. In the vicinity of Plush Hill there are scattered erratic blocks of quartzitic breccia and of coarse dolerite; one small block of white granite was found at the roadside, 40 yd north of Plush Hill cottage [451964]. These deposits show that ice from the low ground to the north surmounted the north-eastern shoulder of the Long Mynd to a height of about 1100 ft. It is thought that this ice provided the melt water which cut the numerous dry valleys in this area.

The Geological Survey Church Stretton No. 4 Borehole [47359653], 550 yd N. 12° W. from Botvyle found the following downward succession in the drift: boulder clay 16 ft; sand and gravel 14 ft; boulder clay 8 ft; solid rocks (Coed-yr-allt Group). The erratics from this borehole are mainly of Palaeozoic rocks, including quartzite and sandstone possibly of Cambrian origin, grit similar to the Hoar Edge Grit, mudstone and siltstone of Silurian type, some soft sandstones possibly of Carboniferous origin, and some specimens of impure limestone. Erratics of Pre-Cambrian rocks include greywackes and siltstones of Longmyndian type, and altered igneous rocks possibly of Uriconian origin. There are also fairly abundant examples of dolerite, including several similar to the Breidden Dolerite. J.E.W.

Clungunford, Craven Arms, Ape Dale, Callaughton. Boulder clay covers most of the low ground on both sides of the Clun valley between Oaker [384816] and Clungunford [399787] although it is patchy around Clungunford. About 8 ft of roughly bedded clayey gravel were seen in a track-side exposure [412807] 1340 yd E. 15° S. of Broome Station and 5 ft of red-brown silty clay with abundant erratics were exposed 800 yd S. 25° E. of Broome Station [403803]. Erratics in this area are predominantly siltstones, calcareous siltstones and limestones of local derivation, with numerous pebbles of Hoar Edge Grit on the west side of the valley. There are, however, scarce pebbles of red-brown rhyolite, purple sandstone, grey grit, grey micaceous sandstone, quartzite and quartz.

In the neighbourhood of Hopesay, boulder clay is confined to the valley bottom, apart from some small patches at 800-900 ft O.D. near Barlow [383840]. The boulder clay is thickest to the north-east of Hopesay village [390834], 10 ft of buff and purple silty clay being exposed in a road cutting [393836] 600 yd N.E. of Hopesay Church, and up to 5 ft of similar material in a lane 100 yd to the east. Erratics include hard grey grits and greywackes of possible Welsh origin, siltstones of local type, much-weathered volcanic rocks and grey slate. The high level patches of boulder clay near

Barlow are thin and of small extent but there are abundant remanié erratics on the surface in this district.

Boulder clay covers the whole of the col between Broome [400810] and Craven Arms [433827] and it appears to be at least 20 ft thick. It is generally very stony and locally passes into a clayey gravel. A temporary pit [420836] 920 yd E. 35° N. of Sibdon Carwood church exposed 16 ft of very sandy, stony boulder clay with erratics of Longmyndian purple grit, Chatwall Sandstone, Silurian siltstones, Hoar Edge Grit and volcanic rocks. Erratics from the fields around Long Meadowend [410823] are rather more varied, and include grey micaceous grit, greywacke and slaty siltstone of possible Welsh origin, Hoar Edge Grit, quartz, quartzite, Longmyndian purple grit, hard buff siltstone of local origin, and weathered feldspar-porphyry. This suite of erratics suggests derivation from the west via the Oaker gap. B.A.H.

About 1000 yd east of Wart Hill [409844] an outlying area of yellowish brown clay, with erratics locally abundant, occurs at about 900 ft O.D. The gentle slopes south of Cheney Longville are covered with brown and yellowish clay with many small erratics, nowhere exposed to a depth of more than 4 ft. Locally abundant debris of the underlying solid rocks indicates that the drift is thin in places. East and north-east of Cheney Longville the boulder clay is largely obscured by alluvial deposits.

In the flat ground south-west of Wistanstow it is difficult to distinguish between boulder clay and terrace gravel in the absence of sections. At a 'Brick and Tile Works' [430857], now abandoned, 300 yd north-west of the church, it appears that the raw material was brown and yellow boulder clay and Wenlock Shales. The flat ground on which Wistanstow stands has the appearance of a high terrace but orange-brown boulder clay is excavated about 100 yd north-north-east of the church [433857]. Some 4 ft of very stony clay were exposed by building operations at the roadside [434854], 240 yd E. 43° S. of the church. In a stream [432865] 985 yd N. 1° E. of the church and in the railway cutting at Gate [439867] there are sections showing 5 ft of purple and grey stony clay. Around Bushmoor boulder clay is seldom exposed to more than 3 ft and it is commonly absent on the steeper slopes. White erratics are usually abundant in the soil, fragments of the underlying Ordovician flags are common locally. On the high ground at Whittingslow [432890] erratics are common in places but no boulder clay is present. D.C.G.

Patches of boulder clay around Lower Dinchope [451843] indicate that ice penetrated the gap in Wenlock Edge at this locality. This is the only occurrence of boulder clay in the southern part of Hope Dale although there is a small patch of clay with abundant erratics at about 760 ft O.D. on the top of Wenlock Edge [460864], 800 yd S.S.E. of Alcaston. A small area of boulder clay [494876] at Dunstan's Lane may have been deposited by ice spilling through the nick in the scarp to the south of Harton. In Ape Dale several streams cut through the thin deposit of boulder clay into solid rocks, and in a series of excavations for pylons between Banner Wood [449877] and Hatton [468903], both boulder clay and the underlying solid rocks were exposed in most instances. In some areas the till is only about 3 ft thick and in general it does not appear to exceed 6 ft. Except for temporary sections there are few exposures of the boulder clay. A 3-ft trench [484893], 200 yd south-east of Harton Road Station showed orange-brown boulder clay varying in texture within a few feet from almost stoneless clay to a clayey gravel. Erratics include purple Longmyndian slate and grit, weathered tuff possibly Uriconian, hard grey limestone and buff mudstone of local origin, micaceous greywacke, fine grey grit and coarse pale grey grit of possible Welsh origin, and vein quartz. Grey-brown clay with similar erratics was seen in trenches [456893] about 250 yd south of Acton Scott Hall and in an excavation [454881] 400 yd E.S.E. of Henley. This suite of erratics suggests a possible western derivation from ice flowing through the Plowden gap.

Boulder clay is absent from the small hill on which Rushbury [513918] stands but it occurs, in increasing thickness, towards the north-east. A few poor sections along the Lakehouse Brook between Rushbury and East Wall [527935] show yellowish brown clay, locally very stony. The most common erratics are grey sandstone and coarse grey grit, with less abundant rhyolite and tuff of Uriconian type and sporadic red sandstone pebbles possibly derived from the Trias. This assemblage suggests a derivation from northern ice. B.A.H.

Northwards from Rushbury the boulder clay cover appears to increase in thickness while the boundaries of the deposit climb from about 600 ft O.D. near Rushbury to about 970 ft O.D. on Yell Bank to the north of Cardington. G.H.M.

There are extensive deposits of boulder clay on the Wenlock and Aymestry escarpments between Wilderhope [545928] and Presthope [582974]. To the south-west of Bourton [595963] the boulder clay reaches a height of about 900 ft O.D. on top of the Aymestry Group ridge.

The Bourton (Burton) boulders have been described by Mackintosh (1879, p. 442) as consisting of Eskdale Granite and Lake District felstone. Specimens collected recently from the vicinity of Bourton village (Specs. E 30140–30156) include tuffs, rhyolitic tuff, subgreywacke, quartzite, granophyre, granite and dioritic rock. Dr. R. Dearnley reports that the granite specimens are similar to varieties of the Eskdale Granite of the Lake District, the granophyres are of possible Uriconian origin, the tuffs could be derived either from the Borrowdale Volcanics or the Uriconian, while the one specimen of subgreywacke is of Longmyndian type. J.E.W.

Large granite boulders occur at three localities at Callaughton [618975] and small erratics are locally abundant at several points around 400 or 500 ft O.D. between Callaughton and Muckleycross [647960]. They include pebbles of grit, quartz, dolerite and granite. R.W.E.

Onibury and Corve Dale. To the west of the river terrace on the right bank of the Onny at Onibury there is a narrow strip of sandy clay with many rounded erratics. It is not certain whether this is boulder clay or an alluvial deposit.

Between Vernolds Common [4780][1] and the southern limit of the Church Stretton district, there are some patches of boulder clay, of indefinite extent, generally on the higher ground. In the stockyard at Walton [466795] about 6 ft of yellow and reddish sandy clay are exposed, with many subrounded pebbles of Silurian limestones and other erratics. This outcrop is restricted to the vicinity of the farm buildings. In other areas, evidence of boulder clay is limited to the occurrence of abundant erratics in brown or reddish clay soil, the erratics consisting mainly of local Silurian or Downton Series rocks. Pebbles in the soil between High Walton [480786] and Stanton Lacy, and again about 1200 yd north-east of Stanton Lacy church [503796], are probably relics of spreads of boulder clay or high river terraces.

Prescott. In a water main trench [657809] about 700 yd west-south-west of Prescott reddish brown clay with many fragments of dolerite and Carboniferous sandstone was exposed to a depth of 4 ft. To the west this clay becomes much less stony and at about 700 yd west of the River Rea [655809] solid rock occurs at the surface. In other directions the small outcrop of clay is limited by alluvial deposits and the Carboniferous outcrop. The distribution of dolerite outcrops and the nature of the clay indicate that it is almost certainly a till. D.C.G.

Glacial Sand and Gravel

Oaker. The Oaker moraine is preserved as a low mound about 200 yd wide [384816] 300 yd west of Oaker. A road cutting exposes about 25 ft of coarse unsorted gravel with a matrix of silty sand. The pebbles vary in shape from rounded to subangular and show evidence of a slight easterly dip, indicating deposition from the west.

[1]Such numbers are four figure National Grid References within the 100-km square SO (32). They refer to general areas rather than specific localities.

Pebbles consist of greywackes, grits, sandstone and shale possibly of Welsh origin, coarse grits of Longmyndian type, coarse tuff and silicified breccia possibly of Uriconian origin, and buff decalcified sandstone of local Ordovician type. Traces of gravelly material occur on the south bank of the Clun, and it appears that the mound was originally a terminal moraine which marked a stage in the westward retreat of a glacier through the gap between Burrow Hill and Clunbury Hill. B.A.H.

Myndmill Bridge. Bedded gravel exposed at the top of the old railway cutting [382885] 150 yd S. 40° W. of Myndmill Farm, and in a road cutting about 30 yd to the west, forms an elliptical mound in the field east of the cutting. This deposit is interpreted as a water-laid deposit terminal to ice which was retreating westwards from Plowden. The gravel was described by Dwerryhouse and Miller (1930, p. 101) as the lateral moraine of a glacier which flowed down the East Onny valley to Plowden. As noted above (p. 291) there is some doubt as to the existence of this glacier, and the stratification of the deposit suggests that it is more in the nature of a kame. No other examples have been observed in this part of the Onny valley. A small patch of probably glacial gravel overlies Hughley Shales on the east side of the road [388886], 1100 yd S. 12° W. of Myndtown church. D.C.G.

Henley to Marshbrook. Areas of sand and gravel occur on both sides of the valley between Henley [451883] and Callow Hollow [437907]. The deposits rest on boulder clay and solid rock, and where they overlie boulder clay the limits of the gravel are commonly defined by a feature and spring line. The gravels are commonly flat-topped and have the general appearance of high terraces, upwards of 50 ft above the valley floor. They are interpreted as early outwash deposits laid down in the northward retreat of the Church Stretton glacier.

In an old sand pit [449893] 500 yd W. 17° S. of Acton Scott church, there is a section in 4 ft of bedded gravel with bands of pinkish brown sand. Pebbles include purple and grey grits, Acton Scott Limestone, grey quartzite, tuffs and felsite. Six feet of unbedded gravel, chiefly Longmyndian debris, are exposed in a sand pit [444892] 1100 yd W. 13° S. of Acton Scott church, and 3 ft of coarse pinkish brown sand are exposed immediately above the quarry in Marsh Wood, 200 yd to the south. To the north of Marshbrook a section in a gravel mound [439905], 880 yd E. 8° S. from the Manor House, Minton, showed 6 ft of sand with bands of gravel. Pebbles from this exposure are mainly Longmyndian sediments but one example of dolerite, similar to the Breidden Dolerite, was noted. B.A.H., D.C.G., J.E.W.

Brockhurst. To the east of Brockhurst Castle [446925] the main line of the Church Stretton valley is blocked by a ridge of gravel up to about 70 ft in height. Just west of the railway line there is a large disused pit in which there are obscured sections in poorly sorted stony gravel, which is roughly bedded at some horizons. The pit section shows a thickness of about 50 ft of gravel with no indication of the underlying deposit, but a small section at the base of the railway cutting immediately east of the pit, indicates that the gravel rests upon boulder clay. Pebbles and cobbles from the gravel include quartz, local Longmyndian sediments, volcanic rocks possibly of Welsh origin, granite of Eskdale type, Uriconian rhyolite fragments, and dolerite, similar to the Breidden Dolerite. In 1959, archaeological excavations in the inner ditch of Brockhurst Castle (information from the late T. H. Whitehead) revealed 13 ft of coarse, poorly sorted gravel, overlain by 1 ft of fine gravel, and this by about 2 ft of stony clay, possibly boulder clay. The gravel ridge is interpreted as a moraine, apparently laid down with some water-sorting, terminal to the northward-retreating Church Stretton glacier. The 2-ft layer of boulder clay in the Brockhurst Castle excavation suggests a minor readvance of the ice across this moraine. The ridge appears to have caused the late-stage diversion of the southward drainage of the Church Stretton valley around the west side of Brockhurst Hill in the Worldsend channel. J.E.W.

Ragleth and Hazler hills. At Coles Wood [455927] and near Plocks Coppice [463932] there are deposits of poorly bedded sand and gravel up to a height of about 830 ft

O.D. Pebbles from the Coles Wood deposit include grits, quartz, and one specimen of grey granite. As indicated above (p. 290) these gravels were probably laid down in a small marginal glacial lake which spilled eastwards at the Sandford Seat col (Fig. 20). Thus they were probably formed before the deposition of the Brockhurst gravel ridge. G.H.M.

Botvyle, Gorseybank, Lower Wood. Between Botvyle [475960] and Brook House [471977] there is a discontinuous gravel ridge which probably marks a halt stage in the retreat of the Church Stretton glacier. Near Cote House [469975] there are exposures of stony gravel composed mainly of well rounded pebbles in a sandy matrix. This deposit shows evidence of water sorting. G.H.M., J.E.W.

Cardington. To the south of Gretton [515952] there is a well marked topographic feature which faces east and forms the eastern boundary of a morainic deposit extending westwards through Cardington village. This deposit has an irregular boundary, and consists of a roughly-bedded coarse clayey gravel which has been exposed around the village in a number of shallow excavations. The greatest recorded thickness is 26 ft, measured from a well section [506954], 270 yd N. 17° W. of Cardington Church. G.H.M.

Glacial Melt Water Channels

Plowden, Asterton, Wentnor. From Plowden to near Asterton a system of drainage channels, now mainly abandoned, is developed between the River Onny and the Long Mynd. At Myndtown [390895] there are three main channels running parallel to the Long Mynd escarpment, with occasional linking valleys approximately at right angles. The main channels appear to have developed successively from east to west at outlets for melt water flowing to the south, along the edge of ice occupying the low ground west of the Long Mynd. The most easterly channel is the only one at present occupied by a stream, and is overdeepened south of Myndtown in consequence of more recent adjustment to the level of the Onny into which it flows. At its northern end, ¼ mile west of Handless [392904] this channel is just over 700 ft above O.D., and in this area the other two channels are about 20 ft and 50 ft lower respectively. Small lakes and marshy hollows occur in the higher parts of the three main channels, perhaps due to the local persistence of stagnant ice after the westward retreat of the main glacier.

Elongate shallow depressions at about 650 ft O.D. occur in places on the southern end of the Long Mynd [398877], between 500 yd and 700 yd west-south-west of Hillend Farm. These features trend approximately north-east and possibly mark the course of drainage marginal to ice filling the valley to the south.

The present valley of the River Onny at Plowden is considered to have originated as a glacial melt water channel. This has been described above in the interpretation of the glacial history of the area (p. 290). D.C.G.

The spur [382923] south of Wentnor is incised by two east–west channels at heights of about 750 ft O.D. and 730 ft O.D. respectively. Their position suggests that the lower part of the East Onny valley was blocked by ice occupying the low ground at Norbury, and that melt water flowing down the East Onny valley was diverted eastwards across the Wentnor spur, cutting the two channels. There are two similar channels near Walkmill farm [376931] to the west of Wentnor, just outside the Church Stretton district. These occur at heights of 875 ft O.D. and 850 ft O.D. respectively, and presumably were formed as a result of drainage diversions at an earlier stage in the glacial retreat.

Hamperley, Cwm Head, Marshbrook, Brockhurst. These channels are more fully described above (p. 289) in the interpretation of the glacial history of the area. They are thought to have been formed in the retreat of the Church Stretton glacier.

All Stretton area. To the north and west of All Stretton there are numerous dry valleys on the eastern slopes of the Long Mynd. These channels occur at various levels between about 1050 ft O.D. and 600 ft O.D. and in some instances long spurs are incised at several levels in an eastward-descending sequence. Figure 22 illustrates the distribution of channels in the area.

It appears that these valleys were cut by melt water issuing from ice which had ridden over the north-east shoulder of the Long Mynd to a height of about 1100 ft O.D. The occurrence of boulder clay in the lower parts of Gogbatch and the Batch valley, All Stretton (p. 293) indicates that the eastern Long Mynd batches were occupied by ice, possibly lateral lobes from the Church Stretton glacier or ice which had flowed down these side valleys from the north-west. At an early stage in the retreat, melt water from the elevated ice front near Plush Hill would have flowed southward towards Church Stretton across the ice-filled Long Mynd batches, cutting valleys in the intervening spurs. With the gradual northward withdrawal of the Church Stretton glacier, the ice in the tributary valleys would sink to progressively lower levels, allowing water from some of the channels to follow the courses of these partly vacated valleys down to the main Church Stretton valley. For instance, in the case of the large dry valley of Cwmdale [452950], south-west of All Stretton (Plate 13c), it is probable that melt water from the upper part of the Batch valley was diverted into the Cwmdale channel by ice standing immediately west of All Stretton to a height of about 900 ft O.D. The outflow from Cwmdale was past the site of the Church Stretton mineral-water works [456947] into the main valley. This implies that, while the ice at All Stretton effectively blocked the mouth of the Batch valley, the level of the glacier had descended sufficiently half a mile to the south to permit an outflow of water from the Cwmdale channel. A similar situation is inferred for the well-marked channel north of Synalds Coppice, near All Stretton. The upper end of this channel is near Woodnall cottage [458958] and, at the time of formation of this channel, ice must have been banked against the steep slopes east of Woodnall to a height of at least 820 ft O.D. It is likely that the outflow from the the Woodnall channel was via the lower part of the Batch valley to All Stretton, only $\frac{1}{4}$ mile south of the ice barrier at Woodnall.

The Cwmdale channel is one of the largest in the area, being about 100 ft deep at the highest point in the valley floor and about 200 ft deep at its lower end. This valley may not have been entirely excavated by melt water, but may have originated as a normal stream tributary or as a nivation hollow, which cut back towards the Batch valley. The later cutting through from the Batch valley, with some deepening of the channel, is ascribed to melt water drainage. Another good example of a melt water channel occurs to the north of Worsley [455962]. At its upper end this channel cuts through the spur leading from Jinlye to Gogbatch and then turns sharply to the south-south-west, following the 900 ft contour for 600 yd before descending into a tributary batch at Worsley.

POST-GLACIAL AND RECENT DEPOSITS

Head Deposits

The head deposits shown on the one-inch geological map include two types, which in some places grade laterally into each other. On the summit of the Long Mynd there is a thick mantle of stony detritus which has moved little, if at all, from its place of formation. In the bottoms of numerous valleys and on the less steep valley sides and escarpments, deposits of head have accumulated by solifluxion processes, the material having been derived from the adjacent higher ground. The period of formation of these head deposits probably covers a considerable interval of time. The unmoved detritus of the Long Mynd plateau may have been formed largely by frost action during a glacial period, when the upper part of the Long Mynd formed a nunatak surrounded by ice on the lower ground. This frost weathering presumably persisted on the Long Mynd in the waning period of the glaciation. The deposits of head on the lower ground probably accumulated under periglacial conditions during and after the withdrawal of the valley glaciers from the district. The precise relationship between these head deposits and boulder clay has not been determined, but it is probable that the boulder clay was laid down prior to the accumulation of soliflucted head on the valley sides and hill slopes. No distinction is made on the geological map between solifluxion deposits and head *in situ*. J.E.W.

DETAILS

Aston-on-Clun area. A thick deposit of head covers the lower slopes on the west side of Hopesay Hill [4084]. It consists of angular blocks and fragments of purple grit in a purple sandy matrix, probably derived by solifluxion from the Wentnor Series rocks of the area. A small stream [398837], 500 yd west of Hopesay Hill summit, is incised to 15 ft in head. There are smaller patches of a similar deposit on the east side of the hill, the stream section south-east of Oldfield [406830] showing up to 7 ft of head overlying Harnage Shales.

Head deposits occur in most of the valley bottoms on the east side of Burrow [381831] and Clunbury Hill [372799]. A section [382842] 250 yd north-west of Barlow shows 3 ft of buff silty clay (recent hillwash) on 7 ft of unsorted siltstone debris. B.A.H.

The Long Mynd. The gently undulating summit plateau of the Long Mynd carries a thick mantle of angular stony debris which is derived from the underlying solid rocks and which has moved very little, if at all. Exposures are scarce but there is a 3-ft section in brown silt with purple shale debris beside the bridle road [410903] 1430 yd east of Handless farm. About 500 yd east of the Boiling Well [426945] and near Wildmoor Pool [426965], the head has been excavated, probably for road metal for local use. These shallow pits show up to 7 ft of angular stony debris with no exposure of solid rock at the base. This material is overlain by a thin peaty soil over much of the upland area. D.C.G., J.E.W.

The head deposits are restricted to the gentle summit slopes and are absent, or nearly so, from the steep sides of the Long Mynd batches. In this connection, an interesting comparison may be made between the north and south-facing sides of the major Long Mynd valleys, e.g. Ashes Hollow and the Cardingmill Valley. The north-facing slopes are mainly smooth and grass covered, with a thin veneer of rock debris and soil. In contrast, the south-facing slopes show many more exposures of solid rock and, particularly along the outcrop of the Synalds Group, are locally rough and craggy. This difference appears to be the effect of insolation, particularly under periglacial conditions, which resulted in more extensive thawing and solifluxion on the south-facing slopes.

In the lower parts of the Long Mynd valleys there is commonly a deposit of angular stony debris which has accumulated partly by solifluxion. In many of the larger valleys this material has been rearranged by the stream, which has developed narrow alluvial flats. These deposits are therefore mapped as alluvium, but at some localities the alluvium is bordered by narrow areas of head, which extends up the valley sides for a short distance. In some of the larger dry valleys, e.g. Cwmdale [451950], the cross section of the valley bottom is concave upwards and there is no development of an alluvial flat (Plate 13C). The drift deposits in these valleys are mapped as head, the material probably having been derived by solifluxion from the higher ground. J.E.W.

On the west side of the Long Mynd, head derived from the upper slopes fills the re-entrants on the escarpment and spreads out westwards from the foot of the steep slope. The material is a brown sandy loam with much Longmyndian debris, and forms very swampy ground between Myndtown and Handless.

View Edge area. At the foot of the View Edge escarpment there are accumulations of head in places and it is probably present also in the narrow valleys on the dip slope west of Aldon [4379]. The material is a yellowish clay with much debris of Ludlow rocks. D.C.G.

Hope Dale. Deposits of head are present below the scarp of the Aymestry Group especially around Lower Dinchope [452843] and Westhope [472864]. Exposures are few and no clear distinction can be drawn between the periglacial solifluxion deposits and more recent hill wash. B.A.H.

Corve Dale. Below the escarpment of the Ditton Series there are local accumulations of head with abundant fragments of Ditton sandstone and cornstone. Much of this material has been transported by scarp streams rather than by solifluxion processes, and in most instances the limits of the deposits are so uncertain as to make the mapping impracticable.

Clee Hills. Extensive spreads of red clay, characteristically with dolerite blocks of various sizes, cover much of the ground below both of the Clee Hills. Head of this type from Titterstone Clee is conspicuous between Roundthorn [5578] and Ripletts [5879]. Apart from a few sandstone ridges it appears to cover most of the outcrop of the Downton Series in this part of the district. Near Ripletts it is exposed to 14 ft but over most of the area it is probably very much thinner.

Between 150 yd and 350 yd east of Kinson [577821] there are sections in ditches and lanes showing up to 6 ft of red clay, with many dolerite blocks derived from Brown Clee. This deposit largely obscures the solid rocks over an extensive area from the Ledwyche Brook [568814], east of Furlong, to Red Hall [586825]. The head is more widespread to the east and south-east of Brown Clee, extending southwards to the gentle slopes up to $\frac{3}{4}$ mile west of Loughton, and to the lower ground north of the road between Loughton and Hinton [6582]. On open ground it is generally 3 to 4 ft thick but in the valley bottoms up to 9 ft are exposed. Here again the head consists of a reddish brown clay with abundant dolerite blocks of various sizes. On the ridge north-west of Wheathill church [622821] dolerite blocks are common locally but no thickness of head appears to be preserved there.

Between the Carboniferous escarpment and the Silvington–Farlow brook the gentler slopes are irregularly covered by yellow and reddish brown sandy clay containing much debris of yellow Carboniferous sandstone and quartz pebbles derived either from the Farlow Series or the Carboniferous outcrops. Locally this deposit exceeds 6 ft in thickness. D.C.G.

To the north-east and north of Brown Clee Hill there is an extensive apron of head, consisting of boulders and stones of local origin (dolerite, Coal Measures sandstone and Clee Series sandstone) set in a matrix of pale brown silty and sandy clay. On the west side of the hill a train of patchy head and scattered boulders can be traced through Abdon [575863] to Tugford [557870], where in the low ground of Corve Dale a delta-shaped area of head has accumulated by solifluxion from the high ground. Near Tugford this deposit is possibly 20 ft in thickness and near Beambridge [532882] the western corner of the area is cut through by the Corve. R.H.H.

River Terraces and Alluvium

The three chief drainage basins in the Church Stretton district are those of the Clun, the Onny with its tributary the Quinny Brook, and the Corve. Along these rivers there are considerable developments of river terraces above the flood-plain alluvium, and on the geological map these terraces are numbered in ascending topographical sequence. It is considered that the oldest terraces are the highest, these being succeeded by younger and lower terraces. The numbering of the terraces has been made with regard to the sequence in each river basin. Thus it does not follow that the First Terrace of the River Clun is the exact stratigraphical equivalent of the First Terrace of the River Onny. However it has been possible to correlate between the basins of the Onny and the Corve and the terraces of these two rivers are numbered in a common sequence.

In the area of the Long Mynd and in the ground to the east and north-east of the Clee Hills, there are several streams, including the Rea Brook, along which alluvial flats have been formed. There is however no marked development of river terraces in these valleys. J.E.W.

DETAILS

River Clun

Fourth and Third terraces. These two terraces occur only as restricted spreads of gravel to the north of Broome [401808]. The heights of the deposits are about 60 ft and 40 ft respectively above the flood-plain alluvium.

Second Terrace. This is the most extensively developed terrace of the Clun within the present district. It is developed downstream from Oaker [384816] and is restricted mainly to the east side of the river. The deposits occur at 20 to 25 ft above the alluvium and consist of poorly sorted gravel with a sandy matrix. Some 5 ft of this material were exposed in the roadside [398788] 380 yd E. 9° N. of Clungunford church.

First Terrace. This terrace lies between 5 ft and 15 ft above the alluvium and occurs downstream from Aston-on-Clun [393817], mainly on the east side of the river, being rather more restricted in extent than the Second Terrace. The terrace deposits consist of poorly sorted sandy gravel.

Flood-Plain Alluvium. The flood plain of the Clun is up to ½ mile wide between Broome and Clungunford, with numerous abandoned meanders. It narrows to about 250 yd at Oaker and Clungunford. The alluvium is a pinkish brown clayey silt with occasional thin bands of sand and gravel. Water boreholes through the alluvium east of Clungunford bridge [394786] showed up to 89 ft of clay, sand and gravel, but this thickness may include some glacial deposits.

River Onny and tributaries

Fourth Terrace. This terrace is developed only along the west side of the Quinny Brook from Felhampton [445783] to Marshbrook Bridge [448888]. South of Felhampton the terrace is degraded and merges with the Third Terrace. B.A.H.

Poorly exposed gravel at the roadside [431852] 450 yd S. 18° W. of Wistanstow church, occurs at a level higher than that of the adjacent Third Terrace. This occurrence appears to have a very limited extent but might be correlated with the Fourth Terrace.

Third Terrace. This is the most extensively developed terrace in the Onny basin within the Church Stretton district. There are wide gravel spreads on each side of the river south of Onibury [445792]. This village stands on the terrace which presents a steep face, up to 10 ft high, towards the flood plain but the height of this feature increases towards the south-east. D.C.G.

Around Stokesay [436817] the Third Terrace lies at about 30 ft above the flood plain but it is not developed between Craven Arms and the junction of the Quinny Brook with the Onny. B.A.H.

Near Wistanstow the Third Terrace reaches a height of about 35 ft above the flood plain in the angle between the Onny and the Quinny Brook. Beside the old road [429851] 630 yd S. 42° W. of Wistanstow church, there is a section showing 5 ft of unsorted earthy gravel with large and small fragments of Longmyndian and Caradoc rocks. Small areas of gravel, referred to the Third Terrace, occur on both sides of the Onny between Glenburrell [413862] and Horderley [409872]. At the 'Old Gravel Pit' [396876], 250 yd W.S.W. of Hillend Farm, the *Pentamerus* Beds are overlain by up to 6 ft of well rounded coarse gravel. This deposit is of limited lateral extent and is probably a relic of the Third Terrace, rather more than 20 ft above the flood plain. To the south-east of Plowden Mill [385871] there is a small gravel terrace at about 25 ft above the alluvial flat of the Onny. This terrace rests just below the level of the pre-glacial surface in the Plowden gap, and its possible correlation with the Third Terrace implies that the latter is of early fluvio-glacial origin, contemporary with the cutting of the Plowden channel. D.C.G.

Along the Quinny Brook the Third Terrace is extensively developed north of Strefford [444857], and continues northwards on the west bank of the stream to near Marsh Farm [448884]. It is less well developed on the east bank but forms a wide feature in the angle between the Quinny and the Byne brooks [450864]. B.A.H.

Second Terrace. This terrace is in general less prominent than the Third. South-east of Onibury it is represented only by a narrow feature 11 ft above the flood plain, but to the north of Craven Arms, on the west side of the Onny, there is a spread of gravel up to 900 yd wide which varies in height from 25 ft to 5 ft above the alluvium. This is probably formed by the deposits of the First and Second terraces which are not separable by a feature in this area. Boreholes through this terrace in Craven Arms [434833] show 45 ft of sand and gravel. On the east side of the Onny the Second Terrace occurs 15 ft to 25 ft above the alluvium from Craven Arms to Grove [436845] and again on either side of the Quinny Brook from Grove to Strefford. About 20 ft of coarse poorly sorted gravel are exposed on the east bank of the River Onny, 1100 yd N. 16° E. of Craven Arms Station [435840].

First Terrace. This terrace is of limited extent and occurs between 5 ft and 10 ft above the flood plain. South-east of Onibury the First Terrace is developed on the north side of the Onny and a tributary stream which flows past Walton [4679] is graded to it. As noted above, the First Terrace merges with the Second in the wide spread of gravel between Craven Arms and Grove. Near Horderley, and about ½ mile upstream, narrow terraces, referred to the First Terrace, occur 5 ft above the flood-plain alluvium.

Flood-Plain Alluvium. Below Wistanstow the Onny flows through an alluvial flood plain about 300 yd wide, and near Onibury this flat is much dissected by abandoned channels. The alluvial deposits of this stretch consist mainly of buff silty clay with sporadic pebbles. Between Wistanstow and Plowden the alluvium is considerably restricted in width and the deposits of this area are mainly gravel. Above Grove along the Quinny Brook the alluvium is generally about 200 yards wide and consists of buff silt with beds of gravel locally. Downstream from the weir at Strefford two levels of alluvium are developed, the lower being adjacent to the stream and incised about 3 ft into the main flood plain. Upstream from Upper Affcot [444865] the alluvium is mainly gravelly. D.C.G., B.A.H.

River Corve

Third Terrace. This is the most extensive terrace of the Corve and, on Ludlow Golf Course [4977] (Ludlow Sheet 181), is continuous with the Third Terrace of the River Onny. The terrace deposits, consisting of coarse gravel with bands of current-bedded sand, are exposed to a depth of 12 ft in old sand pits [496776] ¾ mile south of Stanton Lacy church (Ludlow Sheet 181). The Third Terrace forms a prominent feature about 20 ft above the alluvium on the west side of the Corve above Stanton Lacy. West of Sparchford [496830] it is up to 200 yd wide and rises here to about 30 ft above the alluvium. This terrace continues on the west side of the Corve north-east of Diddlebury but is considerably dissected in this area. Along the Tugford Brook, a tributary of the Corve, a terrace is developed on the north bank [558871] west of Tugford. This is tentatively correlated with the Third Terrace of the Corve.

First Terrace. This terrace occurs at about 5 ft above the alluvium on the west side of the Corve from Stanton Lacy to Culmington [492820]. It is correlated with the First Terrace of the Onny, the equivalent of the Second Terrace of that river not being developed along the Corve. Along a tributary, the Siefton Brook, the First Terrace occurs to the north-west of Culmington as far as New House [483837].

D.C.G., R.H.H., B.A.H.

Flood-Plain Alluvium. The alluvium of the Corve forms a broad flat about 400 yd in width. In the lower reaches of the river the deposits consist of red silty clay with occasional pebbly layers, which is exposed at some localities to a depth of 8 ft. Similar alluvium borders the Pye Brook and the minor tributaries which join the Corve from the east.

D.C.G.

Streams of the Long Mynd and the Church Stretton Valley

To the west of the Long Mynd in the headwaters of the River East Onny and along its tributary the Criftin Brook, there are narrow strips of alluvium, usually consisting of silt and sand with occasional pebbly layers. In one or two localities, low terraces of limited extent occur above the alluvium but these are exceptional. Within the Long Mynd proper, the streams in the larger batches have developed alluvial flats which are restricted laterally owing to the narrowness of the steep-sided valleys. These flats are composed largely of angular stony rubble which probably originated as solifluxion debris from the adjacent slopes and which has undergone only a small amount of stream transport and rearrangement.

In the Church Stretton valley the present watershed occurs at Church Stretton, the Cardingmill Valley stream flowing north to the Cound Brook, while the Townbrook (Fig. 20), which runs to within 200 yd of the Cardingmill Valley, turns southwards to Little Stretton and Marshbrook. In many places these streams have developed small cuspate alluvial flats, but the dominant alluvial features of this area are the three delta fans occuring at the mouths of Ashes Hollow, Little Stretton, the Cardingmill Valley, Church Stretton, and the Batch Valley, All Stretton. At Little Stretton and Church Stretton these fans have spread right across the valley floor modifying the northward and southward drainage systems. At Little Stretton the present drainage from the north runs in a deep ditch cut through the delta fan. It is evident that the accumulation of this delta at Little Stretton caused the formation of a shallow lake or marsh which almost isolated Brockhurst Hill in former times. Shallow trial holes put down for foundation work to the north and south of Brockhurst Hill, show up to 6 ft of peat in some places and in others, peat interbedded with clay and silt. The Church Stretton delta fan has a very gently convex surface the limits of which are difficult to define. Excavations made in 1959 for a new school at Church Stretton [455944] showed about 15 ft of angular stony rubble with no exposure of the base of the deposit. The Geological Survey Borehole No. 3 near Springbank Farm [45719447], 200 yd east of the school, found 6 ft of angular stony detritus overlying 80 ft of glacial deposits.

The present course of the Cardingmill stream is canalized across the delta fan in a straight ditch but this is not its natural course. In a period of flood in the summer of 1957 the stream left the artificial channel and flowed via the site of the old Quarter Houses [455943] and to the south of Springbank Farm towards the railway. In its natural state the Cardingmill stream would probably have been braided across the delta fan so that the direction of flow might have changed from north to south or *vice versa* from time to time. The alluvial fan at All Stretton is similar in form to the two described above, but is less extensive and does not extend right across the main valley.

The late T. H. Whitehead (personal communication) suggested that the angular stony rubble of these delta fans would be more properly classed as solifluxion material which has flowed out of the tributary hollows into the main valley under periglacial conditions. It is agreed here that this may well be true for the greater proportion of the material forming the deltas. The fans are, however, shown on the map as being of alluvial origin since they grade into the alluvial flats of the parent tributary valleys. It has been noted above (p. 299) that the alluvium of the tributary hollows is probably slightly reworked solifluxion debris. J.E.W.

Ledwyche and Dogditch brooks

Alluvial flats are developed along the courses of these streams and their tributaries where they lie within the outcrop of the Downton Series. In places, e.g. on the Ledwyche opposite Crowsleasows [5478], it is difficult to distinguish the alluvial features from the flat clay land formed by the marly solid rocks. The alluvium is brown silty clay, with abundant gravelly bands composed mainly of pebbles of dolerite and Ditton rocks; it is exposed in places up to 5 ft in depth. Within the outcrop of the Ditton Series to the north and north-west the streams are more deeply incised, generally with insignificant alluvial flats.

Rea Brook

The alluvium of the Rea Brook is exposed locally up to 10 ft in thickness and consists mainly of red clay, but it is yellowish at and below the Carboniferous outcrop south of Prescott. In July 1955 the flood plain was inundated by about 5 ft of water, within a few hours after a prolonged thunderstorm in the Bromdon–Farlow area. Small bridges on the Farlow Brook and riverside properties at Prescott were damaged, and red mud was spread widely over the valley floor. The alluvium of the Farlow Brook and its tributaries above Silvington [622798] is composed of red sandy clay with many pebbles of dolerite and Carboniferous sandstone. The alluvium of the Moor Brook and of the stream which joins it from the west near Moorbrook farm [644837], is red silt which is commonly packed with blocks of dolerite. The terraces in the lower part of the Rea Brook and the Farlow Brook are generally about 6 ft above the flood plain. About 700 yd south-west of Hardwickforge [657814] there is evidence of a river terrace at about 80 ft above the alluvium. The soil on this poorly defined feature is a red loamy clay with scattered pebbles of dolerite, quartz and limestone.

Peat and Lacustrine Deposits

Area west of the Long Mynd. Within the flood plain of the Criftin Brook some 1200 yd west-north-west of Myndtown church [380900] about 3 ft of peat is developed in two small areas. At about 1400 yd north-west of the church [381905] about 2 acres of marshy peat occurs on the higher ground east of the brook.
D.C.G.

The relatively flat ground between Bridges and Stedment may have been the site of an ice-dammed lake (Pocock and others 1938, p. 197). Auger holes put down in this area show that the drift deposits consist of a clay with small scattered stones. This clay may be of limited lateral extent as a ditch section [395974] 1000 yd N. 5° E. from Bridges showed about 3 ft of mottled grey and purple boulder clay with numerous erratic stones. At Ratlinghope [403968] there is an old brick pit in clay with scattered pebbles which may be a similar lacustrine deposit.

Summit of the Long Mynd. On the gently sloping summit plateau of the Long Mynd the soil is commonly peaty but it rarely exceeds a few inches in thickness and is wasting in some localities at the present time To the west and south of Wildmoor Pool [425965] there are some thin deposits of peat, commonly overlying silty clay, in the gentle hollows which drain northwards to Colliersford Gutter. The peat there is not more than about 1 ft thick.

Church Stretton Valley. In recent years (1963–4) excavations for a new sewage scheme have been made in the Church Stretton valley from the old sewage works east of Minton [440910] to All Stretton. The late T. H. Whitehead has recorded the sections through the drift deposits revealed by these excavations.

In the vicinity of the old sewage works, excavations for new settling tanks penetrated a layer of clay, about 2 ft thick, with scattered stones, interbedded with angular stony gravel which is probably outwash torrent gravel or solifluxion material from Callow Hollow. One part of these excavations showed a 1-ft layer of peat and peaty clay above 4 ft of gravel which overlies the layer of clay with scattered stones. The trench section [442915] south of Little Stretton exposed 4 ft of brown clay on c.4 ft 7 in of variable gravel with lenses of clay. Just north of Little Stretton [445920] the sewer trench revealed up to 8 inches of dark brown peaty clay bedded in 2 to 3 ft of brown and pale grey stoneless clay which rests upon stony clay and fine angular gravel. The trench excavation [453923] 350 yd south-west of Church Stretton Station, exposed about 3 ft of grey plastic clay with rare stones, resting on clayey gravel. At about 600 yd north of Church Stretton Station the trench excavation [457941] exposed the following upward sequence: grey plastic clay 1 ft 2 in; brown carbonaceous clay 7 in; peat 2 ft 6 in; grey plastic clay, up to 11 in; fine gravel 5 ft; cream-grey plastic clay with dark streaks 8 in; brown clay, rare stones 2 ft 6 in; brown clayey soil, a few stones 1 ft 6 in. The layer of peat is of restricted extent and thins out about 33 ft to the north-west of the above section.

The late T. H. Whitehead considered that these deposits of clay and peat were formed in a lake or chain of lakes which occupied the Church Stretton valley in a late or post-glacial period. The interbedded layers of angular gravel apparently represent spreads of solifluxion material or flood gravel from the adjacent valleys of the Long Mynd, e.g. Callow Hollow and the Cardingmill Valley, which interrupted lacustrine deposition from time to time. The section north of Church Stretton Station [457941] indicates the degeneration of a lake into a peat bog which was overwhelmed by gravel deposition, but the upper layer of clay in the section indicates that lacustrine conditions were later re-established in the area.

Other parts of the trench sections show that the layers of clay and gravel described above, rest upon an irregular surface of boulder clay with some solid rock locally. This suggests a period of erosion between the deposition of the boulder clay and that of the bedded clay and gravel. J.E.W.

Farlow area. A small area of peat about 2 ft thick occurs on the right bank of Ingardine Brook about 300 yd west-north-west of Ingardine farm [626814].

Calcareous Tufa

Calcareous tufa is commonly deposited in the streams of the Ditton Series outcrop, particularly on rocks forming waterfalls. For example it forms a thick layer on the '*Psammosteus*' Limestones beneath the road bridge at The Hope [511786]. It is also developed locally on the faces of old quarries in rocks of the Aymestry Group on Whettleton Bank, Onibury [4480]. No extensive spreads of tufa have been observed. D.C.G.

References

COBBOLD, E. S. 1900. in *Church Stretton.* 1. ed. C. W. Campbell-Hyslop. Shrewsbury.

COWIE, J. W. and ADAMS, P. J. 1957. The Geology of the Cambro-Ordovician Rocks of Central East Greenland. *Medd. om Grønland*, Copenhagen, **153** (1), 1–193.

DWERRYHOUSE, A. R. and MILLER, A. A. 1930. The Glaciation of Clun Forest, Radnor Forest and some Adjoining Districts. *Quart. J. Geol. Soc.*, **86,** 96–129.

MACKINTOSH, D. 1879. Results of a Systematic Survey, in 1878, of the Directions and Limits of Dispersion, Mode of Occurrence, and Relation to Drift-Deposits of the Erratic Blocks or Boulders of the West of England and East of Wales, including a revision of many years' previous Observations. *Quart. J. Geol. Soc.*, **35,** 425–55.

POCOCK, R. W., WHITEHEAD, T. H., WEDD, C. B., and ROBERTSON, T. with contributions by WRAY, D. A., STUBBLEFIELD, C. J., CANTRILL, T. C. and DAVIES, W. M. 1938. Shrewsbury District. *Mem. Geol. Surv.*

WHITTARD, W. F. 1952. A Geology of South Shropshire. *Proc. Geol. Assoc.*, **63,** 143–97.

—— with contributions by BALL, H. W., BLYTH, F. G. H., DINELEY, D. L., JAMES, J. H., MITCHELL, G. H., POCOCK, R. W. and STUBBLEFIELD, C. J. 1953. Report of Summer Field Meeting in South Shropshire, 1952. *Proc. Geol. Assoc.*, **64,** 232–50.

Chapter 11

GEOPHYSICAL INVESTIGATIONS

MORE THAN a century of intensive and varied geological research in the Welsh Borders has been supplemented in recent years by much geophysical investigation. The following account summarizes the results of regional geophysical surveys made within the boundaries of the Church Stretton (166) Sheet and of laboratory investigations on local rocks. Detailed surveys by the Geophysics Department of the Geological Survey in and around the Church Stretton valley are also described. These surveys were designed primarily to obtain information about the sub-drift configuration of the Church Stretton valley and geological structure concealed by drift. Lack of contrasting physical properties between the various Pre-Cambrian and Lower Palaeozoic rocks of the area limits the incidence of anomalies interpretable in terms of deep structures, but useful information has been obtained about the form of the sub-drift surface of the Church Stretton valley.

REGIONAL GEOPHYSICAL SURVEYS

GRAVITY SURVEYS

In 1947, Cook and Thirlaway (1950) made gravity observations in a large area of Wales and the Borders. During this survey, four detailed traverses were made across the Church Stretton Fault Complex in the area between Church Stretton and Trecastle. These traverses (1950, fig. 9, p. 42) show no clear fault anomaly but the line of geological disturbance is characterized by a local gravity increase of up to two milligals. At Trecastle the small positive anomaly persists even though the Old Red Sandstone cover has increased to 900 m, and Cook and Thirlaway (1950, p. 43) suggested that it may be possible to follow any continuation of the line of disturbance south-westward under thick cover. There is a strong regional gradient across the area with gravity values dropping to the south-east into an elongate gravity basin, shown on their anomaly contour map (1950, fig. 1) to be centred west of Ludlow and to enclose the Upper Palaeozoic terrain of the Clee Hills area. The Long Mynd is an area of high gravity field with the anomaly level ten milligals above that on the Lower Palaeozoic outcrops west of Ludlow.

A gravity survey in the country between the Clee Hills and Nuneaton (Cook, Hospers and Parasnis 1952) included the eastern edge of the district under review. From the Old Red Sandstone terrain east of Clee Hill, gravity values increase by about ten milligals northward towards Lower Palaeozoic outcrops in the vicinity of Much Wenlock, and increase by two or three milligals southtward on to the Forest of Wyre Coalfield.

In 1955, Cook and Thirlaway published a paper on the geological results of their earlier gravity survey mentioned above, containing a larger-scale Bouguer anomaly map of the Welsh Borders (1955, pl. iv). The Lower Palaeozoic and Pre-Cambrian areas in the north-west of the Church Stretton district are represented on that map by a Bouguer anomaly level about ten milligals above that over the predominantly Upper Palaeozoic terrain to the south-east. However, the highest anomaly occurs locally east of Much Wenlock extending on to the Old Red Sandstone and Carboniferous strata in the extreme north-east corner of the area under review. A negative anomaly over the bulk of the Old Red Sandstone outcrop in the Church Stretton district is shown to extend on to Silurian strata and to be centred midway between Knighton and Ludlow.

Additional surveys are reported by Mr. P. J. Fenning as follows.

"During June 1964 an extensive regional gravity survey was conducted over the area of the Church Stretton (166) Sheet (by the Geophysics Department) using Worden gravity meter W66. Observations were made at 426 gravity stations, maintaining a density of two stations per square mile. Gravity values were referred to a datum of 981·2650 cm/s² at Pendulum House, Cambridge through acceptance of a value of +13·06 milligals at a gravity base station situated in Church Stretton. Gravity data were reduced to sea level assuming a surface density of 2·70 g/cm³ for Pre-Cambrian, Cambrian, Ordovician and Silurian rocks, 2·50 g/cm³ for Old Red Sandstone rocks, and 2·55 g/cm³ for Coal Measures rocks. The Bouguer anomaly value at each station was corrected for the gravitational effect of the surrounding topography.

The 1930 International Gravity Formula has been adopted as the reference for the Bouguer anomaly values shown in Figs. 24 and 26.

The Long Mynd is characterized by a gravity high elongated in a north-east to south-west direction. Both to south and east of the Long Mynd there is a regular decrease in gravity values towards a negative gravity feature south of Onibury and a pronounced gravity closure at Plaish Park [521968].[1]

In the north-east of the area the gravity gradient south-eastwards from Muckleycross [647961] is interrupted by a positive gravity anomaly centred one mile north of Tugford [556870] and a gravity low trending west-south-west through Cleestanton [574793]. From Cleestanton, the gravity trough may extend eastwards to embrace a subsidiary negative gravity centre at Crumpsbrook [628782], or these two negative anomalies may be separated by a positive residual gravity feature trending south-westwards and situated half a mile west of Silvington [622799]".

MAGNETIC SURVEYS

In October 1953, an aeromagnetic survey was flown over the area around Church Stretton (Henson 1958). The flying height of the survey was 500 ft above ground level. North–south flight lines were spaced at ¼-mile intervals and, in addition, there were two east–west tie lines. Complete coverage of the Church Stretton district is provided by two later aeromagnetic surveys commissioned by the Geological Survey and flown in 1958 and 1960; both of these surveys were flown at a mean terrain clearance of 1000 ft. In the earlier of the two surveys, covering the North Midlands and including the northern half

[1]Such figures are National Grid References within the 100-km square SO (32).

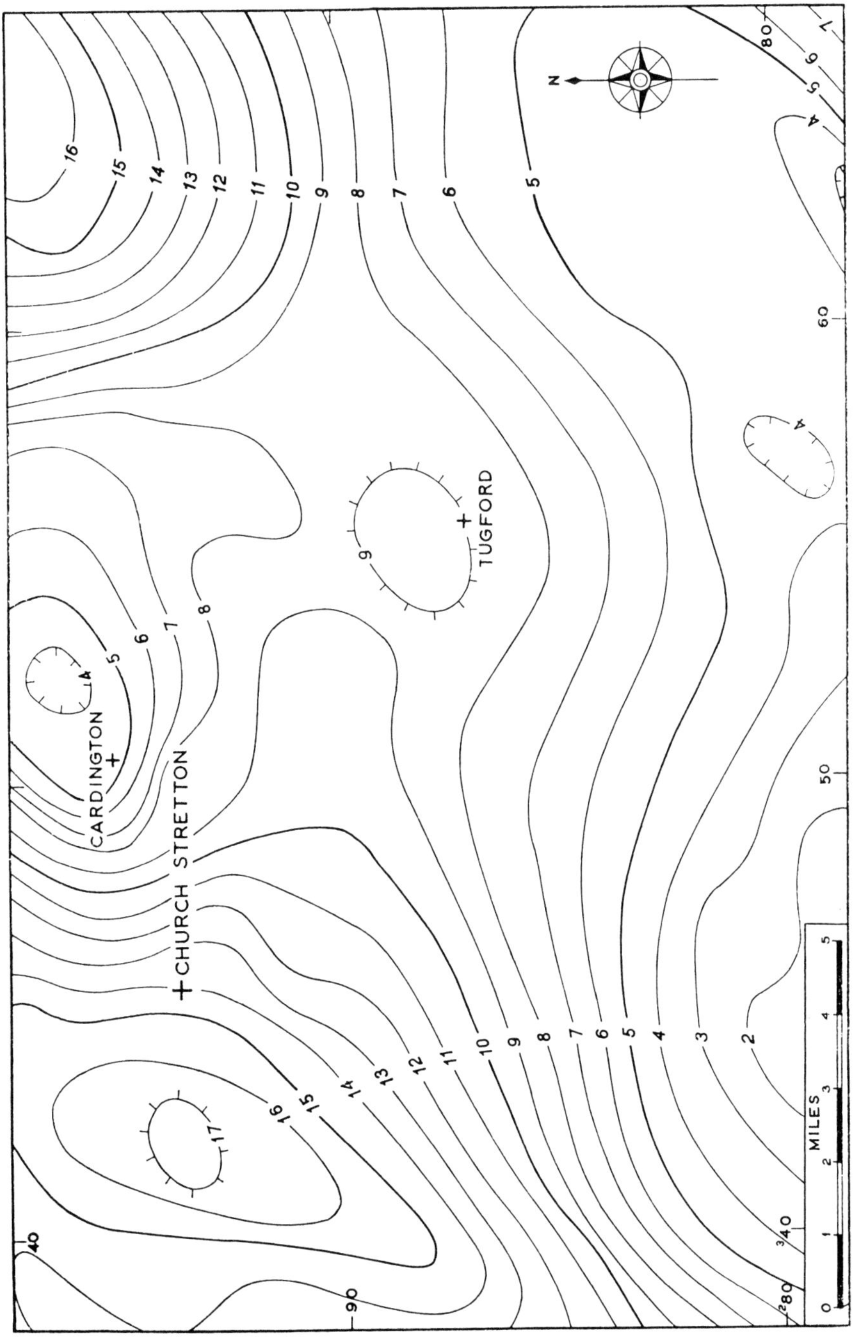

FIG. 24. Church Stretton District: Bouguer anomaly gravity map (Contour interval 1 milligal)

of the Church Stretton district, flight lines were spaced at 2 km along east–west National Grid lines, and north–south tie lines were at 10 km intervals along selected National Grid lines. In the later survey, covering Wales and including the remainder of the Church Stretton district, flight lines were north–south and tie-lines east–west with the same spacings as in the North Midlands survey. Aeromagnetic total field anomalies shown in Fig. 25 represent deviations from a geomagnetic field, calculated from a network of magnetic values derived from observations throughout Great Britain.

The aeromagnetic field over the region shows a large positive anomaly, with an areal extent of about 100 square miles, situated centrally in the Church Stretton district. This anomaly comprises two positive anomaly axes, each about ten miles in length, nearly parallel to the locally predominant north-easterly geological strike. One axis, through Wart Hill and Hope Bowdler, has an amplitude of about 200 gammas and is clearly associated with surface and near-surface Uriconian rocks, for local magnetometer traverses at ground level demonstrate the existence of large positive anomalies over outcropping Uriconian rocks. There is a sharp reduction of over 100 gammas in peak magnetic values between Hope Bowdler village and Caer Caradoc though a narrow strip of positive anomalies, bounded on each side by steep gradients, extends north-eastwards along the line of the Uriconian rocks forming Caer Caradoc and The Lawley. There is no surface geological evidence to account for the lower peak values north of Hope Bowdler but the explanation may be that faulting along the Church Stretton disturbance restricts the width of occurrence of Uriconian rocks at depth in this area. The other positive anomaly axis runs through Bouldon, Holdgate and Shipton across Palaeozoic terrain, with no surface evidence of source material. Mr. P. J. Fenning notes, however, that the positive gravity anomaly noted one mile north of Tugford may well be associated with this positive aeromagnetic anomaly. Peak values occur one mile north-east of Shipton, where the level of the magnetic field exceeds by 200 gammas that in a negative anomaly closure to the north of the positive axis. Again, Mr. P. J. Fenning notes that this negative aeromagnetic closure, situated one mile north-east of Cardington, is coincident with the pronounced negative gravity anomaly which occurs at Plaish Park in an area of exposed Ordovician strata and that these geophysical anomalies may be due either to the presence of a sedimentary basin or to an acid igneous intrusion at depth. To the east, the magnetic field values decrease towards the Forest of Wyre Coalfield.

A method by Vacquier and others (1951) was used to estimate depths to possible causal bodies for the larger anomalies. Calculations were carried out where possible on flight profiles as these permit the most accurate assessment of the parameters on which the method depends. A test on the anomaly centred ½ mile south-east of Ragleth Hill gave a depth of 1050 ft for a source approximated to a vertical-sided prism magnetized by induction in the Earth's field; as the flying height was 1000 ft, this indicates the Uriconian rocks at outcrop. Computations on the anomaly centred north-east of Shipton gave a depth below ground level of approximately 5800 ft for a similar prismatic body. This depth would represent the local Palaeozoic thickness if an anomalous body of the type assumed forms the Pre-Cambrian basement surface in the area.

The Longmyndian and Carboniferous areas north and west of Church Stretton are characterized by the low magnetic relief with field variations

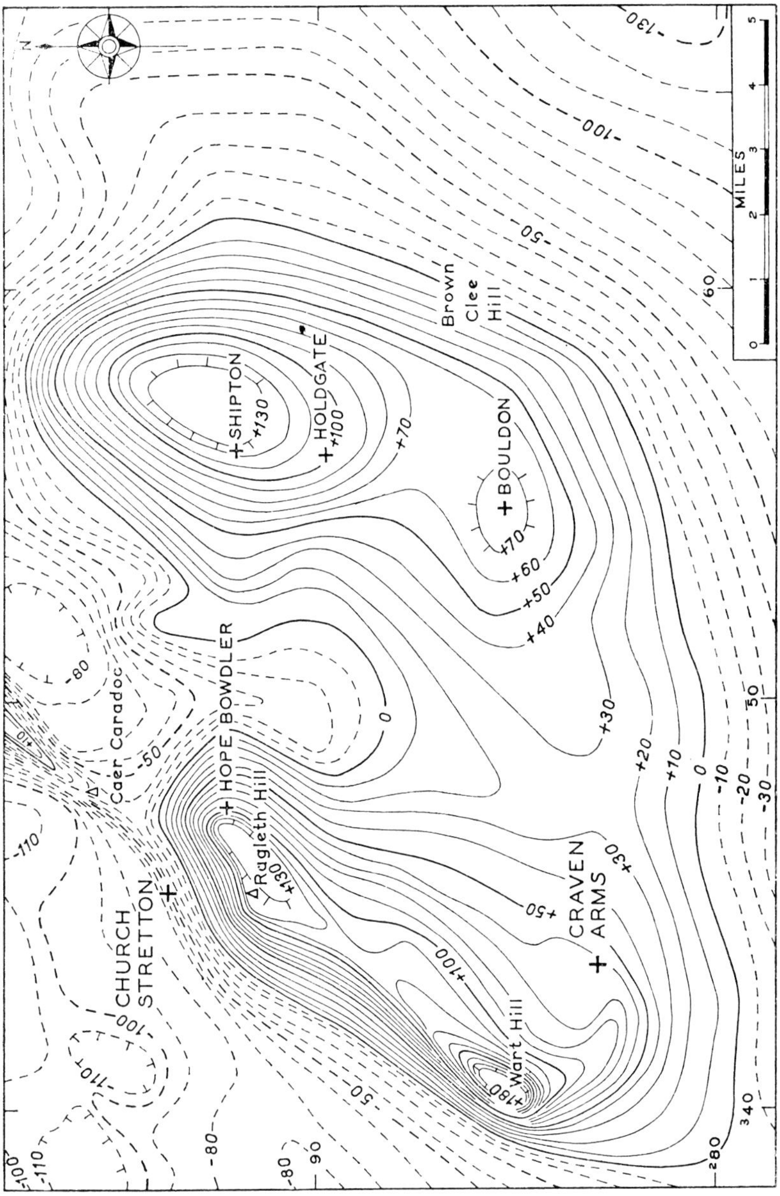

FIG. 25. Church Stretton District: Total field aeromagnetic anomaly map (Contour interval 10 gammas)

normally less than 50 gammas in amplitude. The mean level of the aeromagnetic field in this area is nearly 100 gammas below the geomagnetic datum.

The two small basalt cappings to Brown Clee Hill produce positive anomalies of five gammas on the nearest flight line profile, sited approximately ¼ mile east of the outcrops; these small anomalies are not represented in Fig. 25. Beyond the southern margin of the Church Stretton district, the airborne magnetometer record obtained over Titterstone Clee Hill reveals two positive anomalies with flanking lows, of 60 gammas and 70 gammas peak-to-trough amplitude, associated with the larger basalt occurrence in this area.

Geophysical Investigations in the Church Stretton District

Gravity Survey

During 1961 and 1962 about 550 gravity readings were made along twenty five traverses in the Church Stretton valley and at bench marks and spot heights outside the valley, the latter to outline the regional gravity field. Gravity values were referred to the value of +13·06 milligals at a station in Church Stretton. Data were reduced to sea level using a uniform density of 2·65 g/cm³ and the Bouguer anomalies of Fig. 26 are referred to the International Gravity Formula of 1930.

Results. Both the Bouguer anomaly map (Fig. 26) and the cross-valley traverses show a well-marked negative anomaly coincident with the valley between the Long Mynd and the Caradoc range. This anomaly decreases southwards from near Botvyle, where it is nearly one milligal, to effectively zero in the Marshbrook gap.

Away from the valley, apart from an easterly regional gradient of two milligals per mile and local high anomalies associated with surface Uriconian rocks, the Bouguer anomaly map gives little indication of known geological structure.

Density determinations. Density determinations for use in the interpretation of the survey results were made on a small number of rock samples from local Geological Survey Boreholes using a modification of a method by Parasnis (1952). The results are given below:

TABLE 8

		Borehole No.	No. of Specimens	Saturation density, (g/cm³)	Dry density, (g/cm³)	Porosity (per cent)
Carboniferous	Coed-yr-allt Beds	4	2	2·39 ± 0·01	—	—
Silurian	Wenlock Shale	3	1	2·67	2·57	10
	,,	3	3	2·67	2·61	6
	Hughley Shale	3	3	2·68 ± 0·01	2·61	7
Pre-Cambrian	Stretton Shale	outcrop	2	2·67 ± 0·01	2·61 ± 0·01	6

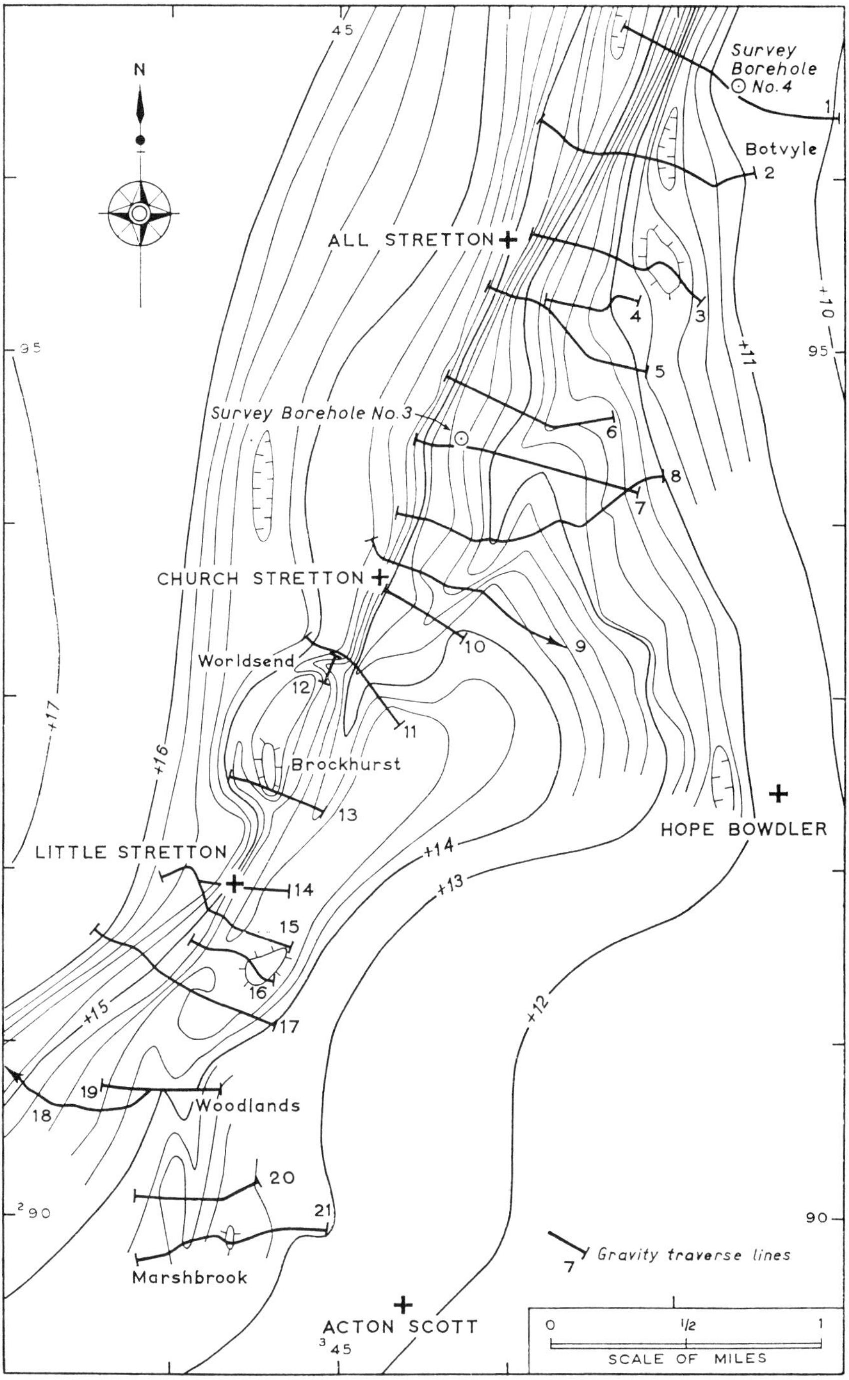

FIG. 26. Bouguer anomaly map of the area around Church Stretton (Contour intervals 0·2 and 1·0 milligal. Gravity values reduced to sea level using uniform density of 2·67 g/cm³.)

Parasnis (1952, p. 266) obtained a saturation density of 2·65 g/cm^3 for a specimen of Wenlock Limestone from Wenlock Edge. These values indicate that Lower Palaeozoic and Longmyndian strata will not be differentiated on the Bouguer anomaly map but that areas of Carboniferous strata or drift will be represented by negative anomalies with respect to the field over Lower Palaeozoic terrain.

Interpretation

(i) Church Stretton Valley. The drift-filled Church Stretton valley is represented on the Bouguer anomaly map by a north-north-easterly trending gravity trough superimposed on a regional gradient, of two milligals per mile, which crosses the area under review. Cross-valley profiles indicate that the gravity trough is of the order of one milligal in amplitude and has a width equal approximately to the width of the valley. In places the limit of drift coincides with an inflexion point on the anomaly profile, suggesting that the sub-drift surface at the side of the valley is steep, but more often the anomaly lies wholly within the mapped limits of drift. Removal of a regional background by inspection of the original profiles successfully isolates the local effect and anomaly profiles thus produced have been used in the following interpretation.

The local anomaly has its largest amplitude of nearly one milligal west of Botvyle where it is due to the combined effects of drift, alluvium and thin Coed-yr-allt Beds. There is a gradual reduction in amplitude southwards to beyond Little Stretton where the anomaly diminishes abruptly. Near Marshbrook there is no obvious anomaly. This pattern suggests a southward thinning of drift and alluvium with solid rock near to the surface in the Marshbrook gap. The position of greatest negative anomaly follows closely the course of present-day streams except (i) the stream detour west of Brockhurst and (ii) south of Little Stretton where the stream turns south to pass through the Marshbrook gap, but the Bouguer anomaly minimum continues south-westward across the line of the railway track before turning into the gap.

Geological Survey Boreholes Nos. 3 and 4, which penetrate the drift within the area of the valley profiles, provide two checks in the accuracy of estimation of drift thickness using the anomaly profiles. At Borehole No. 3 (near Springbank Farm [457945]) the depth of drift is 86 ft; approximating the drift layer to the form of an infinite horizontal slab with a density 0·47 g/cm^3 less than the underlying Lower Palaeozoic and Pre-Cambrian strata, the depth is calculated from the valley anomaly to be 90 ft. At Borehole No. 4 (near Botvyle [474965]) 38 ft of drift are underlain by about 60 ft of Upper Carboniferous. The anomaly of 0·55 milligal compares with a calculated anomaly of 0·44 milligal for horizontal slabs of drift and Carboniferous of the above thicknesses and with densities 0·47 and 0·28 g/cm^3 lower than that of the Lower Palaeozoic strata. The discrepancy is equivalent to the effect of 18 ft of drift.

More accurate thicknesses can be estimated from the maximum anomaly value by approximating the shape of the valley section to either a rectangle (U-shaped valley) or an inverted triangle (V-shaped valley). Keeping the ratio constant between thickness and surface width of drift, the gravity effect of the valley sides can be calculated using a method by Colley (1955) such that it can simply be applied as a correction to a normal Bouguer reduction to give the gravity effect of the structure. Fig. 27 is a longitudinal valley-floor section, based on gravity anomalies which produce the lowest base-of-drift level in any profile and on a V-shaped valley-floor configuration in transverse section. The section indicates that the level of the sub-drift valley floor drops northward from Woodlands and there is some evidence, based on one gravity traverse (Fig. 26, Traverse 21), for a drop in rock head southwards from Woodlands.

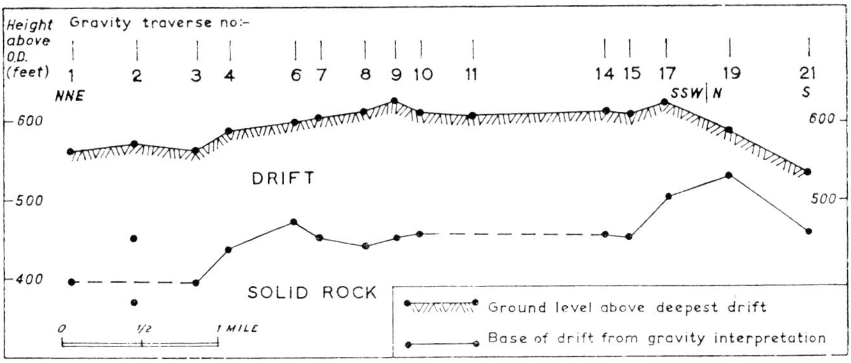

FIG. 27. Longitudinal section along the Church Stretton valley

Irregularities in the interpreted level of the solid rock floor, and in particular the apparent overdeepening between Springbank Farm (Fig. 26, Traverse 7) and Woodlands (Fig. 26, Traverse 19), may be attributed to ice action (see p. 288) or, at least in part, to density variation within the drift succession.

The reliability of the above interpretation is also dependent upon the absence of sharp irregularities of limited lateral extent in the depth to solid rock, for the gravity effect of such irregularities would be small and has not been catered for in the interpretation of the valley traverses. Immediately north of Church Stretton, towards the north-west side of the main valley, solid rock has been exposed within 6 ft of the ground surface at three places [45669416, 45639382 and 45739394] in an excavation running along the valley. These exposures are restricted and in each case the rocks dip steeply out of view. In this area, the local gravity anomaly would be taken to indicate about 80 ft of drift on the simple model postulated above. The absence of local gravity anomalies attributable to the near-surface solid rock is evidence for its restricted lateral extent at shallow depth. This area lies north-west of the maximum gravity anomaly and thus the longitudinal valley section (Fig. 27) is not necessarily affected by it. However the possibility of unidentified rock mounts within the Church Stretton valley, which could have a modifying effect upon the interpretation given, cannot be ruled out.

A gravity traverse (Fig. 26, Traverse 15) across the mouth of Ashes Hollow, which joins the main valley at Little Stretton, shows the presence of a small negative anomaly attributable to valley deposits. Though it is difficult to estimate the background anomaly in the area, it is unlikely that the valley anomaly exceeds 0·15 milligal which represents 25-30 ft of superficial deposits with a density of 2·20 g/cm^3.

From geological evidence it is considered that pre-Glacial drainage in Ashes Hollow continued southwards through Marshbrook, at a level of about 650 ft (see pp. 288–9). In this view, melt water from the snout of a glacier standing in the Church Stretton valley modelled the present-day Marshbrook gap and considerable downcutting to the north of Marshbrook is attributable to erosion during the advance of the glacier.

The rock barrier at Woodlands indicated by the gravity results is incompatible with the view that pre-Glacial drainage was southwards at a level substantially lower than the present valley floor. Further, low-level drainage in the main valley would require that Ashes Hollow be graded to this low level whereas the gravity evidence, though slight, is that there is no great thickness of superficial deposits in the mouth of Ashes Hollow so that this valley probably hangs over the rock floor in the main valley.

(ii) Worldsend and Brockhurst. South of Church Stretton the present-day watercourse through Worldsend and west of Brockhurst appears from the Bouguer anomaly map to be along a well-established channel. Across Worldsend the gravity anomaly is at least 0·5 milligal, which is comparable in size with the main anomaly and indicative of approximately 100 ft of superficial deposits. This anomaly, together with a large anomaly in the low ground west of Brockhurst Castle, suggests an intimate pre-Glacial or Glacial relationship between the two valleys. The new course may be that of an overflow channel which eventually cut down nearly as deeply as a main channel east of Brockhurst. A gravity traverse south of Brockhurst Castle indicates a gravity high separating the present-day drainage valley from the drift-covered area east of Brockhurst, showing that the solid core exposed in Brockhurst Wood continues southwards under the mound of Brockhurst Castle.

MAGNETIC SURVEY

At the same time as the gravity survey was being carried out, about 500 total field magnetic measurements were made on the ground along twenty one traverse lines. Traverses across the Church Stretton valley were designed to locate structure below the valley drift. Further traverses were established in other areas of interest, partly to check the inferred position of geological boundaries and partly to test the magnetic effect of volcanic rocks within the Longmyndian succession.

Results and Interpretation. The cross-valley profiles revealed no interpretable anomaly associated with the drift or with structure under the drift. However, the Longmyndian–Uriconian boundary was found to have an associated anomaly often exceeding 100 gammas, the field increasing sharply on to outcropping Uriconian strata.

Detailed magnetic traverses were made across the Longmyndian–Uriconian boundary on the south-west slopes of Caer Caradoc, below Three Fingers Rock [471947] to check a faulted displacement of the boundary inferred from surface mapping. They show that the Helmeth Grit coincides with a negative anomaly, of between 20 and 200 gammas, between the Stretton Shale field level and the sharp rise on to the Uriconian. Though this negative anomaly is not due necessarily to the Helmeth Grit, but possibly to the Helmeth Grit–Uriconian boundary, nevertheless to the extent that the relationship between anomaly pattern and mapped geology is constant on the five traverses, they support the mapped displacements.

The negative anomaly observed in this area over the Helmeth Grit is a local effect. Of the cross-valley traverses, one coursing the south-west slope of Helmeth Hill and two coursing the west side of Ragleth Hill cross the sequence:

Stretton Shale, Helmeth Grit, Uriconian tuffs, and on one of them is there a recognizable negative anomaly which could be due to the Helmeth Grit or attributed to its boundary with Uriconian rocks. This change in anomaly pattern southwards from below Three Fingers Rock may reflect a change in the mineral content of the Grit or a change in the nature of the Helmeth Grit–Uriconian boundary. The negative anomaly is restricted in occurrence to an area in which, according to Cobbold and Whittard (1935, pl. 29), there is no Helmeth Grit. In this account, however, it is considered that the Grit extends northwards to the Church Stretton Fault (F1).

ELECTRICAL SURVEY

The main purpose of electrical resistivity investigations was to obtain information on the depth to, and configuration of, the base of the drift and to investigate the possibility of outlining boundaries within it. Ten cross-valley traverses using a constant electrode separation were designed to outline the position of deepest superficial cover so that expanding probes along the valley could be sited where they would give information on the maximum thickness of drift. Eight expanding probes were established at various points in the valley, including those at which the constant separation traverses had indicated maximum thickness of drift, to investigate the detailed lithology and thickness of the drift. Electrical measurements were taken in Geological Survey boreholes in the Church Stretton district to supplement the lithological logs and obtain estimates of formation resistivities.

Constant Separation Traverses. The cross-valley profiles typically show an inverse relationship between apparent resistivity and drift thickness, with resistivity values dropping towards the centre of the valley, which implies that the drift resistivity, below the water table, is lower than the solid rock resistivity. This conclusion is upheld by resistivity values measured in the area and given below (p. 319). Assuming that the lowest value of apparent resistivity relates to the position of thickest drift, the resistivity indication of thickest drift always agrees in position with the gravity interpretation to within 50 yd, including a position south-west of Little Stretton where gravity and resistivity minima are both on a steep bank of glacial deposits, 300 yd west of the present stream position.

Electrical measurements support the gravity interpretation that Brockhurst Castle is underlain by a solid core separating the present day drainage valley from the main drift-filled valley east of Brockhurst. Two profiles south of Brockhurst Hill show a resistivity peak of 120 ohm-metres separating the low resistivity values of the two valleys, which coincides with a local gravity high.

Expanding Probes. Three expanding probes were established in the valley on a traverse section through Church Stretton Borehole No. 3 [457945]. Profiles at each probe position are indicated, by comparisons with Mooney and Wetzel interpretation curves (1956), to represent multi-layer configurations. It is impossible to correlate satisfactorily between the profiles or simply to compare the resistivity with the gravity interpretation. Other expanding probes, west of Brockhurst and at positions of maximum depth of drift in the main valley, similarly indicate multi-layer configurations and are difficult to interpret quantitatively. The complexity of the resistivity curves obtained and the lack of correlation between them probably indicate a heterogeneous and locally variable drift succession.

Electrical logs from Church Stretton Boreholes. A programme of borehole investigation, supplementary to the mapping of the Church Stretton district, was implemented during 1961 and 1962. Six boreholes, of which one was inclined to the vertical, were sunk in the area around Church Stretton to obtain local stratigraphical and structural information. Using portable logging equipment, the Geophysics Department recorded electrical resistance and spontaneous potential logs in the five vertical boreholes. The equipment used employs a non-focussing single-electrode system suspending in the drilling fluid and measures the total resistance between this electrode and an earthing electrode situated at the ground surface some distance from the borehole. Formation resistivities cannot be derived from a single-electrode resistance curve by direct numerical calculation. However, records obtained in boreholes elsewhere using the equipment described above, have been compared with those obtained using multi-electrode logging equipment through identical strata. From these tests it has been found possible to construct empirical curves relating single-electrode resistance values and apparent resistivity. For boreholes with the characteristics of those in the Church Stretton area it is estimated that at low resistivities and for thick beds the readings given by the equipment are close to the formation resistivity but that at values exceeding 100 ohm-metres the readings obtained may be less than the formation resistivity by a factor of about two. The results of an analysis of the Church Stretton boreholes are given below.

(i) Borehole No. 1 (near Eaton Farm; [37238978]). The Hughley Shale succession, characterized by a relatively low resistivity of 75 to 90 ohm-metres, is clearly distinguishable from the overlying Wenlock Shale, through which higher and more variable resistance readings were obtained. The Western Longmyndian is marked by a sharp increase to the highest resistances recorded in the hole, with an estimated resistivity of 160 ohm-metres.

(ii) Borehole No. 2 (near Robury Ring; [39409302]). Resistances measured in the Hughley Shale are higher than in Borehole No. 1. The top of the Western Longmyndian is again characterized by a sharp increase in resistance reading.

(iii) Borehole No. 3 (near Springbank Farm; [45719447]). The total resistance encountered was too high to be measured by the equipment. The Hughley Shale and the upper 90 ft of *Pentamerus* Beds are characterized by similar resistance values. Below this there is a slight but well-defined increase in measured resistance. Thin limestones and massive siltstones in the *Pentamerus* Beds are represented by resistance peaks. Again, there is a notable increase of resistance at the top of the Pre-Cambrian succession.

(iv) Borehole No. 4 (near Botvyle; [47359653]). Boring was stopped before Pre-Cambrian strata were reached. The resistance values recorded through the Wenlock Shale are consistently lower than in Borehole No. 1.

(v) Borehole No. 5 (near Hamperley; [42178912]). Estimates of resistivity for the Hughley Shale are lower than those recorded elsewhere and increase downwards through the succession from 50 to 70 ohm-metres. A sandstone and conglomerate at the top of the *Pentamerus* Beds are together represented by an estimated resistivity peak of 250 ohm-metres. Entry into Pre-Cambrian strata is marked by a large increase in resistance.

Casing in the top sections of the boreholes precluded the measurement of resistance in drift deposits but the surface resistivity survey indicated apparent resistivities, for drift successions, in the range 45–700 ohm-metres, the higher values relating to drift above the water table. Constant separation traverses showed a reduction in apparent resistivity towards the centre of the Church Stretton valley indicating that locally the drift resistivity is significantly lower than the solid rock resistivity.

The estimates of formation resistivity from the borehole logs are given below:

TABLE 9

Approximate formation resistivities (ohm-metres)

		B.H. No. 1	2	3	4	5
Silurian	Wenlock Shale	110–150	—	—	50–60	—
	Hughley Shale	75–90	95–105	80*	—	45–70
	Pentamerus Beds	—	—	80–90*	—	120–220
Pre-Cambrian	E. Longmyndian	—	—	125*	—	?240
	W. Longmyndian	160	235	—	—	—

*Dependent upon an assumed Hughley Shale Value.

Spontaneous potential logs were recorded simultaneously with the resistance logs but did not indicate any consistent features of geological interest.

CONCLUSIONS

The geophysical work, and in particular the gravity survey, has fulfilled its primary function of contributing to an understanding of near-surface geology in the Church Stretton valley. The gravity and electrical resistivity surveys give consistent indications of the position of deepest drift, but inadequate knowledge about the heterogeneous drift succession precludes quantitative analysis of the resistivity results obtained. Further useful information about the sub-drift configuration of the Church Stretton valley could be obtained by a seismic refraction survey which would be valuable in checking the gravity interpretation of drift thickness variation.

Magnetic traverses indicate the utility of magnetic methods for locating and outlining the Longmyndian–Uriconian boundary but no interpretable magnetic anomalies have been encountered in the drift-covered ground of the Church Stretton valley.

Resistance measurements in local boreholes have enabled estimates to be made of formation resistivities for local rock types. The variations in electrical properties between boreholes and the overlapping resistivity values encountered for some of the rock groups would limit the scope of geological correlation by electrical logs from uncored boreholes, but the Pre-Cambrian basement normally has a significantly higher resistance than the overlying Lower Palaeozoic strata.

Other Geophysical Work

Palaeomagnetic Studies. The existence, within the area of the Church Stretton Sheet, of well exposed Pre-Cambrian strata and Carboniferous igneous rocks has resulted in considerable local activity by workers in palaeomagnetism.

Creer (1957) measured the direction and intensity of normal remanent magnetization of more than forty samples from twelve localities in the Longmyndian Wentnor Series. After correcting the values for geological dip, assuming the structure to be a synclinal overfold, two distinctly opposed magnetization directions were obtained, interpreted as representing at least one magnetization reversal. The mean inclination of the normal remanence vector was found to be 29° downwards at an inclination of N. 66° W., which is intermediate in position between directions obtained for the Lower and Upper Torridonian (Irving and Runcorn 1957) and thus suggestive of an intermediate age.

Clegg and others (1957) collected specimens of basalt and contact rocks from the Clee Hills for palaeomagnetic measurements. Large differences were noted between the directions of remanent magnetization obtained and those determined by Belshé (1957) for other British Carboniferous rocks. Possible explanations given for the disagreement are (a) secondary magnetization, (b) a physicochemical process resulting in the establishment of magnetization directions oblique to the prevailing field, (c) large scale continental movement relative to the poles during the Carboniferous period, and (d) a Carboniferous magnetic field differing substantially from that of an axial dipole. Close similarity of remanence direction for contact rocks above and below the basalt occurrence on Titterstone Clee Hill supports the view of Marshall (1943) that the body is intrusive. This conclusion is upheld by Everitt (1960) after a wider palaeomagnetic study of Midland basalts which indicated that of all the rocks investigated only the basalt occurrence near Little Wenlock is extrusive. Everitt and Belshé (1960), in a review of palaeomagnetic measurements on Carboniferous rocks, concluded that the Midland basalts investigated exhibit stable primary magnetization and that secondary magnetization of many other Carboniferous rocks during the Triassic period is the cause of disparities in Carboniferous magnetization directions.

Age determination. Using the helium method, Urry obtained an age oı 135 million years for an Upper Carboniferous 'analcime-olivine-dolerite' from the Clee Hills and this age was used in a revised helium time scale (Urry and Holmes 1941, p. 60, table vi). Subsequent investigations by various workers (see Faul 1954, pp. 306–328) established that the escape of helium from rocks commonly led to the assumption of ages too low in value, and the discrepancy between the above helium age and the age of the Upper Carboniferous period in modern time scales (e.g. Kulp 1961) is explicable in terms of helium loss from the dolerite sample. M.B

References

Belshé, J. C. 1957. Palaeomagnetic investigations of Carboniferous rocks in England and Wales. *Advanc. Physics*, **6**, 187–91.

Clegg, J. A., Deutsch, E. R., Everitt, C. W. F. and Stubbs, P. H. S. 1957. Some recent palaeomagnetic measurements made at Imperial College, London. *Phil. Mag. Suppl.* **6**, 219–31.

Cobbold, E. S. and Whittard, W. F. 1935. The Helmeth Grits of the Caradoc Range, Church Stretton; their bearing on part of the Pre-Cambrian Succession of Shropshire. *Proc. Geol. Assoc.*, **46**, 348–59.

Colley, G. C. 1955. Gravity variations in surveys across geological boundaries. *Geophys. Prosp.*, **3**, 403–24.

Cook, A. H., Hospers, J. and Parasnis, D. S. 1952. A gravity survey of the country between the Clee Hills and Nuneaton. *Quart. J. Geol. Soc.*, **107**, 287–302.

—— and Thirlaway, H. I. S. 1950. Recent observations of gravity in Wales and the Borders. *Rep. XVIII. Int. Geol. Congr. Gt. Brit. 1948* Pt. 5, 33–44.

—— 1955. The geological results of measurements of gravity in the Welsh Borders. *Quart. J. Geol. Soc.*, **111**, 47–70.

Creer, K. M. 1957. Palaeomagnetic investigations in Great Britain. IV. The natural remanent magnetization of certain stable rocks from Great Britain. *Phil. Trans. Roy. Soc.* (A), **250**, 111–29.

Everitt, C. W. F. 1960. Rock magnetism and the origin of the Midland basalts. *Geophys. J. Roy. Astr. Soc.*, **3**, 203–10.

—— and Belshé, J. C. 1960. Palaeomagnetism of the British Carboniferous System. *Phil. Mag.*, **5**, 675–85.

Faul, H. 1954. *Nuclear Geology*, 1st Edit. New York.

Henson, F. A. 1958. An aeromagnetic survey of the Church Stretton area. *Proc. Geol. Assoc.*, **68**, 107–14.

Irving, E. and Runcorn, S. K. 1957. Palaeomagnetic investigations in Great Britain. II. Analysis of the palaeomagnetism of the Torridonian Sandstone series of north-west Scotland. *Phil. Trans. Roy. Soc.* (A), **250**, 83–99.

Kulp, J. L. 1961. Geologic Time Scale. *Science*, **133**, 1105–14.

Marshall, C. E. 1943. Field relations of certain of the basic igneous rocks associated with the Carboniferous strata of the Midland counties. *Quart. J. Geol. Soc.*, **98**, 1–25.

Mooney, H. M. and Wetzel, W. W. 1956. *The potentials about a point electrode and apparent resistivity curves for a two-, three- and four-layer earth.* 1st edit. Minneapolis.

Parasnis, D. S. 1952. A study of rock densities in the English Midlands. *Mon. Not. Roy. Astr. Soc. Geophys. Suppl.*, **6**, 252–71.

Urry, W. D. and Holmes, A. 1941. Age determination of Carboniferous basic rocks of Shropshire and Colonsay. *Geol. Mag.*, **78**, 45–61.

Vacquier, V., Steenland, N. C., Henderson, R. G. and Zeitz, I. 1951. Interpretation of Aeromagnetic maps. *Mem. Geol. Soc. Amer*, **47**.

Chapter 12

ECONOMIC GEOLOGY AND WATER SUPPLY

THE CHURCH STRETTON district lies on the south-eastern fringe of the Shelve mining area which formerly yielded considerable amounts of lead, zinc and barytes. Within the district, however, practically no vein-mineral deposits of economic value have been found. Indeed, at the present time there are no important mineral workings of any kind in the area. The following account is mainly of historical interest.

COPPER ORE

Various trials were made for copper in the neighbourhood of the Shelve and one of these, at Medlicott [400947][1], occurs within the Church Stretton district. These workings were described by Murchison (1839, p. 261) who noted that ore was extracted by the 'Snailbatch company' (Snailbeach). Murchison's brief note appears to be the only record of these workings, which consist of a shaft and a level, about 60 yd north of Hall Farm, Medlicott. In the dump material there is vein-quartz and calcite with traces of malachite and chalcocite. Traces of malachite also occur in purple Longmyndian sandstones about 280 yd north-north-west of the old shaft at Medlicott. Murchison (1839, p. 261) referred to unsuccessful trials for copper having been made near Ratlinghope village [403968]. No trace of these works was found in the present survey.

Just west of the Church Stretton district, at Shuttocks Wood [373922] near Wentnor, there are several old shafts sunk in the *Pentamerus* Sandstone, probably for copper. These may be the workings referred to by Murchison (1839, p. 261), as follows: "In that [the Norbury] district, these [copper] veins are of high geological interest, for proceeding upwards from the slaty purple rock [Longmyndian] of these hills they also penetrate the Caradoc [Llandovery] sandstone . . . " According to Dines (1958, p. 41), Murchison's description refers to copper workings on Norbury Hill, 1 mile north-north-west of Norbury village and about 1½ miles north-west of Shuttock's Wood. The fact that the Norbury Hill workings lie within the outcrop of the Bayston–Oakswood Group (Longmyndian) does not agree with Murchison's mention of the 'Caradoc' [Llandovery] sandstone, which does not occur north of Norbury village.

Murchison (1839, p. 222) referred to the working of malachite, probably as diffuse deposits in Ordovician sandstone, in the vicinity of Cardington [507952]. No other record of these workings has been found.

[1]Such numbers are National Grid References within the 100-km square SO (32).

Near Hayton's Bent [518811] there are traces of old workings, probably for copper, within the '*Psammosteus*' Limestones. These were mentioned by Murchison (1839, p. 188) as never having produced any profits and as having been abandoned for 100 years. The workings are reputed to have been re-opened more recently and to have closed again in 1919. No written record of this later venture has been traced.

Barytes

There appear to be no deposits of barytes of economic value within the area under description. Near Squilver Farm [377974], just west of the Church Stretton district, some exploration adits were driven by the Malehurst Barytes Company during the last war and 1357 tons of barytes were recovered from these workings in 1944 (Dines 1958, p. 40).

At the southern end of Adstone Hill [391941] there are traces of an old trial level, 100 yd west of Adstone Farm. There seems to be no record of the the working which is in green and purple sandstone of the Bridges Group (Longmyndian). The trial may have been for barytes or possibly copper, the Medlicott copper workings being about 1300 yd to the north-east.

At the south end of the Long Mynd there is an old quarry 455 yd north-north-east of Hillend Farm [398877], in which grey siltstones and shales of the Synalds Group (Longmyndian) are cut by a number of barytes veins up to 2 inches in thickness. It is thought that the excavation was made for the extraction of barytes but no record of this has been seen. In the same area, 820 yd W. 25° N. of Hillend Farm, a small quarry in massive sandstone of the Lightspout Group (Longmyndian) has been excavated, probably in the working of a vein of barytes between 3 inches and 9 inches in thickness.

Building Stone and Roadstone

Building stone and roadstone are dealt with together, since in many instances it is not now possible to determine for what use a particular quarry was excavated. It is probably true to state that almost every rock type of reasonable hardness has been used locally for building.

Uriconian. Uriconian rocks have been worked in a number of small quarries in the Cardington–Hope Bowdler district, probably for local use mainly as road metal or in rough stone walling.

Longmyndian. The Buxton Rock (fine siliceous tuff) has been quarried to a limited extent at Buxton Quarry, All Stretton [459955] and in the Cardingmill Valley [450943]. The rock has been used locally as a rough building stone and probably also for road metal. Within the outcrop of the Longmyndian, many dolerite dykes were formerly worked in small quarries, probably as local sources of road metal.

Ordovician. The Hoar Edge Grit has been worked in small quarries probably for local building, and at Glenburrell [413862] a calcareous variety has been worked for lime. The Chatwall (Soudley or Horderley) Sandstone has been more widely worked as a good building stone. It has been used extensively in the district and its brown and purple colour banding is a distinctive feature of many of the local buildings. There are fairly extensive quarries in the sandstone at Long Lane [413841] Longville Plantation [412852]. Soudley [477918] and Chatwall Hall [515975]. In the record of a field meeting of the Caradoc and Severn Valley Field Club (1923, p. 62), La Touche is recorded as saying that the *alternata* Limestone has been used for millstones and the Longville Flags for roofing slabs. It is not known from which quarries these materials were obtained. Around Acton Scott [455895] there are numerous small quarries in the Acton Scott Limestone, which has been used in many of the buildings in that village.

Silurian. There are numerous old quarries along the outcrops of the Tickwood Beds and the Wenlock Limestone. These rocks were probably worked mainly as a source of lime but they have also been used for rough building stone and road metal. In the outcrop of the Aymestry Limestone there are a number of disused quarries, e.g. at Brandhill [420790], View Edge [425807], Rotting Wood [446811] near Norton Camp and West Knowles [453828] east of Craven Arms. As in the Wenlock Limestone, the workings were probably for lime but the rock has also been used for rough building and roadstone. The Upper Ludlow Shales have been used for local building and rough stone walling. There are numerous small quarries along the dip slope of these beds and one of the larger quarries is situated at New House [481838] at the southern end of Siefton Batch.

Old Red Sandstone. The Downton Castle Sandstone is well bedded and fairly easily worked. It has been widely used for local building and there are disused quarries at Stokesay Court [445790], Onibury [455793], Norton [460816] and Burley [475815]. Within the outcrop of the Ditton Series there are many small quarries in the '*Psammosteus*' Limestones, cornstones, and cornstone-conglomerates, which were probably worked mainly for lime burning, but cornstone-conglomerates have been used for local building. The Farlow Sandstone has been used more widely as a building stone and has been worked at Farlow Bank [641807] and Oreton [651809]. Cantrill (1908, p. 25) remarked that the Farlow Sandstone was used for the rebuilding of Farlow church in 1862. Howe (1910, p. 136) noted that some beds within the sandstone were known as 'firestones' and were formerly employed for hearthstones and fireplaces.

Carboniferous. The Carboniferous Limestone has been quarried at numerous places along the outcrop from Farlow to Prescott [663807], the most extensive quarry being near Oreton [648806]. Murchison (1839, p. 120) referred to the Oreton quarries as follows: "These quarries have afforded all the finest specimens of the grey oolitic limestone known as the 'Clee Hill Marble'. Columns of great length and breadth have been occasionally extracted. Beautiful examples of these may be seen in the mansions of Oakley Park, Downton Hall and Hoptoun House".

The dolerites of Brown Clee and Titterstone Clee have been quarried extensively in the past mainly for road metal and setts, or concrete aggregate. The workings on Brown Clee are now disused but on Titterstone Clee, large quarries are still active at Clee Hill [597757], just south of the Church Stretton district.

LIME

A calcareous variety of the Hoar Edge Grit has been quarried for lime at Glenburrell [413862] and the Wenlock and Aymestry limestones have been quarried extensively for lime throughout the district. In the Ditton Series the '*Psammosteus*' Limestones have been worked fairly widely for lime as have the cornstone-conglomerates and the Abdon limestones. Cantrill (1908, p. 24) recorded that cornstones were burned for lime at Abdon Lime Works, ½ mile south-east of Abdon, and at Ditton Lime Works, 1 mile south-west of Ditton Priors.

Brick and Tile Clays

There are no deposits of brick clay of any importance within the Church Stretton district. La Touche (1923, p. 62) is recorded as saying that the *Trinucleus* (Onny) Shales "furnished excellent material for brick making", but it is not known where the shales were dug for this purpose.

The Wenlock Shales have been dug for brick making near Plowden [380887] and near Wistanstow [430858]. Marls in the Downton Series have been worked near Norton [466811], Powkesmoor Farm, Ditton Priors [593882], Monkhopton [640942] and Bourton [605956]. In the Ditton Series marls have been dug near Heywood [625828], and Wheathill [602823]. These workings were probably mainly for local use. Boulder clay has been worked at Wistanstow [430858] and in small pits near Harton [473893] and near Minton [438908]. At Ratlinghope [403968] there is a small pit in a clay with some pebbles, which was used in making bricks for a nearby farmhouse. This material may be boulder clay but alternatively it may be of lacustrine origin.

Sand and Gravel

There has been no extensive working of sand and gravel within the Church Stretton district, but there appears to be no reason why the river terrace deposits of the Clun, Onny and Corve (see p. 301) should not be a useful source of this material if the need arises. Near Brockhurst Castle [448925], there is a large disused pit in poorly-bedded morainic gravel. It is not known for what this material was used. In the deposits of glacial sand and gravel around Marshbrook [442898] (Fig. 20) there are several small pits which were probably worked for local use. In Corvedale there is a fairly large pit, ¾ mile south of Culmington [490810] in No. 3 terrace of the Onny–Corve system, and there are a number of smaller pits in the same terrace between the above locality and Bromfield [495770].

Soils and Agriculture

The Silurian shales west of Hopesay Hill [400830] give rise to a rather poorly-drained clayey soil which is used mainly for meadowland, supporting sheep, beef cattle and dairy cattle. North of Plowden [381877] there is a thick cover of boulder clay on the Silurian rocks and this area is also used mainly for grazing.

On the Long Mynd, the Pre-Cambrian sediments carry a thin stony soil which is locally peaty on the poorly-drained upper slopes. This area supports a cover of rough grass, bilberry, and heather which is used for sheep grazing, but the quality is impaired in many places by considerable spreads of bracken. To the south of Minton Batch [420905] a large area on the upper slopes has been planted with conifers by the Forestry Commission. The Uriconian outcrops of the Ragleth, Caer Caradoc and Cardington hills support thin acid soils like those of the Long Mynd and are mainly used for rough grazing for sheep.

The Church Stretton valley is poorly drained and floored with a thick cover of glacial drift, mainly boulder clay. The agricultural land of this area is mainly meadow, used for dairy cattle, some beef cattle and winter grazing for sheep.

North-eastwards from Aston on Clun [392818] and along Ape Dale the Ordovician and Silurian shales and sandstones are largely covered by boulder clay. This area has a generally clayey soil which supports mixed arable and pastoral farming. The poorly-drained lower ground of Ape Dale is dominantly meadowland for beef and dairy cattle with some sheep, but there is a higher proportion of arable ground on the better drained slopes south-east of Hope Bowdler [475925] and south of Whittingstow [432890]. There appears also to be some concentration of arable land on the better-drained glacial and terrace gravel spreads of the Marsh Brook and Quinny Brook between Marshbrook [442898] and Craven Arms [433828].

The prominent scarp slope of Wenlock Edge is largely covered with woodland, mainly oak with some ash; some of the newer plantations however are of conifers e.g. Strefford wood [446850]. Similar use is made of the Aymestry Group scarp and there are large woods of oak and ash below View Edge [425810] and Norton Camp [447820]. Callow Hill plantation [459850] is a more recent coniferous planting.

To the south-east of the Aymestry Group feature, there are practically no glacial deposits of any consequence and on the dip slope of the Silurian the Upper Ludlow Shales give rise to a well-drained calcareous loamy soil. This belt is used for mixed farming, but between Bourton [595963] and Siefton [484830] there is a notably higher proportion of arable land than in the rest of the district. This higher proportion of arable ground is also found on the Upper Ludlow dip slope north and west of Onibury.

In Corve Dale the rocks of the Downton Series give rise to a variable clay soil which is used principally for grass, although at some localities there are considerable areas of arable ground, used mainly for root crops including sugar-beet. The principal land usage is in cattle grazing and some farms are given over entirely to the fattening of store cattle. The rocks of the Ditton Series provide a soil similar to that of the Downton Series, but it commonly contains more rock debris, is very thin locally and is usually more easily drained. Farms are normally small and devoted mainly to the raising of cattle and sheep.

The upper part of Brown Clee Hill above 1000 ft O.D. is formed largely of sandstones of the Clee Group overlain by Coal Measures. This high ground is mainly heathland and is used only for rough pasture. There are some extensive coniferous and deciduous plantations on the east side of the hill. The outcrop of the Cornbrook Sandstone and the Coal Measures on Silvington and Oreton commons also supports mainly rough pasture, but in this area there are a number of small farms with cultivated grassland.

A more detailed account of land usage in this district prior to the last war can be found in the maps of the Land Utilisation Survey of Britain, Sheets 70 (Bishop's Castle) and 71 (Kidderminster), and in the accompanying report for Shropshire by E. J. Howell (1941). J.E.W., D.C.G., B.A.H.

Water Supply

There are no important water-bearing formations in the district, and it is only locally, from drift deposits in the south-west, that ground water is abstracted in any quantity.

The moderate demands of Craven Arms and Church Stretton are supplied from wells and small impounding reservoirs respectively. The East Shropshire Water Board has an agreement with the City of Birmingham Water Department to supply parishes in the south-eastern part of the area with water from the Elan Aqueduct. Elsewhere villages and farms use small individual sources of supply from springs, wells, and boreholes. Obtaining a water supply in this area by drilling remains speculative, as it is dependent, in the solid rocks, upon intersecting a system of water-bearing secondary channels, as for example fissures or fractures. Wells into the solid rocks can normally be left unlined.

Pre-Cambrian. A bore at Aston on Clun is recorded as giving 2400 gallons per hour, but yields from 9 other bores varied from nothing to 600 gal/h.

Cambrian. No wells or bores are recorded, and no ground water in any quantity can be expected.

Ordovician. Records of yields from all the formations except the Onny Shales, for which no data are available, vary between nothing and 2000 gal/h. Successful bores can normally be expected to yield less than 200 gal/h.

Silurian. No wells are recorded in the basal grits and conglomerates, but these beds would probably yield small supplies to bores. Where the Upper Llandovery rocks rest on the Longmyndian [389923], 700 yd south-east of Wentnor church, water from a spring contains 3374 milligrams per litre (mg/l) of sodium chloride. Saline water encountered in a well at Criftin House [380915] may also have come from these beds. In the Llandovery and Wenlock series boreholes commonly yield a few hundred gallons per hour, the highest recorded being 1080 gal/h, from the Hughley Shales at Upper Day House [548967]. In the Ludlow Series yields normally vary between nothing and 1000 gal/h. A borehole at Diddlebury [505856] produces 5000 gal/h, mainly from the Upper Ludlow. Many springs flow from the bases of the Aymestry Group and, to a less extent, of the Wenlock Limestone.

Old Red Sandstone. Bores intersecting sandstones in the Downton Series may yield up to 900 gal/h. Water from such a borehole at Acton Round has a total hardness of 250 (mg/l) and a chloride content of 47 mg/l. Boreholes in the marls yield up to 700 gal/h, but supplies greater than 300 gal/h are rarely to be expected. Some of the water in the Diddlebury Borehole may originate in the Downton Castle Sandstone.

Many small springs issue from the base of the '*Psammosteus*' Limestones and wells into the Limestones should yield up to 500 gal/h. Borings in the lower part of the Ditton Series supply up to 600 gal/h, the quantity depending on the thickness and extent of the sandstone beds intersected. Near Ditton Priors yields up to 2000 gal/h are obtained, and most 6-in bores give at least 300 gal/h. The higher part of the Series has been little exploited but supplies adequate for individual houses and farms should be available from these arenaceous beds. The total hardness of water from the Ditton Series varies between 300 and 350 mg/l.

There are no records of wells into the Clee Group or the Farlow Series. Development of the Clee Group is not likely to be successful, any water within the formation issuing as springs at the level of the Abdon limestones. An attempt was made by Ludlow R.D.C. to use the springs at Bockleton Pool [576832], fed from the Lower Abdon Limestone, but this had to be abandoned as the pool dried out during the dry summer of 1959. Small supplies may be available from the Farlow Series.

Carboniferous. The Carboniferous rocks are not likely to yield to boreholes more than a few hundred gallons a day, but some water may be available from flooded coal workings on Brown Clee Hill.

Superficial Deposits. A well and borehole at Church Stretton, through 89 ft of gravelly boulder clay, yielded 5000 gal/h of water, with a total hardness of only 35 mg/l. In general, however, only small quantities can be expected from boulder clay or moraine. Glacial sand and gravel yield 7500 gal/h to an 89-ft well at Clungunford, and 3000 gal/h to a shaft 24 ft deep at Craven Arms. Other wells and bores in these deposits supply from 120 to 2640 gal/h. Total hardness ranges from 109 to 220 mg/l, partly because of lithological variations between sites and partly because of seasonal changes in groundwater composition. The water levels and the yields fall in these wells in summer.

A shaft at Craven Arms, 31 ft deep in terrace gravels, has yielded 7000 gal/h, the water having a total hardness of 109 mg/l. These gravels and alluvial deposits contribute water to other wells in the Clun, Onny and Corve valleys.

On the Long Mynd spring-flow issues peripherally from the base of the head overlying impermeable Pre-Cambrian strata. A group of small springs 1⅜ miles east of Wentnor [406931] produces at least 7600 gallons a day, but contains 14·4 mg/l of iron and 1·6 mg/l of lead. 600 gal/h are obtained from a borehole through 60 ft of drift at Ashgrove Farm [384946]. H.J.R.

References

Cantrill, T. C. 1908. Chapter I Geology, *in* A History of Shropshire. *The Victoria History of the Counties of England*, 1–46.

Dines, H. G. 1958. The West Shropshire Mining Region. *Bull. Geol. Surv. Gt. Brit.*, No. 14, 1–43.

Howe, J. A. 1910. *The Geology of Building Stones*. London.

Howell, E. J. 1941. Shropshire. *Report of the Land Utilisation Survey of Britain*, **66.**

La Touche, W. M. D. 1923. *in* Field Meeting—Plowden. *Trans. Caradoc and Severn Valley Field Club*, **7**, for 1922, No. 2, 61–3.

Murchison, R. I. 1839. *The Silurian System*. Part I. London.

APPENDIX I

Lists of Fossils Localities and Fossils from the Ordovician

(a) fossil localities

National Grid references are given within square brackets. Geological Survey registered numbers of specimens are given without brackets, e.g. BAH 101–105.

HOAR EDGE GRIT

Locality

1 Coston, old quarry [38608008] 200 yd S. of Coston farmhouse; BAH 101–105, CG 79, DEW 1660–1674, JD 176–208.

2 As Locality 1 but exposure at entrance to quarry; BAH 106–107.

3 Coston, track [38678028] 50 yd E. of Coston farmhouse; DEW 1675–1692.

4 Horderley, old quarry [41188614] 180 yd W. 22° S. of Glenburrell; DEW 1312–1324.

5 Horderley, quarry [41368658] 400 yd N. of Glenburrell; DEW 1452–1458.

6 Cwm Head, old quarries [41888797] 920 yd S. 30° W. of Cwm Head church; WM 2482–2497.

7 Cwm Head, outcrops [41938785] 1020 yd S. 24° W. of Cwm Head church; CG 214–217.

8 Willstone, outcrop [48439520] 1710 yd N. 38° E. of Cwms Farm; DEW 459–475.

HARNAGE SHALES

9 Sibdon Carwood, stream [40648295] 760 yd W. 17° S. of Sibdon Carwood church; DEW 1–13.

10 Sibdon Carwood, stream [40698284] 740 yd W. 27° S. of Sibdon Carwood church; DEW 14–18.

11 Sibdon Carwood, stream [40738424] 935 yd E. 35° S. of Wart Hill summit; DEW 1622–1649.

12 Sibdon Carwood, stream [40958418] 1150 yd E. 31° S. of Wart Hill summit; FGD 3591–3628.

13 Cheney Longville, track [40678538] 950 yd N.E. of Wart Hill summit; JD 57–100.

14 Woolston, outcrop [42088753] 1275 yd S. 11° W. of Cwm Head church; DEW 1431–1451.

15 Little Stretton, quarry [44549136] 620 yd W. 20° N. of The Hough, limestone band c. 8 ft above base of section; Mi 729–730.

16 As Locality 15 but c. 12 ft above limestone band; Mi 731–740.

17 Little Stretton, outcrop [45059148] 340 yd N. 10° W. of The Hough; DEW 337–352.

18 Hope Bowdler, roadside [47399244] 200 yd W. 10° N. of Hope Bowdler church; DEW 107–120.

19 Hope Bowdler, stream [49019304] 870 yd W. 4° S. of Gutter Farm; Mi 654–664.

20 Wall under Heywood, stream [49679301] 140 yd W. 25° S. of Gutter Farm; Mi 634–646.

Locality

21 Wall under Heywood, stream [50299309] 520 yd E. 5° N. of Gutter Farm; Mi 647–648.

22 Caer Caradoc Hill track [48049550] 1800 yd N. 20° E. of Cwms Farm; Mi 416–430.

CHATWALL FLAGS

23 Sibdon Carwood, track [40868335] 515 yd W. 23° N. of Sibdon Carwood church; DEW 33–37.

24 Horderley, R. Onny [41328610] 2130 yd W. 14° N. of Wistanstow church; FGD 1132–1152.

25 Cwm Head, track [42158805] 720 yd S. 15° W. of Cwm Head church; JD 3316–3319.

26 Cwm Head, outcrop at Blakemoor [42598810] 700 yd S. 25° E. of Cwm Head church; JD 3217–3232.

27 Hope Bowdler, old quarry [47119174] 870 yd S. 34° W. of Hope Bowdler church; DEW 121–139.

28 Willstone, road section [48589512] 960 yd E. 17° S. of Caer Caradoc Hill summit, 9 ft above base of section; Mi 379–388.

29 As Locality 28 but 21 ft above base of section; Mi 389–393.

30 As Locality 28 but 42 ft above base of section; Mi 394–405.

31 Cardington, outcrop [49949431] 1160 yd S. 37° W. of Cardington church; DEW 401–419, Mi 371–372.

32 Chatwall Hall, quarry [51379758] 120 yd W. 34° N. of the road junction at Chatwall Hall; RT 588.

33 Chatwall Hall, roadside [51289752] 200 yd W. of the road junction at Chatwall Hall; RT 562–583.

CHATWALL SANDSTONE

34 Sibdon Carwood, Longlane Quarry [41278422] 1130 yd N. of Sibdon Carwood church; DEW 1166–1205.

35 As Locality 34 but north end of quarry, 20 ft below top of section; JD 411–415.

36 As Locality 34 but north end of quarry, 6 ft below top of section; CG 256–257, JD 416–437.

37 Cheney Longville, outcrop [41558578] 540 yd S. 33° E. of Glenburrell; CG 239.

38 Cheney Longville, quarry [41208512] 1280 yd E. 18° N. of Wart Hill summit; DEW 1789–1815, JD 371–384.

39 Cheney Longville, quarry [41548593] 400 yd S. 42° E. of Glenburrell; DEW 1579–1597, JD 333–340.

40 Cheney Longville, old quarry [41018524] 1150 yd E. 27° N. of Wart Hill summit; DEW 1816–1824.

41 Cheney Longville, quarry [41738584] 600 yd E. 37° S. of Glenburrell; DEW 1563–1578.

42 Woolston, old quarry [42228734] 250 yd W. 33° N. of road junction, Woolston; CG 191–193, FGD 1476–1494.

43 Hope Bowdler, Soudley Quarry [47729182] 640 yd S. 24° E. of Hope Bowdler church, 14 ft below *alternata* Limestone; Mi 463–471.

44 As Locality 43 but 6 ft below *alternata* Limestone; Mi 472–475.

45 Little Stretton, railway cutting [44039053] 815 yd north of Marshbrook Station, 382 ft N. of 15-mile post; FGD 3224–3228.

46 As Locality 45 but 348 ft N. of 15-mile post; FGD 3229–3250.

Locality

47 Willstone, old quarry [48989562] 1860 yd W. 15° N. of Cardington church, 15–25 ft below *alternata* Limestone; DEW 376–385.

48 Enchmarsh, old quarry [49789622] 50 yd E. of Folly Bank crossroads; DEW 361–374.

49 Enchmarsh, roadside [50159652] 550 yd E. 39° N. of Folly Bank crossroads; DEW 501–515.

LOWER CHENEY LONGVILLE FLAGS
(INCLUDING ALTERNATA LIMESTONE)

50 Cheney Longville, roadside [41588524] 1810 yd W. 13° S. of Wistanstow church; DEW 1459–1487.

51 Cheney Longville, quarry [42078546] 1240 yd W. 7° S. of Wistanstow church; DEW 1415–1430.

52 Cheney Longville, R. Onny [41828570] 1510 yd W. 4° N. of Wistanstow church, 91 ft downstream from footbridge; FGD 1062–1072.

53 As Locality 52 but 235 ft downstream from footbridge; FGD 1033–1061.

54 Woolston, stream [41868639] 640 yd W. 39° N. of Upper Barn; DEW 1488–1489.

55 Woolston, roadside [42258733] 230 yd W. 33° N. of road junction, Woolston; CG 241–242.

56 Marshbrook, stream [43788974] 1990 yd E. 36° N. of Cwm Head; CG 172–176.

57 Little Stretton, railway cutting [44039050] 775 yd north of Marshbrook Station, 278 ft N. of 15-mile post; FGD 3251–3256.

58 As Locality 57 but 244 ft N. of 15-mile post; FGD 3260–3265.

59 As Locality 57 but 223 ft N. of 15-mile post; FGD 3269–3272.

60 Ragdon, track [45979089] 980 yd E. 16° S. of The Hough; BAH 209–216.

61 Chelmick, debris at reservoir site [46619154] 1400 yd W. 44° S. of Hope Bowdler church; Mi 627–633.

62 Soudley, stream [47789159] 900 yd S. 16° E. of Hope Bowdler church, 4 ft above base of section; DEW 160–175.

63 As Locality 62 but 9 ft above base of section; DEW 176–184.

64 Soudley, stream [47829149] 1030 yd S. 15° E. of Hope Bowdler church; DEW 140–159, Mi 621–626.

65 Soudley, outcrop at barn [48139177] 350 yd W. of Hollies; Mi 560–575.

66 Soudley Quarry [47729182] 640 yd S. 24° E. of Hope Bowdler church, base of *alternata* Limestone; Mi 476–480.

67 As Locality 66 but 0–1 ft above base of *alternata* Limestone; Mi 481–494.

68 As Locality 66 but 1 ft 6 in above base of *alternata* Limestone; Mi 495–508.

69 As Locality 66 but 1 ft 7 in above base of *alternata* Limestone; Mi 509–511.

70 As Locality 66 but 1 ft 10 in–2 ft 3 in above base of *alternata* Limestone; Mi 512–513.

71 As Locality 66 but 2 ft 10 in above base of *alternata* Limestone; Mi 514.

72 As Locality 66 but 3 ft 10 in above base of *alternata* Limestone; Mi 517–529.

73 As Locality 66 but 5 ft 10 in above base of *alternata* Limestone; Mi 530–532.

74 As Locality 66 but 7 ft 10 in above base of *alternata* Limestone; Mi 533.

75 Wall under Heywood, track [49639262] 500 yd E. 15° N. of Common Farm; DEW 386–400.

76 Willstone, N.W. corner of old quarry [48989562] 1860 yd W. 15° N. of Cardington church; DEW 375.

77 Enchmarsh, laneside [50329639] 670 yd E. 17° N. of Folly Bank crossroads; DEW 628–650.

78 Chatwall Hall, roadside [51379741] 100 yd S. 43° W. of Chatwall Hall; RT 584–587, 664–666.

UPPER CHENEY LONGVILLE FLAGS

Locality

79 Cheney Longville, roadside [41908493] immediately E. of motte and bailey; DEW 1281–1298.

80 Cheney Longville, R. Onny [42248542] 1060 yd W. 11° S. of Wistanstow church; FGD 929–955.

81 Woolston, outcrop at junction of road and lane [42508729] 100 yd N.E. of road junction, Woolston; CG 180–183.

82 As Locality 81 but c. 120 ft along lane to N.E.; FGD 1454–1475.

83 Whittingslow, stream [43328783] 1420 yd E. 38° S. of Cwm Head church; CG 194–199.

84 Whittingslow, stream [43868874] 1660 yd E. 3° N. of Cwm Head church; CG 209–210.

85 Whittingslow, old working [43028834] 850 yd E. 27° S. of Cwm Head church; CG 177–179.

86 Whittingslow, track [43718801] 1680 yd E. 25° S. of Cwm Head church; JD 3201–3216.

87 Whittingslow, track [43578922] 1450 yd E. 23° N. of Cwm Head church; JD 3233–3244.

88 Marshbrook, stream [43928972] 1580 yd W. 11° N. of Acton Scott church; CG 200–203.

89 Marshbrook, track and S. end of old quarry [44468901] 1080 yd W. 28° S. of Acton Scott church, 4 ft from base of section; BAH 31–32.

90 As Locality 89 but 34 ft from base of section; BAH 30.

91 As Locality 89, but at N. end of quarry; JD 385–403.

92 Marshbrook, track [44088990] 150 yd W. 20° N. of Marsbrook Station; JD 3124–3137.

93 Marshbrook, stream [44248914] 1240 yd W. 16° S. of Acton Scott church, and for 110 yd upstream; JD 3111–3123.

94 Marshbrook, old quarry and road cutting [44068976] 180 yd W.S.W. of Marshbrook Station; JD 341–369.

95 Marshbrook, roadside [43989002] 370 yd E. 18° N. of New House; CG 165–170; JD 3302–3315.

96 Acton Scott, stream [44848995] 785 yd W. 44° N. of Acton Scott church; BAH 65, JD 2937–2955.

97 Acton Scott, track [45579010] 785 yd N. 15° E. of Acton Scott church; JD 3049–3069.

98 Little Stretton, stream [45079094] 250 yd S. of The Hough; BAH 224–237.

99 Ragdon, stream [45669071] 770 yd E. 37° S. of The Hough; BAH 242–253.

100 Chelmick, outcrop at spring [46759101] 1760 yd S. 29° W. of Hope Bowdler church; Mi 595–616.

101 Soudley, outcrop at Hollies [48459176]; DEW 185–206, Mi 617–620.

102 Ticklerton, stream [48939199] 600 yd S. 26° W. of Common Farm; Mi 545–558.

103 As Locality 102 but 80 ft upstream; Mi 534–544.

104 Cardington, lane [50019571] 900 yd W. 43° N. of Cardington church; Mi 321–332.

105 Chatwall Hall, ditch [51579738] 190 yd S. 40° E. of the road junction at Chatwall Hall; Mi 365–370.

ACTON SCOTT GROUP

106 Cheney Longville, reef on south side of R. Onny [42388538] 940 yd W. 16° S. of Wistanstow church; FGD 749–767.

107 As Locality 106 but from river cliff above reef; FGD 768–783.

Locality

108 As Locality 106 but 25 ft upstream; FGD 784–806.

109 As Locality 106 but 50 ft upstream; FGD 807–840, JD 137–175.

110 As Locality 106 but 300 ft upstream; FGD 841–872.

111 Wistanstow, roadside [42418547] 860 yd W. 9° S. of Wistanstow church; DEW 1533–1562.

112 Wistanstow, outcrop [42478581] 270 yd S. 42° E. of Upper Barn; DEW 1513–1532.

113 Woolston, roadside [42568648] 1200 yd N. 36° W. of Wistanstow church, and for 70 yd to S.E; DEW 1132–1165.

114 Marshbrook, stream [43738973] 1900 yd E. 37° N. of Cwm Head church; JD 3280–3286.

115 Marshbrook, stream [44608992] 400 yd S. of Oakwood; JD 3151–3174.

116 Acton Scott, stream cliff [44808850] 380 yd W. 33° N. of Henley; JD 2956–2976.

117 Acton Scott, old quarry [44968955] 470 yd W. 15° N. of Acton Scott church; JD 3035–3048.

118 Acton Scott, old quarry [45488928] 200 yd S. 35° E. of Acton Scott church; JD 3021–3034.

119 Little Stretton, stream [45059053] 1540 yd S. 21° E. of Little Stretton halt; JD 3085–3097.

120 Hatton, stream cliff [46459013] 1370 yd E. 35° N. of Acton Scott church; 2 ft about base of section; JD 3007–3011.

121 As Locality 120 but 17–20 ft above base of section; JD 3012–3020.

122 Hatton, stream [46459025] 1450 yd E. 39° N. of Acton Scott church; JD 2995–3006.

123 Ticklerton, stream [47929104] 975 yd S. 36° W. of Hollies; DEW 228–239.

124 Cardington, outcrop at vicarage gate [50499520] 135 yd W. 24° N. of Cardington church; Mi 313–319.

125 Plaish, loose blocks [51709630] 1700 yd N. 43° E. of Cardington church; Mi 345–347.

126 Plaish, temporary pit [52889646] 125 yd W. of Plaish Hall; Mi 336–344.

127 Plaish, temporary pit at Plaish Park [51989676]; Mi 333–335.

ONNY SHALES

128 Cheney Longville, river cliff on north bank of R. Onny [42588536] 720 yd W. 22° S. of Wistanstow church, 5-10 ft below unconformity; CG 243, JD 273–306, FGD 435–466.

129 As Locality 128 but 15 ft upstream; FGD 467–505.

130 As Locality 128 but 125 ft upstream; calcareous mudstone ledge across river; FGD 555–603.

131 As Locality 128 but 335 ft upstream and on south bank; FGD 667–707.

132 Upper Affcot, floor of railway cutting [43888666] 1370 yd N. 30° E. of Wistanstow church; FGD 3273–3300.

133 Henley, dug section in railway cutting [45268851] 250 yd N. 40° E. of Hollies; JD 3175–3181.

134 Acton Scott, stream adjacent to Hatton Pool [46228942] 900 yd E. of Acton Scott church; JD 3070–3084.

(b) LISTS OF FOSSILS COLLECTED

The name of each fossil is followed first by the abbreviation of the formation present at the locality, and then by the locality number (see preceding list).

Abbreviations: HEG Hoar Edge Grit, HS Harnage Shales, CF Chatwall Flags, CS Chatwall Sandstone, LCLF Lower Cheney Longville Flags, UCLF Upper Cheney Longville Flags, ASG Acton Scott Group, OS Onny Shales.

SCYPHOZOA

Metaconularia vesicularis (Slater) ASG 109.
M. spp. ASG 115, 127.
Conulariid indet. HS 19.

ANTHOZOA

Anthozoans indet. ASG 117, 119.

CRINOIDEA

Balacrinus aff. *basalis* (McCoy) CF 33.
B. sp. CF 32.
B. ? CF 25–6.
Crinoid columnals HS 9, 11–2, 20. CF 23–4, 27–8, 30. CS 40, 43. LCLF 59. UCLF 82, 86–7, 100. ASG 117, 119, 125.

MACHAERIDIA

Lepidocoleus suecicus Moberg HS 13.

ASTEROZOA

Asterozoan indet. CF 26.

POLYZOA

Batostoma murchisoni Spjeldnaes HEG 6.
'*Favosites fibrilla*' Smith LCLF 78. UCLF 100.
Mesotrypa lens (McCoy) CS 46.
Polyzoans indet. HEG 5. HS 13, 18, 20. CS 35. LCLF 50, 64, 67–70. UCLF 93, 95, 100. ASG 121, 125, 127.

BRACHIOPODA

Bancroftina typa (Whittington) CS 41.
B. aff. *typa* LCLF 59.
B. robusta (Bancroft) LCLF 50, 53–4, 62.
B. cf. *robusta* LCLF 60, 67, 69, 72, 77.
B. spp. LCLF 51, 53, 57, 64.
B. ? CS 47. LCLF 63, 73, 75.
Bicuspina spiriferoides (McCoy). CS 34, 38.
B. sp. HEG 6. CS 42, 44.
'*Camarotoechia*' *spp.* HS 13, 20. CF 30.
Conotreta ? ASG 116.
Craniops sp. ASG 113.
Cryptothyris paracyclica (Bancroft) ASG 116–9, 122.
C. sp. ASG 125.
C. ? ASG 124.
Dalmanella horderleyensis (Whittington) CS 34, 36, 39.
D. cf. *horderleyensis* CS 35, 42, 46, 49.
D. aff. *indica* Whittington CS 34.
D. cf. *indica* CS 38.
D. lepta (Bancroft) CS 45.
D. cf. *lepta* CS 38, 41.
D. unguis (J. de C. Sowerby) UCLF 85, 88, 90, 94, 99, 101–2.
D. wattsi (Bancroft) LCLF 64. UCLF 79, 84, 91, 93, 95, 100, 103, 105.
D. spp. HEG 1–2. HS 12–4, 16, 18–9. CF 31, 33. CS 48. LCLF 51, 53, 56, 65, 67, 77–8. UCLF 87, 92.
D. ? LCLF 61.
Dalmanellids indet. HS 22. CF 28, 31. CS 43. LCLF 59, 68, 70, 72. ASG 112, 126.
Dinorthis flabellulum (J. de C. Sowerby) HEG 3–4.

D. cf. *flabellulum* HEG 8. CF 27.
D. spp. HEG 7. HS 12.
D. ? HS 20. CF 31. CS 44.
Dolerorthis duftonensis (Reed) LCLF 60.
D. cf. *duftonensis* LCLF 51.
D. spp. HEG 7–8. HS 14. UCLF 91. ASG 109.
D. ? HEG 4.
Glyptorthis spp. HEG 6. HS 17–8.
G. ? ASG 121.
Harknessella vespertilio (J. de C. Sowerby) HEG 3. CF 33.
Hedstroemina fragilis Bancroft UCLF 81–2, 86, 89.
H. cf. *fragilis* UCLF 96.
H. robusta Bancroft ASG 117, 124.
H. cf. *robusta* ASG 107, 113, 114.
H. spp. UCLF 97. ASG 110, 115.
Heterorthis alternata (J. de C. Sowerby) LCLF 50, 52–3, 55, 57–8, 61, 66–7, 69–73, 76, 78.
H. patera (Davidson) HEG 3, 5–6.
H. cf. *patera* HS 13.
H. sp. UCLF 83.
H. ? HS 12.
Horderleyella plicata Bancroft HEG 1, 6–7.
Howellites antiquior (McCoy) CS 34–6, 38–9, 42.
H. cf. *antiquior* CS 48.
H. ? CF 26. CS 47.
Kjaerina bipartita (Salter) LCLF 50, 53, 61, 66–8.
K. cf. *bipartita* LCLF 64, 75.
K. geniculata Bancroft LCLF 65, 77.
K. cf. *geniculata* LCLF 56.
K. jonesi Bancroft CS 34, 36.
K. cf. *jonesi* CS 47–9.
K. typa Bancroft LCLF 51, 62.
K. cf. *typa* LCLF 77.
K. ? ASG 116, 120.
Kjerulfina polycyma Bancroft UCLF 80, 82, 86, 92.
K. cf. *polycyma* UCLF 83, 97. ASG 127.
K. trigonalis Bancroft UCLF 101–2.
K. spp. UCLF 89, 91, 94.
K. ? UCLF 100, 104.
Kullervo ? HS 16.
Leptaena salopiensis Williams ASG 123–5.
L. spp. HEG 1, 5. ASG 115, 118–9.
Leptellina ? ASG 109.
Lingula spp. sensu lato. HEG 8. HS 16. CS 36. ASG 106, 113, 116.
Lingulids indet. CS 40. ASG 115.
Macrocoelia expansa (J. de C. Sowerby) CF 24–5, 27–8, 33. CS 40, 43–4.
Macewanella sp. HEG 7.
Nicolella actoniae (J. de C. Sowerby) ASG 111, 118, 123.
N. cf. *actoniae* UCLF 91.
N. sp. UCLF 95.
Onniella cf. *aspasia* Bancroft ASG 108, 110, 112–3, 115, 123.
Onniella broeggeri Bancroft OS 128–31.
O. cf. *grandis* Bancroft ASG 114, 117, 121–2.
O. reuschi Bancroft UCLF 81, 86, 89.
O. cf. *reuschi* UCLF 82–3, 96–7.
O. cf. *soudleyensis* (Bancroft) CF 26, 31.

O. spp. CF 27. LCLF 64, 73. UCLF 80, ASG 106–13, 116, 119–20, 124–7. OS 132–4.
Orbiculoidea sp. HS 22.
Orthids indet. HS 20. CF 33.
Oxoplecia ? HEG 7.
Parastrophinella sp. HS 17.
Pholidops ? HS 12.
Platystrophia spp. UCLF 92. ASG 124.
Rafinesquina spp. CF 30. LCLF 78.
Reuschella bilobata (J. de C. Sowerby) ASG 118, 125–6.
R. cf. *bilobata* ASG 115, 117, 123.
R. horderleyensis Bancroft CF 24, 27.
R. semiglobata Bancroft ASG 109, 121.
R. spp. UCLF 79–80, 83, 85, 94. ASG 113.
Rhynchonelloid indet. ASG 116.
Rostricellula spp. HEG 8. HS 16.
Salopia salteri (Davidson) HEG 8. HS 16–8.
S. ? HEG 1.
Sampo spp. ASG 111, 118.
Schizocrania spp. LCLF 75. UCLF 96. ASG 113.
Sericoidea spp. ASG 106–7, 109–13, 115–6, 120. OS 128–30, 132.
Smeathenella harnagensis Bancroft HS 13, 16–7.
S. strophomenoides Bancroft HEG 1.
Sowerbyella sericea (J. de C. Sowerby) LCLF 50, 52, 55, 57, 61–3, 65–9, 71–2, 77. UCLF 87, 91.
S. cf. *sericea* UCLF 79, 81–3, 86, 93, 95–6.
S. soudleyensis Jones CF 24–5, 27, 33. CS 34–9, 41–2, 46–9.
S. aff. *soudleyensis* HEG 8.
S. spp. HEG 1. HS 14–6. CF 28. LCLF 56, 59, 78. UCLF 92, 97, 100. ASG 108, 111, 117, 119, 123, 125–6.
Strophomena grandis (J. de C. Sowerby) LCLF 50, 53–4, 58, 64, 67–8, 72. UCLF 80, 84, 91–5, 98–101. ASG 117.
S. cf. *grandis* ASG 120.
S. sp. LCLF 62.
'*S.*' *holli* Davidson OS 128–9.
Strophomenid indet. HEG 4. CS 45–6. ASG 116.
Trematis punctata (J. de C. Sowerby) LCLF 50, 52, 61, 78.
Vellamo sp. HEG 6.
Inarticulate brachiopods indet. ASG 109, 115, 117.

GASTROPODA

Archinacella cf. *oblongata* (Portlock) ASG 109.
Ceratotheca ? ASG 106.
Cyclonema crebristria (McCoy) CF 33.
Cyrtolites nodosus (Salter) LCLF 77. UCLF 79.
Holopea striatella (J. de C. Sowerby) UCLF 86.
Hyolithes sp. LCLF 78.
Lophospira spp. CS 43. ASG 123.
Sinuites bilobatus (J. de C. Sowerby) ASG 106–9.
S. cf. *bilobatus* ASG 116.
S. pseudocompressus Reed ASG 107. OS 130.
S. soudleyensis Reed CS 46, 48.
S. spp. CS 34. ASG 113, 122–3. OS 129, 131.
Tropidodiscus acutus (J. de C. Sowerby) ASG 107, 109, 111, 113.
Gastropods indet. CS 34. LCLF 73, 78. UCLF 94. ASG 106, 109, 113.

? SCAPHOPODA

Tentaculites spp. LCLF 50–1, 56, 59–60, 64–5, 77. UCLF 79–80, 84–5, 87–8, 91–5, 97, 99–103. ASG 114–5, 125.

LAMELLIBRANCHIA

Modiolopsis obliqua (J. de C. Sowerby) UCLF 97. ASG 122.
M. orbicularis (J. de C. Sowerby) LCLF 64. UCLF 101.
M. cf. *orbicularis* LCLF 78.
M. spp. LCLF 53. UCLF 81, 92, 95–6.
Orthonota sp. ASG 114.
Orthonota ? ASG 122.
Lamellibranchs indet. UCLF 103. ASG 116.

CEPHALOPODA

Orthocone cephalopod indet. LCLF 65. UCLF 88, 97. ASG 106–9, 121–2. OS 128.

TRILOBITA

Acanthoparypha stubblefieldi (Bancroft) HS 13.
Ampyxella edgelli (Reed) ASG 109, 116.
Asaphids? indet. HEG 1. CF 30.
Broeggerolithus cf. *broeggeri* (Bancroft) HS 12. CF 24, 28, 30–1. CS 43.
B. cf. *harnagensis* (Bancroft). HS 18, 20.
B. cf. *longiceps* (Bancroft) LCLF 68, 72.
B. nicholsoni (Reed) CS 34, 38, 42, 46.
B. cf. *nicholsoni* CS 37.
B. soudleyensis (Bancroft) CF 27.
B. cf. *soudleyensis* CF 26, 31. CS 40, 43.
B. transiens (Bancroft) UCLF 90–4, 96, 101. ASG 117.
B. cf. *transiens* UCLF 97, 102–4.
B. spp. CS 36. LCLF 55–6, 67. UCLF 80, 87, 98.
B. ? LCLF 65.
Brongniartella bisulcata (McCoy) LCLF 53, 57, 62–4, 67–8, 75, 77. UCLF 79, 85, 87–8, 92–3, 95, 101.
B. cf. *bisulcata* CF 30. UCLF 104.
B. minor (Salter) CS 38.
B. minor subcarinata Dean CF 24. CS 36.
B. spp. HS 10. CF 28. CS 41, 43, 47. LCLF 50, 59, 72–3. UCLF 80, 97, 100–3.
Calymenids indet. HEG 1, 5. HS 13, 17, 20. CS 34, 39. UCLF 80, 86, 93–4, 96, 98–100. ASG 109, 115, 117–9, 121, 125.
Calyptaulax actonensis Dean ASG 109.
Chasmops extensa (Boeck) UCLF 79, 82, 84, 87, 91, 96, 101, 104. ASG 110, 121.
Chasmops spp. UCLF 86, 94, 98, 103. ASG 111, 125.
Costonia ultima (Bancroft) HEG 1.
C. ? HEG 6.
Flexicalymene cf. *acantha* Bancroft HEG 1, 2, 6.
F. caractaci (Salter) UCLF 83, 87, 89, 91–2, 96, 101, 103.
F. aff. *caractaci* LCLF 65.
F. cf. *planimarginata* (Reed) CS 34–6, 38–9.
F. ? UCLF 79.
Homalonotid ? CS 48.
Illaenus cf. *fallax* Holm OS 130.
Illaenid indet. ASG 111.
Kloucekia apiculata (McCoy) CS 34–6, 38–9, 42. LCLF 50, 53, 59, 67–9, 72.
K. harnagensis (Bancroft) HS 13.
K. cf. *harnagensis* HS 17.
Lonchodomas pennatus (La Touche) ASG 108–9.
Onnia ? cobboldi (Bancroft) OS 131.
O. gracilis (Bancroft) OS 130.

O. cf. *gracilis* OS 134.
O. superba (Bancroft) OS 128–9.
O. cf. *superba* ASG 113. OS 132–3.
Onnicalymene laticeps (Bancroft) ASG 112, 116, 121.
O. cf. *laticeps* ASG 118.
O. onniensis (Shirley) OS 129–30.
O. cf. *onniensis* OS 134.
O. salteri (Bancroft) ASG 116.
O. cf. *salteri* ASG 115, 121.
Otarion spp. UCLF 89, 92, 97, 102. ASG 119.
Parabasilicus powisi (Murchison) CS 36.
P. cf. *powisi* CS 46.
P. ? CF 33.
Platylichas laxatus (McCoy) ASG 109, 118, 123.
P. spp. ASG 117, 121.
Primaspis caractaci (Salter) ASG 125, 127.
P. harnagensis (Bancroft) HS 14.
Proetidella fearnsidesi Bancroft HS 13, 17.
P. sp. HS 22. CF 24.
Proetids indet. CF 30. UCLF 94.
Remopleurella burmeisteri (Bancroft) OS 128.
Reuscholithus reuschi Bancroft HS 13, 20, 21.
Salterolithus caractaci (Murchison) HS 11.
S. cf. *praecursor* Dean HS 22.
S. spp. HS 14, 18–20.
Triarthrus cf. *linnarssoni* Thorslund OS 128.
Trinucleids indet. HS 12–3, 17, 19–20, 22. CF 33. CS 38. UCLF 79, 85, 103. ASG 119.

OSTRACODA

Primitia spp. HS 20. UCLF 100. ASG 108, 111, 125. OS 129, 132.
Tallinnella scripta (Harper) HEG 1. HS 9, 13–9, 22. CF 24.
T. cf. *scripta* LCLF 74. UCLF 91, 97.
T. spp. HS 12, 20. CS 46. LCLF 73. UCLF 100, 104.
Ulrichia bicornis (Jones) HS 18–20, 22.

GRAPTOLITOIDEA

Amplexograptus spp. HS 9, 19.
Climacograptus spp. CF 33. ASG 115. OS 128, 132.
Diplograptus cf. *leptograptoides* Bulman HS 12.
D. cf. *leptotheca* Bulman CF 23.
? D. leptotheca LCLF 63.
D. of *multidens* Elles and Wood HS 12, 19.
D. spp. HS 12. OS 131.
Diplograptids indet. HS 13. ASG 106, 109.
? Dictyonema sp. HS 14.
Glyptograptus aff. *teretiusculus siccatus* Elles and Wood HS 12.
Orthograptus sp. (of *calcaratus* group) CF 29.
O. of. *leptotheca* type. HS 10.
O. truncatus (Lapworth) CF 32–3.
O. truncatus intermedius Elles and Wood CF 33.
O. truncatus pauperatus Elles and Wood CF 33.
? O. sp. (of *truncatus* type) HS 11.
Graptolites indet. HS 9. CF 29.

APPENDIX 2

LIST OF FOSSIL LOCALITIES AND FOSSILS FROM THE SILURIAN

(a) FOSSIL LOCALITIES

National Grid references are given within square brackets. Geological Survey registered numbers of specimens are given without brackets e.g. CG 205–207.

UPPER LLANDOVERY–PENTAMERUS BEDS

Locality

1 Hampperley, stream [41308876] 220 yd N. 39° E. of Churchmoor Hall; CG 205–207.

2 Whittingslow, stream [43428983] 200 yd W. 24° S. of New House; CG 161.

UPPER LLANDOVERY–HUGHLEY SHALES

3 Hamperley, roadside [42348980] 700 yd N. 30° E. of the crossroads, Hamperley; JD 2220–2228.

4 Wistanstow, R. Onny [42578534] 720 yd W. 22° S. of Wistanstow church; CG 238.

WENLOCK SHALES

5 Edgton, track [38318656] 967 yd N. 23° W. of Edgton church; CG 268.

6 Edgton, track [38568658] 920 yd N. 7° W. of Edgton church; DEW 904–916.

7 Horderley, stream [40758651] 1220 yd E. 5° N. of the Ridgway road junction; FGD 3570–3577.

8 Horderley, stream [40648749] 1310 yd S. 24° W. of Churchmoor Hall; DEW 1739–1757.

9 Horderley, path [40548714] 1250 yd E. 40° N. of the Ridgway road junction; DEW 1325–1350.

10 Horderley, cliff [40808708] 1450 yd E. 31° N. of the Ridgway road junction; DEW 1077–1093.

11 Cwm Head, stream [41718832] 760 yd W. 33° S. of Cwm Head church; CG 234–235.

12 Cwm Head, in Scaldbank Gutter [42148886] 260 yd N. 43° W. of Cwm Head church; CG 233.

13 Cwm Head, outcrop [42438854] 210 yd S. 39° E. of Cwm Head church; JD 2200–2219.

14 Cwm Head, old quarry [42428870] 130 yd E. of Cwm Head church; JD 2192–2199.

15 Cwm Head, old quarry [42248877] 100 yd N. 41° W. of Cwm Head church; JD 2175–2191.

16 Minton, roadside [43189078] 150 yd N. 11° E. of Castle Mound, Minton; FGD 3162–3223.

17 Church Stretton, excavation [45699409] 660 yd N.E. of Church Stretton church; De 466–511, 513–594.

18 Church Stretton, excavation [45659385] 470 yd E. 25° N. of Church Stretton church; D 3999–4098.

19 Church Stretton, excavation [45639385] 450 yd E. 27° N. of Church Stretton church; D 3761–3770.

Locality

20 Clungunford, old quarry [38097869] 995 yd S. 12° W. of New House, near Coston; BAH 113–126, JD 1676–1698.

21 Wistanstow, R. Onny [42778509] 717 yd S. 40° W. of Wistanstow church; DEW 1351–1377.

22 Wistanstow, drive to Grove [43598423] 1530 yd S. 16° E. of Wistanstow church; FGD 3386–3403.

23 Wistanstow, roadside [43028518] 480 yd S. 25° W. of Wistanstow church; FGD 3343–3385.

24 Strefford, track section [44688533] 1550 yd W. 11° N. of the Tower, Callow Hill, base of succession; DEW 5894–5901.

25 As Locality 24 but 25–30 ft above base of succession; DEW 5902–5909.

26 As Locality 24 but 25–30 ft above beds at Locality 25; DEW 5910–5915.

27 As Locality 24 but 12 ft above beds at Locality 26; DEW 5916–5938.

28 Harton, stream cliff [48108870] 740 yd S. 19° W. of Harton Road Station; BAH 77, FGD 2018–2034.

29 Harton, stream cliff [48808908] 600 yd E. 28° S. of Harton Road Station; BAH 75–76.

30 Eaton, track [50348997] 370 yd E. 4° S. of Eaton church and for 130 yd to W.S.W.; JD 2768–2786.

31 Eaton, track [50098999] 100 yd E. 4° S. of Eaton church and for 140 yd to W.N.W.; JD 2787–2815.

32 Rushbury, trench [51329175] 110 yd S. 24° W. of Rushbury church and for 100 yd to S.E.; RH 273–296.

33 Rushbury, trench [51519150] 403 yd S. 23° E. of Rushbury church; RH 297–316.

34 Rushbury, trench [51549147] 440 yd S. 26° E. of Rushbury church; RH 317–325.

TICKWOOD BEDS

35 Whettleton, river cliff [43868250] 970 yd E. 39° S. of Craven Arms Station and for 270 yd to S.; FGD 2540–2551.

36 Eaton, track [50378997] 420 yd E. 2° S. of Eaton church; JD 2756–2764.

37 Longville in the Dale, old quarry [54919383] 410 yd W. 8° N. of Pilgrim Cottage; JD 2156–2174.

WENLOCK LIMESTONE
(including Reef Facies)

38 Church Stretton, old quarry [46089408] 700 yd W. 25° S. of New House Farm; JD 2816–2863.

39 All Stretton, old quarry [46969564] 850 yd E. 14° S. of All Stretton church; DEW 526–527.

40 Whettleton, stream [43868221] 300 yd W. 35° S. of Whettleton; FGD 2552–2579.

41 Lower Dinchope, old quarry [45068508] 1090 yd W. 1° N. of the Tower, Callow Hill, 6 ft above base of section; DEW 5939–5947.

42 As Locality 41, but top 5 ft of section; DEW 5948–5968.

43 Harton, old quarry [48058762] 550 yd N. 43° W. of Burwood; FGD 1981–2000.

44 As Locality 43 but 70 yd to S.W.; FGD 1970–1980.

45 Eaton, old quarry [50579007] 630 yd E. 9° N. of Eaton church; JD 2740–2755.

46 Presthope, Lilleshall Quarry [57439682] 1600 yd W. 34° S. of Presthope station; JD 1716–1763.

EDGTON LIMESTONE
(West of Church Stretton Fault)

Locality

47 Edgton, path [38678620] 510 yd N. 3° E. of Edgton church and for 55 yd to N.N.W.; DEW 879–903.

48 Edgton, stream [38468615] 510 yd N. 23° W. of Edgton church; DEW 873–878.

49 Edgton, old quarry [39648651] 100 yd N. of the Ridgway road junction; DEW 1031–1042.

50 Horderley, old quarry [40448678] 940 yd E. 25° N. of the Ridgway road junction; CG 244–247.

51 Horderley, stream [40298644] 1935 yd E. 25° N. of Edgton church; FGD 3551–3569.

LOWER LUDLOW SHALES

52 All Stretton, stream [46699502] 730 yd N. 2° E. of New House Farm; DEW 577–595.

53 Botvyle, old pond [47679620] 880 yd N. 6° W. of Caer Caradoc Hill summit; DEW 516–525, 1130–1131.

54 The Crossways, old quarry [41117922] 300 yd E. 30° N. of The Crossways; JD 1625–1650.

55 The Crossways, old quarry [41207934] 450 yd E. 40° N. of The Crossways; JD 1597–1624.

56 Brandhill, old quarry [41867975] 660 yd W. 11° N of Gorst Barn; JD 1114–1144.

57 Brandhill, old quarry [41517953] 1020 yd W. 7° S. of Gorst Barn; JD 1090–1113.

58 Stokesay, roadside [42418061] 480 yd N.W. of Newfoundland; BAH 98.

59 Stokesay, quarry [42028032] 880 yd N. 33° W. of Gorst Barn; JD 1044–1089.

60 Stokesay, old quarry [42298078] 700 yd N. 42° W. of Newfoundland; JD 1022–1043.

61 Stokesay, exposure [42218141] 1530 yd W. 13° S. of Stokesay church; BAH 316–319.

62 Stokesay, track [43488072] 1110 yd S. 6° W. of Stokesay church; JD 1372–1389.

63 Stokesay, old quarry [43608085] 1120 yd E. 33° N. of Newfoundland; JD 1340–1371.

64 Stokesay, old quarry [43518053] 880 yd E. 18° N. of Newfoundland; JD 1326–1339.

65 Stokesay, R. Onny [44088087] 950 yd N.W. of Park Farm; FGD 1427–1453.

66 Onibury, R. Onny [44648010] 140 yd S. 25° W. of Park Farm; JD 2314–2332.

67 Onibury, R. Onny [44418035] 1800 yd W. 43° N. of Onibury church; FGD 1344–1347.

68 As Locality 67 but 18 ft downstream; FGD 1348–1352.

69 As Locality 67 but 78 ft downstream; FGD 1356–1359.

70 As Locality 67 but 98 ft downstream; FGD 1360–1362.

71 As Locality 67 but 130 ft downstream; FGD 1365–1368.

72 As Locality 67 but 178 ft downstream; FGD 1369–1379.

73 As Locality 67 but 216 ft downstream; FGD 1380–1383.

74 As Locality 67 but 268 ft downstream; FGD 1387–1394.

75 As Locality 67 but 293 ft downstream; FGD 1395–1403.

76 As Locality 67 but 310 ft downstream; FGD 1404–1407.

77 As Locality 67 but 355 ft downstream; FGD 1408–1414.

78 As Locality 67 but 405 ft downstream; FGD 1415–1426.

Locality

79 Whettleton Bank, track [44368073] 1330 yd S. 38° E. of Stokesay church; FGD 1240–1261.
80 Whettleton, old quarry [44898252] 1200 yd W. 17° S. of Greenway Cross; FGD 2580–2613.
81 Upper Dinchope, track [45918351] 780 yd N. 4° W. of Greenway Cross; FGD 2763–2794.
82 Upper Dinchope, roadside [45608350] 860 yd N. 27° W. of Greenway Cross; FGD 2663–2701.
83 Lower Dinchope, old quarry [45418422] 770 yd W. 7° S. of Siefton Forest; BAH 74.
84 Lower Dinchope, stream [45198476] 1000 yd W. 19° S. of the Tower, Callow Hill; DEW 5969–5991.
85 Westhope, stream [46718589] 460 yd S. 38° W. of Westhope church; FGD 1856–1860.
86 As Locality 85 but 14 yd downstream; FGD 1848–1852.
87 As Locality 85 but 37 yd downstream; FGD 1840–1847.
88 As Locality 85 but 48 yd downstream; FGD 1827–1831.
89 As Locality 85 but 62 yd downstream; FGD 1820–1822.
90 As Locality 85 but 73 yd downstream; FGD 1817–1819.
91 As Locality 85 but 77 yd downstream; FGD 1815–1816.
92 As Locality 85 but 79 yd downstream; FGD 1812–1814.
93 As Locality 85 but 91 yd downstream; FGD 1805–1808.
94 As Locality 85 but 101 yd downstream; FGD 1797–1800.
95 As Locality 85 but 116 yd downstream; FGD 1788–1791.
96 As Locality 85 but 120 yd downstream; FGD 1781–1787.
97 Westhope, stream [46988587] 390 yd S. 1° E. of Westhope church; BAH 84.
98 Burwood, track [48578694] 1170 yd N. 40° E. of Titterel; FGD 1495–1518.
99 Munslow Aston, old quarry [49808678] 1480 yd W. of Munslow Aston chapel; JD 2680–2697.
100 Middlehope, track [50478844] 1795 yd S. 18° E. of Eaton church; JD 2864–2881.
101 Upper Millichope, stream [52058952] 1960 yd N. 2° W. of Munslow church and for 80 yd upstream; FGD 2219–2254.
102 Upper Millichope, stream [52088946] 1900 yd N. 1° W. of Munslow church; JD 1556–1596.
103 Brockton, stream [57369426] 330 yd E. 20° S. of Larden Cottage; JD 2103–2138.

AYMESTRY GROUP

104 All Stretton, old quarry [47219541] 580 yd W. of Caer Caradoc Hill summit; DEW 555–576.
105 Aldon, old quarry [42497978] 170 yd N. 10° E. of Gorst Barn, *C. knightii* band; JD 1310–1312.
106 As Locality 105 b·ıt 15 ft above *C. knightii* band; JD 1313–1325.
107 Brandhill, old quarry [41967985] 600 yd W. 24° N. of Gorst Barn, *C. knightii* band; JD 1149–1151.
108 As Locality 107 but c. 10 ft above *C. knightii* band; JD 1152–1154.
109 As Locality 107 but top 10 ft of quarry face; JD 1155–1161.
110 View Edge Quarry [42608070] 450 yd N. 16° W. of Newfoundland, bottom 9 ft of main quarry face; JD 928–966.
111 As Locality 110 but middle 7 ft of main quarry face; JD 967–988.
112 As Locality 110 but top 8 ft of main quarry face; JD 989–999.
113 As Locality 110 but [42628078] 530 yd N. 11° W. of Newfoundland, top 10 ft of northern quarry face; JD 1000–1021.

Locality

114 Aldon, old quarry [43107968] 680 yd E. 5° N. of Gorst Barn, *C. knightii* band; JD 1293–1299.

115 As Locality 114 but 10 ft above *C. knightii* band; JD 1300–1309.

116 Aldon, cliff [42927979] 520 yd E. 20° N. of Gorst Barn; JD 1285–1292.

117 Aldon, old quarry [43568035] 1510 yd S. 1° W. of Stokesay church; JD 1443–1461.

118 Aldon, old quarry [43658035] 1500 yd S. 3° E. of Stokesay church; JD 1424–1442.

119 Stokesay, cliff [43158085] 1070 yd S. 26° W. of Stokesay church; JD 1408–1423.

120 Stokesay, old quarry [43238057] 1320 yd S. 17° W. of Stokesay church; JD 1402–1407.

121 Stokesay, old quarry [43268057] 1300 yd S. 15° W. of Stokesay church; JD 1390–1401.

122 Onibury, R. Onny [44658007] 1410 yd W. 43° N. of Onibury church, base of Aymestry Group; JD 2333–2342.

123 As Locality 122 but 75 ft upstream, 9 ft above base of Aymestry Group; CG 163.

124 Onibury, old quarry [44688005] 1360 yd W. 43° N. of Onibury church; JD 2303–2313.

125 Onibury, roadside quarry [44738006] 180 yd S. 18° E. of Park Farm; JD 1478–1502.

126 Onibury, old quarry [44708030] 50 to 150 yd N. of Park Farm; JD 1462–1477.

127 Whettleton Bank, old quarry face from [44628086] 700 yd N. 7° W. of Park Farm to [44598060] 420 yd N. 14° W. of Park Farm; FGD 1262–1343.

128 Whettleton Bank, old quarry [44408128] 1020 yd E. 29° S. of Stokesay church, 17 ft below top of section; FGD 3051–3056.

129 As Locality 128 but 14 ft below top of section; FGD 3045–3050.

130 As Locality 128 but 10 ft below top of section; FGD 3038–3044.

131 As Locality 128 but 3 ft below top of section; FGD 3029–3037.

132 As Locality 128 but top of section; FGD 3020–3028.

133 Whettleton Bank, old quarry [44578136] 600 yd W. 33° S. of Campbarn; FGD 3000–3019.

134 Whettleton, cliff [44518186] 1000 yd E. 8° N. of Stokesay church, 51-52 ft below top of section; FGD 2993–2999.

135 As Locality 134 but 47-48 ft below top of section; FGD 2984–2992.

136 As Locality 134 but 32–33 ft below top of section; FGD 2975–2983.

137 As Locality 134 but 29 ft below top of section; FGD 2971–2974.

138 As Locality 134 but 24-25 ft below top of section; FGD 2954–2970.

139 As Locality 134 but 20 ft below top of section; FGD 2944–2953.

140 As Locality 134 but 14 ft below top of section; FGD 2936–2943.

141 As Locality 134 but 8 ft below top of section; FGD 2926–2935.

142 As Locality 134 but top 2 ft of section; FGD 2921–2925.

143 Upper Dinchope, old quarry [45398290] 680 yd W. 11° N. of Greenway Cross; FGD 2531–2539.

144 Upper Dinchope, old quarry [45188345] 1100 yd W. 38° N. of Greenway Cross, 4 ft above base of section; FGD 3057–3061.

145 As Locality 144 but 13 ft below top of section; FGD 3062–3070.

146 As Locality 144 but highest beds of section; FGD 3071–3087.

147 Upper Dinchope, old quarry [45908379] 1100 yd N. 4° W. of Greenway Cross; FGD 2742–2762.

148 Upper Dinchope, old quarry [46368319] 600 N.E. of Greenway Cross BAH 327, FGD 2831–2854.

149 Upper Dinchope, old quarry [46358346] 840 yd N. 30° E. of Greenway Cross; FGD 2855–2857.

Locality

150 Upper Dinchope, old quarry [46068260] 250 yd S. 30° E. of Greenway Cross, W. face; FGD 2479–2486.
151 As Locality 150 but S.E. face; FGD 2487–2508. (See also locality 242).
152 Upper Dinchope, cliff [46068384] 1160 yd N. 5° E. of Greenway Cross, and for 200 yd to the S.S.E.; FGD 2614–2627.
153 Upper Dinchope, outcrop [46148339] 670 yd N. 17° E. of Greenway Cross; BAH 321–322; FGD 3088–3122.
154 Upper Dinchope, outcrop [45968386] 1170 yd N. 1° W. of Greenway Cross; FGD 2702–2741.
155 Upper Dinchope, old workings [45428319] 710 yd W. 37° N. of Greenway Cross; BAH 320.
156 Upper Dinchope, outcrop [46058389] 1200 yd N. 4° E. of Greenway Cross; BAH 323.
157 Hillend, old quarry [46548447] 1150 yd S. 31° W. of Hillend; FGD 1955–1969.
158 Hillend, old quarry [47548530] 460 yd E. 8° S. of Hillend, basal 3 ft of section; FGD 1928–1933.
159 As Locality 158 but 6 ft above base of section; FGD 1934–1941.
160 As Locality 158 but 10 ft above base of section; FGD 1942–1945.
161 As Locality 158 but 12 ft above base of section; FGD 1946.
162 As Locality 158 but 19 ft above base of section; FGD 1947–1954.
163 Hillend, track [46478520] 710 yd W. 16° S. of Hillend; FGD 1533–1561.
164 As Locality 163 but 30 yd to W.N.W.; FGD 1519–1532.
165 Hillend, old quarry [47708477] 940 yd E. 43° S. of Hillend, base of section; FGD 1896–1908.
166 As Locality 165 but top 14 ft of section; BAH 86, FGD 1909–1927.
167 New House, stream [47868463] 830 yd N. 20° W. of the Methodist chapel, 30 yd downstream from footbridge; FGD 1885–1895.
168 As Locality 167 but 80 yd downstream from footbridge; BAH 91–92, FGD 1875–1884.
169 New House, track leading from road [47808468] 1040 yd S.E. of Hillend, 7 ft from road; FGD 1669–1672.
170 As Locality 169 but 11 ft from road; FGD 1674–1675.
171 As Locality 169 but 22 ft from road; FGD 1676–1680.
172 As Locality 169 but 26 ft from road; FGD 1681–1683.
173 As Locality 169 but 30 ft from road; FGD 1684–1693.
174 As Locality 169 but 34 ft from road; FGD 1696–1699.
175 As Locality 169 but 36 ft from road; FGD 1700–1707.
176 As Locality 169 but 40 ft from road; FGD 1708–1717.
177 As Locality 169 but 61 ft from road; FGD 1731–1741.
178 As Locality 169 but 69 ft from road; FGD 1742–1745.
179 As Locality 169 but 70 ft from road; FGD 1746–1755.
180 Hillend, outcrop [47778558] 1120 yd E. 39° S. of Westhope church; BAH 85.
181 Diddlebury, old quarry [49078656] 2300 yd W. 33° N. of Diddlebury church; JD 2627–2636.
182 Diddlebury, old quarry [48878611] 2270 yd W. 19° N. of Diddlebury church; JD 2578–2592.
183 Munslow Aston, old quarry [49878693] 1420 yd W. 7° N. of Munslow Aston chapel, bottom beds of section; JD 2637–2640.
184 As Locality 183 but 5 ft above base of section; JD 2641.
185 As Locality 183 but 11½ ft above base of section; JD 2642–2647.
186 As Locality 183 but 13½ ft above base of section; JD 2648–2652.
187 As Locality 183 but 20½ ft above base of section; JD 2653–2656.
188 As Locality 183 but 23½ ft above base of section; JD 2657–2659.
189 As Locality 183 but topmost beds of section; JD 2660–2663.

Locality

190 Munslow Aston, old quarry [49938711] 1390 yd W. 15° N. of Munslow Aston chapel, bottom beds of section; JD 2711–2719.

191 As Locality 190 but 17-18 ft below top of section; JD 2720–2725.

192 As Locality 190 but 8 ft below top of section; JD 2726–2732.

193 As Locality 190 but topmost beds of section; JD 2733–2739.

194 Munslow Aston, old quarry [50618757] 1080 yd N. 34° W. of Munslow Aston chapel; JD 2921–2936.

195 Munslow Aston, roadside [50598763] 1140 yd N. 33° W. of Munslow Aston chapel; JD 2908–2920.

196 Munslow Aston, old quarry [49928688] 1360 yd W. 5° N. of Munslow Aston chapel, bottom 5 ft of section; JD 2666–2674.

197 As Locality 196 but top of section, c. 20 ft above base; JD 2675–2679.

198 Munslow, old quarry [51728841] 870 yd N. 30° W. of Munslow church, bottom 6 ft of section; FGD 2098–2109.

199 As Locality 198 but top 10 ft of section; FGD 2110–2128.

200 Middlehope, old quarry [50458839] 1840 yd S. 17° E. of Eaton church, c. 6 ft above base of section; JD 2882–2896.

201 As Locality 200 but 2-3 ft below top of section; JD 2897–2907.

202 Upper Millichope, old quarry [51308882] 1490 yd N. 37° W. of Munslow church; FGD 2129–2147.

203 Upper Millichope, old quarry [52718912] 1680 yd N. 22° E. of Munslow church; FGD 2168–2196.

204 Upper Millichope, track [52618886] 1360 yd N. 24° E. of Munslow church; FGD 2337–2368.

205 Upper Millichope, roadside [52898899] 1620 yd N. 30° E. of Munslow church; FGD 2148–2167.

206 Hungerford, old quarry [53608943] 2460 yd N. 39° E. of Munslow church, bottom beds of main quarry; FGD 2369–2395.

207 As Locality 206 but top 4 ft in cut behind main quarry; FGD 2396–2403.

208 Hungerford, old quarry [53278921] 2060 yd N. 36° E. of Munslow church; FGD 2197–2218.

209 Broadstone, old quarry [54629055] 890 yd N. 11° E. of Broadstone church, bottom 8 ft of section; JD 2364–2376.

210 As Locality 209 but 21 ft below top of section; JD 2377–2388.

211 As Locality 209 but 13 ft below top of section; JD 2389–2395.

212 As Locality 209 but 3 ft below top of section; JD 2396–2399.

213 As Locality 209 but top of section; JD 2400–2405.

214 Broadstone, old quarry [54579054] 820 yd N. 6° E. of Broadstone church; JD 2356–2363.

215 Shipton, old quarry at Upper Barn [55269166] 1000 yd W. 11° S. of Shipton church; JD 2416–2421.

216 Brockton, old quarry [56519342] 1460 yd W. 15° S. of Brockton crossroads, below top 5 ft of section; JD 2263–2276.

217 As Locality 216 but top 5 ft of section; JD 2277–2287.

218 Brockton, old quarry [56449353] 1500 yd W. 11° S. of Brockton crossroads; JD 2288–2302.

219 Brockton, old quarry [57309421] 730 yd W. 40° N. of Brockton crossroads, bottom beds of section; JD 2074–2090.

220 As Locality 219 but top 8 ft of section; JD 2091–2102.

221 Brockton, stream [57579400] 370 yd W. 43° N. of Brockton crossroads; JD 2139–2155.

UPPER LUDLOW SHALES

Locality

222 Brandhill, old quarry [42257885] 860 yd S. 14° W. of Gorst Barn; JD 1210–1236.

223 Brandhill, outcrop [42887876] 1040 yd S. 28° E. of Gorst Barn; JD 1162–1175.

224 Brandhill, outcrop at Gorst Barn [42467962]; JD 1237–1264.

225 Brandhill, old quarry [42107889] 880 yd S. 26° W. of Gorst Barn; BAH 99–100.

226 Aldon, quarry [43757896] 200 yd W. 43° S. of Aldon Court; JD 1265–1284.

227 Aldon, lane [43337898] 630 yd W. 11° S. of Aldon Court; JD 1192–1209.

228 Aldon, lane [43167890] 800 yd W. 13° S. of Aldon Court; JD 1176–1191.

229 Onibury, old quarry [44297887] 470 yd E. 28° S. of Aldon Court; FGD 1171–1191.

230 Onibury, outcrops [44188008] 1820 yd W. 31° N. of Onibury church and for c. 120 yd to south-east; FGD 1225–1239.

231 Onibury, track [44028012] 2000 yd W. 30° N. of Onibury church; FGD 1193–1224.

232 Onibury, pondside exposure [44938008] 300 yd E. 32°S. of Park Farm; JD 2343–2355.

233 Onibury, roadside quarry [45257941] 420 yd W. 36° N. of Onibury church, bottom 10 ft of section; JD 1525–1555.

234 Onibury quarry [45117967] 740 yd N. 43° W. of Onibury church; JD 1503–1524.

235 Whettleton, old quarry in Norton Camp [44778186] 350 yd W. 35° N. of Campbarn; FGD 2904–2920.

236 Norton, debris [45188171] 1450 yd S. 34° W. of Greenway Cross; BAH 325–326.

237 Bache, track [46728207] 770 yd W. 36° S. of Medley Park; BAH 329.

238 Bache, old quarry [46908195] 710 yd S. 33° W. of Medley Park; FGD 2816–2830.

239 Burley, old quarry [47438151] 1100 yd S. 10° E. of Medley Park, c. 6 ft below Ludlow Bone Bed; FGD 2795–2800.

240 As Locality 239 but c. 3 ft below Ludlow Bone Bed; FGD 2801–2804.

241 Upper Dinchope, old quarry [45788318] 460 yd N. 24° W. of Greenway Cross; BAH 324; FGD 2441–2478.

242 Upper Dinchope, old quarry [46068260] 250 yd S. 30° E. of Greenway Cross, north-east face; FGD 2509–2530, (See also localities 150–1).

243 New House, track [47768466], leading south-west from road, 1020 yd S. 44° E. of Hillend, 122 ft from road; BAH 87–90, FGD 1562–1566.

244 As Locality 243 but 123 ft from road; FGD 1567–1570.

245 As Locality 243 but 126 ft from road; FGD 1571–1580.

246 As Locality 243 but 128 ft from road; FGD 1581–1587.

247 As Locality 243 but 139 ft from road; FGD 1588–1592.

248 As Locality 243 but 143 ft from road; FGD 1593–1597.

249 As Locality 243 but 146 ft from road; FGD 1598–1600.

250 As Locality 243 but 152 ft from road; FGD 1601–1612.

251 As Locality 243 but 157 ft from road; FGD 1613–1619.

252 As Locality 243 but 162 ft from road; FGD 1620–1632.

253 As Locality 243 but 169 ft from road; FGD 1633–1638.

254 As Locality 243 but 183 ft from road; FGD 1640–1649.

255 As Locality 243 but 196 ft from road; FGD 1650–1655.

256 As Locality 243 but 199 ft from road; FGD 1656–1664.

257 New House, old quarry [47888462] 820 yd N. 18° W. of the Methodist chapel; FGD 1756–1780.

Locality

258 New House, stream [48088396] 80 yd N. 28° W. of the Methodist chapel, 486 ft above road bridge and for 51 ft upstream; FGD 2858–2877.

259 As Locality 258 but 351 ft above road bridge; FGD 2878–2890.

260 As Locality 258 but 283 ft above road bridge; FGD 2891–2897.

261 As Locality 258 but 118 ft above road bridge; FGD 2898–2903.

262 New House, stream [47848458] 790 yd N. 22° W. of the Methodist chapel 100 yd downstream from footbridge; FGD 1861–1874.

263 New House, stream [47988443] 590 yd N. 14° W. of the Methodist chapel; BAH 93.

264 Corfton, lane [49698522] 1230 yd W. 8° S. of Diddlebury church; JD 2546–2555.

265 Diddlebury, old quarry [48898612] 2250 yd W. 20° N. of Diddlebury church; JD 2558–2577.

266 Diddlebury, old quarry [48998597] 2100 yd W. 16° N. of Diddlebury church, bottom beds of section; JD 2593.

267 As Locality 266 but 4½ ft above base of section; JD 2594–2595.

268 As Locality 266 but 7 ft above base of section; JD 2596–2600.

269 As Locality 266 but 19 ft above base of section; JD 2601–2613.

270 As Locality 266 but 25 ft above base of section; JD 2614–2621.

271 As Locality 266 but 31 ft above base of section; JD 2622–2626.

272 Munslow Aston, old quarry [50048666] 1230 yd W. 4° S. of Munslow Aston chapel; JD 2698–2710.

273 Munslow, road cutting [52838833] 1010 yd E. 42° N. of Munslow church, bottom 6 ft of section; FGD 2302–2315.

274 As Locality 273 but 33-35 ft above base of section; FGD 2316–2325.

275 As Locality 273 but 60-66 ft above base of section; FGD 2326–2336.

276 Broadstone, old quarry behind Rock Houses [54769067] 1020 yd N. 17° E. of Broadstone church; JD 2406–2415.

277 Shipton, old quarry [56269199] 165 yd N. 26° E. of Shipton church; JD 2248–2262.

278 Shipton, lane [56299174] 170 yd E. 35° S. of Shipton church; JD 2229–2246.

279 Brockton, roadside [57879384] 110 yd N. 35° E. of Brockton crossroads; JD 2046–2073.

LUDLOW SERIES - UNDIVIDED
(West of Church Stretton Fault)

280 Aston on Clun, old quarry [38718215] 1300 yd S. 8° W. of Hopesay church; DEW 85–106.

281 Beambridge, old quarry [38608100] 450 yd N. 8° W. of The Hollies, Coston; BAH 110–112, DEW 1708–1719.

282 Coston, old quarry [38138027] 500 yd W. of Coston; DEW 1693–1707.

283 Hopesay, track [38258272] 940 yd W. 42° S. of Hopesay church; BAH 296–297, DEW 43–58.

284 Hopesay, outcrop [39478452] 1490 yd N. 23° E. of Hopesay church; DEW 1047–1062.

285 Edgton, old quarry [39308480] 230 yd S. 35° W. of Grist House; DEW 1011–1030.

286 Edgton, outcrop [39108542] 600 yd E. 33° S. of Edgton church; DEW 1206–1225.

287 Edgton, stream [38638605] 385 yd N. 4° W. of Edgton church; CG 263.

288 Edgton, spring [38408562] 290 yd W. 19° S. of Edgton church; CG 262.

289 Edgton, roadside [38298542] 490 yd W. 43° S. of Edgton church; CG 260.

290 Edgton, excavation [38868541] 405 yd S. 35° E. of Edgton church; CG 259.

291 Edgton, lane [38888591] 310 yd E. 40° N. of Edgton church; CG 255.

Locality

292 Edgton, lane [39008578] 380 yd E. 10° N. of Edgton church; CG 249–254, DEW 917–930.

293 Edgton, stream [38638605] 445 yd N. 16° W. of Edgton church; FGD 3630–3652.

294 Edgton, track [39938635] 1500 yd E. 27° N. of Edgton church; DEW 1226–1255.

295 Edgton, track [39488651] 1225 yd N.E. of Edgton church, 428 ft from junction with Horderley–Edgton road; FGD 3423–3443.

296 As Locality 295 but 381 ft from junction with road; FGD 3444–3462.

297 As Locality 295 but 309 ft from junction with road; FGD 3463–3490.

298 As Locality 295 but 63 ft from junction with road; FGD 3491–3527.

299 Edgton, roadside [39648643] 1310 yd E. 36° N. of Edgton church; FGD 3404–3416.

300 Edgton, track [38548614] 480 yd N. 14° W. of Edgton church; FGD 3653–3680.

301 Hopesay, track [38238430] 1600 yd S. 16° W. of Edgton church, 239 ft down from gate; FGD 3681–3688.

302 As Locality 301 but 267 ft down from gate; FGD 3689–3691.

303 As Locality 301 but 340 ft down from gate; FGD 3692–3696.

304 As Locality 301 but 363 ft down from gate; FGD 3697–3702.

305 As Locality 301 but 375 ft down from gate; FGD 3703–3709.

306 As Locality 301 but 407 ft down from gate; FGD 3710–3713.

307 As Locality 301 but 450 ft down from gate; FGD 3714–3718.

308 As Locality 301 but 550 ft down from gate; FGD 3719–3727.

309 As Locality 301 but 577 ft down from gate; BAH 286–295, FGD 3728–3736.

310 As Locality 301 but 717 ft down from gate; FGD 3737–3747.

311 Hopesay, old quarry [39078367] 470 yd N. 19° E. of Hopesay church, main part of quarry; DEW 745–754.

312 As Locality 311 but east side of quarry; DEW 755–773.

313 Hopesay, outcrop [39088377] 570 yd N. 15° E. of Hopesay church; BAH 315.

314 Hopesay, outcrop [39088374] 540 yd N. 15° E. of Hopesay church; BAH 313–314.

315 Edgton, path [38768598] 300 yd N. 23° E. of Edgton church; DEW 858–872.

316 Edgton, roadside [38688577] 50 yd N. 40° E. of Edgton church; DEW 848–857.

317 Edgton, roadside [38528556] 250 yd S. 40° W. of Edgton church and for 250 yd to S.W.; DEW 774–806.

318 Edgton, old quarry [38848442] 1430 yd S. 10° E. of Edgton church; DEW 1000–1010.

319 Edgton, outcrop in barnyard [39058525] 670 yd S. 41° E. of Edgton church; DEW 958–999.

(b) LISTS OF FOSSILS COLLECTED

The name of each fossil is followed first by the abbreviation of the formation present at the locality, and then by the locality number (see preceding list).

Abbreviations: PB *Pentamerus* Beds, HuS Hughley Shales, WS Wenlock Shales, TB Tickwood Beds, WL Wenlock Limestone (including Reef Facies), EL Edgton Limestone, LLS Lower Ludlow Shales, AG Aymestry Group, ULS Upper Ludlow Shales, LS Ludlow Series (undivided).

SCYPHOZOA

Conularia sp. TB 35.

Metaconularia aspersa (Lindström) LLS 102.

HYDROZOA

Labechia conferta (Lonsdale) WL 46.

ANTHOZOA

Coenites sp. WL 46.
Favosites asper d'Orbigny AG 109, 200.
F. gothlandicus Lamarck AG 118.
F. gothlandicus forma *forbesi* Milne Edwards and Haime WL 46.
F. spp. PB 1. EL 47. AG 110–1, 209. LS 282.
Halysites catenularius (Linné) EL 47.
Heliolites interstinctus (Linné) WL 46.
H. sp. AG 163.
Monticulipora cf. *fletcheri* (Milne Edwards and Haime) LLS 62.
Phaulactis angusta (Lonsdale) WL 45.
P. sp. AG 201.
Porpites porpitus (Linné) LLS 66.
Pycnactis mitrata (Schlotheim) AG 111.
P. cf. *mitratus* AG 112.
P. sp. WS 9.
P. ? WS 31. AG 122.
Syringopora sp. WL 45. LS 295.
Solitary corals WS 20. WL 40. EL 47. LLS 80, 98, 101–3. AG 115, 153, 202.

CRINOIDEA

Crinoid columnals WS 10, 17–21, 33–4. TB 36. WL 38, 42, 45–6. LLS 53, 55, 62, 84, 99–100. AG 104–6, 109–10, 112–3, 115–6, 118–21, 125, 130, 146, 166, 183, 187, 190, 201. ULS 224, 226, 271, 277–8. LS 281–4, 294, 304, 311, 313–4, 317–8.

MACHAERIDIA

Lepidocoleus spp. WS 33–4.
? Machaeridia WS 18.

ANNELIDA

Campylites [*Serpulites*] *longissimus* (J. de C. Sowerby) AG 211. ULS 229–30, 233, 238, 240–1, 258–9, 261, 266, 269–70, 272, 274, 276–9, LS 283.
Keilorites ? cf. *coriaceus* (Phillips) AG 184.
K. ? squamosus (Phillips) AG 127, 148.
Worm tubes WS 10, 24. TB 37.

POLYZOA

Ceriopora sp. AG 216.
Monotrypa spp. AG 117, 216.
Ptilodictya lanceolata (Goldfuss) EL 51. LLS 103. AG 110, 116, 136, 143, 190.
Polyzoa indet. WS 13. LLS 55. ULS 223–4, 267.

BRACHIOPODA

Amphistrophia funiculata (McCoy) WL 40. LLS 62–3, 80–1, 98, 101. AG 147, 154. LS 294, 300.
Atrypa reticularis (Linné) HuS 4. WS 13, 22. TB 35. WL 38, 41–4, 46. EL 47, 51. LLS 54–7, 59–60, 62, 64, 66, 68, 72–4, 76–80, 84, 98–101, 103. AG 105–7, 109–21, 123–30, 133, 136, 139–40, 143–4, 146–9, 151–7, 159, 161–6, 168, 172, 181–2, 185–6, 188, 190–1, 194, 196, 199–200, 202–4, 206, 208, 212, 214, 216–9, 221. ULS 231, 235, 242, 253–7, 262. LS 280, 284–6, 294–6, 311.
Brachyprion arenaceus (Davidson) HuS 3. WS 15. TB 35. WL 38. LLS 54–5, 60, 63, 66, 81, 103. AG 118, 121–2, 203.
B. cf. *arenaceus* WS 13. TB 37. WL 43. LLS 64, 99. AG 110, 200.
B. fletcheri (Davidson) TB 36.
B. spp. WS 22. WL 44. LLS 70, 72, 98, 101. AG 113, 123, 126, 147, 163, 204.

'*Camarotoechia*' *nucula* (J. de C. Sowerby) PB 2. WS 10. TB 37. WL 38, 40, 42–3, 45. EL 47, 51. LLS 67, 74, 79–80. AG 104, 109, 115–8, 120, 125–9, 132–3, 135–41, 144–6, 148–52, 155–9, 165–9, 171, 173–5, 177, 179–87, 189–97, 199, 205–21. ULS 222–4, 226, 228–36, 238, 240–6, 250–2, 254, 256–74, 276–9. LS 281–3, 285–6, 295, 309–10, 319.

'*Chonetes*' *lepisma* (J. de C. Sowerby) WL 38, 40. EL 50. LLS 54–60, 63–4, 66, 76–7, 79, 81, 103. AG 106, 125, 127, 129, 131–3, 138, 143–4, 146, 148, 150, 152, 165, 168, 170, 175–6, 190, 194, 196, 198, 214, 221. ULS 222–4, 241–2, 250–2, 255, 262. LS 284, 286, 294, 309.

'*C.*' cf. *lepisma* AG 126.

'*C.*' *minimus* (J. de C. Sowerby) WS 13, 26–7, 30. TB 37. WL 42, 45. LLS 59, 65, 82. AG 169, 179, 208. ULS 222, 227–8, 234, 241, 247, 271. LS 300–2.

'*C.*' cf. *minimus* LLS 99. LS 281, 294, 296.

'*C.*' *novascoticus* (Hall) WS 17.

'*C.*' cf. *novascoticus* WS 18, 21.

Chonetoidea grayi (Davidson) HuS 3. WS 10, 17, 20, 23, 25, 27, 29, 31, 33–4. WL 42. LLS 52, 54–6, 61, 80–1, 85–7, 90, 92–4, 96–7, 102. AG 201. ULS 225, 231–2, 235, 241–2. LS 293–4, 297–8, 315, 317–8.

Conchidium knightii (J. Sowerby) AG 105, 107, 111–4, 118–9, 121–2, 134–5, 147, 164. LS 285, 311, 313.

Craniops implicata (J. de C. Sowerby) WS 24, 27, 32–3. WL 38, 40, 42–3. LLS 55, 57, 59–60, 62, 81, 84, 98–9, 101, 103. AG 115, 125, 153–4, 157, 179, 181, 188, 190, 192, 201–3, 205, 211, 220. ULS 232–3, 235, 239, 241, 258, 264, 269, 278. LS 280, 284–5, 294, 314.

Dalejina hybrida (J. de C. Sowerby) WS 26.

'*Dalmanella*' *spp.* WS 9–10, 18, 20, 25. TB 35. WL 38, 42. EL 49. LLS 59, 94, 102. AG 131, 134–5, 164, 168. ULS 222, 276.

Dayia navicula (J. de C. Sowerby) LLS 55. AG 106, 117, 125, 127, 131, 137–8, 146, 150, 152, 157, 160, 166–8, 170, 173–5, 177–9, 181–2, 192–5, 205–6, 211, 217–21. ULS 222–5, 227–8, 231–2, 235, 241, 243–6, 250–1, 255, 262, 270–1. LS 284–6, 290, 301, 304, 308–10, 314, 319.

Dicoelosia biloba (Linné) EL 47. LS 287–8, 294–5, 297, 314.

Dictyonella capewelli (Davidson) AG 194.

Dolerorthis rustica (J. de C. Sowerby) WL 46.

Eospirifer plicatellus (Linné) LLS 59. AG 201.

E. plicatellus interlineatus (J. de C. Sowerby) WL 46.

E. radiatus (J. de C. Sowerby) WS 13. WL 38. LLS 54, 68. AG 203.

'*Fardenia*' *applanata* (Salter) AG 153.

'*F.*' *pecten* (Linné) LLS 54–5, 59, 63. AG 125.

'*F.*' cf. *pecten* LLS 56. AG 115.

Glassia compressa (J. de C. Sowerby) WS 7, 15, 23. LLS 90, 102.

G. obovata (J. de C. Sowerby) WS 21, 30–1.

G. spp. WS 28. LLS 89, 94, 97.

Gypidula galeata (Dalman) EL 47. LLS 77, 80, 84, 98, 101, 103. AG 112, 135, 147, 153–4, 163, 203–4. LS 297.

G. lata Alexander LLS 54, 62. AG 110, 118, 200. LS 285.

G. spp. LLS 68, 79.

Howellella elegans (Muir-Wood) TB 36. WL 41. LLS 56, 59, 70, 75, 99, 101. AG 125–7, 129, 131, 138–9, 144, 146–7, 149–54, 157, 163, 166, 185, 187, 194, 208, 216–20. ULS 231, 242, 264, 279. LS 290, 297, 300.

H. cf. *elegans* LS 294.

H. spp. WS 27. WL 46.

Isorthis orbicularis (J. de C. Sowerby) WL 43. LLS 56, 90, 98, 101, 103. AG 104, 106, 109–10, 113, 117–9, 124–7, 130, 132–3, 137, 139–42, 144–6, 148, 152–3, 156, 158, 160, 165–7, 169, 174, 177–81, 187–8, 190–3, 195, 198–9, 202, 206, 208–10, 213, 216–20. ULS 241–2, 245, 250–1, 254–5. LS 285–6, 304, 308.

I. cf. *orbicularis* LLS 66. AG 112, 171, 185, 200–1, 207, 214. LS 294, 319.
Leangella segmentum (Lindström) HuS 3. WS 13, 17–8, 20, 30. TB 36. LS 288.
Leptaena rhomboidalis (Wilckens) WS 13, 22, 27, 30. WL 42, 44–6. EL 47. LLS 56, 62, 66, 72–4, 77–81, 99–103. AG 110–3, 116, 118, 121–2, 125, 127, 135, 147, 153–4, 163, 166, 182, 194–7, 201, 203, 205, 212–4, 217, 221. ULS 222, 231, 235, 245–9, 252–4, 256–7. LS 284–5, 294, 297, 304, 309, 311, 314.
Leptostrophia filosa (J. de C. Sowerby) WL 44. LLS 55, 74–5, 80, 84, 103. AG 106, 113–4, 117, 121, 125–6, 133–4, 138, 147, 152–3, 157, 163, 166, 168–9, 181, 183, 189–91, 193, 196, 202–3, 205–6, 208, 215, 217–8.
L. cf. *filosa* AG 111.
L. spp. PB 2. TB 37. WL 41–2. EL 47. LLS 54, 64, 67, 78, 100. AG 119, 123. ULS 265.
Lingula lata J. de C. Sowerby LLS 62–3.
L. lewisii J. de C. Sowerby LLS 100. AG 112, 127, 130, 159, 163, 166, 171–2. ULS 241, 257.
L. cf. *lewisii* LLS 66. AG 124, 151, 213.
L. minima J. de C. Sowerby ULS 222–4, 226, 229, 231–3, 238, 258–9, 264, 269, 271–3, 277–8.
L. striata J. de C. Sowerby LLS 54, 102.
L. symondsii Davidson WS 15, 17, 30. EL 49. LLS 57, 60, 63, 65, 75, 80–2. AG 115, 117, 125–6, 133, 142, 148, 158, 173. ULS 235, 249, 256, LS 285–6, 297, 305.
L. cf. *symondsii* LLS 59, 76. AG 169, 194.
L. spp. LLS 64, 103. AG 210, 221. ULS 261. LS 280, 301, 318–9.
Meristina obtusa (J. Sowerby) EL 47.
Orbiculoidea cf. *forbesii* (Davidson) WS 17.
O. rugata (J. de C. Sowerby) LLS 54, 63, 82, 100. AG 132, 175, 181, 194, 205, 210–1, 221. ULS 222–3, 226, 229–30, 233, 236, 241, 258, 262, 269, 272–4, 277–9. LS 283, 285, 294, 319.
Pentamerus oblongus J. de C. Sowerby PB 1.
Plectodonta cf. *duvalii* (Davidson) WS 33.
P. transversalis (Davidson) WS 16. LS 292, 294.
P. cf. *transversalis* WS 32. LLS 99.
Protochonetes ludloviensis Muir-Wood AG 104, 138, 152, 166, 169, 172, 175, 180–1, 192, 206–7, 209–10, 215, 218, 220–1, ULS 226, 228–30, 232–3, 238–41, 257–65, 267–70, 272–9. LS 281–3, 311.
P. cf. *ludloviensis* LS 318.
Resserella basalis (Dalman) WS 27.
R. elegantula (Dalman) WS 21, 24, 27, 30–1. WL 38. EL 47. LLS 56, 84. AG 125. ULS 224. LS 280, 285.
R. cf. *elegantula* WS 33. LLS 53, 102.
Salopina lunata (J. de C. Sowerby) AG 116–7, 177, 181–2, 194–6, 205, 210–1, 217–8, 221. ULS 223–4, 226, 229–30, 233–4, 238–40, 243–4, 252, 258–62, 264, 272–3, 275, 277–9. LS 281–3, 319.
Schizocrania striata (J. de C. Sowerby) WS 27. AG 127, 194. ULS 230, 256–7, 271, 276.
Shaleria ornatella (Davidson) AG 157, 176, 182, 194–6, 205, 207, 212–4, 221. ULS 222, 224, 231, 242, 244–5, 247–53, 262, 265.
Skenidioides lewisii (Davidson) HuS 3. WS 27, 33–4. LLS 65. LS 284, 287, 294, 297–8.
S. cf. *lewisii* WS 20.
Sphaerirhynchia davidsoni (McCoy) WL 20. LLS 66, 103.
S. cf. *davidsoni* LLS 98.
S. wilsoni (J. Sowerby) WL 41, 43. LLS 53–6, 58, 66, 71, 73, 75–6, 78–81, 99–101, 103. AG 104, 106, 110, 112–3, 117, 119, 124–33, 138–43, 145–55, 157, 159, 162–4, 166, 169, 171, 173, 175–6, 180–1, 186, 189–91, 198–200, 202–4, 208–10, 216–21. ULS 242, 244–5, 262. LS 282, 286, 301, 308, 319.

Strophonella euglypha (Hisinger) WL 40, 42. LLS 54–7, 59, 62, 67, 69, 72, 77–8, 100, 103. AG 110–3, 116, 118–9, 121–2, 147, 153–4, 164, 200, 203–4. ULS 262. LS 311.
S. cf. *euglypha* WL 38.
Whitfieldella canalis (J. de C. Sowerby) AG 124, 126, 148, 152, 156, 190, 209.
W. sp. AG 145.

GASTROPODA

Bellerophon cf. *globulus* Lindström ULS 276.
Bembexia lloydii (J. de C. Sowerby) LLS 56.
Bucanopsis expansa (J. de C. Sowerby) LLS 102. AG 214. ULS 228, 270.
Cyrtostropha corallii (J. de C. Sowerby) AG 166, 215–6, 220. ULS 232.
C. cf. *corallii* ULS 228.
Ecculiomphalus laevis (J. de C. Sowerby) WS 17. LLS 102. ULS 265.
Euomphalopterus alatus (Wahlenberg) AG 110, 112–3, 120, 187, 216. ULS 276.
E. cf. *alatus* AG 115, 118.
E. subundulatus (Salter) AG 192.
Gyronema octavia (d'Orbigny) *glabrum* (Lindström) LLS 57, 59. AG 157.
G. spp. LLS 67. AG 152.
Liospira cf. *striatissima* (Salter) AG 200.
L. spp. WS 32–3.
Loxonema sinuosum (J. de C. Sowerby) LLS 63. AG 150, 169. ULS 254.
L. cf. *sinuosum* WS 17, 34. LLS 59, 102.
L. spp. WS 21, 31. LLS 103.
Murchisonia articulata (J. de C. Sowerby) LLS 102.
M. cingulata Hisinger LS 317.
Platyceras sp. WS 21, 31. LLS 76. ULS 244.
Platyschisma sp. LS 318.
'Pleurotomaria' spp. WS 8, 18, 26. LLS 57, 60, 102. AG 201, 219. ULS 222–3, 228.
Poleumita globosa (Schlotheim) WL 46. AG 132.
Theca forbesi (Sharpe) WS 27. ULS 228, 258.

?SCAPHOPODA

Tentaculites spp. WS 30. WL 39. EL 49. LLS 81. AG 205. ULS 277.

LAMELLIBRANCHIA

Actinopterella tenuistriata (McCoy) AG 141, 153, 171, 197, 207, 209, 213–4, 217, 221. ULS 241, 268, 271, 276, 278. LS 312, 319.
Ambonychia sp. LS 317.
Anodontopsis cf. *lucina* Salter LLS 55.
Cardiola interrupta J. de C. Sowerby WS 5, 7, 18. LLS 54–7, 59, 87, 92, 95, 102. AG 154, 199, 220. ULS 222. LS 307, 312, 317, 319.
C. cf. *interrupta* WS 17.
Ctenodonta cf. *deltoidea* (Phillips) LLS 56.
C. spp. WS 17–8, 20, 31. LLS 55, 65. LS 317.
Cypricardinia subplanulata Reed LLS 81, 98, 101, 103. AG 182, 194, 214. ULS 246. LS 296, 300.
Dualina cf. *elegans* (Salter) WS 23.
D. longiuscula Barrande LLS 102.
Fuchsella amygdalina (J. de C. Sowerby) AG 166–8, 177, 181, 193, 196, 205–6, 217–8, 220–1. ULS 222, 226–9, 233–5, 237, 241, 249–51, 257–8, 264–5, 268–71, 273, 277–9. LS 282.
Goniophora cymbaeformis (J. de C. Sowerby) LLS 56. AG 179, 206. ULS 226, 229, 233, 238, 258, 261, 264, 268–9, 274. LS 281.
? Grammysia macadamensis McLearn LLS 86.
Lunulacardium acuminatum Barrande LS 317.
Lyrodesma sp. LLS 64.
Myalina sp. LS 317.

Nuculites anglicus (d'Orbigny) WS 21. LLS 82.
N. cf. *anglicus* WS 17, 33.
N. antiquus (J. de C. Sowerby) ULS 271.
N. cf. *antiquus* LLS 63.
N. coarctatus (Phillips) ULS 267–8.
N. ovatus (J. de C. Sowerby) ULS 278.
N. spp. ULS 222, 230, 279. LS 281, 315, 317.
Orthonota impressa (J. de C. Sowerby) AG 203.
O. cf. *impressa* WS 18.
'*O. rigida* (J. de C. Sowerby) LLS 61, 98.
O. cf. *rigida* ULS 228. LS 280.
Palaeopecten danbyi (McCoy) LLS 79.
Paracyclas perovalis (Salter) WS 21. LLS 59, 63. AG 182, 195–6, 206, 219–20. LS 293.
P. cf. *perovalis* WS 18.
Plethomytilus [*Mytilarca*] *mytilimeris* (Conrad) LLS 56, 103.
P. [*M.*] cf. *mytilimeris* LS 294.
Pteritonella exasperata (Reed) LLS 62.
P. reticulata (Hisinger) AG 187.
P. retroflexa (Wahlenberg) AG 127, 152, 197, 205–6, 210, 215, 219, ULS 229, 257, 265, 278–9.
Salweyia subobliquata Reed LLS 100.
Tancrediopsis ludensis Reed LS 293, 317.

CEPHALOPODA
Cyrtoceras sp. LLS 63.
Dawsonoceras annulatum (J. Sowerby) WS 17. LLS 48, 102. LS 319.
Kionoceras angulatum (Wahlenberg) WS 21, 27. LLS 55, 79. ULS 228. LS 285, 292.
Lituites ibex J. de C. Sowerby AG 197. ULS 227, 229, 232, 234, 269. LS 319.
Michelinoceras bullatum (J. de C. Sowerby) LLS 55–7, 102. ULS 222–3, 226, 233–4, 276–8. LS 282–3.
M. cf. *bullatum* LLS 59–60.
M. ibex (J. de C. Sowerby) LLS 55–6, 59.
M. lineatum (Hisinger) *tenuistriatum* (Portlock) WS 17.
M. ludense (J. de C. Sowerby) AG 192.
M. cf. *originale* (Barrande) WS 17.
M. cf. *primaevum* (Forbes) LS 317.
M. subundulatum (Portlock) WS 8, 31, 34. ULS 227. LS 292.
M. cf. *subundulatum* WS 17. LS 286, 315, 317.
M. tennuiannulatum (McCoy) ULS 222.
Ophioceras articulatum (J. de C. Sowerby) LLS 56, 59.
'*Orthoceras*' *spp.* HuS 3. WS 9, 14, 16, 18, 32. LLS 52, 54. ULS 224, 272, 279. LS 292, 312, 316, 318.

? OTHER MOLLUSCA
Cornulites serpularius Schlotheim LLS 103. AG 182, 203, 207, 210, 218, 220. ULS 228, 278.

TRILOBITA
Acaste downingiae (Murchison) ULS 229, 278.
Acidaspis coronata Salter LLS 102. LS 284.
A. cf. *coronata* LLS 63.
Ampyx spp. WS 17, 33. LLS 65, 87.
Calymene blumenbachii Brongniart WS 16, 21, 28.
C. cf. *blumenbachii* LLS 59.
C. lata Shirley. LLS 81.
C. neointermedia R. and E. Richter ULS 231, 265.
C. cf. *neointermedia* ULS 262.

C. nodulosa Shirley WS 33–4.
C. spp. WS 20, 34. LLS 54, 62, 103. AG 128, 150, 190, 204. ULS 254.
Cheirurus bimucronatus (Murchison) WS 20.
Dalmanites caudatus (Brünnich) WS 14, 17–9, 21, 23, 25, 27–8, 31–3. WL 40, 45. EL 47, 49. LS 297.
D. cf. *caudatus* WS 30. WL 44. EL 50.
D. myops (König) LLS 54–6, 58, 61, 63, 65–6, 78, 80, 82, 85–8, 92–6, 98–103. AG 122, 153–4, 164, 201, 204. LS 285, 291, 293–4, 317.
D. cf. *myops* LLS 53, 59–60. AG 119.
D. cf. *weaveri* (Salter) WS 18.
D. spp. PB 2. WS 6, 8–9, 13, 17–8, 24, 32, 34. WL 38, 42. EL 48. LLS 62. AG 123, 135. LS 284, 296, 298.
Encrinurus spp. EL 50. LLS 69, 80, 84, 98. AG 144, 202. ULS 231, 235, 254.
Homalonotus knightii König ULS 279.
Lichas sp. WS 20.
Otarion elegantula (Lovén) WS 30.
O. megalops (McCoy) WS 13. AG 206. LS 284, 294.
O. spp. WS 18. TB 37. WL 38, 40. AG 205.
Phacops stokesii (Milne Edwards) WS 17, 20. EL 50. LLS 82, 85.
Proetus spp. WS 13, 18. WL 38, 45. LLS 60, 66. AG 106, 110, 118, 179, 214, 221. ULS 231.

OSTRACODA

Beyrichia kloedeni McCoy WS 20–1, 28, 34. WL 45. EL 48. LLS 55, 59, 84. AG 201. ULS 222, 272. LS 318–9.
B. cf. *kloedeni* WS 33. LLS 103.
B. torosa T. R. Jones AG 221. ULS 223–4, 227, 229, 232, 236.
Entomis impendens Haswell ULS 227.
Hemsiella [*Beyrichia*] *maccoyiana* (T. R. Jones) WS 8, 13, 17–20, 27, 31. TB 37. LLS 56, 60–1, 65, 103. AG 200.
H. [*B.*] cf. *maccoyiana* ULS 222, 227.
Neobeyrichia lauensis (Kiesow) ULS 241.
N. cf. *lauensis* ULS 231–2.
Primitia nana T. R. Jones LLS 84.
P. sp. LLS 99.

GRAPTOLITOIDEA

Gothograptus nassa (Holm) WS 9.
Monograptus bohemicus (Barrande) LS 292.
M. chimaera (Barrande) LLS 83. LS 292.
M. cf. *chimaera* LLS 65, 103.
M. chimaera salweyi Lapworth LS 316.
M. cf. *chimaera salweyi* LS 292.
M. colonus (Barrande) LLS 52, 87, 91, 95.
M. cf. *colonus* LLS 86, 89. LS 315.
M. colonus compactus Wood LLS 52. LS 315, 317.
M. dubius (Suess) LS 312, 316.
M. cf. *dubius* WS 6–7, 10, 12, 28, 32. LLS 55, 63, 82, 86. LS 292.
M. aff. *dubius* LLS 56.
M. flemingii (Salter) WS 11, 21, 29.
M. cf. *flemingii* WS 20, 32.
M. leintwardinensis Lapworth ULS 222, 224. LS 303–4, 306, 309.
M. cf. *leintwardinensis* AG 104, 176. ULS 231, 246, 256. LS 305, 307–8, 318–9.
M. leintwardinensis incipiens Wood LS 286.
M. cf. *leintwardinensis incipiens* AG 221. ULS 231.
M. cf. *priodon* (Bronn) WS 15–6, 23, 28, 31–2, 34.
M. tumescens Wood LLS 54–5, 57, 59, 63. LS 312.

M. cf. *tumescens* LLS 56.
M. varians Wood LLS 102.
M. cf. *varians* LLS 52. LS 289, 315.
M. cf. *vomerinus* (Nicholson) WS 32.
M. cf. *vulgaris* Wood LLS 82.
Spinograptus spinosus (Wood) LS 317.

PISCES

Thelodont denticles ULS 240.

APPENDIX 3

LIST OF GEOLOGICAL SURVEY PHOTOGRAPHS

(CHURCH STRETTON SHEET 166)

Copies of these photographs are deposited for reference in the library of the Geological Survey and Museum, South Kensington, London, S.W.7. Prints and lantern slides may be supplied at a fixed tariff. An asterisk indicates that a colour transparency is also available.

All numbers belong to Series A.

URICONIAN

8890 Crag of rhyolitic tuffs, Three Fingers Rock, 900 yd S.S.W. of Caer Caradoc summit.
8891 Detail of crags of rhyolitic tuffs, Three Fingers Rock.
8893 Andesite and rhyolite crags forming Caer Caradoc summit.
8897 Rhyolite crags, 300 yd south of Caer Caradoc summit.
8905 Syncline in tuffs on Caer Caradoc Hill. Viewpoint, 400 N E of Cwms Farm.
8909 Uriconian rhyolite, quarry, 250 yd north of Hope Bowdler church.
8910 Ditto.
8911 Gorge and quarry in Uriconian rhyolite. Viewpoint, 500 yd N.N.E. of Woodgate Cottage.
8918 Rhyolite crags, The Gaer Stone, ½ mile N.W. of Hope Bowdler church.
8921 View of Hope Bowdler Hill from the north end of the hill.
8923 View of Caer Caradoc Hill from Cwms Farm.
8926 Crags on the west side of Caer Caradoc Hill. Viewpoint, near quarry, 300 yd S.W. of Botvyle.
*9465 Wart Hill, from north-east of Grist House, Hopesay.
*9571 Detail of coarse tuff, 50 yd S.S.W. of summit of Wart Hill.

LONGMYNDIAN

6199 Lightspout waterfall, over massive beds in the Lightspout Group, Cardingmill Valley, Church Stretton.
6200 Close view of Lightspout waterfall.
*9421 Western scarp slope of Long Mynd. Viewpoint, Myndmill Wood, Myndtown.
*9423–4 Topographic feature formed by 30-ft dolerite dyke in Longmyndian. Viewpoint, 100 yd S.W. of The Villa, High Park Hollow.
*9425 Upper part of Batch valley, All Stretton. Viewpoint, 550 yd W. 15° S. of Jinlye.
*9435 View down New Pool Hollow, Cardingmill Valley, Church Stretton.
*9448 Lower part of Nut Batch from Prior's Holt.
*9449 Lower part of Wooler's Batch from near Churchmoor.
*9450 Upper part of Mount Gutter, near Horderley.
*9451 Lower part of Mount Gutter, near Horderley.
*9489 Hollow in skyline on outcrop of shaly Synalds Group, looking N.N.E. Viewpoint, 2000 yd E. 9° S. of Boiling Well.

*9490 Ditto, looking S.S.W. Viewpoint, 2000 yd E. 9° S. of Boiling Well.
*9558 Steeply dipping flaggy siltstones of Bridges Group. Roadside, Bridges.
*9559 Detail of lower rib of Stanbatch Conglomerate. Roadside, 1900 yd E. 36° S. of Manor House, Ratlinghope.
*9560 Double feature formed by the two conglomeratic ribs of the Stanbatch Conglomerate. Viewpoint, roadside, 600 yd W.S.W. of Upper Darnford farm (Pl. 7B).
*9561 The Huckster (Huxter) Stone, crag of Huckster Conglomerate, the basal member of the Portway Group. Long Mynd, above Myndtown.
*9562 Ditto.
*9563 Minor tear fault in Synalds Group. Pike Hollow, south end of the Long Mynd.
*9564 Bedding and cleavage in Synalds Group, Mount Gutter, Horderley (Pl. 7A).
*9565 Lineation on bedding planes, Synalds Group. Between Pike Hollow and Mount Gutter, south end of Longmynd.
9566 Cardingmill Grit, cut by quartz veins along joints. Mount Gutter, Horderley.
*9567 Crags formed by Cardingmill Grit. Devil's Mouth, Burway Road, near Church Stretton.
*9568 Crags formed by Cardingmill Grit. Opposite Batch Cottage, Batch valley, All Stretton.
*9569 Close view of Buxton Rock, Buxton Quarry, All Stretton.
*9570 General view of Buxton Quarry with Buxton Rock, All Stretton.

CAMBRIAN

4856 Lower Comley Sandstone overlain by *Olenellus* Limestone, Comley Quarry.
4857 Excavation in Lower and Middle Cambrian beds, 200 yd south of Comley Quarry.
8899 Cambrian quarzite crags of Hill End. Viewpoint, 600 yd N.N.W. of Stoneacton.
8901 Cambrian quartzite crags, 1100 yd S.S.W. of Cardington Church.
8902 Cambrian quartzite crags, Hill End quarry, 700 yd S. 9° E. of Cardington church.

ORDOVICIAN

8886 Six-inch Neptunian dyke of Harnage Shales in Uriconian andesite, Hazler Quarry.
8887 View of Uriconian, Harnage Shale and Chatwall Sandstone features. Viewpoint 300 yd E.S.E. of Dryhill.
8888 Colour-banded Chatwall (Soudley) Sandstone, overlain by *alternata* Limestone. Soudley Quarry, 500 yd W.S.W. of Hope Bowdler church (Pl. 9A).
8903 Dip slope of Chatwall Flags and scarp of Chatwall Sandstone. Viewpoint, Folly Bank, 1 mile N.W. of Cardington Church.
8907 Dip slope of Cheney Longville (Chelmick) Flags; Ralgeth Hill on skyline. Viewpoint 300 yd S.E. of Hope Bowdler.
8908 Basal conglomerate of Harnage Shales on Uriconian rhyolite, Upper House, Hope Bowdler (Pl. 7C).
8915 Folded Harnage Shales, quarry, 600 yd S. 20° E. of Little Stretton church.
8916 Detail of junction of Harnage Shales with grit of ? Wentnor Series, quarry, 600 yd S. 20° E. of Little Stretton church.
8917 Folding in Harnage Shales, quarry, 600 yd S. 20° E. of Little Stretton church.
*9549 Cheney Longville Flags, road cutting, Cheney Longville.

*9550 Cheney Longville Flags, north-west of Marshbrook.
*9551 Cheney Longville Flags, Marshbrook Bridge, 750 yd N.N.W. of Henley.
*9552 Exposure of *alternata* Limestone, railway cutting, north of Marshbrook.
*9553 Colour banding and current bedding in Horderley Sandstone, Glenburrell, Horderley.
*9554 Ditto.
*9555 Basal beds of Hoar Edge Grit, south side of River Onny, Glenburrell, Horderley.
9556 Calcareous Hoar Edge Grit, south side of River Onny, Glenburrell, Horderley.

SILURIAN

4196 Ballstone in Wenlock Limestone, Presthope Quarry, Wenlock Edge.
8924 Steeply dipping Aymestry Group, quarry, 300 yd S.W. of Botvyle.
8925 Weathered Aymestry Group, near quarry, 300 yd S.W. of Botvyle.
*9455 Escarpment of Tickwood Beds, Ridgway Hill, Horderley.
*9459 Burrow Hill. Viewpoint 225 yd E.N.E. of The Hollies, Coston.
*9471 Wenlock and Aymestry scarps from Wart Hill.
*9472 View Edge, Aymestry Group scarp from Rowton, Clungunford.
*9473 Wenlock Edge and Callow Hill. Viewpoint, 1050 yd north of Sibdon Carwood church.
*9474 Norton Camp, Aymestry Group scarp. Viewpoint, 1050 yd north of Sibdon Carwood church.
*9475 View Edge, Aymestry Group scarp. Viewpoint, 1050 yd north of Sibdon Carwood church.
*9477 Wenlock and Aymestry scarps. Viewpoint, 225 yd E.N.E. of The Hollies, Coston.
*9492 Wenlock Edge and Stoke Wood. Viewpoint, 450 yd north of Affcot House.
*9493 View along Hope Dale between the Wenlock and Aymestry scarps. Viewpoint, 50 yd N.E. of The Pole, Stanway (Pl.8A).
*9495 Callow Hill, Aymestry Group scarp. Viewpoint, 450 yd E.S.E. of Middle Westhope (Pl. 8B).
*9496 Siefton Batch, a valley through the Aymestry scarp. Viewpoint, 100 yd W.S.W. of Westhope church.
*9497 Hope Dale, between the Wenlock and Aymestry scarps. Viewpoint, 600 yd east of Morwood.
*9498 Dry valley in dip slope of Upper Ludlow Shales. Viewpoint, 950 yd south of Hillend, Westhope.
*9530 Section in Upper Ludlow Shales, Aldon, Onibury.
*9531 Slump bedding in Upper Ludlow Shales, 150 yd E.N.E. of Bache Cottage (Pl. 10B).
*9532 Evenly bedded Upper Ludlow Shales, Newhouse Quarry, 650 yd E.N.E. of Culmington Manor.
*9533 Aymestry Group, quarry, 250 yd north of Fernhall Mill.
*9535 Band of *Conchidium knightii* in Aymestry Group, Viewedge Lime Works (Pl. 10A).
*9536 Quarry face in Aymestry Group, Viewedge Lime Works.
*9537 Flaggy Lower Ludlow Shales, River Onny, 1½ miles above Onibury.
*9538 Flaggy Lower Ludlow Shales, Stokewood Cottage quarry, Stokesay.
*9539 Stream section in Lower Ludlow Shales, 300 yd N.N.W. of Upper Millichope.
*9540 Bedded Wenlock Limestone, 700 yd S.W. of Morwood (Pl. 9B).
*9541 Bedded, rubbly Wenlock Limestone, quarry, 1100 yd S.W. of Stanway Manor.
*9542 Reef development in Wenlock Limestone, Littleshall Quarry, 1 mile S.E. of Hughley.
*9543 Ditto (Pl. 9C).

*9544 Crags of massive reef facies of Wenlock Limestone, 300 yd north of Plough Inn, Hilltop.
*9545 Tickwood Beds, quarry, west end of Ridgway Hill, Horderley.
*9546 Tickwood Beds, quarry, Harton Hollow Way, 750 yd south of Harton.
*9547 Wenlock Shale with calcareous bullions, The Bank, Horderley.
*9548 Unconformity between rocks of the Caradoc and Llandovery series, Onny River section, south-west of Wistanstow.
*9557 Llandovery shore line features, Wooler's Batch, Churchmoor.

OLD RED SANDSTONE

*9501 '*Psammosteus*' Limestone escarpment from near Stanton Lacy.
*9502 Valley through sandstone escarpment of the Ditton Series from near Lodge Farm, Stanton Lacy.
*9503 Fault valley in Ditton Series, Brown Clee in background, from Meesons, Hopton Cangeford.
*9504 Weston Hill, the scarp of a sandstone band in the Ditton Series, from Lesser Poston, Hopton Cangeford.
*9505 Nordy Bank Camp and Abdon Burf from The Thrift, Stoke St. Milborough.
*9506 Clee Burf, Brown Clee, from The Thrift, Stoke St. Milborough.
*9507 Slopes below Brown Clee from west of Loughton.
*9510 Titterstone Clee from near The Moor, Stoke St. Milborough.
*9511 Titterstone Clee from near Meesons, Hopton Cangeford.
*9521 Flaggy Farlow Sandstone, Prescott, near Stottesdon.
*9522 Pebbly Farlow Sandstone, Prescott, near Stottesdon.
*9523 Sandstone wedge in cornstone-conglomerate of the Ditton Series, Hoptongate, Hopton Cangeford.
*9524 Cornstone-conglomerate, of the Ditton Series, quarry, Hoptongate, Hopton Cangeford.
*9525 Sandstone wedge in cornstone-conglomerate of the Ditton Series, Lesser Poston, Hopton Cangeford (Pl. 11B).
*9526 Waterfall over '*Psammosteus*' Limestone, The Hope, Stanton Lacy.
*9527 'Fish eyes' in marl of the Downton Series, The Hope, Stanton Lacy.
*9528 Current bedding in Downton Castle Sandstone, quarry at Onibury (Pl. 11A).

CARBONIFEROUS

8467 Contrast between vegetation cover on dolerite and Coal Measures, East side, Green Lea, Brown Clee.
8468 Contrast between vegetation cover an Old Red Sandstone and Coal Measures, 400 yd N.W. of Green Lea, Brown Clee.
*9508 Farlow Bank and Silvington Common from The Down, Farlow.
*9509 Farlow Bank from north of Silvington (Pl. 12B).
*9512 Dolerite cap of Titterstone Clee from Bedlam, near Bitterley.
*9513 Dip slope of Carboniferous Limestone, Oreton Bank.

GLACIAL, POST-GLACIAL AND RECENT

8882 Glacial drainage channel between Helmeth and Hazler hills, Church Stretton.
8906 View west across Hope Bowdler up glacial drainage channel. Viewpoint 300 yd S.E. of Hope Bowdler.
*9416 Glacial drainage channels west of River East Onny. Viewpoint, 720 yd N. 23° E. of Wentnor church.

*9417–8 Marshy ground between Bridges and Stedment, possible site of glacial lake. Viewpoint 60 yd west of bridge over river at Bridges.
*9422 Glacial drainage channel, Worsley. Viewpoint, 300 yd E. 28° S. of Jinlye.
*9427 Upper end of glacial drainage channel of A 9422. Viewpoint, 550 yd N. 24° E. of Jinlye.
*9428 Glacial drainage channel south of Castle Hill, All Stretton.
*9429–30 Glacial drainage channel, Synalds Coppice, All Stretton. Viewpoint 640 yd E. 33° S. of Jinlye.
*9431 Glacial drainage channel, Cwmdale, All Stretton. Viewpoint, 1550 yd S. 2° W. of Jinlye (Pl. 13C).
*9433 Glacial drainage channel at Golf Club House, Church Stretton.
*9434 Upper part of channel of A 9433.
*9436 Incised cross section of Ashes Hollow. Viewpoint 1 mile E. 20° S. of Narnell's Rock.
*9437 View down Ashes Hollow showing incised character. Viewpoint, 1120 yd E. 16° S. of Narnell's Rock.
*9438 Glacial drainage channel west of Brockhurst Hill, Church Stretton valley. Viewpoint, 620 yd N. 7° E. of Little Stretton church.
*9439–41 Panoramic view of terminal moraine at Brockhurst Castle, Church Stretton and 700-ft pre-glacial erosion level at Little Stretton. Viewpoint 650 yd N.E. of Little Stretton church.
*9442–3 Panoramic view to north-east of 700 ft pre-glacial erosion level at Little Stretton. Viewpoint 950 yd west of Little Stretton church.
*9444–7 Panoramic view to west of 700-ft pre-glacial erosion level from Little Stretton to Minton. Viewpoint, 530 yd S. 22° E. of Little Stretton church (9444 and 9445 are shown as Pl. 13B).
*9458 Moraine near Oaker. Viewpoint 350 yd N.W. of Oaker.
*9466 Marshbrook valley, alluvial flats. Viewpoint, 800 yd N.N.W. of Henley.
*9467 River terraces at Marsh Farm. Viewpoint, 200 yd S.S.E. of Marsh Farm.
*9469 Alluvium of the Quinny Brook from Affcot Bridge.
*9476 Alluvium of the River Clun. Viewpoint, 650 yd south of post office, Aston-on-Clun.
*9478 Alluvium of the Eaton Brook. Viewpoint, 400 yd south of Harton Road Station.
9480 View west across Church Stretton valley to head of glacial drainage channel north of Brockhurst Castle. Viewpoint, 1430 yd N. 40° E. from Little Stretton church.
*9514 Coarse terrace gravel on Downton Castle Sandstone, stream section, 100 yd S.E. of Newhouse Farm.
*9515 Coarse alluvial gravel in the Batch valley, All Stretton, 560 yd S. 35° W. of Jinlye.
*9516 Coarse terrace gravel in Onny valley, road section south-west of Wistanstow.
*9519 Road cutting in moraine gravel, 250 yd west of Oaker.
*9520 Section in boulder clay, farmyard, Walton Farm, Onibury.

GENERAL VIEWS

6195 Cardingmill Valley, Church Stretton.
6196 Cardingmill Valley above the Carding Mill, Church Stretton.
6197 Townbrook Valley, Church Stretton.
6198 Church Stretton valley, Caer Caradoc and the Lawley in the background.
6202 View across Church Stretton valley to the Long Mynd.
6203 View across Church Stretton valley to the Long Mynd from Caer Caradoc.
6204 The Lawley, Hoar Edge and The Wrekin from Caer Caradoc.
6205 Ditto.

6206 View east-south-east from Caer Caradoc.

6209 Caer Caradoc, Helmeth and Hope Bowdler hills from the foot of Ragleth Hill.

6210 View across Bodbury Ring from the Bur Way to Caer Caradoc and The Lawley.

6211 View across Townbrook Valley from the Bur Way to Caer Caradoc and The Lawley.

6912 View west over east end of Clee Hill Carboniferous syncline. Viewpoint, 300 yd east of Prescott Mill.

8883 Caer Caradoc and Hope Bowdler hills from ¼ mile east of Hazler Hill.

8884 Caer Caradoc and Helmeth hills from Hazler Hill.

8889 View west across Church Stretton valley towards the Long Mynd, Church Stretton.

8894 View across The Cwms to Hope Bowdler Hill from Caer Caradoc Hill.

8896 View from Caer Caradoc towards Battle Stones.

8898 The Lawley, Hoar Edge and The Wrekin from the north shoulder of Caer Caradoc.

8900 Battle Stones and Caer Caradoc from Sharpstones.

8904 Caer Caradoc Hill with Ordovician rocks in foreground. Viewpoint 400 yd N.E. of Cwms Farm.

8912 View east to Wenlock Edge from Uriconian hills west of Woodgate Cottage.

8913 View south-east to Wenlock Edge, Brown Clee and Titterstone Clee from Uriconian hills west of Woodgate Cottage.

8922 View south-east to Wenlock Edge and Titterstone Clee from Willstone Hill.

8928 View across Church Stretton valley to north end of the Long Mynd. Viewpoint, near quarry, 300 yd S.W. of Botvyle.

8929 Ditto.

*9419–20 Western scarp slope of the Long Mynd, Wentnor ridge and Stiperstones ridge. Viewpoint, 700 yd N. 35° E. of Asterton chapel.

*9432 Cardingmill Valley. Viewpoint 400 yd S. 35° E. of Carding Mill.

*9452 The Long Mynd at Minton from Whittingslow.

*9453 Southern end of the Long Mynd from Wart Hill.

*9454 Silurian escarpment of Ridgway Hill with pre-glacial erosion bench, from near Hillend Farm.

*9460 Burrow hill and Hopesay valley from Wart Hill.

*9461 View north-north-east along Church Stretton valley from Whittingslow.

*9462 Ditto from near Ridgway, Edgton.

*9463 Ditto from Wart Hill (Pl. 6B).

*9464 View northwards along the Church Stretton Fault. Viewpoint 225 yd E.N.E. of The Hollies, Coston.

*9468 The Long Mynd from near Wolverton. Viewpoint, 600 yd north of Wolverton.

*9470 The Cardington hills. Viewpoint, 300 yd S.S.E. of Manor Farm, East Wall.

*9479 The Wrekin and the North Shropshire Plain from Wenlock Edge. Viewpoint 350 yd N.N.E. of Plough Inn, Hilltop.

*9481–2 Caer Caradoc, The Lawley and The Wrekin from the Long Mynd. Viewpoint, Jinlye.

*9483–5 Panoramic view of north end of the Long Mynd across the Church Stretton valley. Viewpoint, 650 yd N.W. of Hill House.

*9486–8 Panoramic view of Ragleth, Hazler and Helmeth hills across Church Stretton. Viewpoint, Links Road, Church Stretton.

*9491 Cardington Hill and the Long Mynd from Wenlock Edge. Viewpoint 350 yd N.N.E. of Plough Inn, Hilltop.

*9494 '*Psammosteus*' Limestone escarpment and Titterstone Clee. Viewpoint 1300 yd S.W. of Stanway Manor.

INDEX